*Methods of
Experimental Physics*

VOLUME 22

SOLID STATE PHYSICS: SURFACES

METHODS OF
EXPERIMENTAL PHYSICS

Robert Celotta and Judah Levine, *Editors-in-Chief*

Founding Editors

L. MARTON
C. MARTON

Volume 22

Solid State Physics: Surfaces

Edited by

Robert L. Park

Department of Physics and Astronomy
University of Maryland
College Park, Maryland

Max G. Lagally

Department of Metallurgical and Mineral Engineering
and Materials Science Center
University of Wisconsin
Madison, Wisconsin

1985

ACADEMIC PRESS, INC.
(Harcourt Brace Jovanovich, Publishers)

Orlando San Diego New York London
Toronto Montreal Sydney Tokyo

COPYRIGHT © 1985, BY ACADEMIC PRESS, INC.
ALL RIGHTS RESERVED.
NO PART OF THIS PUBLICATION MAY BE REPRODUCED OR
TRANSMITTED IN ANY FORM OR BY ANY MEANS, ELECTRONIC
OR MECHANICAL, INCLUDING PHOTOCOPY, RECORDING, OR
ANY INFORMATION STORAGE AND RETRIEVAL SYSTEM, WITHOUT
PERMISSION IN WRITING FROM THE PUBLISHER.

ACADEMIC PRESS, INC.
Orlando, Florida 32887

United Kingdom Edition published by
ACADEMIC PRESS INC. (LONDON) LTD.
24–28 Oval Road, London NW1 7DX

Library of Congress Cataloging in Publication Data
Main entry under title:

Solid state physics.

(Methods of experimental physics ; v. 22)
Includes index.
1. Solids--Surfaces--Optical properties.   2. Solid
state physics.   I. Park, Robert L.   II. Lagally, Max G.
III. Series.
QA176.8.S8S65   1985   530.4'1   84-12427
ISBN 0-12-475964-5 (alk. paper)

PRINTED IN THE UNITED STATES OF AMERICA

85 86 87 88   9 8 7 6 5 4 3 2 1

CONTENTS

CONTRIBUTORS	xiii
FOREWORD	xv
PREFACE	xvii
LIST OF VOLUMES IN TREATISE	xix

1. Work Function Measurements
by L. W. SWANSON AND P. R. DAVIS

1.1. Introduction	1
1.2. Work Function Theory	2
1.2.1. Basic Concepts	2
1.3. Electron Emission Methods	5
1.3.1. Thermionic Method	5
1.3.2. Photoelectric Emission	7
1.3.3. Field Electron Emission	8
1.4. Retarding-Potential Methods	11
1.4.1. The Shelton Triode	11
1.4.2. Electron-Beam Method	14
1.4.3. Field Emission Retarding-Potential Method	15
1.5. Capacitance Methods	17
1.5.1. The Kelvin–Zisman (Vibrating Capacitor) Method	17
1.5.2. Static Capacitor Methods	19
1.6. Comparison of the Various Measuring Techniques	19
1.7. Experimental Embodiments of the Various Techniques	21
1.8. The Future of Work Function Measurements	22

2. Vibrations in Overlayers
   by W. Henry Weinberg

   2.1. Introduction 23
   2.2. Electron Energy Loss Spectroscopy 31

       2.2.1. Introduction 31
       2.2.2. Electron Energy Loss Spectrometers 33
       2.2.3. Theoretical Considerations 46
       2.2.4. Experimental Results: Dissociation of NO
              and Coadsorption of NO and CO on Ru(001) . . . 66

   2.3. Reflection IR Spectroscopy 75

       2.3.1. Introduction 75
       2.3.2. Experimental IRAS 76
       2.3.3. Theoretical Considerations 81
       2.3.4. Experimental Results: The Adsorption of CO on
              Pd 84

   2.4. Inelastic Electron Tunneling Spectroscopy 95

       2.4.1. Introduction 95
       2.4.2. Experimental Details 96
       2.4.3. Theoretical Concepts 101
       2.4.4. Experimental Results 103

   2.5. Conclusions 113

       2.5.1. Introduction 113
       2.5.2. CO Chemisorption on Ru(001) 113
       2.5.3. Tabulation of Previous EELS and IRAS Results . . 121
       2.5.4. Update 124

3. Photoemission Spectroscopy of Valence States
   by Giorgio Margaritondo and John H. Weaver

   3.1. Introduction: Three-Step Model, Escape Depth, Relevant
        Parameters 127

       3.1.1. Photoemission: A Versatile Probe
              of the Valence States 127

3.1.2.	The Three-Step Model	129
3.1.3.	Escape Depth	130
3.1.4.	The Important Parameters	132

3.2. Instrumentation 133

    3.2.1. Electron-Energy Analyzers 134
    3.2.2. Photon Sources 140

3.3. Conventional Photoemission 144

    3.3.1. Density of States, Joint Density of States, and Energy Distribution of the Joint Density of States 145
    3.3.2. Transition Probabilities 146
    3.3.3. Interpreting the Energy-Distribution Curves . . . 147
    3.3.4. Processes without k Conservation 148
    3.3.5. Processing the Photoemission EDCs 149
    3.3.6. Experimental EDCs and Theoretical Curves: Bulk States 150
    3.3.7. Surface-States Studies in the EDC Mode 152
    3.3.8. Adsorption States 154

3.4. Polarized-Photon Photoemission 156

    3.4.1. Photon-Polarization Selection Rules 156
    3.4.2. Experimental Problems in the Photon-Polarization Techniques 158
    3.4.3. Photon-Polarization Effects in Angle-Resolved Photoemission: The Hermanson Rule 159

3.5. Angle-Resolved Photoemission: The Band-Mapping Technique 161

    3.5.1. The Surface Reciprocal-Lattice Vectors 162
    3.5.2. Mahan Cones 163
    3.5.3. Physical Information from the Angular Distribution of Photoelectrons 164
    3.5.4. Band-Structure Mapping for Two-Dimensional Crystals 165
    3.5.5. Band Structure of Adsorbed Overlayers 168
    3.5.6. Three-Dimensional Band-Structure Mapping . . . 169

3.6. Synchrotron Radiation Techniques 171

    3.6.1. Constant-Final-Energy-State Spectroscopy and Partial-Yield Spectroscopy 173

3.6.2. Constant-Initial-State Spectroscopy	177
3.7. Conclusions and Future Prospects	182
3.7.1. Future Prospects.	183

4. Core-Level Spectroscopies
by ROBERT L. PARK

4.1. Introduction	187
4.2. The Core-Level Structure of Atoms	188
4.2.1. Binding Energies.	189
4.2.2. Auger Yields and Lifetime Broadening.	190
4.3. The Interaction of Electrons with a Solid.	192
4.3.1. The Secondary-Electron Energy Distribution	193
4.3.2. The Loss Spectrum.	195
4.3.3. The Emission Spectrum.	197
4.3.4. The Inelastic Scattering Mean Free Path	198
4.4. Appearance-Potential Spectroscopy	198
4.4.1. Core-Hole Excitation.	199
4.4.2. Background Suppression	202
4.4.3. Resolution	205
4.4.4. Soft-X-Ray Appearance-Potential Spectroscopy	207
4.4.5. Auger Electron Appearance-Potential Spectroscopy	210
4.4.6. Disappearance-Potential Spectroscopy	213
4.5. X-Ray Photoelectron Spectroscopy	214
4.5.1. Core-Hole Excitation.	215
4.5.2. Electron Spectroscopy	218
4.5.3. Chemical Shifts	223
4.6. Comparison of Binding Energy Measurements	225
4.7. Electron-Excited Auger Electron Spectroscopy	226
4.7.1. The Auger Transition Energies	227
4.7.2. The Auger Line Shape	230
4.7.3. AES and Surface Composition	231

4.8. Extended Fine Structure Analysis of Surfaces 232
    4.8.1. Surface-Extended X-Ray Absorption Fine Structure 233
    4.8.2. Extended Appearance-Potential Fine Structure . . 235

5. Diffraction Techniques
by MAX G. LAGALLY

5.1. Introduction 237
5.2. Elements of Diffraction Theory 240
    5.2.1. Diffraction from Surfaces 242
    5.2.2. Surface Defects 245
5.3. The Measurement of Diffracted-Intensity Distributions . . 258
    5.3.1. Sensitivity 259
    5.3.2. Resolving Power 260
5.4. Surface Crystallography Measurements 267
    5.4.1. Diffraction Patterns 267
    5.4.2. Equilibrium Position Determinations 268
    5.4.3. Structural Defects 272
    5.4.4. Thermodynamics and Kinetics 277
5.5. Instrumentation and Sample Preparation 280
    5.5.1. Electron Guns 281
    5.5.2. Detectors 283
    5.5.3. Goniometers 288
    5.5.4. Sample Preparation 289
5.6. Representative Experimental Results 291
5.7. Conclusions 297

6. Ion Scattering and Secondary-Ion Mass Spectrometry
by W. HEILAND AND E. TAGLAUER

6.1. Ion Scattering Spectrometry 299
    6.1.1. Fundamental Aspects 299

6.1.2. Experiment. 312
6.2. Secondary-Ion Mass Spectrometry 328
    6.2.1. Fundamental Aspects. 328
    6.2.2. SIMS Experiments. 337
6.3. Synopsis. 347

7. High-Field Techniques
by J. A. PANITZ

7.1. Field-Electron-Emission Microscopy 349
    7.1.1. The Fowler–Nordheim Equation. 350
    7.1.2. The Field-Electron-Emission Microscope. . . . 353
    7.1.3. The Magnification of a Point-Projection
           Microscope. 357
    7.1.4. The Resolution of a Point-Projection Microscope . 358
    7.1.5. Adsorption Studies Using the FEEM 359
    7.1.6. Field Emission Energy Distributions. 361
7.2. Applications of Field-Electron-Emission Microscopy . . . 363
    7.2.1. Surface Diffusion 363
    7.2.2. Sputtering, Nucleation, and Electron Sources . . . 366
    7.2.3. Electrical Breakdown in High Vacuum. 368
    7.2.4. Molecular Imaging. 370
    7.2.5. Single-Atom Imaging. 374
7.3. Field-Ion Microscopy 376
    7.3.1. Field Ionization 378
    7.3.2. Field-Ion Energy Distributions 380
    7.3.3. The Hopping Gas Model 383
    7.3.4. The Low-Temperature Field-Ion Microscope . . . 384
    7.3.5. The Magnification of a Field-Ion Image 388
    7.3.6. The Modern Field-Ion Microscope 389
    7.3.7. Microchannel-Plate Image Intensification 390
    7.3.8. Surface Diffusion Studies Using the FIM 392
7.4. Field Evaporation. 397
7.5. Field-Ion Mass Spectroscopy 399
    7.5.1. Field Ionization Sources 401

7.5.2.	Liquid-Metal Field Desorption Sources	402
7.6.	Atom-Probe Mass Spectroscopy	404
	7.6.1. The Voltage-Pulsed Atom-Probe	404
	7.6.2. The Pulsed-Laser Atom-Probe	406
	7.6.3. Atom-Probe Measurements	407
	7.6.4. First-Layer Composition	407
	7.6.5. Composition of One Atomic Layer	408
	7.6.6. Average Composition of the Near-Surface Region	411
7.7.	Field-Desorption Microscopy	412
	7.7.1. The 10-cm Atom-Probe	414
	7.7.2. The Imaging Atom-Probe	414
7.8.	Molecular Imaging with Ions	416

8. The Thermal Desorption of Adsorbed Species
by JOHN T. YATES, JR.

8.1.	Introduction	425
8.2.	Early Studies of Desorption from Polycrystalline Substrates	427
	8.2.1. Introduction	427
	8.2.2. Material Balance Equation for Thermal Desorption in a Vacuum System	430
8.3.	Thermal Desorption from Single Crystals	431
	8.3.1. Introduction	431
	8.3.2. Preparation, Mounting, and Temperature Programming of Single Crystals	432
	8.3.3. Isothermal Desorption Measurements	436
	8.3.4. Isosteric Heat of Adsorption	438
	8.3.5. Absolute Coverage Measurements	441
	8.3.6. Detectors for the Study of Thermal Desorption	444
8.4.	Treatment of Experimental Desorption Data	447
	8.4.1. Desorption Kinetics Using Gas Evolution Measurements (Constant Rate of Heating)	447
	8.4.2. Desorption Kinetics Using Gas Evolution Measurements (Variable Rate of Heating)	449

8.4.3. Coverage Measurements Made during Programmed Desorption 451
8.4.4. Chemical Methods for Measuring Desorption Kinetics 453

8.5. Theories of Thermal Desorption 454

8.5.1. The Mobile Precursor Model in Adsorption and Desorption 454
8.5.2. Statistical Thermodynamics of Adsorption and Desorption 457

8.6. Concluding Remarks. 464

9. **Experimental Methods in Electron- and Photon-Stimulated Desorption**
by THEODORE E. MADEY AND ROGER STOCKBAUER

9.1. Theory and Mechanisms of Electron-Stimulated Desorption and the Relationship between ESD and PSD . 465

9.1.1. Experimental Observations 465
9.1.2. Mechanisms of Ion Formation and Desorption . . 467

9.2. Experimental Methods in ESD and PSD 473

9.2.1. Direct Detection of Desorbing Species 473
9.2.2. Evaluation of ESD and PSD Data: Cross Sections and Electron-Induced Surface Damage 498
9.2.3. Methods Based on Changes in Surface Properties: Detection of ESD and PSD Damage in Adsorbed Monolayers. 502
9.2.4. Measurements of Electron-Beam-Induced Damage in Thin Films and Bulk Samples 505
9.2.5. Electron-Stimulated Adsorption 511

9.3. Conclusions 513

AUTHOR INDEX 515
SUBJECT INDEX 531

CONTRIBUTORS

Numbers in parentheses indicate the pages on which the authors' contributions begin.

P. R. DAVIS (1), *Oregon Graduate Center, Beaverton, Oregon 97006*

W. HEILAND (299), *Universität Osnabrück, 4500 Osnabrück, Federal Republic of Germany*

MAX G. LAGALLY (237), *Department of Metallurgical and Mineral Engineering and Materials Science Center, University of Wisconsin, Madison, Wisconsin 53706*

THEODORE E. MADEY (465), *Surface Science Division, National Bureau of Standards, Gaithersburg, Maryland 20899*

GIORGIO MARGARITONDO (127), *Department of Physics, University of Wisconsin, Madison, Wisconsin 53706*

J. A. PANITZ (349), *Surface Science Division, Sandia National Laboratories, Albuquerque, New Mexico 87185*

ROBERT L. PARK (187), *Department of Physics and Astronomy, University of Maryland, College Park, Maryland 20742*

ROGER STOCKBAUER (465), *Surface Science Division, National Bureau of Standards, Gaithersburg, Maryland 20899*

L. W. SWANSON (1), *Oregon Graduate Center, Beaverton, Oregon 97006*

E. TAGLAUER (299), *Max-Planck-Institut für Plasmaphysik, 8046 Munich, Federal Republic of Germany*

JOHN H. WEAVER (127), *Synchrotron Radiation Center, University of Wisconsin, Madison, Wisconsin 53706, and Department of Chemical Engineering and Materials Science, University of Minnesota, Minneapolis, Minnesota 55455*

W. HENRY WEINBERG (23), *Division of Chemistry and Chemical Engineering, California Institute of Technology, Pasadena, California 91125*

JOHN T. YATES, JR.* (425), *Surface Science Division, National Bureau of Standards, Washington, D.C. 20234*

* Present address: Department of Chemistry, University of Pittsburgh, Pittsburgh, Pennsylvania 15260.

FOREWORD

This volume is the first treatment of the field of surface physics in this treatise. Surface preparation techniques were mentioned in Volume 6A, Solid State Physics, which also contained a description of the methods used to grow single crystals and produce thin films. Volume 14, Vacuum Physics and Technology, describes the methods of obtaining the vacua necessary to the preparation and maintenance of well-characterized surfaces. The advances described there were of course instrumental in the explosive growth in the field of surface physics.

Surface physics has become an extremely important part of the scientific landscape. This is due not only to its high degree of relevance to major industrial technologies, but also to the challenge posed by a discipline in which many fundamental questions remain to be decided. The editors and authors have done a fine job of presenting the main methods used by the surface scientist to answer these questions, making this volume a valuable addition to the treatise.

R. J. CELOTTA

PREFACE

The importance of the properties of surfaces is increasingly felt in a wide variety of technologies. Although some surface properties, such as chemical interactions leading to corrosion, have been recognized for many hundreds of years, the use of surfaces for specific applications is a recent phenomenon that is fueled by our increasing understanding of their properties. The technological success of devices and processes involving surfaces has in turn driven the development of increasingly sophisticated instrumentation designed to improve further our understanding of surface properties. This volume attempts to review, in a tutorial fashion, techniques that allow a definition of the major static properties of a surface. These can be classified as chemical or compositional, structural, and electronic. Because this volume is one in a treatise on experimental methods, the emphasis is deliberately on the techniques, their use, and the evaluation of measurements, rather than principally on the surface properties themselves.

The book is organized as follows. Part 1 deals with measurement of the work function and Part 2 with measurement of vibrational states of surface or adsorbed-layer atoms. Parts 3 and 4 describe the spectroscopy of energy levels in surface and near-surface atoms, with Part 3 emphasizing valence-level structure and Part 4 core-level structure. The use of such measurements to determine both surface composition and surface electronic properties is discussed. In Part 5, diffraction techniques, the major means for crystallographic analysis of surfaces, are discussed. Part 6 deals with ion-scattering methods from the point of view of using them as both structural and compositional analysis techniques. Part 7 discusses a specialized but powerful set of techniques, field-emission and field-ion microscopy, that permit atomic resolution in the image. Parts 8 and 9 cover two types of desorption of species from a surface, thermally induced desorption and electron- and photon-stimulated desorption. Both techniques give information on the nature of the chemical bonding at the surface.

It will be noted that no effort has been made to organize the parts into the major categories of surface properties mentioned earlier. This is deliberate: most of the techniques give information about surfaces that fit into more than one of the categories. The interweaving of parts on techniques whose principal functions differ emphasizes the interdisciplinary nature of research into the properties of surfaces and serves to accentuate the multiple secondary functions of a technique.

ROBERT L. PARK
MAX G. LAGALLY

METHODS OF EXPERIMENTAL PHYSICS

Editors-in-Chief
Robert Celotta and Judah Levine

Volume 1. Classical Methods
Edited by Immanuel Estermann

Volume 2. Electronic Methods, Second Edition (in two parts)
Edited by E. Bleuler and R. O. Haxby

Volume 3. Molecular Physics, Second Edition (in two parts)
Edited by Dudley Williams

Volume 4. Atomic and Electron Physics — Part A: Atomic Sources and Detectors; Part B: Free Atoms
Edited by Vernon W. Hughes and Howard L. Schultz

Volume 5. Nuclear Physics (in two parts)
Edited by Luke C. L. Yuan and Chien-Shiung Wu

Volume 6. Solid State Physics — Part A: Preparation, Structure, Mechanical and Thermal Properties; Part B: Electrical, Magnetic, and Optical Properties
Edited by K. Lark-Horovitz and Vivian A. Johnson

Volume 7. Atomic and Electron Physics — Atomic Interactions (in two parts)
Edited by Benjamin Bederson and Wade L. Fite

Volume 8. Problems and Solutions for Students
Edited by L. Marton and W. F. Hornyak

Volume 9. Plasma Physics (in two parts)
Edited by Hans R. Griem and Ralph H. Lovberg

Volume 10. Physical Principles of Far-Infrared Radiation
By L. C. Robinson

Volume 11. Solid State Physics
Edited by R. V. Coleman

Volume 12. Astrophysics — Part A: Optical and Infrared Astronomy
Edited by N. Carleton
Part B: Radio Telescopes; Part C: Radio Observations
Edited by M. L. Meeks

Volume 13. Spectroscopy (in two parts)
Edited by Dudley Williams

Volume 14. Vacuum Physics and Technology
Edited by G. L. Weissler and R. W. Carlson

Volume 15. Quantum Electronics (in two parts)
Edited by C. L. Tang

Volume 16. Polymers — Part A: Molecular Structure and Dynamics; Part B: Crystal Structure and Morphology; Part C: Physical Properties
Edited by R. A. Fava

Volume 17. Accelerators in Atomic Physics
Edited by P. Richard

Volume 18. Fluid Dynamics (in two parts)
Edited by R. J. Emrich

Volume 19. Ultrasonics
Edited by Peter D. Edmonds

Volume 20. Biophysics
Edited by Gerald Ehrenstein and Harold Lecar

Volume 21. Solid State: Nuclear Methods
Edited by J. N. Mundy, S. J. Rothman, M. J. Fluss, and L. C. Smedskjaer

Volume 22. Solid State Physics: Surfaces
Edited by Robert L. Park and Max G. Lagally

1. WORK FUNCTION MEASUREMENTS

By L. W. Swanson and P. R. Davis

Oregon Graduate Center
Beaverton, Oregon

1.1. Introduction

The determination of the surface work function is an important part of the overall characterization of a surface. Any change in the chemical or physical state of the surface, such as adsorption or geometric reconstruction, can be observed through a work function modification. In the case of cathode and related studies, the work function may be the single most important surface parameter to be measured. In this part we shall discuss a variety of methods used for these measurements. Most of the discussion will be devoted to those techniques that may be used for routine measurements of the work function to complement other surface analysis methods. Thus, particular attention will be given to the thermionic and field emission retarding-potential techniques as well as the capacitor methods, which give accurate, reproducible results with experimentally simple arrangements. On the other hand, several other methods will be given very little consideration. Diode, magnetron, and surface ionization techniques have limited applicability and will not be discussed at all. The reader is referred to review articles that discuss these techniques.[1-3]

We begin with a short discussion of work function theory for metal surfaces. The reader may wish to consult one or more of the numerous detailed articles on this subject.[4-7] The technniques of work function measurement will be divided into three categories. The first deals with methods involving the cathode of an electron emission system and includes thermionic, photoelectric, and field emission techniques. The second category also involves

---

[1] G. A. Haas and R. E. Thomas, *Tech. Met. Res.* **6**, Part 1, 91 (1972).
[2] J. C. Rivière, *Solid State Surf. Sci.* **1**, 180 (1969).
[3] L. Dobretsov, "Electronic and Ionic Emission" (NASA Tech. Transl. F-73, 1973).
[4] C. Herring and M. H. Nichols, *Rev. Mod. Phys.* **21**, 185 (1949).
[5] F. I. Itskovich, *Sov. Phys.—JETP (Engl. Transl.)* **23**, 945 (1966); **24**, 202 (1967).
[6] R. V. Culver and F. C. Tompkins, *Adv. Catal.* **11**, 67 (1959).
[7] A. L. Reimann, "Thermionic Emission." Chapman & Hall, London, 1934.

electron emission, but the work function being measured is that of the anode; these methods will be referred to as retarding-potential methods. Consideration will then be given to capacitive contact potential methods. Finally, experimental embodiments will be discussed and the various techniques will be compared.

1.2. Work Function Theory

1.2.1. Basic Concepts

When a large number of atoms are brought together to form a crystal, the potential inside the solid assumes the periodicity of the lattice, a periodicity that is terminated at the surface, as shown in Fig. 1. The electronic work function ϕ is influenced by two factors.[4] The first, the chemical potential μ, is determined solely by the bulk properties of the solid and is independent of the surface. The second, the surface potential χ, is very sensitive to the properties of the surface and is related to the electrostatic potential energy at the surface of the metal. The latter effect decays slowly with distance away from the surface. (As used here, the term *surface potential* refers to the difference between the electrostatic potentials inside and outside the metal and will be written $\chi = V_{out} - V_{in}$.) There are undoubtedly many contributions to the surface potential; some of the principal ones that have been considered theoretically are given here.

(1) Of primary importance is the penetration of the electrons beyond the surface of the conductor. This is a quantum effect that leads to the formation of a dipole layer (negative side outward) at the surface and an increase in the magnitude of the surface potential.

(2) Accompanying this "spill-over" of electrons is the decrease in the surface potential due to the decrease in the density of electrons within the metal. This energy is called the correlation energy and is due to the mutual repulsion of the electrons through Coulomb interactions and the Pauli exclusion principle.

(3) Still another contribution to the surface potential is the ion-core potential, which will vary with crystal plane. From what has been said, it is reasonable to expect that an adsorbed layer would alter the surface potential, especially if the adsorbed molecules exist in a charged or partially charged state on the surface.

The state of any solid, liquid, or gaseous phase containing electrons can be described thermodynamically by the *electrochemical potential* $\bar{\mu}$, where for a solid of essentially fixed volume v at constant temperature T, $\bar{\mu}$ is defined by

$$\bar{\mu} = (\partial F/\partial n)_{T,v}, \quad (1.2.1)$$

1.2. WORK FUNCTION THEORY

where $F = U - TS$ is the Helmholtz potential. The ambiguity associated with the way in which the entropy S depends upon the number of electrons is removed by requiring that $(\partial S/\partial n)_{T,v} \to 0$ as $T \to 0$. This is equivalent to defining $\bar{\mu}$ as the work done in bringing an electron from infinity and adding it isothermally to a metal of fixed volume. Thus any change in the potential at the surface of the metal will alter the electrochemical potential $\bar{\mu}$.

It is seen readily from Fig. 1 that the work function measured in electron volts can be written as

$$\phi = V_{\text{out}} - \bar{\mu}/e. \tag{1.2.2}$$

If we define the *chemical potential* as

$$\mu = \bar{\mu} - eV_{\text{in}}, \tag{1.2.3}$$

then

$$\phi = \chi - \mu/e. \tag{1.2.4}$$

Defined as it is, μ is a quantity dependent solely upon temperature and the bulk properties of the material, such as electron density. Consequently, changes in ϕ due to adsorption are directly related to changes in the surface potential χ.

The work function of a metal is changed as a result of the adsorption,[6] and one can view the adsorbed particles as having a discrete dipole moment that tends to modify the total dipole layer at the surface and consequently change the work function. A dipole layer of σ dipoles per unit area, all identically oriented normal to the surface, may be treated in first approximation as a parallel-plate capacitor with plate separation d. The plates then have a uniform charge density σe and the product of d and the electric charge e is the effective dipole moment μ_e. The external field of this layer is zero and the change in work function $\Delta\phi$ is equal to the change in surface potential $\Delta\chi$.

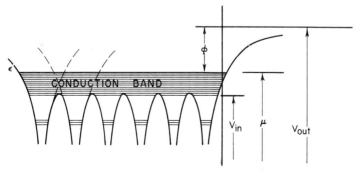

FIG. 1. Schematic energy diagram of a metal surface.

Thus,

$$\Delta\phi = \Delta\chi = 4\pi\,\Delta P, \tag{1.2.5}$$

but $\Delta P = \pm\sigma\mu_e$, so $\Delta\phi = \pm 4\pi\sigma\mu_e$.

If we write the coverage $\theta = \sigma/\sigma_0$ (σ_0 is the number of adsorption sites), then

$$\Delta\phi = \pm 4\pi\theta\sigma_0\mu_e. \tag{1.2.6}$$

We note immediately from this that the sign (positive for negatively charged ions) of μ_e is dependent upon the charge q of the adsorbed molecules and the separation d of the two charge layers (i.e., $\mu_e = qd$). The three cases corresponding to van der Waals, ionic, and covalent bonding are depicted in Fig. 2. An example of van der Waals adsorption would be inert gas adsorption in which little if any charge transfer with the substrate occurs and the dipole moment is contained within the adsorbate. Cesium adsorbed on refractory metals is an example of ionic adsorption that involves charge transfer to the substrate. Covalent adsorption involves localized orbital overlap and charge transfer between the adsorbate and surface atoms, e.g., oxygen on metals.

The magnitude of the dipole moment per adsorbed atom varies owing to a reduction in μ_e with decreasing lateral spacing (i.e., increasing θ). Although several mechanisms may contribute to the coverage dependence of μ_e, the most popular mechanism is that of Topping[8] which assumes mutual depolarization of adsorbed particles. This leads to $\mu_e(\theta)$, given by

$$\mu_e(\theta) = \mu_{e0}\,[1 + 9\alpha\,(\theta\sigma_0)^{3/2}]^{-1}, \tag{1.2.7}$$

where μ_{e0} is the dipole moment per adsorbed particle at $\theta = 0$ and α its effective polarizability. If we substitute Eq. (1.2.7) into Eq. (1.2.6), we can predict the well-known maximum or minimum in the $\Delta\phi$ versus θ curve.

Equating the work function measured by emission methods and the true work function must be done with care. One must be cognizant of the com-

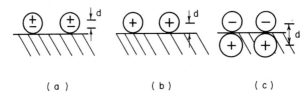

FIG. 2. Generalized diagram of charge separation for different types of adsorbate bonding: (a) van der Waals, (b) ionic, (c) covalent.

[8] J. Topping, *Proc. R. Soc. (London)* **A114**, 67 (1927).

plications associated with (1) the temperature dependence of the work function, (2) electronic band structure effects, and (3) patch fields. The patch field problems are eliminated when emission is measured from single crystal faces and the product of the applied field F and face size l exceeds the work function change between adjacent patches, i.e., $\Delta\phi < Fl$. The problems associated with items (1) and (2) have been discussed in detail.[4,5] Itskovich discusses the band structure effects and concludes that the photoelectric, field emission, and thermionic work functions ϕ_{PE}, ϕ_{FE}, and ϕ_{TE}, respectively, may not be lower than, but may be higher than the true or contact potential work function.

1.3. Electron Emission Methods

1.3.1. Thermionic Method

In thermionic emission, the metal sample is heated and electrons are emitted from the surface because the tail of the electron energy distribution within the metal extends above the vacuum level. The current density of electrons emitted from such a system is shown theoretically to be dependent upon the temperature of the cathode through the so-called Richardson equation:[4]

$$J = A(1 - r)T^2 \exp(-e\phi/kT) \qquad (1.3.1)$$

where ϕ is the work function of the metal, $A = 120 \text{ A cm}^{-2} \text{ deg}^{-2}$, and r is the reflection coefficient of electrons arriving at the work function barrier. Note that a graph (called a Richardson plot) of $\ln(J/T^2)$ against $1/kT$ is a straight line with slope equal to $e\phi$. Thus, a change in the work function of the cathode will show up as a change in the slope of the Richardson plot.

Experimentally, Richardson plots are generally found to be straight lines, although the intercept is usually not equal to the theoretical constant and ϕ is often temperature dependent. Thus, an experiment designed to measure the work function changes occurring during adsorption must include careful temperature control and measurement, as well as emission current measurements. The Richardson equation is reasonably valid for single crystal surfaces, but if patches of different work functions are present on the surface, emission will arise predominantly from one patch or another and lead to an average work function weighted toward the lowest value present.[3] Additional difficulty arises in attempting to measure changes in the work function due to adsorption because of the possibility of thermal desorption, an effect that is exaggerated if the adsorbate tends to increase the work function, because then even higher temperature is required to obtain emission. However, it is possible to measure the change in the work functions due to adsorption of alkali metals on refractory metals by this method because significant electron currents are emitted at relatively low temperatures.[7] Also, since the electric field must be uniform for the Richardson equation to be valid, precautions must be taken to provide a geometry amenable to this requirement. Generally speaking, plane-parallel geometries are the most convenient for this application.

The externally applied electric field must be high enough to overcome space-charge buildup between the emitting surface and the anode. However, this applied field causes an increase in the emission current density because of the Schottky effect, changing the Richardson equation to

$$J = A(1 - r)T^2 \exp\{-e[\phi - (eF)^{1/2}]/kT\}, \qquad (1.3.2)$$

where F is the applied field. This effect is normally eliminated by extrapolating the linear portion of a $\ln J$ versus $F^{1/2}$ plot to zero field at constant T and using the zero-field current density values in determining the Richardson slope. Thus, a single work function determination requires measurement of J for several values of F at fixed T and repetition of the procedure at several different values of T.

A complication arises if the work function is temperature dependent, either intrinsically or through temperature-induced changes in the surface layer. This effect may cause the Richardson plot to be nonlinear or linear over some range, with slope dependent upon $d\phi/dT$. The apparent work function ϕ will be related to the true work function ϕ_0 by

$$\phi = \phi_0 - T(d\phi/dT). \qquad (1.3.3)$$

For a well-defined emitting area, such as the active surface of a cathode, the emitted current density can be calculated simply by dividing the total emission current by the emitting area. Otherwise, a collector/guard-plate assembly can be used to measure a fraction of the emitted current, with the known area of the collector being used to calculate the current density. In any case, the collector should be of material that can be cleaned to prevent thermal- and electron-stimulated desorption of gases during measurements. For low-work-function emitters, collector overheating must be prevented by water cooling or by pulsing the accelerating voltage with low duty cycle and monitoring peak voltage and emission current.

Despite its limitations, the thermionic method of work functions determination has widespread application. The experiment is simple to set up and perform and requires no specialized electron gun, probe, or electronics. In addition, measurements of emission current density changes during changing surface conditions (such as desorption of an adsorbate) are sometimes very informative. Other work function measurement techniques are often difficult to apply with the sample at elevated temperature.

1.3.2 Photoelectric Emission

When electromagnetic radiation of sufficiently high frequency is directed on a metal surface, electrons are emitted in a process known as photoelectric emission.[9] The threshold frequency for emission ν_0 is related to the work function by the expression

$$\phi = h\nu_0/e. \qquad (1.3.4)$$

Hence, changes in work function of the metal surface are given by

$$\Delta\phi = h(\nu - \nu_0)/e. \qquad (1.3.5)$$

Because of competition from thermal emission, the threshold for photoemission is not always easily determined, and it has been found useful to apply a method devised by Fowler to determine ν_0. The basic equation in the Fowler analysis[10] is written

$$\ln(I/T) = B + F(X), \qquad (1.3.6)$$

where I is the photocurrent for a fixed photon flux, B is a constant independent of the frequency, $F(X)$ is a function having the same form for all metals, and

$$X = h(\nu - \nu_0)/kT. \qquad (1.3.7)$$

Specifically,

$$F(X) = \ln\left\{\frac{\pi^2}{6} + \frac{1}{2}X^2 - \left[\exp(-X) - \frac{\exp(-2X)}{2^2}\right.\right.$$
$$\left.\left. + \frac{\exp(-3X)}{3^3} - \cdots\right]\right\}, \qquad (1.3.8)$$

where $X \geq 0$.

Thus, a graph of $\ln(I/T^2)$ versus $h\nu/kT$ will yield a curve identical to the theoretical $F(X)$ versus X curve except that it will be shifted in the vertical direction by an amount B and in the horizontal direction by an amount $h\nu_0/kT$. The latter shift is used to obtain ν_0. The threshold measurement is thus reduced to determination of the spectral distribution of photoelectric yield near ν_0 and comparison with the function $F(X)$. This technique requires a photon source (light source with monochromator or filters) of calibrated spectral intensity and variable energy in the range around ν_0.

For measurements made on patchy surfaces having more than one work function, the photoelectric method is subject to the same problems as the

---

[9] G. L. Weissler, *Handb. Phys.* **21**, 304 (1956).
[10] R. H. Fowler, *Phys. Rev.* **38**, 45 (1931).

thermionic emission method. Other difficulties that are inherent in the photoelectric method of measuring the work function are (1) in order to obtain a precise value for ϕ, currents of the order 10^{-14} A must be measured in the region of ν_0; (2) for work functions of 5 V or more, the threshold frequency is in the ultraviolet region, which introduces numerous experimental problems. In principle, however, the photoelectric method involves the simplest experimental arrangement of any method discussed here.

1.3.3. Field Electron Emission

The principles and experimental adaptations of high-field phenomena are discussed in detail elsewhere.[11] However, since field emission measurements have contributed greatly to our current understanding of the surface work function, some background discussion here is warranted.

Field emission is a quantum mechanical phenomenon with no classical analogue. When a sufficiently high electric field is applied to the surface of a metal or semiconductor, the surface potential barrier is deformed to provide a finite length through which an electron within the metal can "tunnel." This phenomenon can be formulated mathematically by considering a Fermi sea of electrons within the metal impinging on the surface.[11] If we multiply this impingement rate by the appropriate quantum mechanical transmission coefficient and then multiply by the electronic charge and integrate over the electron energy limits $-\infty$ to ∞, we obtain an expression for the current density of the field emitted electrons, the Fowler–Nordheim formula, which can be rewritten in terms of the directly measurable field emission current I and applied field F in the form

$$I = A_f F^2 \exp(-a/F). \qquad (1.3.9)$$

In view of this relation and $F = \beta V$ it follows that a *Fowler–Nordheim* (FN) *plot* of the current–voltage relationship [$\ln(I/V^2)$ versus $1/V$] yields a straight line having a slope a and an intercept A_f. The linearity of a FN plot is normally regarded as adequate proof that the emission is due to stable field emission. When the Fowler–Nordheim law is satisfied, it can be shown that ϕ is related to the slope a by the expression

$$a = 2.96 \times 10^7 \phi^{3/2} s(y)/\beta, \qquad (1.3.10)$$

where $s(y)$ is a tabulated function equal to 0.95 ± 0.009 over the range of current densities normally encountered. If the work function ϕ_0 of a clean surface is used as a reference, then the work function ϕ_1 of the surface when

---

[11] R. H. Good, Jr. and E. W. Müller, *Handb. Phys.* **21**, 176 (1956).

1.3. ELECTRON EMISSION METHODS

coated with an adsorbate can then be determined from

$$\phi_1 = \phi_0 \left[\frac{a_1 s(y_0)}{a_0 s(y_1)} \right]^{2/3}, \qquad (1.3.11)$$

where a_0 and a_1 are the slopes of the corresponding Fowler–Nordheim plots. Absolute determination of the work function requires a knowledge of the geometric factor β, which can be determined from an electron micrograph of the emitter profile with an accuracy of about 15%. If the work function of the clean surface is well known, both β and the work function of the coated surface can be determined with good accuracy.

The properties of the clean or contaminated emitter surface become apparent only through their effects on the electron emission. The variation in the electron emission over the surface of a smooth hemispherical and atomically clean field emitter is caused by variation of the work function with crystallographic direction. The use of Eq. (1.3.11) to obtain the work function change due to adsorption is strictly correct only if F and ϕ are constant over the emitting region, which is seldom the case. In general the electron emission is averaged over a variety of crystal faces with widely varying work functions. It can be shown that the slope of a FN plot yields an average value of $\phi^{3/2}/\beta$, which is related to the individual regions by

$$\phi^{3/2}/\beta = \sum_i f_i \phi_i^{3/2}/\beta_i, \qquad (1.3.12)$$

where f_i is the fraction of the total current carried by the ith emitting region.

In practice, the curvature of the FN plot over the voltage range is negligible, so that its slope yields a constant $\phi^{3/2}/\beta$. It can be seen from Eq. (1.3.12) that the experimentally obtained value is weighted heavily in favor of the highly emitting low-work-function regions and may in practice be almost identical to the lowest ϕ encountered on the emitter surface. Thus, in utilizing Eq. (1.3.11) to determine work function changes on adsorption it should be remembered that average values are obtained, and when adsorption greatly alters the emission distribution from that of the clean surface, these values cannot always be related directly to contact potentials.

The difficulties inherent in the determination of averages can be overcome by measuring the electron current from individual crystal planes by suitably designed current probes. A tube that allows measurement of the emission from a single plane is shown in Fig. 3. By utilizing the average work function versus coverage relationships and the results of probe-tube measurements, it is possible to establish the relationship between the work function of individual crystal planes and average coverage.

The major problem with work function determination by the field emission method is that a highly specialized sample configuration must be used. Only at a very sharp point (radius $\sim 10^{-5}$ cm) can sufficiently high fields be maintained for emission measurements. This requirement precludes the use of other surface analysis techniques. On the other hand, for specialized applications, field emission may be the only method or the most convenient method available. For example, it has been possible to measure directly the work function change induced by the adsorption of a single atom on a tungsten plane.[12]

At very low adsorbate coverages a small correction to the field emission work function due to adsorbate dipole moment discreteness should be considered. This problem has been discussed by Gomer.[13] In addition, a small correction due to the finite temperature can be included. In this case Eq. (1.3.11) becomes

$$\phi_1 = \phi_0 \left[\frac{a_1(1+\delta)}{a_0 s(y_1)} s(y_0) \right], \quad (1.3.13)$$

where

$$\delta = V(1 - \pi p \cot \pi p) \quad (1.3.14)$$

and

$$p \approx kT(8m\phi)^{1/2}/\hbar eF. \quad (1.3.15)$$

It has been pointed out that the true work function of the surface could be obtained by a combination of measurements including the slope of the FN plot [Eq. 1.3.9] and the slope of a' of the tail of the total energy distribution.

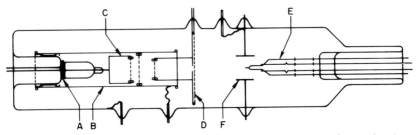

FIG. 3. Diagram of probe tube used for measuring field emission currents from cesium layers on single crystallographic planes. A, guard ring; B, suppressor grid support; C, collector; D, platinum anode containing a 2.5-mm hole and coated with a fluorescent screen; E, emitter assembly; F, quadrupole deflection plates.

[12] C. J. Todd and T. N. Rhodin, *Surf. Sci.* **42**, 109 (1974).
[13] R. Gomer, *J. Chem. Phys.* **21**, 1869 (1953).

Unfortunately, band structure effects for nonfree electron metals, which have little effect on the FN slope, dramatically alter a', thus rendering this method somewhat suspect for most transition metals. However, Vorburger et al.[14] have shown how the band structure problems can be partially overcome by proper analysis, thereby making the method more applicable.

1.4. Retarding-Potential Methods

In each of the techniques discussed so far, the work function is determined from the electron emission properties of a surface, with emission produced by one of several possible mechanisms. In this chapter, we consider methods based upon the low-energy electron collection properties of the surface being studied. These retarding-potential methods have the inherent advantage that the surface under study need not be heated nor be subjected to high fields or irradiation. On the other hand, the experimental arrangements tend to be somewhat more complicated than those used for emission methods.

1.4.1. The Shelton Triode

Consider a plane-parallel two-electrode system consisting of a cathode and an anode, with the cathode heated to a temperature T and possessing a uniform work function ϕ_c and the anode at a low temperature and having a uniform work function ϕ_a. If a potential difference is applied to this system such that a small accelerating field exists at the cathode surface, the emission current density is given by the Richardson equation [Eq. (1.3.1)]. In this case the barrier to emission is simply the cathode work function ϕ_c (Fig. 4a). However, if a negative potential V of sufficient magnitude is applied to the anode, the barrier to emission becomes $\phi_a - V$ (Fig. 4b), and it can be readily shown[3] that the emission current density is given by

$$J_e = AT^2 \exp[e(-\phi_a + V)kT]$$
$$= J_0 \exp[e(\phi_c - \phi_a + V)/kT]. \qquad (1.4.1)$$

Implicit in the derivation of Eq. (1.4.1) is the conservation of the Maxwellian normal energy distribution.

In practice, the electron energy distribution in this dioide arrangement is modified by space-charge effects and by field distortion at sample edges. In the technique devised by Shelton,[15] a third electrode consisting of a thin plate with a central aperture is inserted between the emitting and collecting elec-

---

[14] T. V. Vorburger, D. Penn, and E. W. Plummer, *Surf. Sci.* **48**, 417 (1975).
[15] H. Shelton, *Phys. Rev.* **107**, 1543 (1957).

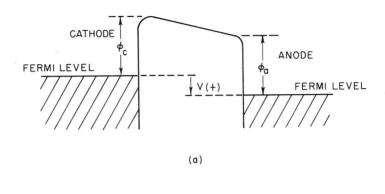

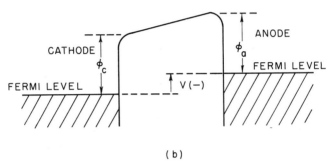

FIG. 4. Electron motive diagram for a diode in (a) accelerating-potential mode, $V \geq \phi_c - \phi_a$, and (b) retarding-potential mode, $V \leq \phi_c - \phi_a$.

trodes. This intermediate electrode is operated at a potential sufficiently high to overcome space charge. In addition, a strong uniform axial magnetic field is applied to preserve the Maxwellian normal energy distribution and to define the area for electrons emitted from the cathode and passing through the aperture in the intermedite electrode. The electrode spacing must be such that all electrons passing through the aperture are collected at the anode.

Under these conditions, a plot of ln J_e, or more conveniently ln I, as a function of anode voltage V results (a) in a constant value independent of V for $V \geq (\phi_c - \phi_a)$ or (b) in a linear variation of ln I with V, giving a slope $1/kT$ for $V \leq (\phi_c - \phi_a)$ (see Fig. 5). The voltage at the point of intersection of these two straight lines, the so-called knee of the curve, is just equal to the difference in the work functions of the two surfaces (their contact potential).

Nonuniform work functions and deviations in geometry from the plane-parallel case (i.e., effects that disturb the conservation of normal energy) affect the I–V curve primarily by rounding the knee. The position of the

1.4. RETARDING-POTENTIAL METHODS

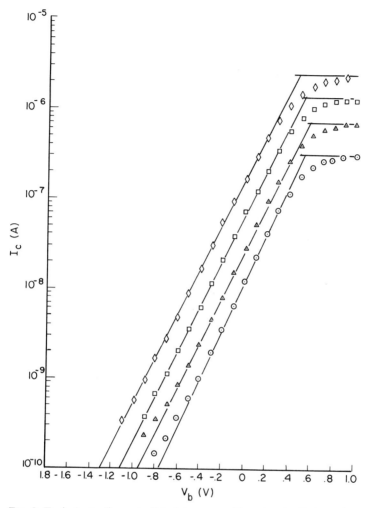

FIG. 5. Typical retarding-potential plots measured by using the Shelton triode.

knee under these circumstances can be obtained by extrapolation of the two straight lines of cases (a) and (b) until they intersect.

Increasing the temperature of the cathode increases the current at any value of V. Plotting the current as a function of temperature in the form of a Richardson plot [$\ln(I/T^2)$ versus $1/T$] for fixed V permits determination of the work function of the cathode for case (a) and of the anode for case (b) from the slopes of the resulting straight lines. Thus, the Shelton method combines both retarding-potential and thermionic emission measurements in one experiment. Even the cathode temperature need not be measured, in principle, since it may be determined from the slope of ln I versus V for $V \le (\phi_c - \phi_a)$.

A variation of work function with temperature affects the $I-V$ curves in two ways: (1) the position of the knee is shifted along the V axis, and (2) the measured thermionic constants A_m and ϕ_m for the heated cathode will differ from the constants A and ϕ in Eq. (1.3.1). For example, if we assume $\phi_c = \phi_{oc} + \alpha T$, then $\alpha = \Delta V/\Delta T$, where ΔV, is the shift in the knee for a temperature change ΔT. Equation (1.3.7) thus becomes

$$J_0 = AT^2 \exp[-e(\phi_{oc} + \alpha T)/kT]$$
$$= AT^2 \exp(-e\alpha/k) \exp(-e\phi_{oc}/kT) \qquad (1.4.2)$$

if it is assumed that other complications such as reflection are negligible in comparison with the temperature effect. Thus, $\phi_m = \phi_{oc}$, and the temperature coefficient of the work function α may be obtained from

$$A_m = A \exp(-\alpha/k). \qquad (1.4.3)$$

The Shelton method is limited in application primarily by the bulky magnet maintaining the axial field. It is experimentally complex to adapt the triode assembly to a movable collector configuration that would enable adsorption experiments and sputter cleaning. When low-work-function surfaces are studied, photoemission from the collector surface to the intermediate electrode may be a serious problem. Finally, material evaporation during thermal cycling may cause cross-contamination of the electrode surfaces, thus making the technique suitable primarily for the study of refractory materials. However, in its limited range of application the Shelton technique can be very useful.

1.4.2. Electron-Beam Method

This technique, originally used by Anderson,[16] has broader application and greater flexibility than the Shelton method but involves the same physical principles. In this case the cathode and intermediate electrode are combined to form an electron gun, which generally has other lenses as well and which produces a parallel electron beam of small cross section. This beam impinges on the sample surface (anode) at normal incidence, and a curve of collected current versus cathode–anode potential difference V is plotted. In general, the cathode work function is not known, so the contact potential provides no new information. However, if the anode surface is changed by modifying the surface or by examining a different sample, and the character-

[16] P. A. Anderson, *Phys. Rev.* **47**, 958 (1935).

1.4. RETARDING-POTENTIAL METHODS

istic curve is replotted, then the shift in the curve, or change in contact potential, is equal to the change in anode work function. This method, like all contact potential methods, yields relative work function values rather than absolute values.

Several conditions must be met if accurate results are to be obtained. The cathode work function must remain constant throughout the experiment since the cathode is effectively a reference surface. Thus, adsorption of gases on the anode must be done with care to prevent contamination of the cathode. In practice, this means operating the cathode at a temperature high enough to prevent adsorption or thermally cleaning the cathode before each measurement. The measured $I-V$ curves must have the same shape and be parallel over their entire range. Therefore, the cathode temperature must be constant, the anode position highly reproducible, and the beam accurately normal to the anode surface. If these criteria are met, the electron-beam method provides an accurate, convenient means of measuring work function on a routine basis. The technique is ideally suited for use in conjunction with the other methods of surface analysis discussed throughout this book.

1.4.3. Field Emission Retarding-Potential Method

The field emission retarding-potential (FERP) method of work function measurement is also highly compatible with other surface analysis techniques. It has most of the advantages of the previously discussed electron-beam method but has the added benefit of providing an absolute measure of the sample work function. In addition, the resolution of the technique is excellent.

In the FERP method, originally developed by Holscher[17] and further developed and discussed in detail by Strayer et al.,[18] electrons emitted from a field emission cathode are focused and collimated in an electron lens system and made to strike the analyzed surface at normal incidence. From the potential energy diagram for this arrangement (Fig. 6), it is clear that the threshold for collection of the emitted electrons is the work function of the collector. This is true since, in field emission, electrons are emitted with energies near the Fermi energy of the cathode with a total energy spread of only a few tenths of an electron volt, as shown theoretically by Young and Müller.[19]

Figure 7 shows a schematic diagram of a FERP setup. Typical gun-element potentials are 0 V for the emitter, 1000 V for the anode, 100 V for the lens, 8–20 V for the mesh, V_c for the sample, variable from fully retarding to

---

[17] A. A. Holscher, *Surf. Sci.* **4**, 89 (1966).
[18] R. W. Strayer, W. Mackie, and L. W. Swanson, *Surf. Sci.* **34**, 225 (1973).
[19] R. D. Young and E. W. Müller, *Phys. Rev.* **113**, 115 (1959).

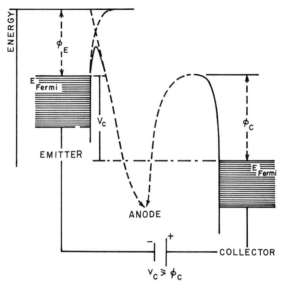

FIG. 6. Potential-energy diagram for FERP setup.

~20 V above the collection threshold. By using tip modulation and phase-locked detection (not shown), the derivative of the total current can be measured to give the energy distribution of collected electrons. Theoretically, this distribution is described by[20]

$$P(E) = C \frac{\exp(E/d)}{\exp[(E - E_F)/kT] + 1}, \quad (1.4.4)$$

where

$$d = \hbar eF/[(2m\phi_e)^{1/2}t((e^3F)^{1/2}\phi_e)], \quad (1.4.5)$$

where ϕ_e is the emitter work function, e electronic charge, $2\pi\hbar$ Planck's constant, k Boltzmann's constant, C constant, and t a slowly varying elliptic function, and where an emitted electron has energy $E = 0$ at infinity. At an emitter temperature T of 0 K the maximum in the energy distribution corresponds to the Fermi energy E_F of the emitter. Thus, the potential V_c (Fig. 6) required to collect electrons at the distribution maximum is equal to the work function ϕ_c of the collector. At practical temperatures, the distribution will be broadened and its maximum shifted because of population of states above the Fermi energy. The effect of temperature may be taken into

[20] R. D. Young, *Phys. Rev.* **113**, 110 (1959).

1.5. CAPACITANCE METHODS

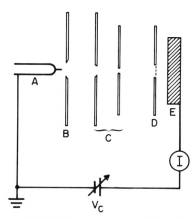

FIG. 7. Schematic diagram of FERP electron gun. A, emitter; B, anode; C, lens assembly; D, mesh screen; E, sample.

account, but even without this compensation the method can measure ϕ_c absolutely to within 1%, in principle.

In practice, an instrumental broadening is folded into the true energy distribution to yield the observed distribution curve. It is possible to extract the actual distribution from the data, as discussed by Young and Kuyatt[21] and Reifenberger et al.[22] For typical instruments, the effect of broadening is an increase in apparent ϕ_c by ~ 50 mV. The overall uncertainty in the absolute value of ϕ_c is of about the same magnitude.

Since the energy of the emitted electrons is referenced to E_F rather than ϕ_e, work function and electric field changes of the emitter have little effect on the measured value of ϕ_c. However, because of the increase in beam noise due to cathode sputtering at $P \gtrsim 1 \times 10^{-9}$ Torr, the measurements must be performed in a good vacuum.

1.5. Capacitance Methods

1.5.1. The Kelvin–Zisman (Vibrating Capacitor) Method

The contact potential difference between two metals connected electrically is just equal to the difference in their respective work functions. The contact potential difference arises from the transfer of electrons from the metal of lower work function to the second metal until their Fermi levels line up as illustrated in Fig. 8. If the two metals are arranged as the two plates of a

---

[21] R. D. Young and C. E. Kuyatt, *Rev. Sci. Instrum.* **39**, 1477 (1968).
[22] R. Reifenberger, H. A. Goldberg, and M. J. G. Lee, *Surf. Sci.* **83**, 599 (1979).

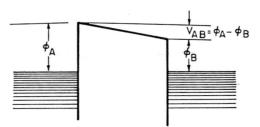

FIG. 8. Potential energy diagram for Kelvin probe.

capacitor, then in the absence of an applied emf, an electric field exists in the region between the plates. By the application of a compensating potential, the field between the plates can be reduced to zero. In other words, the applied potential necessary to reduce the electric field between the plates to zero is just equal to the contact potential difference.

Experimentally, one of the capacitor plates is used as a reference electrode of known work function and the work function of the second plate is determined by measuring the applied potential required between the plates. Alternatively, the reference electrode may be mounted on a rotatable mount in a manner identical to the experimental sample. Then, from the sample-to-measuring-electrode and reference-to-measuring-electrode contact potential differences, the work function of the sample may be deduced.

The actual measurement is made by causing the measuring (probe) electrode to vibrate at a fixed frequency. When there is a nonzero electric field between the two electrodes, an alternating current will flow in the external circuit. When the electric field between the plates is reduced to zero by the applied potential, the ac current likewise reduces to zero. For low noise and maximum sensitivity, phase-locked detection is usually employed in the measuring circuit, and a feedback loop may be used for automatic nulling of the field between the electrodes. Care must be taken with shielding around the measuring electrode to prevent stray electric and magnetic fields from interfering with the measurements. A clever, compact Kelvin probe design, completely compatible with ultrahigh-vacuum requirements and incorporating feedback nulling and continuous readout has been discussed Desplat.[23]

The major difficulty encountered in use of the Kelvin probe method is the necessity of maintaining a reference surface of known work function. This problem may be lessened by using an inert surface like gold whose work function change is small due to adsorption of residual gases. Alternatively, the reference surface may be cleaned before each measurement, which

[23] J.-L. Desplat, *NASA [Contract. Rep.] CR* **NASA-CR-152272** (1979).

makes the apparatus more complicated unless the previously mentioned three-electrode technique is used.

1.5.2. Static Capacitor Methods

A related method for determining the zero-electric-field condition between two parallel plates is by measuring directly the tendency for charge to flow as a result of *changing* the work function of one of the surfaces such as might occur during adsorption. In this static method, the tendency for charge to flow through the external circuit is detected electronically and used to establish very quickly the correct value for the compensating potential. The sensitivity of this method is of the order of 30 mV.[24]

1.6. Comparison of the Various Measuring Techniques

The measurement of surface work function means determination of the energy difference between the Fermi level and the top of the potential energy barrier. We expect to measure different apparent values of ϕ by the various techniques due to changes in the barrier height, band structure effects, reflection problems, and so forth. For example, the Schottky effect must be corrected for in thermionic emission measurements, and thermal effects must be considered in field emission and photoemission measurements. The effects of patch fields at polycrystalline surfaces have been discussed elsewhere[1,2,4] and will not be discussed here, except to note that emission measurements on a polycrystalline surface will yield an "average" work function strongly weighted toward the lowest work function patches. Capacitance and retarding methods, on the other hand, will yield geometrically weighted average values.

For a single crystal surface, the capacitance methods of work function measurement should yield true values of contact potential difference between the sample under study and the reference surface. These methods do not depend upon detailed knowledge of electron emission mechanisms, nor are they affected by surface reflection of slow electrons or other similar properties. They simply measure the difference in surface potential between two electrodes whose Fermi energies are equalized. Because this type of measurement is usually made at room temperature, thermal effects are generally not serious. There is a major drawback, however, in these methods. The work function of the reference surface must be precisely known if absolute determination of the sample work function is to be made. Even when only relative changes are to be measured, as in adsorption experiments,

---

[24] T. Delchar, A. Eberhagen, and F. C. Tompkins, *J. Sci. Instrum.* **40**, 105 (1963).

the reference surface work function must be stable, either unaffected by adsorption or cleanable before each measurement.

The various emission techniques depend upon precise knowledge of the emission mechanism involved. The theories of thermionic emission, field emission, and photoemission are based upon idealized models and do not take into account the detailed band structure in the region of the Fermi level (field emission) or near the top of the emission barrier (thermionic emission and photoemission). As discussed by Itskovich,[5] the emission (and consequently the barrier heights measured by the emission methods) may correspond exactly with theory only in the free-electron limit. Deviation from the free-electron case, as observed to a greater or lesser extent in all metals, can result in an apparent work function greater than the true barrier height when measured by the emission methods. One would expect the three emission methods to deviate in different ways, so that their resulting measured work functions may differ. Nevertheless, emission methods have proved to be extremely useful and are usually convenient to set up. In cathode work, an "effective" thermionic work function may be all that is desired. In detailed surface studies, however, it is desirable to use other techniques in conjunction with, or in place of, one or another of the emission methods.

The retarding-potential methods have perhaps the broadest applicability of all those discussed here. A variety of experimental arrangements may be used, including the scanning low-energy probe technique (SLEEP) developed by Haas and Thomas,[25] which can be used to study work function variation over a nonuniform surface. When the field emission retarding-potential (FERP) technique is used, absolute work function values may be determined. In general, the emission properties of the electron source may be checked independently to verify cleanliness, but, to first order, this is unnecessary in the FERP case.

The retarding-potential measurements are expected to be affected by sample band structure in analogy to the emission methods, by application of the principle of detailed balance at the surface barrier. Thus, the energy distribution of collected electrons may differ from the energy distribution in the incident beam. For example, strong reflections near the onset of electron collection (top of the energy barrier) that have been observed for (111) LaB_6.[26] and (111) Ir[18] are believed to be due to band gaps at these energies. In addition, periodic variations in reflection occur at higher energies for various materials[18] owing to Bragg and other reflection mechanisms. These observations constitute the basis for the low-energy electron reflection

---

[25] G. A. Haas and R. E. Thomas, *Surf. Sci.* **4**, 64 (1966).
[26] L. W. Swanson, D. R. McNeely, and M. Gesley, *NASA [Contract. Rep.] CR* **NASA-CR-159649** (1979).

(LEER) technique developed by Haas et al.,[27] by which it is possible in the case of single crystal surfaces to separate the bulk (μ/e) and surface (χ) contributions to the work function. Thus, reflection measurements may provide additional information on the surface, complementary to the work function determinations, without use of any additional equipment. This is a rarely exploited advantage of the retarding-potential methods.

1.7. Experimental Embodiments of the Various Techniques

In this section we highlight applications of the various techniques already discussed, concentrating on experiments that are either particularly innovative, typical of a large variety of possible arrangements, or of special interest for some other reason. We begin with the field emission microscope,[28] but, since a thorough discussion appears in Part 7 by Panitz, we shall make only a few comments here. Much of the early understanding of adsorption was obtained with this device. The instrument yields a highly magnified image of a metal point, displaying the work function topography of the various crystal faces. Thus, relative ordering of the work functions of the different planes is apparent. Using a probe technique to collect current only from a given plane, the variation of work function with adsorption on that plane can be determined. The technique has proved to be very adaptable and has yielded a large amount of data on single crystal work functions.

Both thermionic emission and photoemission have also been adapted to microscope form. The thermionic emission microscope[29] yields information on emitted current distributions from nonuniform surfaces and has found particular application in the study of cathode emission properties. When used in conjunction with techniques yielding information on surface composition, this microscope allows areas of high emission to be pinpointed and their chemical nature determined so that the physical processes leading to efficient electron emission can be understood.[30,31]

Retarding-potential methods have been used in a variety of adaptations, including the scanning arrangement (SLEEP) mentioned earlier.[25] The SLEEP method has a wide range of applicability and complements other analysis techniques such as scanning electron microscopy (SEM) and scanning Auger microscopy (SAM). It is particularly useful for examining metal-

[27] G. A. Haas, A. Shih, and R. E. Thomas, *J. Appl. Phys.* **47**, 5400 (1976).
[28] R. Gomer, "Field Emission and Field Ionization." Harvard Univ. Press, Cambridge, Massachusetts, 1961.
[29] W. Heinze and S. Wagener, *Z. Tech. Phys.* **20**, 16 (1939).
[30] R. A. Tuck, *Appl. Surf. Sci.* **2**, 128 (1979).
[31] R. E. Thomas, T. Pankey, J. W. Gibson, and G. A. Haas, *Appl. Surf. Sci.* **2**, 187 (1979).

lographic specimens and other nonuniform surfaces. By using synchronous scanning of the beam and an oscilloscope trace, the work function topography of a surface may be studied. Alternatively, the beam may be moved in finite steps, with the relative work function being determined by the beam voltage required to maintain a constant collected current. Comparison with a known standard surface (clean tungsten single crystal) allows a plot of the absolute work function distribution to be made. Spatial resolution of this technique is of the order of several micrometers, and the work function resolution may be a few tens of milli-electron-volts.

Another device for work function analysis of nonuniform surfaces is the reflected electron microscope. In one form,[32] this device is similar to the thermionic emission microscope except that the imaged electrons are reflected rather than emitted from the surface being analyzed so that cold samples may be studied. A broad incident beam is used and the contact potential difference between the electron source and sample surface is adjusted so that electrons are reflected from patches with ϕ greater than a predetermined value and collected by patches with ϕ smaller than this value. An alternative reflection microscope consists of a scanning electron gun and a reflected electron collector.[33] This device is similar to the SLEEP except that reflected rather than collected electrons are used to form the oscilloscope image or to make point-by-point work-function determinations.

In a study by Butz and Wagner[34] a high-resolution Kelvin probe was constructed to measure the diffusion rate of an adsorbed layer.

1.8. The Future of Work Function Measurements

Measurement of the work function of a surface is an important part of overall surface characterization. The development of other surface analysis techniques, e.g., probes of surface composition and electronic and geometrical structure, does not decrease the importance of work function measurements. Indeed, these developments have encouraged expansion of work function studies to include, for example, SLEEP and LEER and have stimulated improvement of FERP, Kelvin probe, and other techniques for better precision and spatial resolution. Continued development of the large variety of surface electron spectroscopies will likely cause further evolution of work function techniques. The retarding-potential methods should be particularly affected because of the wealth of untapped information simultaneously available in easily accessible electron reflection measurements.

[32] L. Mayer, *J. Appl. Phys.* **26**, 1228 (1955).
[33] K. Bauer, *Z. Angew. Phys.* **20**, 294 (1966).
[34] R. Butz and H. Wagner, *Appl. Phys.* **13**, 37 (1977).

2. VIBRATIONS IN OVERLAYERS

By W. Henry Weinberg

Division of Chemistry and Chemical Engineering
California Institute of Technology
Pasadena, California

2.1. Introduction

One of the most important areas of current scientific research is unquestionably the general field of surface and interfacial science. The reason for this significance is that the applications of research in this field are both diverse and extremely important to modern technology. For example, the following three major applications come to mind immediately: (1) chemisorption, heterogeneous catalysis, and catalyst promotion and poisoning; (2) properties of materials that are related to surface phenomena, such as corrosion, adhesion, lubrication, friction and wear, protective coatings, two-dimensional clustering and nucleation, alloying, and epitaxial-growth mechanisms; and (3) the long-term stability and reliability of thin-film microelectronic devices and optical detectors. Of these, it is certainly clear that research in surface chemistry has direct and crucial applications to the fields of catalysis and materials science. The relevance to long-life microelectronic devices is perhaps not so obvious. However, the failure of these devices [e.g., metal–oxide–semiconductor (MOS) devices] is related to the segregation of impurities at the interfaces, which can lead to high local electric fields in addition to the applied field across the device. This can lead to an increased current flow, heating, and eventual device failure through breakdown. A very appealing aspect of research in surface chemistry then is its application to a wide variety of important technological problems. Specifically, the present chapter will deal primarily with research into the static and dynamic properties of the surfaces of heterogeneous catalysts.

Fundamental research in heterogeneous catalysis is similar in its intent to early research in homogeneous chemistry. Its purpose is to enable the chemist to predict reaction pathways (selectivity) and relative rates (activity) for heterogeneously catalyzed surface reactions. Thus, much current research in this field is directed toward identification of molecular structures present on solid surfaces during reaction and relating the observed species to reaction mechanisms. However, a heterogeneously catalyzed chemical conversion occurring under realistic industrial conditions is very complicated indeed. A "real" catalytic reaction involves five steps: (1) diffusion of the reactants to the surface of the catalyst, (2) adsorption of at least one and often more than one of the reactants on the surface of the catalyst, (3) chemical reaction on the catalyst surface, (4) desorption of the products of the reaction from the surface of the catalyst, and (5) diffusion of the reaction products from the catalyst. Although the problems of mass transport (steps 1 and 5) are not well suited for study by the surface chemist, it is certainly true that the surface-chemical reaction as limited by steps 2–4 may be investigated incisively by using modern surface analysis techniques.

The research that will be discussed in this review is primarily concerned with the adsorptive and catalytic properties of the surfaces of metallic single crystals that are well characterized chemically and structurally. However, a brief comparative discussion will also be given concerning both homogeneous catalytic clusters and reduced metallic clusters bound to oxide supports and thin polycrystalline films. Although the conditions of temperature, gas pressure, surface cleanliness, and surface structure under which such experiments are performed are far removed from the conditions prevailing in a real catalytic system, these experiments allow careful control of all the variables involved. An understanding of a well-defined, well-controlled catalytic surface is obviously a necessary fundamental step toward understanding the vastly more complex surface of a real operating catalyst.

Experimentally, a limiting factor in controlling the chemical composition of a surface is the residual gas pressure in the experimental apparatus. The relationship between background pressure and surface characterization can be appreciated best by realizing that at a pressure of 10^{-6} Torr (an impingement flux of approximately 5×10^{14} cm^{-2} s^{-1} for a room-temperature gas of molecular weight 28 amu) a surface is contaminated totally by a monolayer of adsorbed gas (1 monolayer $\sim 10^{15}$ admolecules cm^{-2}) within approximately 2 s if the sticking probability of the gas impinging on the surface is assumed to be unity. For a background pressure of 10^{-10} Torr, the comparable time is nearly 3 hr. The sticking probability of many gases and vapors on metallic surfaces is close to unity, and so fundamental chemisorption and catalysis experiments on well-characterized metallic surfaces require the use of an ultrahigh-vacuum (UHV) chemical reactor. On insulator or semiconductor surfaces, the sticking probability is often much less than unity, and so fundamental investigations of chemisorption may be possible in favorable cases in high-vacuum systems with a background pressure on the order of 10^{-6} Torr. This requires merely the judicious choice of the gas–surface

2.1. INTRODUCTION

system under examination. Thus, although it is possible to conduct definitive fundamental experiments in a high-vacuum system, the most versatile apparatus is clearly a UHV one.

A comprehensive view of the solid surface can be provided by four general types of information:[1] (1) an *in situ* compositional analysis of the surface region, both qualitative and quantitative, (2) an analysis of the geometrical structure of the surface, (3) an analysis of the electronic structure of the surface, and (4) an analysis of the vibrational structure of the surface. Although there are many possible experimental techniques that may be used to determine this information,[2-4] (see also other parts of this volume) the methods of Auger electron spectroscopy (AES),[5-9] low-energy electron diffraction (LEED),[10-14] and x-ray and UV photoelectron spectroscopy (XPS and UPS, respectively)[9,15,16] are the most commonly used means of obtaining the first three types of information needed to characterize the surface.

Prior to approximately 1977, methods of measuring the vibrational structure of admolecules at solid surfaces were extremely limited. Presently, however, the field is flourishing. Vibrational information was first available from the optical spectroscopies of transmission–absorption or reflection–

---

[1] C. B. Duke and R. L. Park, *Phys. Today* **25** 23, (1972).

[2] J. T. Yates, Jr., *Chem. Eng. News* Aug. 26, p. 19 (1974).

[3] G. Ertl and J. Küppers, "Low-Energy Electrons and Surface Chemistry." Verlag Chemie, Weinheim, 1974.

[4] H. Ibach, ed., "Electron Spectroscopy for Surface Analysis." Springer-Verlag, Berlin and New York, 1977.

[5] C. C. Chang, *Surf. Sci.* **25**, 53 (1971).

[6] H. P. Bonzel and R. Ku, *Surf. Sci.* **33**, 91 (1972).

[7] H. P. Bonzel and R. Ku, *J. Chem. Phys.* **58**, 4617 (1973).

[8] H. P. Bonzel and R. Ku, *J. Chem. Phys.* **59**, 1641 (1973).

[9] *Proc. Int. Conf. Electron Spectrosc., Namur, Belg., 1974*, published in *J. Electron Spectrosc. Relat. Phenom.* **5** (1974).

[10] J. B. Pendry, "Low-Energy Electron Diffraction." Academic Press, New York, 1974.

[11] C. B. Duke, "Dynamic Aspects of Surface Physics" (Proceedings of the International School of Physics "Enrico Fermi," Course LVIII), pp. 52–249. Academic Press, New York, 1974.

[12] C. B. Duke, *Adv. Chem. Phys.* **27**, 1 (1974).

[13] C. B. Duke, *Proc. 2nd Int. Conf. Solid Surf., Kyoto*, p. 641 (1974).

[14] S. Y. Tong, *Prog. Surf. Sci.* **7**, Part 2 (1975).

[15] C. R. Brundle, *in* "Electronic Structure and Reactivity of Metal Surfaces" (E. G. Derouane and A. A. Lucas, eds.), Nato Advanced Study Institute, p. 389. Plenum, New York, 1976.

[16] *Proc. Int. Symp. Photoemiss., Noordwijk, Neth., 1976* (R. F. Willis, B. Feuerbacher, B. Fitton, and C. Backx, eds.), published in *Eur. Space Agency [Spec. Publ.] ESA SP* **ESA SP-118**, rev. 1 (1976).

absorption infrared (IR) spectroscopy[17-25] as well as laser Raman spectroscopy.[26] High-resolution electron energy loss spectroscopy (EELS),[27-47] inelastic electron tunneling spectroscopy (IETS),[48-51] and field emission energy distribution (FEED) measurements[52] have all been developed more recently to derive vibrational structures. Since FEED measurements have severe restrictions in their application, the discussion in this chapter will be restricted to EELS, IR spectroscopy, and IETS, with emphasis on the EEL experiment.

Infrared transmission spectroscopy has been used extensively to probe chemisorption and heterogeneous catalysis over approximately the past

[17] A. M. Bradshaw, J. Pritchard, and M. L. Sims, *Chem. Commun.* p. 1519 (1968).
[18] L. H. Little, "Infrared Spectra of Adsorbed Species." Academic Press, New York, 1966.
[19] M. L. Hair, "Infrared Spectroscopy in Surface Chemistry." Dekker, New York, 1967.
[20] N. J. Harrick, "Internal Reflection Spectroscopy." (Wiley Interscience), New York, 1966.
[21] A. M. Bradshaw and O. Vierle, *Ber. Bunsenges. Phys. Chem.* **74**, 630 (1970).
[22] J. Pritchard and M. L. Sims, *Trans. Faraday Soc.* **66**, 427 (1970).
[23] M. A. Chesters and J. Pritchard, *Surf. Sci.* **28**, 460 (1971).
[24] H. G. Tompkins and R. G. Greenler, *Surf. Sci.* **28**, 194 (1971).
[25] J. T. Yates, Jr., R. G. Greenler, I. Ratajczykowa, and D. A. King, *Surf. Sci.* **36**, 739 (1973).
[26] P. J. Hendra, *in* "Chemisorption and Catalysis" (P. Hepple, ed.), p. 80. Inst. Pet., London, 1970.
[27] F. M. Propst and T. C. Piper, *J. Vac. Sci. Technol.* **4**, 53 (1967).
[28] H. Ibach, *J. Vac. Sci. Technol.* **9**, 713 (1971).
[29] H. Ibach, K. Horn, R. Dorn, and H. Lüth, *Surf. Sci.* **38**, 433 (1973).
[30] H. Froitzheim, H. Ibach, and S. Lehwald, *Phys. Lett. A* **55**, 247 (1975).
[31] S. Andersson, *Solid State Commun.* **21**, 75 (1977).
[32] H. Froitzheim, H. Ibach, and S. Lehwald, ref. 16, p. 277.
[33] H. Froitzheim, H. Ibach, and S. Lehwald, *Phys. Rev. Lett.* **36**, 1549 (1976).
[34] H. Froitzheim, H. Ibach, and S. Lehwald, *Phys. Rev. B* **14**, 1362 (1976).
[35] H. Froitzheim, H. Ibach, and S. Lehwald, *Surf. Sci.* **63**, 56 (1977).
[36] H. Froitzheim, H. Hopster, H. Ibach, and S. Lehwald, *Appl. Phys.* **13**, 147 (1977).
[37] H. Ibach, *Surf. Sci.* **66**, 56 (1977).
[38] J. F. Wendelken and F. M. Propst, *Rev. Sci. Instrum.* **47**, 1069 (1976).
[39] S. Andersson, *Solid State Commun.* **20**, 229 (1976).
[40] H. Hopster and H. Ibach, *Surf. Sci.* **77**, 109 (1978).
[41] J. C. Bertolini, G. Dalmai-Imelik, and J. Rousseau, ref. 16, p. 285.
[42] J. C. Bertolini, G. Dalmai-Imelik, and J. Rousseau, *Surf. Sci.* **68**, 539 (1977).
[43] C. Backx, B. Feuerbacher, B. Fitton, and R. F. Willis, ref. 16, p. 291.
[44] C. Backx, B. Feuerbacher, B. Fitton, and R. F. Willis, *Phys. Lett. A* **60**, 145 (1977).
[45] C. Backx, B. Feuerbacher, B. Fitton, and R. F. Willis, *Surf. Sci.* **63**, 193 (1977).
[46] C. Backx, R. F. Willis, B. Feuerbacher, and B. Fitton, *Surf. Sci.* **68**, 516 (1977).
[47] A. Adnot and J. D. Carette, *Surf. Sci.* **74**, 109 (1978).
[48] J. Lambe and R. C. Jaklevic, *Phys. Rev.* **165**, 821 (1968).
[49] A. L. Geiger, B. S. Chandrasekhar, and J. G. Adler, *Phys. Rev.* **188**, 1130 (1969).
[50] B. F. Lewis, M. Mosesman, and W. H. Weinberg, *Surf. Sci.* **41**, 142 (1974).
[51] M. G. Simonson, R. V. Coleman, and P. K. Hansma, *J. Chem. Phys.* **61**, 3789 (1974).
[52] L. W. Swanson and L. C. Crouser, *Surf. Sci.* **23**, 1 (1970).

2.1. INTRODUCTION

25 years. Infrared absorption occurs by means of an interaction between incoming infrared radiation (4000–200 cm^{-1}, where 1 meV = 8.066 cm^{-1}) and the dynamic charge (transition dipole moment) of an oscillating dipole. The experiment has been applied both to chemisorption on oxide and supported metal catalysts by using infrared transmission–absorption spectroscopy (ITAS) and to chemisorption on metallic ribbons, films, and single crystals by using infrared reflection–absorption spectroscopy (IRAS). The former method can be used to study dispersed catalysts with a high surface area. Thus, the total number of oscillators adsorbed within the area of the radiation probe is large enough that good signal-to-noise ratios can be achieved rather easily with commercial instruments; however, the size (typically on the order of 10^3–10^4 cm^2) and dispersed nature of the adsorbent necessarily mean that the surface cannot be characterized in detail. The IRAS experiment can be applied to study chemisorption on well-characterized surfaces, but it is handicapped by interrelated technological limitations in sensitivity, noise, and spectral range. Both types of experiments are capable of achieving resolution on the order of approximately 3 cm^{-1}.

The early application of IRAS to low-area well-characterized surfaces is exemplified by work in the laboratory of David King[25] in the case of chemisorption of CO on polycrystalline W in which two states of molecularly adsorbed CO were observed *in situ* and correlated with previous electron stimulated desorption and thermal desorption data. However, the sensitivity of the measurements was only on the order of 1×10^{14}–2×10^{14} molecules, i.e., approximately 4% of a total monolayer of CO, and the weakly bound α-CO states were not resolved fully. Moreover, the strongly held β-CO states, which represent dissociated CO, were not visible at all due to the low excitation energies of chemisorbed carbon and oxygen adatoms, i.e., less than 1000 cm^{-1}.[32,35,46,47] However, in the regions of the spectrum that were visible, the resolution of the commercial spectrometer used was quite good, and peak positions were accurate to within approximately 5 cm^{-1}. More recently, in very elegant work, Bradshaw and Hoffmann[53,54] have combined LEED and reflection IR in the investigation of CO chemisorbed on the (100), (111), and (210) surfaces of Pd. By using polarization modulation, a sensitivity of approximately 0.003 monolayer has been obtained.[55] These authors were able to distinguish easily between bridge-bonded and linearly bonded CO, to measure clearly frequency shifts with coverage due to adsorbate–adsorbate interactions, and to correlate the frequency shifts with ordering phenomena

---

[53] F. M. Hoffmann and A. M. Bradshaw, *Proc. 3rd Int. Conf. Solid Surf.*, Vienna, p. 1167 (1977).
[54] A. M. Bradshaw and F. M. Hoffmann, *Surf. Sci.* **72**, 513 (1978).
[55] F. M. Hoffmann, personal communication, (1979).

observed via LEED. This work, together with reflection IR spectroscopy in general, will be discussed in detail in Chapter 2.3.

One of the most promising techniques being used to obtain information concerning vibrations in overlayers is EELS. In this technique, a highly monochromated beam of low-energy (1 – 10-eV) electrons is reflected from a surface, and the energy of the scattered electrons is analyzed. Some of the incident electrons (for this low range of kinetic energies) can lose a discrete amount of energy upon reflection, which can be associated with excitation of a vibration at the surface. Although state-of-the-art resolution [defined as the full width at half-maximum (FWHM) $\Delta E_{1/2}$ of the elastically scattered intensity distribution] is only approximately 60 cm^{-1}, its sensitivity is on the order of 0.1% of a monolayer, i.e., 10^{10} admolecules for a typical electron-beam area of approximately 1 mm^2. Most important, by using EELS it is possible to observe vibrational excitations at an energy as low as approximately 240 cm^{-1}, the energy limited only by interference due to the elastically scattered beam. Furthermore, in principle, with the same instrumentation one may tune the incident electron energy in order to observe electronic excitations of the valence electrons as well as core-level excitations.[38,56,57] Taken together, this information allows important progress to be made toward a better understanding of molecular interactions including catalytic reactions on solid surfaces.

The experimental technique of high-resolution EELS as applied to solid surfaces was pioneered by Franklin Propst at the University of Illinois in 1966[28] and is currently practiced by nine research groups around the world: Ibach and co-workers in Jülich, West Germany; Andersson in Göteborg, Sweden; Willis and co-workers in Noordwijk, the Netherlands; Bertolini and co-workers in Lyon, France; Carrette and co-workers in Quebec; Weinberg and co-workers in Pasadena, California; Sexton in Warren, Michigan; Wendelken and Uléhla in Oak Ridge, Tennessee; and Plummer and co-workers in Philadelphia, Pennsylvania. The experimental work published by these groups to date is summarized in Table IV in Chapter 2.5. Of these nine, the groups of Ibach *et al.*, Andersson, Bertolini *et al.*, Carrette *et al.*, Weinberg *et al.*, and Sexton are capable of a resolution of 80 cm^{-1} (10 meV) or less. A detailed discussion of EELS is presented in Chapter 2.2.

Inelastic electron tunneling spectroscopy may be viewed as a technique that is complementary both to reflection IR and to EELS. The IETS measurement involves detecting the fraction of electrons that tunnel inelastically between two metal electrodes (in the form of films) in a metal–insulator–

---

[56] H. Froitzheim and H. Ibach, *Z. Phys.* **269**, 17 (1974).
[57] G. E. Thomas and W. H. Weinberg, *Rev. Sci. Instrum.* **50**, 497 (1979).

metal tunnel junction. The second derivative of the tunneling current with respect to applied bias voltage is proportional to the oscillator strength of vibrational modes that are excited; hence, the second derivative of the tunneling current with respect to voltage as a function of voltage is the IETS analog of a reflection IR or of an EEL spectrum. Inelastic electron tunneling spectroscopy has been most effectively used to deduce the vibrational structure ($\sim 240-4000$ cm^{-1}) of molecules chemisorbed on Al_2O_3 surfaces in $Al-Al_2O_3-Pb$ tunneling junctions, and three reviews deal extensively with this topic.[58-60] In Chapter 2.4, emphasis will be placed on more recent applications of IETS: chemisorption on reduced metallic clusters ($\sim 30-80$ Å in diameter), which are "supported" on the Al_2O_3 substrate, and the characterization and catalytic reactions of inorganic cluster compounds (homogeneous catalysts), which are anchored to the Al_2O_3 substrate. It is in these two areas that IETS is likely to make a major impact in view of the difficulty of obtaining similar information on such systems using more conventional techniques.

As would be anticipated intuitively, each of the three different probes of the vibrational structure of overlayers has its own set of advantages and disadvantages. Although some of the "disadvantages" will undoubtedly be overcome as the field matures, a listing of the various pros and cons of the three techniques, as currently practiced, is given in Table I. This set of comparisons will be useful to keep in mind while reading the next three chapters in which each is discussed in more detail.

To conclude, there are various types of information to be gained from studies of the vibrational properties of overlayers.

(1) *Molecular structural determination:* The identification of complex molecular adsorbates, products of dissociative adsorption, and chemical reaction intermediates or products.

(2) *Evidence for surface-site specificity:* The identification of chemically identical admolecules that are adsorbed at adsites of different symmetry on the same surface and are distinguished by different vibrational energies and the observation of competition for adsites in coadsorbed overlayers.

(3) *Lateral interactions in the adlayer:* The observation of frequency shifts and line-shape changes as a function of surface coverage, which can be correlated with such interactions.

[58] P. K. Hansma, *Phys. Rep.* **30C**, 145 (1977).
[59] W. H. Weinberg, *Annu. Rev. Phys. Chem.* **29**, 115 (1978).
[60] T. Wolfram, ed., "Inelastic Electron Tunneling Spectroscopy." Springer-Verlag, Berlin and New York, 1978.

2. VIBRATIONS IN OVERLAYERS

TABLE I. Advantages and Disadvantages of EELS, IRAS, and IETS

Spectroscopy	Advantages	Disadvantages
EELS	High sensitivity: The detection limit is ~0.1% of a monolayer and the sampling area is ~1 mm². Wide spectral range: Features are observable from ~240 cm^{-1} upward.	Low resolution: The FWHM of the elastically reflected beam is typically 80 cm^{-1}. Relative difficulty of construction. Limited to use in low-pressure systems ($\leq 10^{-6}$ Torr).
IRAS	High resolution: The typical resolution is ~3 cm^{-1}. Relative ease of construction. Can be used at high ambient pressures to simulate conditions of real catalysts.	Rather low sensitivity: The ratio of the scattering cross sections for EELS and IRAS has been estimated to be ~200:1,[38] although detection of 0.1% of a monolayer of CO can be achieved (Section 2.3.2). Narrow spectral range: Observable features are typically limited to the region above 1500 cm^{-1} by solid-state IR detector technology and sensitivity, noise and interference.
IETS	High sensitivity: The detection limit is ~0.1% of a monolayer, and the sampling area is ~1 mm². Wide spectral range: Features are observable from ~240 cm^{-1} to above 8000 cm^{-1}. High resolution: The typical resolution is ~3 cm^{-1}. Relative ease of construction. Ability to simulate and study real catalytic systems: In its utility for examining adsorption on reduced supported metallic clusters and supported homogeneous cluster catalysts, this technique is complementary to IRAS.	Sample limitations: This probe cannot be applied to adsorbates on bulk metallic substrates. It is limited to examination of amorphous oxide surfaces, or metallic particles or inorganic cluster compounds supported on the oxide. Sample preparation technique: This may be thought of as a matrix isolation technique in which dynamic surface phenomena cannot be monitored in real time.

(4) *Geometrical orientation of the adlayer:* Observation of the vibrational modes that are selectively excited, which can be used to infer crudely the orientation of the admolecule with respect to the surface normal.

In Chapter 2.2 EELS will be discussed, and in Chapters 2.3 and 2.4 reflection IR spectroscopy and IETS will be described. Finally, in Chapter 2.5, a synopsis is given, which includes a discussion of CO chemisorbed on

2.2. ELECTRON ENERGY LOSS SPECTROSCOPY

Ru(001), which has been studied both by EELS[61,62] and by IRAS.[63,64] Also, tabulations are given of the various adsorbate–substrate systems that have been investigated by the techniques of EELS (Table IV) and IRAS (Table V).

2.2. Electron Energy Loss Spectroscopy

2.2.1. Introduction

Some of the early experimental results in the field of EELS will be presented here to lay the foundation for the following discussion of a particular EELS spectrometer, theoretical considerations for EELS, and selected experimental results. The development of EELS as a spectroscopic technique began in 1966, when Propst and Piper[28] measured the vibrational energies of H_2, H_2O, CO, and N_2 chemisorbed on the W(100) surface. Their apparatus consisted of an electron gun with a 127° cylindrical monochromator used to form a 4.5-eV electron beam with a resolution of 400 cm^{-1} (50 meV). Similarly, the spectrum of the scattered electrons was monitored with a 127° cylindrical analyzer. The vibrational spectrum of each of the adsorbates was measured as a function of coverage, and chemical structures were proposed. It was concluded that hydrogen adsorbs dissociatively on W(100) since the characteristic H_2 vibrational energy of 4400 cm^{-1} was not observed. They found instead two vibrational energies, one at 1080 cm^{-1} and the other at 550 cm^{-1}.

A more systematic analysis of their H_2 chemisorption data was carried out by Ibach[65] on the basis of a surface-phonon calculation by Dobrzynski.[66] Ibach concluded that the fourfold symmetric binding site, in which atomic hydrogen is shared by four tungsten atoms, was unlikely since the calculated vibrational energies $\hbar\omega_\parallel$ and $\hbar\omega_\perp$ were both too high to explain the low-frequency vibration at 550 cm^{-1}. He further concluded that the bridge site is possible if the high-frequency parallel and perpendicular vibrations appear together near 1080 cm^{-1}, and the low-frequency parallel vibration appears at

---

[61] G. E. Thomas and W. H. Weinberg, *J. Chem. Phys.* **70**, 954 (1979); see also G. E. Thomas and W. H. Weinberg, *in* "Vibrations in Adsorbed Layers" (H. Ibach and S. Lehwald, eds.), Conference Records Series of KFA, p. 97. Jülich, Fed. Rep. Ger., 1978.
[62] G. E. Thomas and W. H. Weinberg, *J. Chem. Phys.* **70**, 1437 (1979).
[63] H. Pfnür, D. Menzel, F. M. Hoffmann, A. Ortega, D. Menzel, and A. M. Bradshaw, *Surf. Sci.* **93**, 431 (1980).
[64] F. M. Hoffmann, A. Ortega, H. Pfnür, D. Menzel, and A. M. Bradshaw, *J. Vac. Sci. Technol.* **17**, 239 (1980).
[65] H. Ibach, *Festkoerperprobleme* **11**, 135 (1971).
[66] L. Dobrzynski, *Surf. Sci.* **20**, 99 (1970).

550 cm^{-1}. Although some subsequent measurements suggested that the 550-cm^{-1} feature might be due to slight contamination of the surface by CO,[33] more recent studies confirm the assignment of the 1080-cm^{-1} loss to the symmetric stretching mode of atomic hydrogen in the twofold bridge site[32,33,45,67-71] and support Ibach's assignment of the 550-cm^{-1} loss energy as due to a bending mode.[68-71]

Other chemisorption studies were carried out by Ibach et al.[28-31] for adsorption of O_2 and H_2 on Si. Ibach also used a cylindrical analyzer and monochromator but shortened them to 119° to correct for fringing fields. In their study of H_2 on Si, Froitzheim et al.[30] reported operating resolutions of 80 cm^{-1}. In the study of oxygen on the Si(111) surface, Ibach et al.[28,29] proposed a peroxide structure. Measurements by Ludeke and Koma[72] of energy losses due to electronic transitions of oxygen chemisorbed on a Ge(100) surface suggested a reinterpretation of the surface silicon oxide vibrational spectrum in terms of a $>$Si=O structure, but this disagreement has yet to be resolved fully.

Approximately three years ago, the focus of research using EELS turned from an investigation of semiconductor surfaces back to a study of admolecules on metallic substrates. Ibach and co-workers investigated the chemisorption of oxygen, hydrogen, and CO on a W(100) surface,[32-35] as well as the chemisorption of CO on a Pt(111) surface.[32,36,40] These experiments were conducted by using an improved monochromator and analyzer that were 119° cylindrical condensors that gave an overall resolution of 60 cm^{-1} (7.5 meV) at a primary beam current of 10^{-10} A.[73] This work of Ibach et al. demonstrated unequivocally the unique value of EELS insofar as investigating chemisorption on metals is concerned. For example, a long-standing controversy concerning the nature of CO chemisorbed on a W(100) surface was resolved by making EELS measurements as a function of surface coverage.[32,35] At low coverages, the appearance of excitations at 540 and 620 cm^{-1} were interpreted as being caused by the vibrations of isolated carbon and oxygen atoms, respectively, giving evidence for the dissociative adsorption of the β-CO adstate. At higher coverages, two additional modes at 360 and 2081^{-1} were observed also; these are due to the vibration of the CO molecule against the W surface and the C−O stretching vibration, respectively. This demonstrated that the more weakly bound α-CO adstates are molecular in

[67] A. Adnot and J. D. Carette, *Phys. Rev. Lett.* **39**, 209 (1977).
[68] M. R. Barnes and R. F. Willis, *Phys. Rev. Lett.* **41**, 1729 (1978).
[69] W. Ho, R. F. Willis, and E. W. Plummer, *Phys. Rev. Lett.* **40**, 1463 (1978).
[70] W. Ho, R. F. Willis, and E. W. Plummer, *Surf. Sci.* **95**, 171 (1980).
[71] R. F. Willis, W. Ho, and E. W. Plummer, *Surf. Sci.* **80**, 593 (1979).
[72] R. Ludeke and A. Koma, *Phys. Rev. Lett.* **34**, 1170 (1975).
[73] H. Froitzheim, H. Ibach, and S. Lehwald, *Rev. Sci. Instrum.* **46**, 1325 (1975).

2.2. ELECTRON ENERGY LOSS SPECTROSCOPY

nature. Also, Ibach et al.[32,36] were able to correlate their EELS measurements for CO chemisorbed on Pt(111) with LEED data of Ertl et al.[74] in order to gain a detailed perception of the filling of two different adsites by CO on the surface as a function of coverage. At low surface coverages, only on-top site bonding states are filled [leading to a $(\sqrt{3} \times \sqrt{3})R30°$ LEED structure at $\frac{1}{3}$ monolayer], whereas at higher coverages, a mixture of on-top sites and twofold bridge sites are populated [leading to a $c(4 \times 2)$ LEED structure at $\frac{1}{2}$ monolayer].

In Sweden, Andersson[31,39] used a cylindrical-mirror electron-kinetic-energy analyzer to examine the chemisorption of both oxygen and CO on a Ni(100) surface. He found two different vibrational modes for oxygen adatoms at 424 and 312 cm^{-1} corresponding to $p(2 \times 2)$ and $c(2 \times 2)$ LEED structures, respectively. He was able to distinguish also between linearly bonded and bridge-bonded CO on this Ni surface on the basis of the EELS measurements and was able to correlate these results with observed LEED structures as a function both of surface temperature and surface coverage.[32]

Subsequently, Backx et al.[43-46] in the Netherlands and Bertolini et al.[41,42] in France reported EELS measurements using hemispherical monochromators and electron-energy analyzers (discussed in detail in Section 2.2.2). The former group considered the chemisorption of H_2 and C_2H_2 on the (110) and (111) surfaces of W,[43-46] whereas the latter group investigated the chemisorption of benzene (including deuterated benzene) on the (100) and (111) surfaces of Ni.[41,42] The work of Bertolini et al. is of particular note since it demonstrates explicitly the potential importance of EELS applied to fundamental studies of heterogeneous catalysis.

Within the past three years, the EELS technique has begun to flourish with the entrance of several additional research groups to the field. Selected examples of the more recent work are discussed later in this chapter. As is customary in surface science, the least equivocal measurements are those which combine EELS with other surface-sensitive probes such as LEED, UPS, AES, and thermal-desorption mass spectrometry. Most of the next generation of EELS machines will undoubtedly be "multitechnique" instruments in which many of these other measurements may be made *in situ*.

2.2.2. Electron Energy Loss Spectrometers

In principle, an electron energy loss spectrometer is very simple indeed. It consists of two components: a monochromator that focuses low-energy electrons onto the surface under investigation and an energy analyzer that measures the kinetic energy of the electrons reflected from the surface. As will be brought out in Section 2.2.3, a long-range dipolar scattering mechanism

---

[74] G. Ertl, M. Neumann, and K. M. Streit, *Surf. Sci.* **64**, 393 (1977).

generally dominates the electron–phonon coupling, and hence the greatest intensity of the reflected electrons is peaked sharply in the specular direction. From this point of view, having a fixed monochromator with an energy analyzer fixed in the specular direction would be sufficient. However, as discussed in Section 2.2.3, it is possible to obtain much additional information (e.g., all normal modes and adsorbate geometry) if short-range "impact" scattering can be observed. This short-range interaction is best viewed away from the specular angle (in order to minimize the contribution of dipolar scattering), and it occurs with characteristic angular distributions. Hence, it is quite useful to have an electron-energy analyzer that is *rotatable in situ*.

There are three different types of monochromator–analyzer pairs that have been used in low-energy electron-scattering experiments on surfaces: concentric 180° hemispherical deflectors,[45,57,75] 127° cylindrical deflectors,[28,38,56,76–80] and two segments of a sliced cylindrical-mirror analyzer.[31,39,81] These three basic geometries are shown in Figs. 1–3, respectively. Of the six research groups engaged in EELS that report a resolution below 10 meV, two are using 180° hemispherical deflectors,[57,75] three are using 127° cylindrical deflectors,[28,56,76–80] and one is using a modified cylindrical-mirror analyzer.[31,39,81] The choice among these three types of monochromator–analyzer pairs is rather arbitrary. The best resolution to date (6-meV FWHM) has been achieved with the cylindrical mirror, which, however, does not have the desirable feature of being rotatable.[31,39,81] The 127° cylindrical deflectors have typically a simpler series of lens elements than the 180° hemispherical deflectors and, consequently, are easier to focus. It is possible to rotate both the 127° cylindrical[38,56,76–79] and the 180° hemispherical[57,70,71,82] deflectors *in situ*. However, the added complexity of the lens system usually associated with the 180° hemispheres may be used to advantage; it is possible to tune the impact energy up to several kilovolts to make measurements of core-level (e.g., Auger) excitations as well as of electronic (valence-band) excitations.[57] Furthermore, the addition of a rare-gas discharge lamp to the UHV system permits angularly resolved ultraviolet photoelectron spectroscopy (ARUPS) measurements to be made with the rotatable 180° hemispherical electron energy analyzer.[57,83] However, the

---

[75] J. C. Bertolini, G. Dalmai-Imelik, and J. Rousseau, *J. Microsc. Spectrosc. Electron.* **2**, 575 (1977).

[76] D. Roy, A. Delage, and J. D. Carette, *J. Phys. E* **8**, 109 (1975).

[77] D. Roy and J. D. Carette, ref. 4, p. 13.

[78] H. Froitzheim, H. Ibach, and S. Lehwald, *Rev. Sci. Instrum.* **46**, 1325 (1975).

[79] H. Ibach, H. Hopster, and B. Sexton, *Appl. Surf. Sci.* **1**, 1 (1977).

[80] B. A. Sexton, *J. Vac. Sci. Technol.* **16**, 1033 (1979).

[81] S. Andersson, *Surf. Sci.* **79**, 385 (1979).

[82] W. Ho, Ph.D. Thesis, Univ. of Pennsylvania, Philadelphia, 1979.

[83] P. A. Thiel, Ph.D. Thesis, California Inst. Technol., Pasadena, 1980.

2.2. ELECTRON ENERGY LOSS SPECTROSCOPY

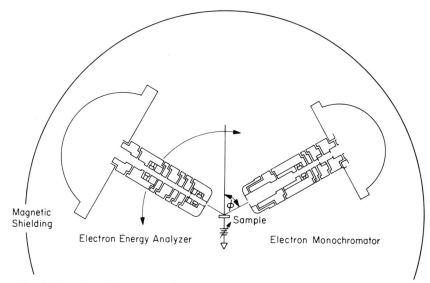

FIG. 1. Top view of the Kuyatt–Simpson type of spectrometer used in the author's laboratory, showing the two sets of lenses in the scattering plane. For a vertical cross-sectional view showing all the lens elements, see Fig. 4.

127° cylindrical deflectors and cylindrical-mirror deflectors may be constructed to be sufficiently compact so as to fit in a UHV system that contains also LEED and AES.[40,80,81]

The choice as to type of spectrometer then is largely one of taste. A detailed discussion will be given concerning the version of the 180° hemispherical monochromator–analyzer used in the author's laboratory, and the reader is referred elsewhere for further information concerning either the cylindrical mirror or the 127° cylindrical monochromator–analyzers.[31,39,77,81] However, at least a part of the discussion that follows (e.g., electron-lens calculations and discussion of the data-acquisition system) is obviously relevant to any type of spectrometer.

The electron spectrometer is of the Kuyatt–Simpson type[84-86] with hemispherical energy-dispersing elements and electron optics consisting of cylindrical tube lenses. Hemispherical elements were chosen because of their convenient geometry,[77,85] since the hemispherical system is used most easily with circular apertures rather than the rectangular slits used frequently in cylindrical systems. Thus, the scattering intensity, in the case of a circular

---

[84] J. A. Simpson, *Rev. Sci. Instrum.* **35**, 1698 (1964).
[85] C. E. Kuyatt and J. A. Simpson, *Rev. Sci. Instrum.* **38**, 103 (1967).
[86] C. E. Kuyatt and E. W. Plummer, *Rev. Sci. Instrum.* **43**, 108 (1972).

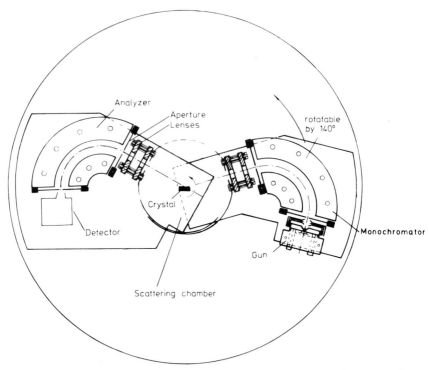

FIG. 2. High-resolution electron spectrometer of the 127° cylindrical deflector type. (From Ibach.[79] Copyright North-Holland Physics Publishing, Amsterdam, 1977.)

beam, is interpreted more easily in terms of a partially integrated cross section than is the scattering intensity of a rectangular beam.

The scattering system was designed to be used so that the sample remains in approximately the same position for the various measurements. The electron optics, ultraviolet-discharge lamp,[83] ion gun, and directional gas doser[87] are all directed at the sample at the center of the stainless-steel bell jar so that a minimum of alignment difficulties is encountered in the course of an experiment. The sample is positioned at the scattering center with a manipulator mounted on a side port of the bell jar. The manipulator allows the sample to move through two angles, which are not strictly orthogonal, and to be translated in three orthogonal directions in the scattering region. The ultraviolet light source, the ion gun, and the directional (adsorbate) beam doser are mounted on side ports that face the sample at various angles. The electron spectrometer is attached to a 35.6 cm i.d. Wheeler flange, which

[87] D. W. Goodman, T. E. Madey, M. Ono, and J. T. Yates, Jr., *J. Catal.* **50**, 279 (1977).

2.2. ELECTRON ENERGY LOSS SPECTROSCOPY

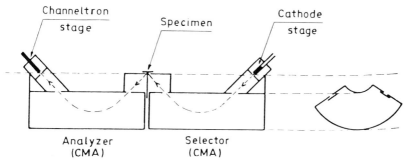

FIG. 3. High-resolution electron spectrometer of the cylindrical-mirror analyzer type. (From Andersson.[81] Copyright North-Holland Physics Publishing, Amsterdam, 1979.)

serves to close off the top of the bell jar, the base pressure of which is below 5×10^{-11} Torr. The spectrometer consists of a fixed electron monochromator (Fig. 4a) and a rotating energy analyzer (Fig. 4b). The angle of incidence of the electron beam may be changed either by rotating and resealing the top of the bell jar or by rotating the crystal. The last element of the monochromator is 20 mm from the sample, with the distance taken along the center of the beam trajectory. The entrance aperture of the analyzer is 19 mm from this point.

The scattering system is lined with Conetic shielding, which serves to reduce the magnetic field to < 30 mG throughout the region occupied by the electron spectrometer. The relatively high magnetic field is quite compatible with the high-resolution electron spectrometer, although commonly used design criteria would suggest otherwise.[76]

The flange that supports the spectrometer contains the feedthroughs necessary for the electrical and mechanical connections to the spectrometer. The axis of rotation of the analyzer (Fig. 4b) is the center line of the bell jar and is defined by two Cu–Be bearings in the spectrometer frame. The shaft that holds the analyzer in these bearings is hollow to accommodate the multiplier collection lead and supports a 15.2-cm-diameter spur gear parallel to the flange face. Angle markings, engraved in 2° increments on the gear face, are visible through a window on the top Wheeler flange. This allows the angular position to be measured ($\pm 0.2°$) without consideration of the backlash in the rotary feedthrough and off-axis spur gear that turns the energy analyzer.

The electron optics are supported by a stainless-steel frame that is suspended from the top flange. The electrical connections were made to the rotating analyzer with flexible shielded cables to avoid the noise associated with sliding contacts.

The electron optics, including the hemispheres, are machined from Cu–Be alloy (chiefly Berylco 10) with the exception of the cathode and anode, which are Mo. The Cu–Be pieces were treated with a nonmagnetic Electroless nickel plate, approximately 3.8 μm thick. This was followed by a layer of electroplated gold of approximately 3.3-μm thickness. Although the reflectivity of gold is thought to be undesirably high for use in electron spectrometers, a treatment that would provide chemical stability in a wide range of experiments and that would not be degraded with use at high electron energies was sought. This appears to have been successful in practice.

The hemispheres of the monochromator define a central trajectory 5.08 cm in diameter. The hemispheres of the energy analyzer define a central path 7.62 cm in diameter. The monochromator hemispheres are small due to the difficulty of space-charge spreading[85] imposed by the low-pass energy and relatively high-beam currents in this region.

The two sets of hemispheres were machined to within 0.001 cm of true hemispheres and positioned with sapphire balls on mounting plates. The hemispheres and other optics were positioned with dowel pins and alignment grooves during machining, and further alignment was not necessary. Deflector plates positioned throughout the optics proved superfluous, although in a similar spectrometer they seem to be essential.[82]

The electron optics of this instrument are similar to several reported previously.[84–86,88,89] A metal–oxide dispenser cathode obviates the necessity of accurate filament placement for high-resolution work. The current for the indirect heater is delivered through four wires arranged in four holes of an alumina rod in such a way that both the dipolar and quadrupolar components of the magnetic field set up by the current cancel sufficiently far from the rod.

The beam-defining apertures of the spectrometer were chosen to satisfy the resolution requirements of vibrational energy loss measurements. The beam-defining apertures of the gun (element 4, Fig. 4a) are approximately 0.03 cm in diameter. Two additional apertures at the analyzer entrance are 0.05 cm in diameter and define an acceptance cone from the surface that subtends a full-angle of 4.5°, independent of electron energy.

The Spiraltron electron multiplier is positioned after the energy defining aperture of the analyzer in such a way that the fixed collector can lie on the axis of rotation of the analyzer. The pulse from the electron multiplier travels a short path (25 cm) through a high-voltage feedthrough and decoupling capacitor before reaching the amplifier–discriminator, thus minimizing losses and reflections.

---

[88] A. Chutjian, *J. Chem. Phys.* **61**, 4279 (1974).
[89] A. Chutjian, *Rev. Sci. Instrum.* **50**, 347 (1979).

2.2. ELECTRON ENERGY LOSS SPECTROSCOPY

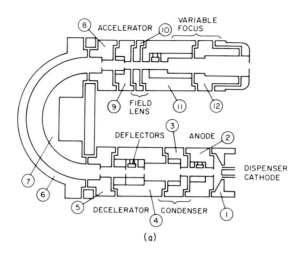

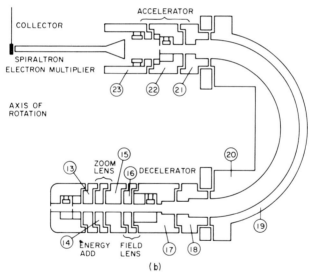

FIG. 4. (a) Schematic drawing of electron monochromator showing the function of the various optical stages. The diameter of the central trajectory between the hemispheres is 5.1 cm. The monochromator element nearest the sample is at ground potential. (b) Electron energy analyzer. The average diameter of the hemispherical electrodes is 7.6 cm. The analyzer element nearest the sample is at ground potential. (From Thomas and Weinberg.[57])

The hemispherical energy dispersing elements have the property of first-order angle focusing. If the optics at the hemisphere entrance are designed so that the angular divergence α of the beam is small enough, then the resolution of the analyzer takes the form

$$\Delta E_{1/2}/E_0 = r/R_0 \quad \text{for} \quad \alpha^2 = r/4R_0, \tag{2.2.1}$$

where R_0 is the radius of the central trajectory between the hemispheres; $\Delta E_{1/2}$ the overall spectrometer resolution (FWHM); E_0 the kinetic energy of an electron on that trajectory, i.e., the pass energy; and r the radius of the beam at the hemisphere entrance.[85,90] The Kuyatt–Simpson spectrometer incorporates electron optics at the entrances and exits of the hemispherical elements both for the gun and for the analyzer. This means that the impact energy E_i of the electrons and the energy E_a of the scattered electrons that reach the detector may be varied independently of E_0 and that E_0 may be selected without regard for the properties of the cathode or electron multiplier. The gun current is limited by the space charge of the electrons in the monochromator and is found to vary as $(\Delta E_{1/2})^{5/2}$. The design of each of the lens systems has been described by both Chutjian[88,89] and Kuyatt.[91]

A very brief summary of the optical principles used in the design of the spectrometer will be given here. A computer program written by C. E. Kuyatt was updated to include the results of lens calculations[92] and used to calculate the properties of the relevant four-tube-lens systems.

In the approximation of geometrical optics, the space charge of the electron beam is neglected. In this approximation, an electron beam may be described by limiting rays that are straight lines in regions of constant potential. A cylindrically symmetric monochromatic beam in such a region may be defined completely by its radius r, the angle θ made by the limiting ray with the axis of symmetry, and the kinetic energy of the beam.

A tube lens consists of two closely spaced cylinders held at different potentials. Since the beam in each drift tube is characterized by a column vector of three elements, the properties of the lens may be described by a 3 × 3 matrix. The potential energy zero is chosen so that the kinetic energy of the electron is equal to the negative of its potential energy. Then the kinetic energy of the electron is fixed by the lens potential, and the lens may be described by a 2 × 2 matrix.

It is customary to use tube lenses in which the gap between the drift tubes is $\frac{1}{10}$ the diameter of the lens, since the properties of these lenses are well known.[94] The four elements of a lens matrix, e.g., elements 8–11 of Fig. 4a,

---

[90] J. K. Rice, Ph.D. Thesis, California Inst. Technol., Pasadena, 1969.
[91] C. E. Kuyatt, unpublished notes of a lecture series on spectrometer design.
[92] S. Natali, D. DiChio, E. Uva, and C. E. Kuyatt, *Rev. Sci. Instrum.* **38**, 103. (1967).

2.2. ELECTRON ENERGY LOSS SPECTROSCOPY

are then defined completely by the potential ratio V_{11}/V_{10}, where V_x is the voltage applied to element x.

The beam may be defined as a group of trajectories in $r-\theta$ space. In the small-angle approximation, this area is proportional to $V^{-1/2}$ times the phase-space area in $r-p_r$ space. According to Liouville's theorem, therefore, the quantity $V^{1/2}$ times the area in $r-\theta$ space, called the normalized-beam emittance, is conserved.

In beams of high current and low energy, the space charge of the electrons becomes important. The maximum current through an enclosed volume, such as a drift tube or a hemispherical monochromator, is limited at a given energy by the charge of the electrons. Kuyatt and Simpson[86] estimated the maximum current through the monochromator with a model in which the deflectors were treated as a straight tube, and the universal space-charge curve for a laminar beam[93] was used to calculate beam spreading. The focusing properties of the hemispheres were approximated by two thin lenses in the tube. The current actually obtained was a factor of 5–10 less than calculated, depending on the deflector energy. This factor must be taken into account in estimating the gun current when designing the spectrometer.

According to Eq. (2.2.1), the resolution of the spectrometer should improve directly with increasing hemisphere size. That this does not work out in practice is thought to be due to space-charge focusing in the monochromator[94] and/or to the effect of an uneven work function in both the analyzer and monochromator.[95] Accordingly, as mentioned earlier, the monochromator was chosen to have a radius of 2.54 cm and the analyzer to have a radius of 3.81 cm.

The design criteria for the lens systems include the following. For an entering beam defined by the positions and sizes of two circular apertures, the exit beam may be required to have image apertures with positions and sizes close to specified values. This condition must satisfy conservation of normalized beam emittance in order to be valid. Alternatively, the exit beam may be required to illuminate fully a pair of apertures, which, in turn, define a new beam. As the beam traverses the system of tube lenses, it is desirable that the limiting rays not approach the tube walls too closely, in order to minimize aberrations. For the same reason, the minimum lens spacing should be approximately one lens diameter.

The first set of lenses, between the cathode and the monochromator (Fig. 4a), serves to accelerate and condense the beam from the cathode that illuminates two beam defining apertures. The beam is then decelerated and focused at the entrance to the hemispheres. The anode and condenser are

[93] K. R. Spangenberg, "Vacuum Tubes." McGraw-Hill, New York, 1948.
[94] C. E. Kuyatt, personal communication, 1975.
[95] E. W. Plummer, personal communication, 1975.

designed to deliver a beam current in excess of the space-charge limit to the monochromator. The 10:1 decelerator is calculated to image the first beam-defining aperture at the hemisphere entrance with an approximate magnification of unity and beam angle of zero. Since the criteria for this part of the gun are the same as those used by Chutjian, the same arrangement was used for the condenser and anode.[88,89]

The second lens system of the gun was designed to deliver a beam of uniform resolution in the energy range between 1 and 10 eV. With the crystal treated as a mirror, the beam was calculated to illuminate fully the analyzer apertures over this energy range. The beam emittance matches the acceptance of the analyzer at 10 eV, and so at lower energies some of the beam will be lost. It was found that in order to achieve a narrow beam at the analyzer entrance at low energies, a focusing element, labeled 11 in Fig. 4a, was necessary. The potential on this element is varied so that the kinetic energy of the electrons in that element is always twice the impact energy. Then the potential on element 12 (Fig. 4a) is varied to keep the final image position constant during a sweep of the impact energy, and the potential on element 10 (Fig. 4a) is varied to control the beam angle. It should be noted that at low energies the beam will be spread by space charge. However, since the area in $r-\theta$ space is larger at low energies, the problem of matching the beam emittance to analyzer acceptance can be solved experimentally by searching for the maximum transmitted current.

At the entrance to the analyzer, shown schematically in Fig. 4b, the beam is redefined by two apertures that can accept electrons from an area of the crystal approximately 2 mm in diameter, depending on the crystal orientation, and with an angular resolution of about 4.5°. The first element is followed by optics designed to allow an experimental spectrum to be measured at constant resolution < 10 meV over the range 1–10 eV. The optics at the exit to the analyzer hemispheres serve to accelerate the electrons into the energy-defining aperture and into the electron multiplier. The collector optics are designed so that the collector for the electron multiplier is on the axis of rotation of the analyzer and, consequently, can remain fixed during rotation.

The lens calculations have been made for three different modes of operation: low-energy vibrational-loss measurements ($E_i = 1-10$ eV), medium-energy UV photoemission ($h\nu = 16-41$ eV) and electronic energy loss measurements ($E_i = 10-100$ eV), and high-energy Auger electron spectroscopy measurements ($E_i = 2000$ eV). The potentials on all elements in Fig. 4 remain fixed with the exception of elements 10, 11, 12, 14, and 16. These variable potentials are applied by independent power supplies controlled by a minicomputer. Consequently, the nonlinear variation of these potentials causes no special difficulties.

An outline of the spectrometer electronics is given in Fig. 5. The system is

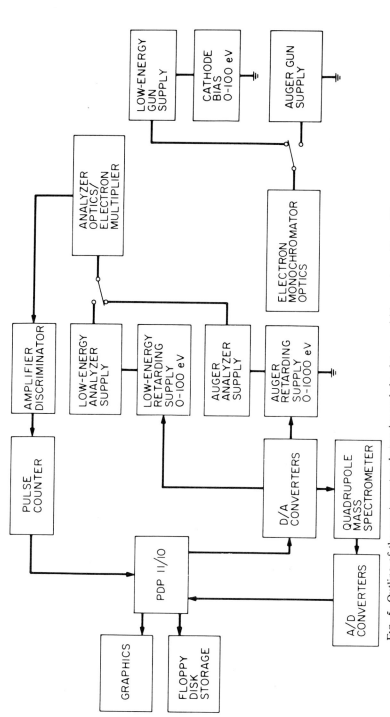

FIG. 5. Outline of the spectrometer electronics and the data acquisition system. (From Thomas and Weinberg.)[57]

under the control of a minicomputer (PDP 11/10) that is supervised by a software package designed for data acquisition and reduction.[96] The software system controls the spectrometer by interpreting a string of statements once during each clock period of 16 ms. The relatively slow clock reduces the machine overhead to the point at which time sharing is possible.

The kinetic energy of the electrons that are counted is selected by a power supply programmed by a 12-bit digital-to-analog converter. The low-noise requirements of high-resolution measurements were satisfied by Kepco PCX supplies. A separate set of high-voltage supplies is used for analyzer

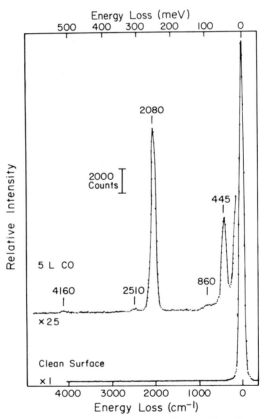

FIG. 6. Energy-loss spectra of CO on the Ru(001) surface at 100 K. The angle of incidence is 62° from the surface normal; the analyzer is centered on the specularly reflected beam. The incident energy is approximately 4.5 eV. The exposure is reported in units of langmuirs, where 1 langmuir ≡ 1 L ≡ 1 × 10^{-6} Torr s. (From Thomas and Weinberg.[57])

[96] J. L. Taylor, Ph.D. Thesis, California Inst. Technol., Pasadena, 1978.

operation to 1000 eV and for incident beam energies up to 2 keV, chiefly for AES. The lens potentials are set with potentiometers. The focusing error encountered in retardation does not appear to be serious in the vibrational energy loss experiment, presumably due to the narrow energy range involved. Two additional D/A channels and power supplies were provided for nonlinear corrections to the field lens and zoom lens of the energy analyzer, although these corrections have proved not to be necessary.

Pulses from the amplifier–discriminator are accumulated in a 16-bit counter that is read and cleared by the computer every 16 ms. This allows for a maximum count rate of 4 MHz. The maximum rate encountered in energy loss experiments is about 2×10^6 counts s^{-1}. The amplifier–discriminator (PAR 1120) has a 10-ns dead time and the first stages of the counting string are Schottky-clamped flip-flops, and so good counting statistics obtain even at high rates. The spectrometer electronics are isolated completely from the computer with optical isolators. Although the spectrometer is quite versatile, the most demanding application is obviously the measurement of vibrational energy loss spectra. A typical spectrum for CO chemisorbed near saturation coverage on the Ru(001) surface is shown in Fig. 6. The resolution (FWHM) of this spectrum is 11 meV. Typical electrode potentials employed in the electron energy loss measurements are listed in Table II. Some care in

TABLE II. Voltages Applied to Elements of Electron Monochromator and Analyzer to Produce and Deflect Electrons with a Kinetic Energy of 4.5 eV.[a]

Electrode	Potential (V)	Electrode	Potential (V)
Cathode	−5.9	Sample	0.0 (ground)
1	−5.9	13	34.2
2	16.2	14	22.7
3	125.0	15	18.5
4	−1.1	16	68.5
5	−3.9	17	22.7
6	−3.6	18	−3.3
7	−4.1	19	−3.0
8	−1.1	20	−3.6
9	0.5	21	−3.3
10	6.7	22	5.1
11	48.9	23	150
12	14.2		

[a] The voltages are given relative to the sample potential. The contact potential difference between the sample and the cathode, using the equations for ideal hemispherical deflectors,[86] is 1.4 ± 0.2 eV.

tuning is required to eliminate spurious peaks similar to those described previously.[36,78] The vibrational energy loss data are accumulated typically in 20 min by using incident currents of $1 \times 10^{-10} - 3 \times 10^{-10}$ A. The count rate in the elastic channel in the specular direction from the clean surface is typically 3×10^5 counts s^{-1}, although under optimal conditions a counting rate of 5×10^5 counts s^{-1} at a resolution of 8 meV is attainable.[97]

2.2.3. Theoretical Considerations

2.2.3.1. *Long-Range Dipolar (Multipolar) Scattering.* The dominant mechanism for electron scattering in the specular direction in EELS has been observed experimentally to be dipolar in nature. Hence, the scattering mechanism, as discussed also in Chapter 2.3, is similar to that in IRAS, and for such long-range scattering the same selection rules are apparently applicable in both experiments. For example, consider a CO molecule adsorbed perpendicular to a metallic surface through the carbon atom as shown in Fig. 7. For the moment, all interactions between neighboring molecules are ignored. As shown in Fig. 7, the adsorbed CO has two stretching modes [v_1 (the carbon–oxygen stretch) and v_2 (the center of mass of the CO molecule vibrating against the surface)] and two doubly degenerate bending modes (v_3 and v_4 as shown in Fig. 7). In this example, the two stretching modes are perpendicular to the surface, whereas the bending modes are parallel to the surface. As will be seen, dipolar scattering theory would predict only an excitation of the v_1 and v_2 modes, and indeed these are the only two modes observed in the spectrum of Fig. 6 at 2080 and 445 cm^{-1} for CO adsorbed on Ru(001).

The more general case of an adsorbed diatomic molecule that is not oriented perpendicularly to the surface is shown in Fig. 8. If a perfect image dipole is assumed to be created in the metal, then all components of the adsorbate dipole that are parallel to the surface cancel exactly. Moreover, there is an image of the approaching electron created in the bulk metal (Fig. 8).

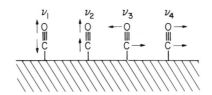

FIG. 7. Vibrational modes of adsorbed CO.

---

[97] P. A. Thiel, unpublished work. (1979).

2.2. ELECTRON ENERGY LOSS SPECTROSCOPY

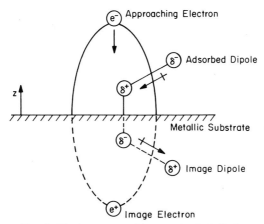

FIG. 8. Electron approaching an adsorbed dipole.

Considering only the long-range coulombic interactions of the electron, the resulting electric field is perpendicular to the surface near the dipole, and only those vibrational modes of an adsorbed molecule with a component of the dipole moment perpendicular to the surface will be excited, e.g., the v_1 and v_2 modes of CO in Fig. 7. This is the essence of the so-called "dipole-normal selection rule" and is identical to that associated with reflection IR.[37] With this physical view of the scattering mechanism, the scattering cross section has been computed both semiclassically[98-100] and quantum mechanically within the Born approximation[37,101,102] with the same result.

An important consideration in the interpretation of energy loss data is the assignment of the various energies to vibrational modes and the use of the dipole-normal selection rule in this interpretation.[79] Energy loss measurements of hydrogen adsorbed on the W(100) surface show that this selection rule may be broken for electrons scattered out of the specular direction.[59-71] As these results indicate, if dipole-allowed and dipole-forbidden transitions may be identified separately in the angular distribution of inelastically scattered electrons, EELS will be even more powerful in the interpretation of molecular structures on metal surfaces.

The possible mechanisms(s) for excitation of dipole-forbidden transitions is a controversial subject. It has been proposed that such excitations may

[98] F. Delanaye, A. Lucas, and G. D. Mahan, *Surf. Sci.* **70**, 629 (1978).
[99] D. M. Newns, *Phys. Lett. A* **60**, 461 (1977).
[100] Z. Lenac, M. Sunjic, D. Šokčević, and R. Brako, *Surf. Sci.* **80**, 602 (1979).
[101] E. Evans and D. L. Mills, *Phys. Rev. B* **5**, 4126 (1972).
[102] B. N. J. Persson, *Solid State Commun.* **24**, 573 (1977).

occur through a temporary negative ion resonance[69-71,103] or through impact or thermal diffuse scattering, in which the electron interacts with the short-range potential of the ion cores[104] (Subsection 2.2.3.2). In electron scattering from gas molecules, resonance scattering is an important mechanism,[105,106] and calculations have been made of angular distributions expected for resonance scattering from oriented molecules.[103] The effect of the metal surface on the temporary negative ion mechanism apparently remains unresolved. However, there is another possible mechanism for the excitation of dipole-forbidden transitions in which the electron scatters from the periodic potential set up by the surface vibration as it does in dipole scattering. In this picture, vibrations parallel to the surface appear at higher order in the potential and scatter more weakly than dipole-allowed transitions. This treatment will be discussed in detail in this subsection.

Only recently have theoretical calculations appeared that attempt to treat dipole-forbidden transitions in inelastic electron scattering. Several authors have calculated the inelastic scattering cross section for scattering from a single dipole on a surface[99,102,107] or for scattering from a dipole array[101,108] in the Born approximation. The effect of metal plasmons in moderating the image potential at high electron energies has also been calculated.[109] As discussed earlier, the assumption of high surface reflectivity and of a perfect imaging plane leads to the result that the electron is scattered only from vibrations that have a component of the dynamic effective-charge tensor perpendicular to the surface, and the inelastically scattered electrons are peaked sharply in the specular direction. If the approximation of high reflectivity or of a perfect imaging plane is relaxed, then parallel vibrations are allowed, but the electrons remain sharply peaked in the specular direction.[102,107] A general expression for the differential scattering probability, including both the dipolar and quadrupolar contribution to the potentials, will be derived below.[110] Pure dipolar scattering is then a special case of this more general result, and, moreover, the relative magnitude of quadrupolar contributions may be assessed. If the quadrupolar term in the potential is included, then vibrational modes that are strictly parallel to the surface may be excited also, and *the electrons are scattered relatively isotropically.*[110]

The periodic potential of an array of dipoles and quadrupoles on a metal surface is calculated as follows.[101] First, the potential due to the charge

---

[103] J. W. Davenport, W. Ho, and J. R. Schrieffer, *Phys. Rev. B* **17**, 3115 (1978).
[104] V. Roundy and D. L. Mills, *Phys. Rev. B* **5**, 1347 (1972).
[105] G. J. Schulz, *Rev. Mod. Phys.* **45**, 378 (1973).
[106] G. J. Schulz, *Rev. Mod. Phys.* **45**, 423 (1973).
[107] D. Šokčević, Z. Lenac, and R. Brako, *Z. Phys. B* **28**, 273 (1977).
[108] E. Evans and D. L. Mills, *Phys. Rev. B* **7**, 853 (1973).
[109] F. Delanaye, A. Lucas, and G. D. Mahan, *Surf. Sci.* **70**, 629 (1978).
[110] G. E. Thomas and W. H. Weinberg, *J. Chem. Phys.* **70**, 1000 (1979).

2.2. ELECTRON ENERGY LOSS SPECTROSCOPY

distribution at each lattice site $x_{\|}$ together with its image is calculated from

$$\Phi(\mathbf{r}) = 2 \sum_{x_\|} \left[\frac{P_z r_z}{|\mathbf{r} - \mathbf{x}_\||^3} + \frac{Q_{xz} r_x r_z + Q_{yz} r_y r_z}{|\mathbf{r} - \mathbf{x}_\||^5} \right]. \quad (2.2.2)$$

Here, $\Phi(\mathbf{r})$ is the potential far from the surface, P_z is the z component of the dipole moment of the charge distribution, and Q_{xz} and Q_{yz} are the corresponding elements of the quadrupole-moment tensor. Since r_z appears in each term, $\Phi = 0$ at the image plane, and Q_{zz} is zero to satisfy the requirement that the quadrupole-moment tensor be traceless.

Equation (2.2.2) may be expanded to first order in displacements, which are assigned a phase factor $\exp(i\mathbf{q}_\| \cdot \mathbf{x}_\|)$ for a surface wave of wave vector $\mathbf{q}_\|$. Sufficiently far from the surface, the sum in Eq. (2.2.2) may be replaced by an integral, and the term that leads to inelastic scattering is

$$V(\mathbf{r}) = 4\pi n_0 \exp(i\mathbf{q}_\| \cdot \mathbf{x}_\| - q_\| z)$$

$$\times \sum_\phi \left[e^*_{z\phi} + \frac{i}{3} \left(P^*_{xz\phi} q_x + P^*_{yz\phi} q_y \right) \right] e_\phi(\mathbf{q}_\|, \omega), \quad (2.2.3)$$

where the sum is over the Cartesian components and e_ϕ is the displacement operator of the normal mode of interest. The z-axis is defined to be perpendicular to the surface, and the reciprocal of the unit cell area is n_0. The quantities $e^*_{z\phi}$, $P^*_{xz\phi}$, and $P^*_{yz\phi}$ are defined by

$$e^*_{z\phi} \equiv \partial P_z / \partial u_\phi \quad \text{and} \quad P^*_{\beta z\phi} \equiv \partial Q_{\beta z} / \partial u_\phi, \quad (2.2.4)$$

where u_ϕ is a displacement vector. This may be generalized to more than one atom per unit cell by summing over each atom.

The scattering probability may be calculated to first order using plane waves reflected from a barrier as the unperturbed wave functions. The "golden rule" may be employed if the wave functions are properly normalized Bloch waves parallel to the surface,[107] or the scattered wave function may be calculated directly.[101] The resulting scattering probability per unit solid angle is

$$\frac{dS}{d\Omega} = \frac{2n_0(n_s + 1)e^2 m^2}{M\hbar^3 \omega(q_\|) \cos \phi} \left\{ \left(\sum_\phi e^*_{z\phi} \xi_\phi \right)^2 \right.$$

$$\left. + \left[\frac{1}{3} \sum_\phi (P^*_{xz\phi} q_x + P^*_{yz\phi} q_y) \xi_\phi \right]^2 \right\} \left(\frac{k_f}{k_i} \right) |f(q_\|, k_{iz}, k_{fz})|^2, \quad (2.2.5)$$

where

$$f(q_\|, k_{iz}, k_{fz}) = \frac{1}{q_\| + i(k_{iz} + k_{fz})} - \frac{R_f}{q_\| + i(k_{iz} - k_{fz})}$$

$$- \frac{R_i}{q_\| - i(k_{iz} + k_{fz})} + \frac{R_i R_f}{q_\| - i(k_{iz} - k_{fz})}. \quad (2.2.6)$$

Here, ξ_ϕ is the ϕ component of the normal coordinate of the vibration, n_s is $[\exp(\hbar\omega/k_B T) - 1]^{-1}$, e is the electronic charge, and m the electronic mass. The surface oscillator has a frequency $\omega(\mathbf{q}_\parallel)$ and a reduced mass M. The wave vectors of the incident and scattered wave vectors are $\mathbf{k}_i$ and $\mathbf{k}_f$, respectively, and R_i and R_f are the corresponding reflection coefficients. Under the experimental conditions, the approximation $(k_{iz} + k_{fz}) \gg q_\parallel$ usually holds, even at large scattering angles. If it is assumed that $R_i \approx R_f \approx R$, then

$$|f(q_\parallel, k_{iz}, k_{fz})|^2 = \frac{4|R|^2 q_\parallel^2}{q_\parallel^2 + (k_{iz} - k_{fz})^2}. \tag{2.2.7}$$

The shape of $f(q_\parallel, k_{iz}, k_{fz})$ and the role of this approximation has been discussed in detail by Šokčević et al.[107] Since the full periodic potential was used [Eq. (2.2.3)], the requirement of momentum conservation, $\mathbf{k}_i - \mathbf{k}_f = \mathbf{q}_\parallel$, is obtained also.

The first term in braces of Eq. (2.2.5) has been published by a number of authors, and perhaps it is worthwhile summarizing these results. Within the approximation $k_f \approx k_i$, Eq. (2.2.7) is consistent with the partially integrated cross section for small-angle scattering reported by Newns.[99] It differs from the result of Evans and Mills by a factor of two, which seems to be due to the definition of reduced mass in Eq. (2.2.5).[101] It agrees with the cross section of Šokčević et al. within a factor $1/\cos\phi$[107] and with the result of Persson within a factor of $4\pi^2$.[102] These factors are largely insignificant in a model calculation of this type, but when experimental intensities are interpreted, *different values of e^* will be obtained depending on which expression is used*. Note that in the derivation of Eq. (2.2.5) the potential due to images was included explicitly, and $e^*_{z\phi}$ and $P^*_{\phi z\beta}$ are defined so as to be compared (where possible) directly with gas-phase values.

The first term of the differential scattering probability depends on the quantity $\Sigma_\phi e^*_{z\phi} \xi_\phi$. In many cases, it appears that the dynamic effective charge tensor is diagonal, and only normal modes that involve motion perpendicular to the surface can cause scattering in the dipole approximation.[101] The second term in Eq. (2.2.5), which depends on the quadrupole derivatives $p^*_{xz\phi}$ and $P^*_{yz\phi}$, is then the first term, which in general excites modes that are strictly parallel to the surface. Since the product $q_\parallel f(q_\parallel, k_{iz}, k_{fz})$ is approximately constant with scattering angle when $|q_\parallel| > 0$, *the electrons scattered by the parallel vibrations will be scattered through all angles*.

To illustrate the influence of the quadrupole potential on the inelastic cross section, energy loss measurements by Ho et al.[69] of the v_1 state of H adsorbed on the W(100) surface may be considered. In this work, an intense energy loss feature at 130 meV, observed also by other workers,[32,33,44,67,68] is ascribed to the dipole-allowed normal vibration (v_1). Additional features at 80 and 160 meV are assigned to the weak (v_2) and strong (v_3) parallel modes of atomic hydrogen in a bridged site. Within this picture, the parameters $e^*_{z\phi}$ and $P^*_{\phi z\beta}$ can be estimated, and the importance of the quadrupolar term can be assessed.

The plane of incidence is along the (001) direction, and it is the angular dependence of scattering in this plane that is of interest. It will be assumed that v_2 and v_3 are strictly parallel to the surface and that the normal coordinate of each parallel mode is in the (001) direction for half of the H atoms on the surface. This is to be expected for H occupying all possible bridge sites as proposed by Estrup and Anderson.[111] Then, the term in braces in Eq. (2.2.5) may be simplified. For the v_1 mode, scattering in the $x-z$ plane depends only on e^*_{zz} and P^*_{xzz}. For the v_2 and v_3 modes, the scattering probability depends only on P^*_{xzx}, where the x axis is in the (001) direction.

The differential scattering probability can be compared directly with experiment after integration over the acceptance aperture angle of the spectrometer. An analytic expression has been given for the partially integrated dipole scattering probability in the specular direction,[101,112] and an expression for the quadrupolar term can be developed in the same way. Since the major concern is primarily with scattering out of the specular direction, however, $f(q_\parallel, k_{iz}, k_{fz})$ can be taken as a slowly varying function over the analyzer aperture, and the scattering probability is well approximated by $(dS/d\Omega)\Delta\Omega$. This quantity is shown in Fig. 9 as a function of scattering angle for (a realistic value of) $\Delta\Omega = 1 \times 10^{-4}$ sr. The dynamic effective charge e^* and the quadrupole derivative P^*_{zxz}, which appear in the scattering probability of v_1, are 0.17 and 4.5 a.u., respectively. The quantity P^*_{xzx}, which determines the intensity of the modes v_2 and v_3, is taken as 2.7 a.u. These values are chosen arbitrarily to match approximately the experimental intensities for scattering close to the specular direction and at 30° off specular. The angle of incidence ϕ is 23°, and the primary energy is 9.65 eV.

The quadrupole terms P^*_{xzx} and P^*_{xzz} seem somewhat large when compared with fixed dipole moments, but the same considerations apply in estimating these quantities as in estimating e^*. Unfortunately, P^*_{xzx} and P^*_{xzz} are not so amenable to independent measurement as e^*. As a point of reference, the quadrupole derivative of H_2 has been estimated from calculations to be 1.13 a.u.[112]

The comparison with the experimental results of Ho et al.[69] does not imply that the detailed angular intensities are explainable purely in terms of this potential. The calculated scattering probability is only weakly dependent on energy, and rather sharp features in the scattering probability as a function of energy have been observed experimentally.[70,71,82] Ho et al.[70,71,82] interpret

---

[111] P. J. Estrup and J. Anderson, *J. Chem. Phys.* **45**, 2254 (1966).
[112] E. L. Breig and C. C. Lin, *J. Chem. Phys.* **43**, 3839 (1965).

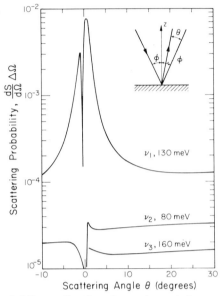

FIG. 9. Scattering probability as a function of scattering angle from assumed values of e^* and $P^*_{\phi z\beta}$ as given in the text. (From Thomas and Weinberg.[110])

their observations in terms of the trapping of an incident electron in a two-dimensional band gap (a surface "resonance") on W(100) and have correlated their findings convincingly with modulation in the elastic reflectivity of the W(100) surface. Nevertheless, the calculation presented above serves two purposes: (1) It quantifies the concept of a differential scattering probability for dipolar scattering, and (2) it suggests that an additional mechanism for diffuse scattering may exist and may be detectable at the limit of sensitivity of present instruments. This dipolar-forbidden scattering occurs at second order and may be expected to be two or three orders of magnitude less intense than dipolar scattering in the specular direction, if a quadrupole derivative of 1–3 a.u. is assumed.

2.2.3.2. Negative Ion Resonance Scattering. When a low-energy electron scatters inelastically from an adsorbed atom or molecule, the possibility of forming a negative ion resonance represents a possible nondipolar scattering mechanism in EELS. Davenport et al.[103] have considered this problem theoretically by assuming *oriented diatomic molecules* (CO, H_2, N_2) as targets but *in the absence of a surface*. Their calculation employed both the $X\alpha$ multiple scattering formalism, in which the exchange-correlation potential is assumed to be proportional to the $\frac{1}{3}$ power of the charge density and spherical averaging of nonoverlapping spheres about each nucleus with an

2.2. ELECTRON ENERGY LOSS SPECTROSCOPY

outer sphere placed about the entire molecule (the "muffin-tin" approximation).

In the gas phase, resonance effects are known to dominate electron scattering cross sections in the 1 – 10-eV range of impact energies.[105,106] A temporary negative ion resonance occurs when the incident electron is trapped temporarily in a virtual bound state of the target molecule. For example, H_2 has a σ_u resonance near 3 eV (the antibonding level of H_2),[113] CO a π resonance near 1.7 eV,[114] and N_2 a π_g resonance near 2.3 eV.[115,116] Davenport et al.[103] have calculated both the elastic and the vibrationally inelastic cross sections of oriented CO, H_2, and N_2 as a function both of impact energy and scattering angle. Figure 10 shows the total inelastic scattering cross section for CO as a function of impact energy for excitation from the ground state to the first excited vibrational state (lower panel) and from the ground state to the second excited vibrational state (top panel). The experimental results of Ehrhardt et al.[114] are shown as dashed lines, and good qualitative

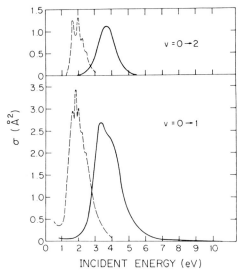

FIG. 10. Total cross section for the $0 \rightarrow 1$ (bottom) and $0 \rightarrow 2$ (top) vibrational transitions of CO as a function of the incident energy. The dashed curves are the measurement by Ehrhardt et al.[114] with the absolute cross sections obtained from integrating the differential scattering cross sections and normalizing to absolute elastic cross sections. (From Davenport et al.[103])

[113] F. Linder and H. Schmidt, Z. Naturforsch., A **26A**, 1603 (1971).
[114] H. Ehrhardt, L. Langhans, F. Linder, and H. S. Taylor, Phys. Rev. **173**, 222 (1968).
[115] D. E. Golden, Phys. Rev. Lett. **17**, 847 (1966).
[116] S. K. Srivastana, H. Chutjian, and S. Trajmar, J. Chem. Phys. **64**, 1340 (1976).

agreement is apparent. The differences between the experimental and theoretical cross sections are undoubtedly due to the potential employed in the theoretical calculation, which should be somewhat intermediate between the potential of a ground state molecule and that of the negative ion. Figure 11 shows the theoretically predicted angular distribution for negative ion resonance scattering from H_2 at a 4-eV impact energy and an angle of incidence of 45° from the axis of the molecular H_2 (see insert). Figure 11a shows the angular distribution for elastic scattering, and Figure 11b shows the angular distribution for vibrationally inelastic scattering from the ground state to the first vibrationally excited state. It is important to note that the angular distribution is certainly not peaked sharply in the forward direction as it would be for long-range dipolar scattering. Hence, it seems likely that an analysis of the angular distribution of resonantly scattered electrons from an adlayer could lead to an assignment of the geometrical position of the adatom or admolecule.

In extremely elegant work, Andersson and Davenport[117] have examined, both experimentally and theoretically, vibrationally inelastic scattering of CO adsorbed on Ni(100) and OH on NiO(111). The latter was an epitaxially grown (300-L O_2 at 296 K) oxide, which was moderately well ordered on the Ni(100) surface as judged by a relatively sharp LEED pattern (1L = 1 Langmuir = 10^{-6} Torr s). Figure 12 shows the variation with impact energy of the first and second (overtone) vibrational loss intensities normalized by the intensity of the elastically scattered electron beam for the 2057-cm^{-1} stretching frequency of the CO molecule adsorbed in a c(2 × 2) overlayer structure on Ni(100). The resonance loss intensity (dashed curve) was calculated with the theory described earlier,[103] and the energy scale was shifted arbitrarily by 2 eV so that the calculated resonance energy aligns itself with the experimental value for CO in the gas phase. Even if a surface reflectivity of 10% (quite high) is used, the calculated intensity of the loss feature is clearly too large. Moreover, the resonance theory does not reproduce the qualitative variation of the loss intensity with the impact energy, and resonance theory predicts an intensity I_2 of the overtone feature, which is on the order of 10% of the fundamental I_1. This is obviously not borne out experimentally. Hence, it must be concluded that CO on Ni(100) does not exhibit significant resonance scattering (bearing in mind that the measurements were made in the specularly reflected direction). On the other hand, the dipolar scattering theory describes the variation in intensity with impact energy of both the fundamental and the overtone very well indeed, as may be seen in Fig. 12. This level of agreement was achieved by using a dipole derivative (or dynamic effective charge) of $(0.61 \pm 0.09)e$ in the calculation

[117] S. Andersson and J. W. Davenport, *Solid State Commun.* **28**, 677 (1978).

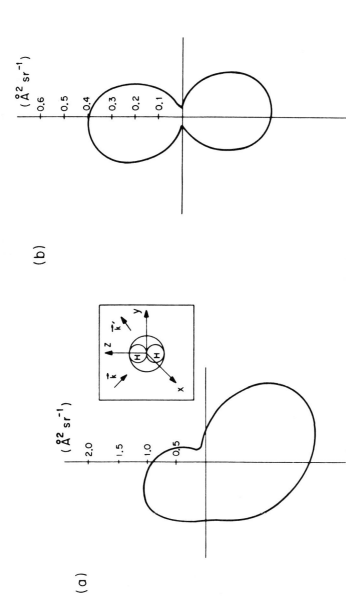

FIG. 11. Polar plots for the angular distribution from oriented H$_2$ at 45° incidence and $E = 4$ V. (a) Elastic scattering. (b) $0 \to 1$ vibrational excitation. The incident energy is chosen to correspond to the peak in the total vibrational excitation in the gas phase. Note the difference in the scale. (From Davenport et al.[103])

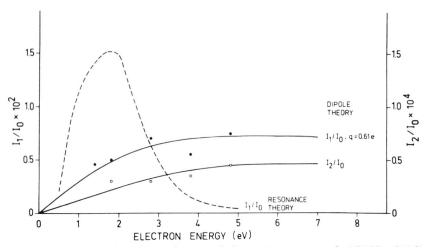

FIG. 12. Relative loss intensity as a function of primary-electron energy for Ni(100) c(2 × 2) CO derived from measured fundamental (●) and overtone (○) loss spectra. Also shown are the resonance (dashed) and dipole theories (solid curves). (Reprinted with permission from *Solid State Communications,* Vol. 28, S. Anderson and J. W. Davenport.[117] Energy dependence of vibrational excitation of adsorbed CO and OH. Copyright 1978, Pergamon Press, Ltd.)

of I_1 [the value for gas-phase CO is $(0.64 \pm 0.01)e$[118]], and the overtone matrix element of CO in the gas phase $\mu_2 = (0.257 \pm 0.004) \times 10^{-2} ea_0$, where a_0 is the Bohr radius.[118] It must be concluded that dipolar scattering is dominant for CO on Ni(100) in an ordered half-monolayer overlayer and that both the fundamental and overtone matrix elements are very nearly the same as for CO in the gas phase.

In the case of OH groups on the NiO(111) surface, the situation is completely different, as shown in Fig. 13.[117] Calculations based on the dipolar theory are somewhat less firm in this case, since the nature of the OH group on the surface is less well defined (as is its coverage, which is assumed be a half-monolayer). However, for a free OH group the matrix elements of the fundamental and the overtone losses are $-1.41 \times 10^{-2} ea_0$ (corresponding to a dynamic effective charge of $-0.11e$), and $-0.45 \times 10^{-2} ea_0$, respectively;[119] for an OH⁻ the matrix element of the fundamental is $3.73 \times 10^{-2} ea_0$ (a dynamic effective charge of $0.29e$).[120] When these values are used, no agreement between theory and experiment is obtained, as is apparent from Fig. 13. No detailed calculations of resonance scattering of the OH groups have been carried out, but two features of the experimental data are at least suggestive of such a mechanism: (1) the characteristic decrease in inten-

---

[118] R. A. Toth, R. H. Hunt, and E. K. Plyler, *J. Mol. Spectrosc.* **32,** 85 (1969).
[119] F. H. Mies, *J. Mol. Spectrosc.* **53,** 150 (1974).
[120] P. E. Cade, *J. Chem. Phys.* **47,** 2390 (1967).

2.2. ELECTRON ENERGY LOSS SPECTROSCOPY

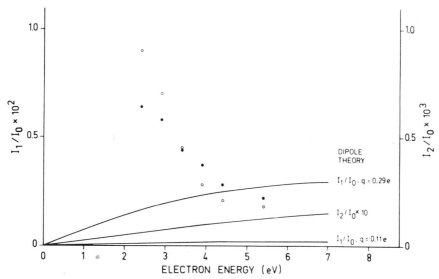

FIG. 13. Relative loss intensity as a function of primary-electron energy for NiO(111)–OH derived from measured fundamental (●) and overtone (○) loss spectra. Solid curves are dipole predictions as explained in the text. (Reprinted with permission from *Solid State Communications*, Vol. 28, S. Anderson and J. W. Davenport.[117] Energy dependence of vibrational excitation of adsorbed CO and OH. Copyright 1978, Pergamon Press, Ltd.)

sity with increasing impact energy, and (2) the fact that the overtone loss is approximately 10% of the fundamental, as expected in resonance scattering. This is far more intense than expected in dipolar scattering. Finally, intuitively it is not difficult to imagine that an incident electron occupying the 1π state in the OH would form a 4σ resonant state coupled to the continuum.[117] It would obviously be of great interest to measure the *angular* distribution of electrons that have lost energy in the excitation of the OH mode to test the validity of the hypothesis of negative ion resonance as well as to determine the geometry of the OH groups on the surface.

2.2.3.3. Determination of Adsite Symmetry. Depending on the symmetry of the surface, it is easy to imagine that an adsorbed molecule may occupy various sites of different symmetry. The question then arises as to whether it is possible to delineate the binding site of the admolecule as well as its (approximate) orientation based on the number of vibrational modes observed in EELS and the frequencies of those modes. For example, in Fig. 14 a diatomic AB molecule is shown in four different site symmetries. It would be expected that the AB-bond excitation energy as well as the MA-bond excitation energy would vary, depending on the adsite symmetry, i.e., depending on the nature of the local bonding. An outstanding example of this line of reasoning is the case of CO, either as an adsorbate on a surface or

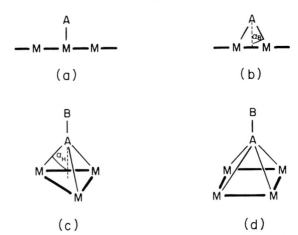

FIG. 14. Four possible adsite symmetries for bonding of a diatomic AB molecule to a surface: (a) the linear (on-top) site, (b) the twofold bridging site, (c) the threefold bridging site, and (d) the fourfold hollow site.

as a ligand in a multimetallic inorganic cluster compound. Lore has it that a CO stretching frequency above approximately 2000 cm^{-1} corresponds to linear bonding (Fig. 14a), a frequency between approximately 1850 and 2000 cm^{-1} corresponds to twofold bridge bonding (Fig. 14b), and a frequency below approximately 1850 cm^{-1} corresponds to more highly coordinated bonding (Fig. 14c,d). This rule of thumb is a consequence of increased back-donation of metallic electron density into the $2\pi^*$ antibonding orbital of the CO as the coordination of the CO to the surface or the cluster increases. These types of arguments, however, should be made with great caution, especially when considering adsorption on an extended surface.

In a provocative calculation, Efrima and Metiu[121] have considered the influence of electrodynamic interactions between a single chemisorbed molecule and a metallic surface in shifting the vibrational frequency of the diatomic molecule. In their model, the interaction potential between the bonding atoms is represented by a Morse potential, and the electrodynamic properties of the chemisorbed molecule are simulated. The two atoms in the chemisorbed diatomic molecule are assumed to have two different effective atomic charges, which are dependent on the internuclear separation. The use of effective atomic charges to represent the admolecule, as opposed to an oscillating point dipole used in previous theories to describe frequency shifts

---

[121] S. Efrima and H. Metiu, *Surf. Sci.* **92**, 433 (1980); **108**, 329 (1981); **109**, 109 (1981).

2.2. ELECTRON ENERGY LOSS SPECTROSCOPY

induced by lateral interactions,[122-128] should be a major improvement insofar as describing the interaction of an admolecule with a metallic surface is concerned. This approach has proved useful in the theory of IETS as well.[129] The forces which must be considered in this model calculation are due to both the Morse potential and the electrodynamic interaction of the two partial charges (of the two atoms in the diatomic adsorbate) and the induced dipole with the metallic substrate: the two partial charges and the dipole moment polarize the metal and then interact with the polarization charges. These authors conclude that the use of a model that represents an admolecule as a point dipole and considers its interaction with a metallic surface via an electrostatic image force (either with or without the inclusion of Thomas–Fermi screening of the charge–metal interaction) appears to be too simplistic and may even lead to erroneous conclusions. This point of view should be kept in mind as well when lateral interactions are discussed in Subsection 2.2.3.4. However, Richardson and Bradshaw[130,131] have modeled localized vibrations of admolecules via inorganic cluster compounds and have found encouraging agreement between theory and experiment.

In the case of atomic hydrogen chemisorbed on the Pt(111) surface, the observation of both short-range and long-range scattering has allowed a firm conclusion to be drawn concerning the bonding geometry of the hydrogen on the surface.[132] For scattering in the specular direction, a single, weak loss feature is observed near 550 cm^{-1}. The intensity of this feature is approximately an order of magnitude less than that observed for chemisorbed CO: for hydrogen on Pt(111), the differential cross section at an impact energy of 5 eV is 9×10^{-16} cm^2 sr^{-1} with a dynamic effective charge of $0.018e$. This value of the dynamic effective charge is a factor of three less than that observed for hydrogen chemisorbed dissociatively on the W(100) surface.[37] However, when the scattered beam was energy analyzed away from the specular direction (angle of incidence of 55.7° and angle of reflection of 70°,

---

[122] R. M. Hammaker, S. A. Francis, and R. P. Eischens, *Spectrochim. Acta* **21**, 1295 (1965).

[123] R. A. Shigeishi and D. A. King, *Surf. Sci.* **58**, 484 (1976).

[124] G. D. Mahan and A. A. Lucas, *J. Chem. Phys.* **68**, 1344 (1978); see also A. A. Lucas and G. D. Mahan, in "Vibrations in Adsorbed Layers" (H. Ibach and S. Lehwald, eds.), Conference Records Series of KFA, p. 1. Jülich, Fed. Rep. Ger., 1978.

[125] F. Dalanaye, M. Schmeitz, and A. A. Lucas, *J. Chem. Phys.* **69**, 5126 (1978).

[126] M. Moskovits and J. E. Hulse, *Surf. Sci.* **78**, 397 (1978).

[127] J. C. Campuzano and R. G. Greenler, *Surf. Sci.* **83**, 301 (1979).

[128] M. Scheffler, *Surf. Sci.* **81**, 562 (1979).

[129] J. R. Kirtley, D. J. Scalapino, and P. K. Hansma, *Phys. Rev. B* **14**, 3177 (1976).

[130] N. V. Richardson and A. M. Bradshaw, in "Vibrations in Adsorbed Layers" (H. Ibach and S. Lehwald, ed.), Conference Records Series of KFA, p. 2. Jülich, Fed. Rep. Ger., 1978.

[131] N. V. Richardson and A. M. Bradshaw, *Surf. Sci.* **88**, 255 (1979).

[132] A. M. Baró, H. Ibach, and H. D. Bruchmann, *Surf. Sci.* **88**, 384 (1979).

both with respect to the surface normal), an additional loss feature was observed at 1230 cm^{-1}. In this off-specular direction at an impact energy of 7.25 eV, the differential cross section for the 1230-cm^{-1} mode is 1.4×10^{-18} cm^2 sr^{-1}. Both these bands shift down in frequency by a factor of 1.37 on adsorption of deuterium, which demonstrates that they are both derived from chemisorbed hydrogen. Moreover, both of the observed loss features are associated with the same adsite on the surface since they develop simultaneously from the lowest measurable exposures (2 L) up to saturation exposure (~ 60 L). With this information, it may be concluded that the 550-cm^{-1} mode possesses a small perpendicular component of a dipole moment since weak dipolar scattering by this mode is observed. On the other hand, the 1230-cm^{-1} mode is only excited by a short-range impact mechanism, and, consequently, its vibration produces no net changes in the dipole moment perpendicular to the surface.

The observation of these two loss features with different scattering mechanisms permitted Baró et al. to locate the geometrical position of the hydrogen adatoms on the surface.[132] As shown in Table III (compare Fig. 14a–c), only the three highest symmetry sites on the (111) surface were considered: (a) the linear (on-top) site, (b) the twofold bridging site, and (c) the threefold hollow site. Both the on-top and the twofold bridging sites may be excluded immediately. For the on-top site, two modes should be observed, but the A_1 (stretching) mode should be at a higher frequency than the E (bending) mode, contrary to the experimental observation. The bridging site may be excluded for two reasons. (1) Three loss features should be observed due to one dipolar allowed A_1 (symmetric stretching) mode and two dipolar forbidden (a B_1 asymmetric stretching and a B_2 bending) modes. (2) The observed dipolar allowed mode (550 cm^{-1}) is too low in frequency, considering the expected (e.g., hard-sphere) Pt–H bond length and, consequently, the angle α_B, which is defined in Table III and Fig. 14. At this point, one is forced to conclude that the hydrogen atoms occupy the threefold hollow sites if they occupy high-symmetry sites on the Pt(111) surface. As shown in Table III, it is possible to solve the following two equations simultaneously:

$$\hbar \omega_H^S \cos \alpha_H = 550 \quad \text{cm}^{-1}, \qquad (2.2.8)$$

$$\hbar \omega_H^S \sin \alpha_H = 1230 \quad \text{cm}^{-1}, \qquad (2.2.9)$$

to obtain values for $\hbar \omega_H^S$ (the effective linear Pt–H stretching vibration for H adsorbed in the hollow site) and α_H (the bond angle defined in Table III and Fig. 14). These quantities are found to be $\hbar \omega_H^S = 1347$ cm^{-1} and $\alpha_H = 65.9°$, which corresponds to a Pt–H bond length of 1.76 Å and an adlayer–surface-layer spacing of 0.71 Å. These values are physically quite reasonable: a frequency near 2100 cm^{-1} has been observed with transmission IR of

TABLE III. Expected Modes for Atomic Adsorption in the Three Highest Symmetry Positions on an fcc (111) Surface.[a]

A. On-Top C_{6v}[b]			
Vibration	Mode	Frequency	Intensity for specular reflection
Symmetric stretching	$A_1(\perp)$[c]	ω_T^s [d]	Strong
Bending	$E(\parallel)$	$\omega_T^b < \omega_T^s$	Weak

B. Bridge C_{2v}[e]			
Symmetric stretching	$A_1(\perp)$	$\omega_B^s \cos\alpha_B$ [f]	Strong
Asymmetric stretching	$B_1(\parallel)$	$\omega_B^s \sin\alpha_B$	Weak
Bending	$B_2(\parallel)$	$\omega_B^b < \omega_B^s$	Weak

C. Hollow C_{3v}[g]			
Symmetric stretching	$A_1(\perp)$	$\omega_H^s \cos\alpha_H$	Strong
Asymmetric stretching	$E(\parallel)$	$\omega_H^s \sin\alpha_H$	Weak

[a] Only the first surface layer is considered.
[b] See Fig. 14a.
[c] The symbols $\perp$ and $\parallel$ refer to modes perpendicular and parallel to the surface, respectively.
[d] Superscripts s and b refer to stretching and bending modes, respectively.
[e] See Fig. 14b.
[f] Angles α are defined in Fig. 14b and c.
[g] See Fig. 14c.

supported Pt,[133,134] which is attributed to a terminal Pt–H bond, and it would be expected that the corresponding frequency at a hollow site would be considerably lower; and the hard-spheres bond length of the Pt–H bond is 1.69 Å (1.39 Å for Pt + 0.30 Å for H), in close agreement with the deduced value of 1.76 Å. By using this type of analysis, it is not possible to distinguish between the two inequivalent types of threefold sites on Pt(111), one having a vacancy in the second layer below the top-layer hollow site and the other having a Pt atom in the second layer. For comparison, a detailed LEED analysis of hydrogen adatoms on the Ni(111) surface showed that the hydrogen adatoms occupy *each type* of threefold site on the surface with a Ni–H bond length of 1.84 Å, corresponding to a H–Ni interlayer spacing of

---

[133] W. A. Pliskin and R. P. Eischens, *Z. Phys. Chem.* **24**, 11 (1960).
[134] J. P. Candy, P. Fouilloux, and M. Primet, *Surf. Sci.* **72**, 167 (1978).

1.15 Å.[135] Nevertheless, the results of Baró et al.[132] demonstrate very nicely the possibility of extracting geometrical information from EELS even in the absence of lengthy calculations of the angular distributions of loss features resulting from impact scattering.

2.2.3.4. Frequency Shifts with Surface Coverage. Although coverage-dependent frequency shifts are observed for adsorbates other than chemisorbed CO, much of the discussion in the literature concerning frequency shifts with coverage centers around that admolecule. Hence, that point of view will be followed in this account as well. Possible mechanisms that might lead to frequency shifts include (a) vibrational coupling, (b) dipole–dipole interactions, and (c) local bonding variations. Although an entirely satisfactory explanation of adsorbate frequency shifts is not yet available, that such a shift occurs is irrefutable. For example, a shift in the CO stretching vibration as a function of coverage on Ru(001) from 1980 to 2080 cm^{-1} is shown in Fig. 15.[62] In an effort to understand better this common observation, each of the three effects mentioned earlier will be discussed separately.

2.2.3.4.1. VIBRATIONAL COUPLING. There are two types of vibrational coupling, which may be thought of as intramolecular coupling and intermolecular coupling, that can cause shifts in vibrational frequencies. Each of these topics will be developed separately.

Intramolecular vibrational coupling involves a coupling between the vibrations in an admolecule and the vibrations of the substrate–admolecule bonds. This coupling, which occurs even for a *single* admolecule, can cause significant shifts from gas-phase frequencies.[31,136,137] As a specific example, consider CO bonded in a twofold bridging site between two metallic surface atoms denoted as M in Fig. 16. As defined in this figure, $\hbar\omega_1$, $\hbar\omega_3$, and $\hbar\omega_2$ are the excitation energies corresponding to the symmetric and asymmetric MC stretching vibrations and the CO stretching vibration, respectively. A classical mechanical analysis yields the set of equations

$$\omega_1^2 + \omega_2^2 = k_1\left(\frac{1}{m_C} + \frac{1}{m_O}\right) + k_2\left(\frac{1}{m_M} + \frac{2}{m_C}\cos^2\alpha\right), \quad (2.2.10)$$

$$\omega_1^2\omega_2^2 = k_1 k_2\left(\frac{1}{\mu m_M} + \frac{2}{m_C m_O}\cos^2\alpha\right), \quad (2.2.11)$$

$$\omega_3^2 = k_2\left(\frac{1}{m_M} + \frac{2}{m_C}\sin^2\alpha\right), \quad (2.2.12)$$

---

[135] K. Christmann, R. J. Behm, G. Ertl, M. A. Van Hove, and W. H. Weinberg, *J. Chem. Phys.* **70**, 4168 (1979).
[136] J. W. Gadzuk, *Phys. Rev. B* **19**, 5355 (1979).
[137] T. Okawa, M. Soma, H. Bandow, and K. Uchida, *J. Catal.* **54**, 439 (1978).

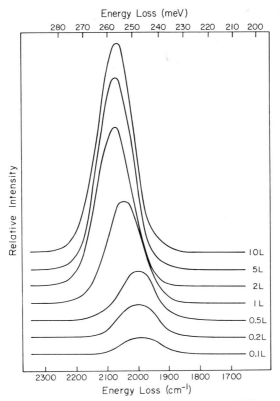

FIG. 15. Detail of carbon–oxygen stretching region as a function of CO exposure on Ru(001) at 100 K. (From Thomas and Weinberg.[62])

where k_1 and k_2 are the stretching force constants of the CO and MC bonds, respectively; the bond angle α is defined in Fig. 16; and m_O, m_C, m_M, and μ are the masses of the oxygen atom, carbon atom, metallic surface atom and reduced mass of the CO molecule, respectively.[138] Okawa et al.[137] have evaluated the relationship between these vibrationally coupled and the uncoupled frequencies for the simpler case of a diatomic molecule linearly bound to a single metal atom. These authors pointed out that such coupling causes the higher energy A–B stretching mode to shift upward and the lower energy M–A stretching mode to shift downward in frequency relative to the uncoupled frequencies, even if the A–B bond strength does not change on adsorption. In the case of CO adsorption on Ni, the calculated shift upward

[138] G. Herzberg, "Infrared and Raman Spectra." Van Nostrand, Princeton, New Jersey, 1945.

2. VIBRATIONS IN OVERLAYERS

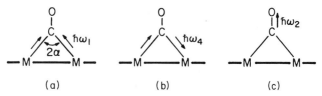

FIG. 16. Normal modes of bridge-bonded CO to be considered in intramolecular vibrational coupling; (a) symmetric M–C, (b) asymmetric M–C, and (c) C–O.

in ν_{CO} for CO bound linearly to one Ni atom is on the order of 40 cm^{-1} above the gas-phase value of 2143 cm^{-1}.[137] Since all CO frequencies observed experimentally to date are below the gas-phase value, the bonding of the molecule to the surface weakens the CO bond.

Intermolecular vibrational coupling of adsorbed CO molecules on adjacent metallic surface atoms may also induce an upward shift in the CO frequency.[126] The physical situation, shown schematically in Fig. 17, may be thought of in terms of the Cotton–Kraihanzel[139] model for metal carbonyls. The vibrational potential energy of interaction in this model is given by

$$V = \tfrac{1}{2}(k_1 q_1^2 + k_1 q_2^2 + k_3 q_1 q_2), \qquad (2.2.13)$$

where q_i is the displacement from the equilibrium CO internuclear separation for the ith molecule ($i = 1, 2$ as shown in Fig. 17), k_1 the primary CO force constant, and k_3 the force constant describing the interaction between the two CO molecules. The extent of the calculated frequency shift depends entirely on the value chosen for the parameter k_3. In order to reproduce the shift of 100 cm^{-1} seen in Fig. 15, a value of approximately 0.3 mdyn Å^{-1} is required.[62] This would appear to be unreasonably large in view of the fact that the experimentally determined interaction force constant for the cluster compound $Mn_2(CO)_{10}$ is 0.07 mdyn Å^{-1}.[126,140] Although intermolecular vibrational coupling must lead to a part of the experimentally observed frequency shifts, it appears not to be the dominant contributing mechanism.

2.2.3.4.2. DIPOLE–DIPOLE INTERACTIONS. Dipole–dipole interactions

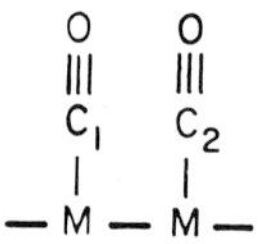

FIG. 17. Model for intermolecular vibrational coupling of adsorbed CO, where C_1 and C_2 can be, for example, different isotopes of carbon.

[139] F. A. Cotton and C. S. Kraihanzel, *J. Am. Chem. Soc.* **84**, 4432 (1962).
[140] H. Haas and R. K. Sheline, *J. Chem. Phys.* **47**, 2996 (1967).

also lead to a predicted upward shift in the frequency of adsorbed CO with coverage.[122-124,126,128,141] The mathematical (functional) relationship describing dipole–dipole interactions is the same as that describing intermolecular vibrational coupling with k_3 replaced by $a_i a_j/R_{ij}^3$, where a_i is the dipole derivative of the ith molecule, and R_{ij} is the separation distance between neighboring CO molecules. The dipole moment (used in the evaluation of the a_i) contains both a static part and a perturbation part (determined by the polarizability and the electric field). By including screening of the dipole from its self-image, Scheffler[128] was able to account theoretically for the complete frequency shift observed experimentally for CO on Pt and Pd. However, the results of Scheffler[128] should be viewed with caution in view of his use of point dipoles rather than the more realistic use of partial charges as described by Efrima and Metiu.[121] The precise contribution of dipole–dipole interactions to the observed frequency shift remains controversial.

2.2.3.4.3. LOCAL BONDING VARIATIONS. It may well be that changes in metal–adsorbate bond strength and adsorbate site symmetry play a decisive role in the observed frequency shifts of adsorbed CO. The chemisorption of CO on transition metals may be considered to be similar to the bonding in metal carbonyl cluster compounds. Blyholder[142] has described this bonding in terms of molecular orbital theory in which CO bonds to the metal by donation of electrons from the 5σ level to the metal with back-donation of metallic d electrons to the antibonding $2\pi^*$ orbital of CO. The back-donation weakens the CO bond, which lowers the CO vibrational frequency. As surface coverage increases, one might expect less back-donation into the $2\pi^*$ orbital, which strengthens the CO bond and causes the observed upward shift in the vibrational frequency. However, calculations by Campuzano and Greenler[127] indicate that, although this mechanism may be important at low surface coverages (below approximately 0.1 monolayer), it cannot cause the entire frequency shift observed for CO on Ni(111).

It would be expected that the local adsorption site of CO changes with increasing surface coverage, and there is considerable LEED evidence to support this expectation.[143] According to the Blyholder model,[142] at higher coordination numbers there is increased back-donation of metal electrons into the antibonding $2\pi^*$ orbital. The subsequent weakening of the C–O bond would then manifest itself in a reduced C–O stretching frequency. It is important to note that the concept of a localized adsorption site, in connection with the term-coordination number, does not mean that for two-, three-,

---

[141] A. Crossley and D. A. King, *Surf. Sci.* **68**, 528 (1977).
[142] G. Blyholder, *J. Phys. Chem.* **68**, 2772 (1964); G. Blyholder and M. Allen, *J. Am. Chem. Soc.* **91**, 3158 (1969); G. Blyholder, *J. Vac. Sci. Technol.* **11**, 865 (1974); G. Blyholder, *J. Phys. Chem.* **79**, 756 (1975).
[143] E. D. Williams and W. H. Weinberg, *J. Chem. Phys.* **68**, 4688 (1978).

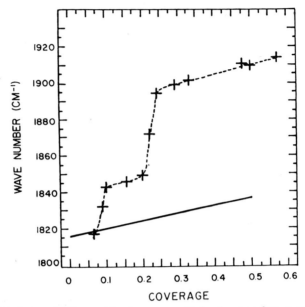

Fig. 18. Variation of CO stretching frequency with fractional surface coverage of CO on Ni(111). Crosses represent IRAS data, and the solid line is the calculated shift assuming dipole–dipole interactions between CO admolecules. (From Campuzano and Greenler.[127] Copyright North-Holland Physics Publishing, Amsterdam, 1979.)

and fourfold sites, strong σ bonds to each of the coordinating substrate atoms need to be formed. Campuzano and Greenler[127] have argued that for CO on Ni(111) at a fractional surface coverage $\theta \lesssim 0.20$, the CO is adsorbed at threefold hollow sites. As θ increases from 0.20 to 0.25, the CO admolecules shift to twofold bridging sites until a coverage of 0.50 is reached. Above $\theta = 0.50$, the admolecules begin to shift to on-top sites. This picture is consistent with LEED information for this system, as well as the variation of CO stretching frequency with surface coverage shown in Fig. 18: the sharp increase in $\hbar \omega_{CO}$ near $\theta = 0.25$ could well be a hollow-to-bridge surface site transformation.

2.2.4. Experimental Results: Dissociation of NO and Coadsorption of NO and CO on Ru(001)

To demonstrate explicitly the very useful surface chemical information that may be obtained from EELS, the case history of the adsorption and dissociation of NO as well as the coadsorption of NO and CO on the close-

2.2. ELECTRON ENERGY LOSS SPECTROSCOPY

packed (001) surface of Ru will be considered in some detail.[144-146] This is an instructive example since it demonstrates several of the extremely valuable aspects of EELS: (1) the ability to distinguish easily two fundamentally different forms of chemisorbed molecular NO that exist simultaneously on the surface, (2) the ability to observe a surface chemical reaction, in this case the dissociation reaction of adsorbed NO, and (3) the ability to use a chemical probe to deduce adsite symmetries via competitive adsorption and site poisoning in coadsorptive systems.

Energy loss spectra of NO adsorbed on Ru(001) are shown in Fig. 19. The NO exposure ϵ_{NO} is reported in units of Torr seconds where the pressure refers to the NO pressure in a gas-storage bulb behind a glass capillary of a directional beam doser,[87] and the time refers to the duration of the exposure. These spectra, for various coverages of NO at crystal temperatures between 120 and 180 K, are consistent with a model in which molecular NO is adsorbed in two sites of different symmetry. The energy loss feature at

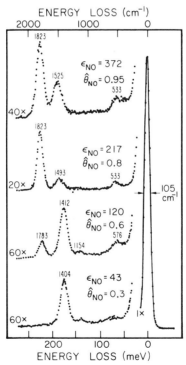

FIG. 19. Electron energy loss spectra of NO adsorbed on the Ru(001) surface at 120 K < T < 180 K for various coverages, where $\hat{\theta}_{NO}$ is the coverage of NO relative to saturation. The energies of individual loss features are given in inverse centimeters and milli-electron volts. (From Thiel et al.[146])

1404–1525 cm^{-1} is assigned to the N–O stretching vibration of NO adsorbed at a site of two- or threefold coordination, in analogy with doubly and triply coordinated nitrosyl ligands in homogeneous metal clusters.[144,147,148] The loss feature at 1783–1823 cm^{-1} can similarly be assigned to NO bound linearly in an on-top site.[144,149] These two forms will be referred to as "bridged" and "linear" NO, respectively. At low coverages, the NO adsorbs in the more strongly bound bridged form; at coverages greater than approximately $\frac{1}{3}$ of saturation, the on-top site becomes populated.

A transient feature at approximately 1154 cm^{-1} occurred in some spectra and apparently represents an adsorbate that is unstable at 120 K ≤ T ≤ 180 K on Ru(001). In several cobalt complexes, the presence of the hyponitrite ligand, N_2O_2, is characterized by infrared bands between 925 and 1195 cm^{-1}, depending on the nature of the anion and other ligands.[150] The exact nature of the transient species on the Ru(001) surface will require further clarification, however.

A low energy feature at 533–576 cm^{-1} is observed also. The frequency range for the Ru–O stretch of atomic oxygen on Ru(001), determined separately,[71,151] is 516–596 cm^{-1}. However, in organic complexes the Ru–N vibration of those nitrosyl ligands with N–O frequencies above 1800 cm^{-1} is reported to occur in the range 570–638 cm^{-1}.[152,153] In Fig. 19 the intensity of the low-energy feature correlates with the appearance and growth in intensity of the loss feature above 1780 cm^{-1}, which is due to linear NO. This suggests strongly that at T ≤ 180 K the vibrational feature at 533–576 cm^{-1} is due to the Ru–NO stretch of linear NO. This assignment is supported by the fact that at T ≤ 150 K Umbach et al.,[154] using x-ray photoelectron spectroscopy, see no evidence for dissociation of NO on Ru(001). At higher temperatures, for which dissociation products and molecular NO are both present on the Ru(001) surface, the Ru–NO loss feature is unresolved from the Ru–O feature in the EEL spectra.

Experimentally, thermally induced dissociation of NO occurs at higher

---

[144] G. E. Thomas and W. H. Weinberg, *Phys. Rev. Lett.* **41**, 1181 (1978).
[145] P. A. Thiel, W. H. Weinberg, and J. T. Yates, Jr., *Chem. Phys. Lett.* **67**, 403 (1979).
[146] P. A. Thiel, W. H. Weinberg, and J. T. Yates, Jr., *J. Chem. Phys.* **71**, 1643 (1979).
[147] J. R. Norton, J. P. Collman, G. Dolcetti, and W. T. Robinson, *Inorg. Chem.* **11**, 382 (1972).
[148] J. Müller and S. Schmitt, *J. Organomet. Chem.* **97**, C54 (1975).
[149] R. Eisenberg and C. D. Meyer, *Acc. Chem. Res.* **8**, 26 (1975).
[150] H. Toyuki, *Spectrochim. Acta, Part A* **27A**, 985 (1971); B. F. Hoskins and F. D. Whillans, *J.C.S. Dalton* p. 607 (1973); see also K. G. Caulton, *Coord. Chem. Rev.* **14**, 317 (1975), and references therein.
[151] G. E. Thomas and W. H. Weinberg, *J. Chem. Phys.* **69**, 3611 (1978).
[152] E. E. Mereer, W. A. McAllister, and J. R. Durig, *Inorg. Chem.* **5**, 1881 (1966).
[153] M. S. Quinby and R. D. Feltham, *Inorg. Chem.* **11**, 2468 (1972).
[154] E. Umbach, S. Kulkarni, P. Feulner, and D. Menzel, *Surf. Sci.* **88**, 65 (1979).

2.2. ELECTRON ENERGY LOSS SPECTROSCOPY

temperatures as the relative coverage increases. Within the framework of the two-site model of NO adsorption, this can be understood as follows: On heating, the bridged NO produces dissociation products that occupy multiple-coordination sites on the Ru(001) surface. This is accompanied by conversion of the bridged form of NO to the linear form. At high NO coverages, dissociation is impeded due to the site competition between bridged molecular NO and its dissociation products. The two-site model is thus consistent with the observed coverage dependence of the activated dissociation of NO. It has been inferred from UPS data that there is a coverage-dependent activation barrier for dissociation of molecular NO on the (111) surface of Ni also.[155]

Energy loss spectra are shown in Figs. 20 and 21 in which the Ru(001) surface was first exposed to NO at 150 K, then heated to a temperature sufficient to dissociate the adsorbed NO, and finally exposed again to NO at 150 K. The initial relative (to saturation) coverage of NO in spectrum (a) of Fig. 20 is $\hat{\theta}_{NO} = 0.15$. The peak at 1379 cm^{-1} is due to bridged NO. Spectrum (b) of Fig. 20 follows heating the surface to 250 K. In this spectrum, the peak at 516 cm^{-1} is due to chemisorbed oxygen. When this surface was then exposed to 43 Torr s NO at 150 K, spectrum (c) of Fig. 20 resulted, with peaks at 1412 and 524 cm^{-1}. It can be seen on the basis of energy loss spectra, by comparison with Fig. 21 spectrum (a), that the presence of the dissociation products resulting from the small initial NO coverage does not significantly influence the subsequent adsorption of NO. The dissociation products do not prevent the NO from being adsorbed into the bridged state, where it would be adsorbed also at low coverages on the clean Ru(001) surface (Fig. 19).

At a higher initial coverage of NO, $\hat{\theta}_{NO} = 0.3$, the results are quite different. The vibrational spectrum of the initial surface is shown in spectrum (a) of Fig. 21, with peaks at 1395 and 1113 cm^{-1}. Spectrum (b) of Fig. 21 follows heating to 311 K, which again leaves a surface with only chemisorbed nitrogen and oxygen atoms present. Spectrum (c) of Fig. 21 was measured after the surface with dissociation products had been exposed to 43 Torr s of NO at 150 K. It can be seen that the higher coverage of nitrogen and oxygen atoms resulting from the higher initial coverage of NO prevents almost completely subsequent adsorption of molecular NO into the bridged state. Rather, adsorption of linear NO occurs after this exposure, in contrast to the behavior shown in spectrum (c) of Fig. 20. Further exposure of this surface to NO leads only to a growth in the intensity of the feature at 1791 cm^{-1}, which is due to linear NO, as shown in spectrum (d) of Fig. 21.

The experimental data of Figs. 20 and 21 provide strong support for the two-site model of adsorption of NO. The dissociation products of NO oc-

---

[155] H. Conrad, G. Ertl, J. Küppers, and E. E. Latta, *Surf. Sci.* **50**, 296 (1975).

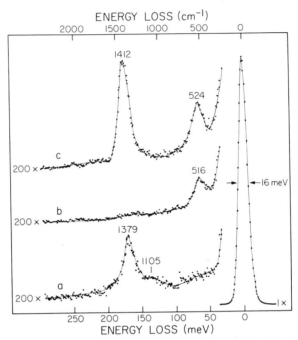

FIG. 20. A series of energy loss spectra of molecular NO and its dissociation products adsorbed on Ru(001). (a) After an exposure of ϵ_{NO} of 22 Torr s ($\hat{\theta}_{NO} = 0.15$) at $T = 150$ K on clean Ru(001). (b) After heating the surface represented in (a) to 250 K. (c) After exposing the surface represented in (b) to 43 Torr s NO at 150 K. (From Thiel et al.[145] Copyright North-Holland Physics Publishing, Amsterdam, 1979.)

cupy the same sites on the Ru(001) surface as the bridged state of molecular NO. When the initial coverage of dissociation products is sufficiently high, as in Fig. 21, the surface is poisoned with respect to subsequent adsorption of bridged NO. At a lower initial coverage of N and O adatoms (Fig. 20), the NO can be adsorbed first into the (vacant) bridged sites, which are energetically preferred. These conclusions are reminiscent of those drawn from previous UPS studies of the adsorption of NO on Ru(100) at 370 K, in which dissociation of NO was shown to precede adsorption of molecular NO in coverage.[156] This adsorption sequence has also been postulated to occur on polycrystalline Rh at room temperature on the basis of CO titration experiments.[157]

At an initial relative coverage of NO of 0.15, dissociation of the bridged NO is complete by 250 K, as shown by spectrum (b) of Fig. 20. If one assumes that the (coverage-independent) value of the pre-exponential factor of the

[156] H. P. Bonzel and T. E. Fischer, Surf. Sci. 51, 213 (1975).
[157] C. T. Campbell and J. M. White, Appl. Surf. Sci. 1, 2996 (1978).

2.2. ELECTRON ENERGY LOSS SPECTROSCOPY

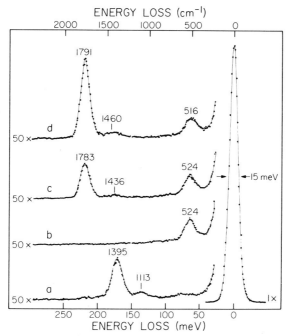

FIG. 21. A series of energy loss spectra of molecular NO and its dissociation products adsorbed on Ru(001). (a) After an exposure of ϵ_{NO} of 55 Torr s ($\theta_{NO} = 0.32$) at 150 K on clean Ru(001). (b) After heating the surface represented in (a) to 311 K. (c) After exposing the surface represented in (b) to 43 Torr s NO at 150 K. (d) After exposing the surface represented in (c) to 43 Torr s NO at 150 K. (From Thiel et al.[145] Copyright North-Holland Physics Publishing, Amsterdam, 1979.)

rate coefficient is 10^{13} s^{-1}, then the activation energy for dissociation of this state is approximately 16 kcal mole^{-1}.

Coadsorption of nitric oxide and carbon monoxide yields further insight into the adsorption of each of these molecules separately on the clean Ru(001) surface and the nature of their interaction when they are coadsorbed on this surface. Previous LEED data for CO on Ru(001)[145] indicate that adsorbed CO forms coincidence lattices at absolute coverages θ_{CO} greater than $\frac{1}{3}$, in which many of the CO molecules are in low-symmetry adsorption sites. A ($\sqrt{3} \times \sqrt{3}$)R30° LEED pattern is observed at $\theta_{CO} = \frac{1}{3}$, however, and presumably corresponds to CO molecules adsorbed in sites of high symmetry. The energy loss and infrared spectra for this system will be discussed more fully in Section 2.5.2, and representative EEL spectra are shown in Figs. 6 and 15. They show a single C-O stretching frequency, which shifts with increasing coverage from 1980 to 2080 cm^{-1} (Subsection 2.2.3.4), and an energy-loss feature at 444 cm^{-1}, which is assigned to the

Ru–CO stretching vibration. The C–O vibration occurs in the frequency range commonly assigned to CO bound linearly to a single metal atom, in spite of the observation of coincidence lattices in the LEED structure.

In Fig. 22, vibrational spectra are shown for four experiments in which the Ru surface, with four different initial coverages of CO, was exposed to various amounts of NO. It is apparent that for initial absolute coverages of CO with $\theta_{CO}^0 < 0.3$ (Fig. 22a and b), the subsequent adsorption of NO occurs with an attenuation of intensity of the linear NO loss feature (1791–1831 cm^{-1} in Fig. 22a and b) relative to the loss spectra of NO on clean Ru(001), which were obtained after equal NO exposures (Fig. 19). Qualitative comparison of the intensities of the loss features in Fig. 22a and b suggests that adsorption of NO may be accompanied by some displacement of CO from the surface for $\theta_{CO}^0 \le 0.3$. The observed variations in energy of the vibrational loss features may be due to changes in coverage of the individual adsorbates and/or interactions between different adsorbates. At $\theta_{CO}^0 \ge 0.3$, adsorption of NO occurs only in the bridged sites (1395–1460 cm^{-1} in Fig. 22c and d). The fact that NO can adsorb into the linear sites only below $\theta_{CO}^0 \sim 0.3$ is evidence that the CO molecules in the $(\sqrt{3} \times \sqrt{3})R30°$ lattice are adsorbed in the on-top sites, as is the case also for adsorption of CO on Pt(111).[36,40]

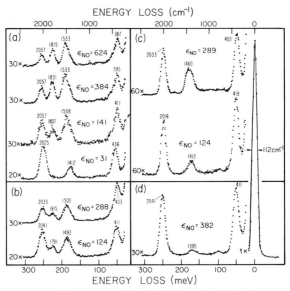

FIG. 22. Electron energy loss spectra of the Ru(001) surface following exposure to CO and subsequent exposure to NO at 120 K $\le T \le$ 180 K. (a) $\theta_{CO}^0 = 0.12$, (b) $\theta_{CO}^0 = 0.24$, (c) $\theta_{CO}^0 = 0.31$, and (d) $\theta_{CO}^0 = 0.65$ (saturation). The energies of the individual loss features are given in inverse centimeters. (From Thiel et al.[146])

In Fig. 23, vibrational spectra are shown for three experiments in which exposure to NO was followed by exposure to CO, the reverse of the exposure sequence of Fig. 22. The spectrum of Fig. 23a is to be compared with the spectrum of Fig. 19, which follows a 120 Torr s NO exposure. It can be seen in Fig. 23a that exposure to CO causes displacement and/or conversion of the NO, which would have yielded the loss feature at 1783 cm^{-1} (linear NO) in the absence of exposure to CO. In Fig. 23b, the conversion/displacement is obviously more difficult due to the higher initial coverage of NO relative to the saturation coverage of NO on the clean surface ($\hat{\theta}_{NO}^0 \equiv 1$ at saturation). Qualitative comparison of the relative peak intensities of Figs. 23a and b implies also that less CO adsorption occurs for the higher value of $\hat{\theta}_{NO}^0$.

Finally, the energy loss spectra of Fig. 23c show clearly that *conversion from linear to bridged* NO *occurs during adsorption of* CO. Whether displacement of NO by CO from the surface occurs also cannot be determined on the basis of these data. The surface represented by the spectrum at the bottom of Fig. 23c was prepared by exposing the Ru(001) crystal to NO at 280 K. At an exposure temperature between 120 and 180 K, this 120 Torr s exposure would have resulted in the spectrum of Fig. 19 and a relative surface coverage of $\hat{\theta}_{NO}^0 = 0.6$. Exposure at the higher temperature allows

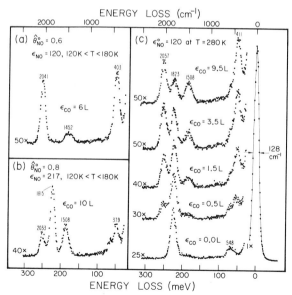

FIG. 23. Electron energy loss spectra of the Ru(001) surface following exposure to NO at (a) $\hat{\theta}_{NO}^0 = 0.6$, $\epsilon_{NO} = 120$, 120 K $< T <$ 180 K, (b) $\hat{\theta}_{NO}^0 = 0.8$, $\epsilon_{NO} = 217$, 120 K $< T <$ 180 K, and (c) $\epsilon_{NO}^0 = 120$, $T = 280$ K and subsequent exposure to CO at the temperatures indicated. The energies of the individual loss features are given in inverse centimeters. (From Thiel et al.[146])

dissociation of bridged NO to occur, which results in the exclusive population of the linear molecular NO state at this coverage. Exposure to CO of the surface of Fig. 23c, where only linear NO and dissociation products are present, results in the growth of the 1508-cm^{-1} loss feature (bridged molecular NO), the growth of the 2057-cm^{-1} feature (adsorbed CO), and a decrease in the intensity of the 1823-cm^{-1} feature (linear molecular NO) relative to the other two. The 548-cm^{-1} feature (Ru–O and Ru–NO stretching vibrations) becomes almost obscured by the 411-cm^{-1} feature (Ru–CO vibration) as the CO exposure increases.

The observed competitive adsorption of CO and NO on Ru(001) has a close analogy in the reactions of two metal-cluster compounds. The compound $Os_3(CO)_{12}$, which has 12 linear, singly coordinated CO ligands, reacts with gaseous NO to form $Os_3(CO)_9(NO)_2$, in which three of the carbonyl

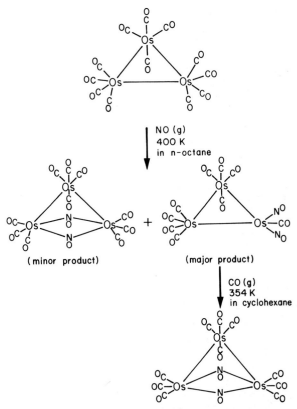

FIG. 24. Illustration of the reaction of $Os_3(CO)_{12}$ with NO(g) and subsequent reaction with CO(g). (From Thiel et al.[148] See also Bhaduri et al.[159])

ligands on one Os atom have been replaced by two singly coordinated nitrosyl ligands. This compound can react further with CO(g) to form $Os_3(CO)_{10}(NO)_2$, in which both nitrosyl ligands are now twofold coordinated (bridged) and an Os–Os bond has been broken.[158] The reaction sequence is illustrated in Fig. 24. This reaction proceeds similarly for the $Ru_3(CO)_{12}$ cluster.[159] The conversion of the nitrosyl ligand by CO(g) from a single to a multiple coordination state with the metal is thus seen in both heterogeneous and homogeneous Ru systems.

To summarize, EELS has been used to show that two forms of adsorbed NO exist on the Ru(001) surface, a bridged and a linear admolecule. Bridged NO dissociates with an activation energy of approximately 16 kcal mole^{-1} at low relative coverages. At a sufficiently high coverage of dissociation products, the Ru(001) surface is poisoned with respect to subsequent adsorption of bridged NO, and adsorption into the linear state occurs. Adsorbed CO competes directly with NO for the adsites that are occupied by the linear form of NO. Adsorption of NO, followed by exposure to gaseous CO, causes at least partial conversion from the linear to bridged form of NO as adsorption of CO occurs. Adsorption of CO, followed by exposure to gaseous NO, prevents adsorption of linear NO at $\theta^0_{CO} \geq 0.3$. The fact that some NO can adsorb into the linear sites only below $\theta^0_{CO} = 0.3$ is evidence that the CO molecules in the $(\sqrt{3} \times \sqrt{3})R30°$ overlayer are adsorbed in the on-top sites.

2.3. Reflection IR Spectroscopy

2.3.1. Introduction

Both the questions posed and the theoretical concepts concerning IR reflection–absorption spectroscopy (IRAS) are quite similar to those related to EELS discussed in detail in the preceding chapter. Consequently, a somewhat less inclusive discussion of IRAS is necessary. This section is organized in the following way: In Section 2.3.2, experimental considerations are reviewed; in Section 2.3.3 a discussion of the interaction of radiation with a metallic surface is presented, including a comparison of predicted cross sections for IRAS and EELS; and in Section 2.3.4 a discussion is given of experimental results for CO chemisorbed on the (100), (111), and (210) surfaces of Pd.[53,54] This will serve to illustrate the power of IRAS when applied to judiciously chosen adsorbate–substrate systems.

---

[158] S. Bhaduri, B. F. G. Johnson, J. Lewis, D. J. Watson, and C. Zuccaro, *J.C.S. Chem. Commun.* p. 477 (1977).

[159] B. F. G. Johnson, personal communication (1979).

2.3.2. Experimental IRAS

The application of IRAS to investigate adsorption on single crystals has been difficult in the past for several reasons.

(1) Low intensity of the monochromatic radiation (typically 10^{-8} W at resolutions of a few wave numbers) and low sensitivity of thermal detectors operated at room temperature. Intensity is further reduced by losses at mirrors in the optical system (typical reflectivity is 97%), by losses at windows in the UHV cell, and by reflection at small single crystals (typically 6 mm $\times$ 10 mm) imposed by the necessarily high angle of incidence (reflectivity for some metals can be as low as 50% at 2000 cm^{-1}).

(2) The weakness of the IR absorption due to vibrational excitation of the adsorbed molecule. For a monolayer of CO on transition metals the absorption is typically on the order of one percent, but is much less for hydrocarbons [e.g., 0.25% for C_2D_6 on $Cu(110)$[160]]. Consequently, a very low noise-to-signal ratio on the order of 0.01% is necessary for the detection of adsorbed molecules at lower coverages.

(3) Drift problems, e.g., those associated with source intensity changes, spectrometer atmosphere changes, and crystal movement.

(4) Absorption bands due to atmospheric water in the wavelength region below 2000 cm^{-1}, which can be several orders of magnitude larger than absorption bands due to the adsorbate.

The above-mentioned restrictions thus impose certain design criteria for an experimental IRAS system. The components of a typical system (Fig. 25)[55] consist of an infrared source, a modulation system (e.g., beam chopper, rotating polarizer, or vibrating mirror), a monochromator, the UHV chamber with the sample, and a detector. The vacuum chamber is equipped with UHV-compatible infrared transmitting windows (e.g., CaF_2). Furthermore, if ordered adlayers are to be investigated, it is advantageous to arrange the optical system so that the IR reflection at the crystal occurs in front of LEED optics, thus allowing simultaneous *in situ* LEED and IR measurements. Keeping these requirements of an IRAS system in mind, the various parts of an optical system are discussed in the following.

The infrared source is usually a thermal emitter operated at high temperatures. Since the emitted power of blackbody radiation is, according to the Stefan-Boltzmann law, proportional to the fourth power of the absolute temperature, operation at higher temperatures is desirable.

The use of blackbody radiation has yet another consequence expressed by the Planck radiation law, whereby the monochromatic emissive intensity I

[160] K. Horn and J. Pritchard, *Surf. Sci.* **52**, 437 (1975).

2.3. REFLECTION IR SPECTROSCOPY

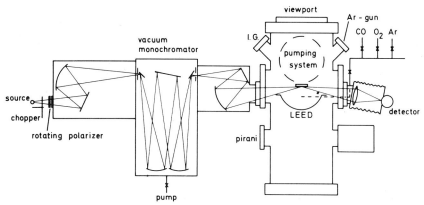

FIG. 25. Schematic diagram of an IRAS system. (From Hoffmann.[55])

varies with the wavelength λ according to

$$I \propto \frac{a}{\lambda^5[\exp(b/T\lambda) - 1]} \, d\lambda, \qquad (2.3.1)$$

where a and b are constants.

This implies a decrease in monochromatic source intensity with increasing wavelength — a serious limitation for the application of thermal emitters in the high-wavelength region above 15 μm (below 660 cm^{-1}). In practice, the infrared source is usually a Globar (SiC rod of approximately 4-mm × 25-mm glowing area) operated at temperatures near 1700 K in a water-cooled housing. For IRAS the Globar source is often favored over the Nernst glower (ZrO$_2$ · $\frac{1}{2}$O$_3$ rod), which is commonly used in commercial IR spectrometers due to the superior stability of the Globar.

An essential improvement may be achieved with the future development of IR lasers that can be tuned conveniently over larger wavelength regions.[161] Provided they possess the necessary stability, they would have higher effective monochromatic power by several orders of magnitude, as compared to thermal emitters (10^{-8} W at a resolution of a few wave numbers). In addition, they would allow for increased spectral resolution (presently 1–5 cm^{-1}); however, this advantage is probably unnecessary in view of the fact that inherent absorption bandwidths of CO adsorbed on metals are typically 6–10 cm^{-1}.

The radiation of the thermal emitter is monochromatized in the IR by means of diffraction gratings. Often in the past, commercial double-beam

---

[161] H. Walther, "Laser Spectroscopy." Springer-Verlag, Berlin and New York, 1976.

instruments, such as those used for chemical analysis, have been modified and adapted to UHV chambers. More recently, however, smaller, evacuable $f/4$ monochromators (e.g., Edinburgh Instruments or Spex Industries) have come into use. These are preferred for several reasons.[162,163] First, evacuation of the optical path is highly desirable to reduce or eliminate absorption due to atmospheric gases. Especially for the frequency region below 2000 cm^{-1}, absorption by atmospheric water vapor can be bothersome or make detection of small absorption bands virtually impossible, even when using dry nitrogen flushing or more sophisticated modulation methods such as double beams. A second advantage of these monochromators is their physical flexibility, an advantage that should not be underestimated in view of the fact that modern UHV systems require adaptation of commercial optical systems, and the optical system must be partially disassembled for bakeout procedures. For a scanning range between 2200 and 1750 cm^{-1}, diffraction gratings of 100–150 grooves mm^{-1} and blaze wavelengths of approximately 5 μm (=2000 cm^{-1}) are used at typical slit widths of 0.2–1 mm, yielding resolutions in the range of 1–6 cm^{-1}. Typical scan speeds are 1 cm^{-1} s^{-1}.

Extending the spectral range necessitates the use of different gratings blazed to the corresponding region to be investigated and different filters to cut off higher-order diffraction from the gratings. To match the typical EEL spectral range of 4000–250 cm^{-1}, which seems to be technically feasible now in IRAS, one needs three different gratings and filters. This difficulty, however, can be overcome by the use of fast Fourier transform IR spectroscopy (FFTS).[164,165] When using this technique, one would not benefit from the luminosity advantage of an arrangement as shown in Fig. 25, but, nevertheless, this technique has very real practical advantages in the avoidance of filter and grating interchanges and in the ability to vary the resolution without altering the slit width.

Another perhaps promising possibility,[166] the feasibility of which, however, has not yet been explored fully, lies in the application of "variable wavelength filters" (Optical Coatings, Ltd.) in monochromatizing the incident radiation. These devices, originally developed as filters, are circular sections of optical coatings with radially increasing thickness that block radiation except in a small range with a typical half-width of 15–25 cm^{-1}. A single small disc allows monochromatization of a wavelength region between approximately 2500 and 800 cm^{-1}. Use of these in a double-pass or tandem configuration to reduce the band-pass half-width should comprise a

[162] J. Pritchard, ref. 130, p. 114.
[163] F. M. Hoffmann and A. Ortega, ref. 130, p. 128.
[164] K. R. Betty and G. Horlich, *Appl. Spectrosc.* **30**, 23 (1976).
[165] A. Sovitsky and M. J. E. Golay, *Anal. Chem.* **36**, 1627 (1964).
[166] W. G. Golden, D. S. Down, and J. Overend, *J. Phys. Chem.* **82**, 843 (1978).

2.3. REFLECTION IR SPECTROSCOPY

strikingly simple optical arrangement that circumvents the usual grating and filter changes needed for wide spectral scans.

Usually, the monochromatized radiation is focused as an image that is typically 1 × 10 mm in area at the crystal with grazing incidence (typically 83° ± 6°), and the reflected light is focused finally on the IR detector. Generally, thermal detectors (e.g., thermocouples or golays) do not meet the signal-to-noise requirements of IRAS. Consequently, liquid-nitrogen-cooled detectors of the photoconductive type (e.g., InSb, PbSnTe, or PbSe) are often used, with a noise level (NEP) typically 1×10^{-11} W Hz$^{-1/2}$. However, these detectors have the disadvantage of restricted spectral range. This has been overcome by the use of a liquid-He-cooled germanium bolometer,[162] which gives comparable performance over the whole spectral range of interest. Further improvement is possible through using cooled filters to reduce the limiting effects of the noise generated by room-temperature background radiation. A more extensive discussion of the use of detectors with IR spectrometers is given elsewhere.[167]

The use of different modulation systems is of fundamental importance. Conventional single-beam systems are inherently more subject to the drift problems associated with source intensity and spectrometer-atmosphere changes than are double-beam spectrometers. Consequently, the latter have been used[127,168] in a configuration in which the IR beam is divided between a sample beam reflected at the crystal and a reference beam passing through the UHV cell, which results in a modulated signal alternating between these two components at typically 13 Hz. However, the problems introduced by using the low modulation frequency, the signal of which is subject to $1/f$ noise, and the problems of matching sample and reference beams at their different optical paths tend to cancel the advantages gained by the double-beam system. These problems have been overcome partially by using polarization modulation or wavelength modulation of a single beam. Polarization modulation exploits the fact that only p-polarized radiation interacts vibrationally with adsorbed molecules on surfaces (compare Section 2.3.3). Consequently, the s component can provide a reference beam following *identically the same optical path*, resulting in a pseudo-double-beam technique. (Here, s and p denote components perpendicular and parallel to the plane of incidence, as indicated in Fig. 26a.) This principle has been followed in the elegant ellipsometric spectroscopy of Dignam and co-workers[169] and in the simpler methods adopted by Hoffmann and Bradshaw[53,54] and Overend *et al.*[166] In practice, this can be performed by rotating a polarizer in the optical

---

[167] H. A. Strobel, "Chemical Instrumentation," 2nd Ed. Addison-Wesley, Reading, Massachusetts, 1973.

[168] H. J. Krebs and H. Lüth, *Appl. Phys.* **14**, 337 (1977).

[169] R. W. Stobie, B. Rao, and M. J. Dignam, *Surf. Sci.* **56**, 334 (1976).

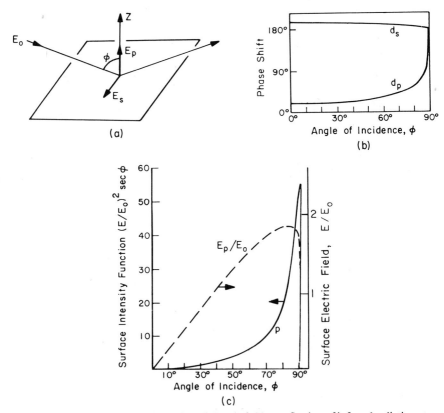

FIG. 26. (a) Schematic representation of electric field on reflection of infrared radiation at a metal surface. (b) Phase shifts d of the electric-field components as functions of incident angle ϕ. (From Pritchard.[162]) (c) Absorption intensity and intensity for the p component of the electric field for $n = 3$, $k = 30$ as functions of incident angle ϕ. (From Pritchard.[162])

path; this yields a modulated signal (typically 150 Hz) of $I_p - I_s$[53] or $(I_p - I_s)/I_p$ when combined with static polarizers.[169] Polarization modulation thus offers the advantage of using a single optical path and an optimal reduction of the effects of water-vapor absorption and intensity drift. In wavelength modulation,[160] an oscillating mirror in the optical path allows measurement of the derivative of the intensity $dI/d\lambda$ at the oscillation frequency (~ 250 Hz). Subsequent integration of the recorded derivative spectrum averages much of the noise and yields a normal spectrum with an improved noise-to-signal ratio (typically 0.005%[160,170]).

Through the use of these modulation methods, the noise-to-signal ratio

[170] K. Horn, ref. 130, p. 140.

2.3. REFLECTION IR SPECTROSCOPY

can be reduced to 0.01% or less, making IRAS sensitive to small fractions of a monolayer of hydrocarbons[160,170] as well as to strongly absorbing species such as CO for which approximately 1/1000 of a monolayer is presently detectable.[63] The high resolution of IRAS offers the possibility of detailed investigations of ordering processes in adsorbed layers and of intermolecular interactions, as will be shown in the following section. However, this method is restricted in present practice to a smaller spectral region than EELS. This may be overcome by technological improvements presently under way,[171,172] but the ease with which large spectral ranges can be scanned in EELS will be difficult to match in the near future. Finally, it should be noted that one of the most important advantages this technique offers is the ability to obtain spectra without restriction on the ambient pressure. In contrast to electron spectroscopies such as EELS, which are restricted to pressures below approximately 10^{-6} Torr, this method can operate even at pressures higher than 1 atm. This is indicative of its importance for investigations under conditions of real catalytic reactions.

2.3.3. Theoretical Considerations

The mechanism of vibrational excitation of molecules with IRAS is based upon the interaction of the electromagnetic field of the reflected light with the dipole moment of an adsorbed molecule. The absorption of radiation induces a change in the dipole moment of the molecule, which leads to an excited vibrational state. Theoretical considerations of IRAS to date are based entirely on classical electrodynamics and have been expressed several times[173-175] since their original introduction by Greenler.[176,177] It has been demonstrated both experimentally[23] and theoretically[176,177] that for IR reflection–absorption spectroscopy of adsorbed molecules on metal surfaces the following considerations are important.

(1) The incident light should be reflected at the surface with grazing incidence.

(2) The validity of a strict dipole selection rule perpendicular to the surface requires that (a) the incident light should have a p-polarized component E_p (compare Fig. 26a) and (b) only vibrational modes with a compo-

---

[171] A. M. Bradshaw, personal communication (1979).
[172] J. Pritchard, personal communication (1979).
[173] J. D. E. McIntyre and D. E. Aspnes, *Surf. Sci.* **24**, 417 (1971).
[174] J. Pritchard, *DECHEMA-Monogr.* **78**, 231 (1975).
[175] H. A. Pearce and N. Sheppard, *Surf. Sci.* **59**, 205 (1976).
[176] R. G. Greenler, *J. Chem. Phys.* **44**, 310 (1966).
[177] R. G. Greenler, *J. Vac. Sci. Technol.* **12**, 1410 (1975).

nent of the dipole-moment change perpendicular to the surface will be excited.

The common reason for these conditions stems from the electrodynamic properties of metallic substrates in the frequency range of IR, as will be shown in the following subsection.

2.3.3.1. Reflection from Clean Metal Surfaces. Following the early work of Francis and Allison,[178] reflection of IR radiation from a clean metallic substrate can be treated theoretically by using Fresnel's equations. Reflection from a metal, as characterized by its complex index of refraction $\tilde{n} = n - ik$, results in phase shifts d_s and d_p at the surface. Results for a typical metal show that the s-polarized component is reversed in phase on reflection, thus resulting in a vanishing surface electric field (destructive interference). The phase change of the p component is strongly dependent on the angle of incidence (Fig. 26b), and at high incident angles a large resultant electric field E_p is established. This is shown in Fig. 26c for a typical metal, copper, at a frequency of 2000 cm^{-1}.[177] As a result, p-polarized radiation can interact strongly with vibrational modes containing a change of dipole moment perpendicular to the surface. The importance of this is that, due to the reflecting properties of metals in the IR, it is impossible to excite vibrations with s-polarized light irrespective of the orientation of a molecular dipole oscillation at the surface. Following Pritchard,[162] the intensity of absorption due to E_p can be assumed to depend on E_p^2 and on the area (i.e., the number of molecules) over which the field is exerted, thus giving an intensity function proportional to $E_p^2 \sec \phi$ (Fig. 26c), based on the reflecting properties of the clean metal surface. This parallels closely a more accurate angular dependence of absorption by a thin surface film computed by Greenler[162,176,177] and results in a maximum absorption predicted at 88°, a result experimentally observed for CO on evaporated copper.[22]

2.3.3.2. Reflection from a Metal Surface with an Adsorbate. This situation can be treated theoretically using a classical dielectric description assuming macroscopic dielectric constants ($\epsilon = \tilde{n}^2$) for the metal substrate, adphase, and gas phase. In the following, it should be kept in mind, however, that the use of such a classical, macroscopic description may be adequate to obtain optical properties of multilayer films, but, clearly, for monolayer and fractional monolayer chemisorption systems, there is a need for a microscopic theory.[128] The three regions of interest are shown in Fig. 27.[176] According to McIntyre and Aspnes[173] and Greenler,[176,177] for radiation polarized perpendicular to the plane of incidence, the change in reflectivity upon

---

[178] S. A. Francis and A. H. Allison, *J. Opt. Soc. Am.* **49**, 131 (1959).

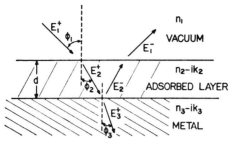

FIG. 27. Model of a thin layer on a metal surface used to formulate the reflection problem as a boundary-value problem. (From Greenler.[176])

absorption can be approximated by

$$\frac{\Delta R_s}{R_s} = \frac{8\pi l}{\lambda} \cos\phi \, \text{Im}\left(\frac{\epsilon_2 - \epsilon_3}{\epsilon_1 - \epsilon_3}\right), \quad (2.3.2)$$

where l is the thickness of the adlayer, λ the wavelength of the radiation, ϕ the angle of incidence, and ϵ_i the complex dielectric constant of the ith phase (as defined in Fig. 27). A somewhat more complicated expression is required to express the change in reflectivity of the p component parallel to the plane of incidence, namely,

$$\frac{\Delta R_p}{R_p} = \frac{8\pi l}{\lambda} \cos\phi \, \text{Im}\left\{\frac{\epsilon_2 - \epsilon_3}{\epsilon_1 - \epsilon_3}\right.$$
$$\left. \times \left[\frac{1 - (\epsilon_1/\epsilon_2\epsilon_3)(\epsilon_2 + \epsilon_3)\sin^2\phi}{1 - (1/\epsilon_3)(\epsilon_1 + \epsilon_3)\sin^2\phi}\right]\right\}. \quad (2.3.3)$$

For CO adsorbed on Pt(111), values of the complex dielectric constant for the adlayer and for the substrate have been found to be $\epsilon_2 = 1 + 2i$ and $\epsilon_3 = -120 + 340i$, respectively.[179] Examination of the expressions for the change in reflectivity shows that as the magnitude of ϵ_3 becomes much greater than that of ϵ_2, the perpendicular component approaches zero, and the expression for the p-polarized component reduces to

$$\frac{\Delta R_p}{R_p} = \frac{8\pi l}{\lambda} \sin\phi \tan\phi \, \text{Im}\left(-\frac{1}{\epsilon_2}\right). \quad (2.3.4)$$

This expression again implies that the maximum enhancement in the ab-

[179] J. Pritchard and T. Catterick, in "Experimental Methods in Catalytic Research" (R. B. Anderson and P. T. Dawson, eds.), Vol. 3, p. 281. Academic Press, New York, 1976.

sorption of radiation by vibrations with a component of the dynamic effective charge normal to the surface occurs at large angle of incidence ϕ.

The IR radiation field that interacts with an adsorbate dipole is periodic in time. At the particular frequency that matches the eigenfrequency of a normal mode, resonant energy absorption occurs. The energy extracted from the local electric field vector is dissipated eventually throughout the adlayer by anharmonic coupling to other vibrational and rotational eigenmodes of the system. This interaction can be approximated by

$$\hat{q}_\| = q_\perp \sin \phi, \quad (2.3.5)$$

where $\hat{q}_\|$ is the wave vector of the surface wave and $q_\perp$ is the value of the wave vector normal to the surface. For single reflections and large angles of incidence ($\phi > 80°$), the surface wave vector results in a coupling cross section of 2×10^{-18} cm^2 molecule^{-1} for the case of CO adsorbed on Pt.[37]

The corresponding wave vector interaction in EELS is given by

$$\hat{q}_\| = (2E_i)^{-0.5} \sin^{-1} \phi, \quad (2.3.6)$$

where E_i is the impact energy of the incident electron beam. For CO adsorbed on Pt, this results in a coupling cross section of 4.6×10^{-16} cm^2 molecule^{-1}.[37] Consequently, there is an advantage in sensitivity in EELS of a factor of greater than 200 compared to IRAS.

2.3.4. Experimental Results: The Adsorption of CO on Pd

2.3.4.1. CO on Pd(100). The Pd(100) surface offers a unique advantage for studying ordered adsorption of CO with IRAS, since both the adsorption site and the coverage can be determined by LEED.[180,181] As ordering takes place, a c(4 $\times$ 2)R45° structure is observed at $\theta = 0.5$, which is followed by a series of uniaxial out-of-registry compression patterns as coverage increases. As shown in Fig. 28, only twofold bridge sites can be occupied at $\theta = 0.5$ when the occupation of equivalent adsorption sites is assumed.[180,181] Intensity measurements by LEED and multiple scattering calculations[182] confirm further the bridge-bonding site yielding a Pd–C bond length of 1.93 ± 0.07 Å and a C–O spacing of 1.15 ± 0.1 Å, with a bulk Pd–Pd spacing.

Figure 29 shows the development of the IR absorption band with increasing coverage at 100 K.[183] A band appears at 1895 cm^{-1}, shifting continuously

---

[180] R. L. Park and H. H. Madden, *Surf. Sci.* **11**, 188 (1968).

[181] J. C. Tracy and P. W. Palmberg, *J. Chem. Phys.* **51**, 4852 (1969).

[182] R. J. Behm, K. Christmann, G. Ertl, M. A. Van Hove, P. A. Thiel, and W. H. Weinberg, *Surf. Sci.* **88**, L59 (1979).

[183] A. Ortega, A. Garbout, F. M. Hoffman, W. Stenzel, R. Unwin, K. Horr, and A. M. Bradshaw, in: *Proc. JCSS*-4 and EC05S-3, Cannes, 1980, p. 335.

2.3. REFLECTION IR SPECTROSCOPY

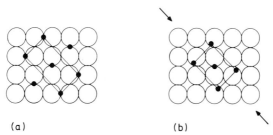

FIG. 28. For CO on Pd(100), (a) one of the two possible domains of the c(4 × 2)R45° structure with $\theta = 0.5$, showing the two unit cell descriptions and (b) the direction of compression of the c(4 × 2)R45° structure (shown by arrows) with $\theta > 0.5$ for the same domain. The filled circles represent CO molecules but are not intended to convey their size relative to the metal atoms. (From Bradshaw and Hoffmann.[54] Copyright North-Holland Physics Publishing, Amsterdam, 1978.)

to 1949 cm^{-1} at $\theta = 0.5$ [as evidenced by the c(4 × 2)R45° structure in LEED].

The ordering of the adlayer, which takes place just below $\theta = 0.5$, can be monitored better at higher temperatures (330 ≤ T ≤ 370 K), for which the surface coverage can be varied sensitively by changing the equilibrium pressure of CO between 10^{-6} and 10^{-8} Torr.[53,54] Figure 30 depicts the coverage range between $\theta \sim 0.45$ and $\theta \sim 0.55$. Below $\theta = 0.5$, the absorption band has a broad asymmetric shape, indicating that it consists of two bands. At $\theta = 0.50$, the band sharpens at 1949 cm^{-1}, which corresponds to the ordering of the adlayer in the simultaneously observed c(4 × 2)R45° LEED pattern. A further increase of the coverage leads to the formation of domains where the c(4 × 2)R45° is compressed (Fig. 29b), as evidenced by the growth of a second IR peak that is shifted slightly to higher wave numbers. At still higher coverages, the c(4 × 2)R45° peak vanishes upon completion of the transition to the compression overlayer. During compression, the frequency of the IR band shifts at an increased rate toward higher frequencies, as may be seen in Fig. 31. This indicates the possible importance of intermolecular repulsion within the adlayer as an additional mechanism contributing to the frequency shift. Saturation coverage at room temperature is reached at an equilibrium pressure of CO of 10^{-6} Torr. This corresponds to a fractional surface coverage of approximately 0.61 and a CO stretching frequency of 1983 cm^{-1}.[53,54] At 100 K, the IR band shifts further to 1997 cm^{-1} at $\theta \sim 0.8$.[183]

As was discussed in Chapter 2.2, there has been a controversy over the past few years concerning the nature of the coverage-dependent shift of the C–O stretching frequency illustrated by the data of Figs. 15 and 31. The technique

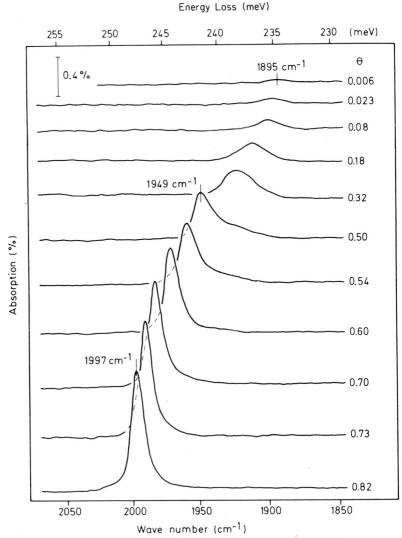

FIG. 29. IR absorption band for the C–O stretching frequency in the system CO/Pd(100) at 100 K as a function of increasing coverage. (From Ortega et al.[184] Copyright North-Holland Physics Publishing, Amsterdam, 1982.)

of IRAS possesses a unique advantage over EELS in its ability to resolve clearly features arising from different isotopically labeled CO admolecules. The vibrational spectra of such isotopic mixtures ($C^{12}O^{16}/C^{13}O^{16}$) have been studied to gain insight into the importance of dipolar coupling and intermolecular vibrational coupling relative to the other types of effects that have been postulated to account for the observed frequency shifts (Subsection

2.3. REFLECTION IR SPECTROSCOPY

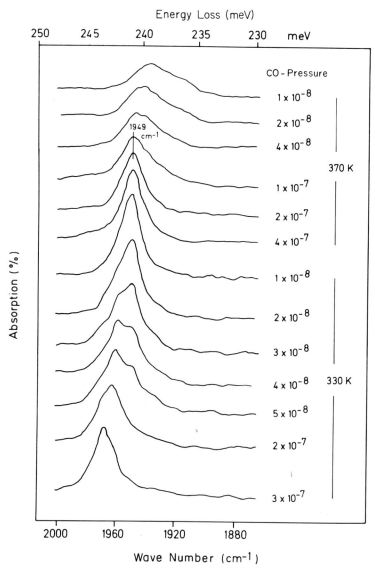

FIG. 30. The IR sequence for equilibrium pressures of CO for CO/Pd(100) at 330–370 K. The spectra show ordering of the c(4 × 2)R45° structure at 1949 cm^{-1} (370 K) and subsequent compression (330 K) of the ordered layer, resulting in the appearance of a second peak. (From Ortega et al.[184] Copyright North-Holland Physics Publishing, Amsterdam, 1982.)

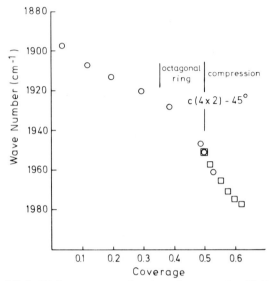

FIG. 31. The shift in IR frequency as a function of coverage for CO/Pd(100) at 300 K. Coverage from LEED; O, coverage from IR intensities. (From Bradshaw and Hoffman.[54] Copyright North-Holland Physics Publishing, Amsterdam, 1978.)

2.2.3.3). On Pd(100)[184] and Cu(111),[185] such studies indicate that both types of interactions (coupling and chemical–electrostatic effects) are important, whereas for linearly bound CO on Pt(111), King and co-workers[123,141] have suggested that dipolar coupling predominates. This illustrates the value of the quantitative assignments that are becoming possible as a result of IRAS studies of isotopic mixtures.

2.3.4.2. CO on Pd(111). Beyond the fact that the fcc (111) surface possesses threefold coordination sites [in contrast to the (100) surface], the CO/Pd(111) system is quite interesting for other reasons. For example, the (111) surfaces of several different fcc transition metals show identical LEED behavior during CO adsorption,[186–190] namely, the formation of an ordered ($\sqrt{3} \times \sqrt{3}$)R30° structure. In addition, it has been predicted theoretically[191] that this surface should be rather "smooth" with respect to variation in

[184] A. Ortega, F. M. Hoffmann, and A. M. Bradshaw, Surf. Sci. **119**, 79 (1982).
[185] P. Hollins and J. Pritchard, Surf. Sci. **89**, 489 (1979).
[186] G. Ertl and J. Koch, Z. Naturforsch. A **25A**, 1906 (1970).
[187] G. Ertl, M. Neumann, and K. M. Streit, Surf. Sci. **64**, 393 (1977).
[188] K. Christmann, O. Schober, and G. Ertl, J. Chem. Phys. **60**, 4719 (1974).
[189] J. Küppers and A. Plagge, J. Vac. Sci. Technol. **13**, 259 (1976).
[190] J. Pritchard, J. Vac. Sci. Technol. **9**, 895 (1972).
[191] G. Doyen and G. Ertl, Surf. Sci. **43**, 197 (1974).

2.3. REFLECTION IR SPECTROSCOPY

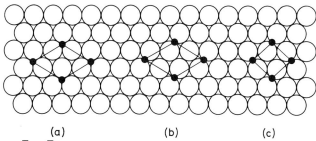

(a) (b) (c)

FIG. 32. (a) ($\sqrt{3} \times \sqrt{3}$)R30° structure for CO on Pd(111) with sites of threefold coordination. (b) One of the three possible domains of the ($\sqrt{3} \times \sqrt{3}$)R30° structure with sites of twofold coordination (bridge-sites); the other two domains are formed by shifting the unit cell into the other two kinds of bridge sites available. [In a sense these domains are only antiphase domains as opposed to the rotational domains of the c(4 × 2) structure.] (c) One of the three possible domains of the c(4 × 2) structure with sites of twofold coordination. (From Bradshaw and Hoffmann.[54] Copyright North-Holland Physics Publishing, Amsterdam, 1978.)

adsorption energy for different adsorption sites. It is indicated by LEED that the ordering of the adsorbed layer in a ($\sqrt{3} \times \sqrt{3}$)R30° structure is $\theta = \frac{1}{3}$.[186] If we assume identical adsorption sites, such an ordered overlayer can be arranged with *all* molecules in threefold or in twofold coordinated bridge sites as indicated in Fig. 32a and b. If one increases the CO coverage on Pd(111), a c(4 × 2) structure at $\theta = 0.5$ is formed by continuously compressing the ($\sqrt{3} \times \sqrt{3}$)R30° unit cell. In this structure however, the molecules can be arranged only in equivalent twofold bridge sites (Fig. 32c). Figure 33 shows the IR absorption sequence for adsorption of CO at 300 K. The initial absorption band occurs at 1823 cm^{-1}, 70 cm^{-1} lower than for CO on Pd(100). With increasing coverage, the IR band increases in intensity at rather constant wavelength up to $\theta = \frac{1}{3}$, where the ($\sqrt{3} \times \sqrt{3}$)R30° superstructure is complete. The constancy of the IR frequency as a function of coverage and the early appearance of fractional order LEED beams has been attributed to island formation at this stage.[53,192] The coverage region between $\theta = \frac{1}{3}$ and $\theta = \frac{1}{2}$, i.e., the region where a transition from the ($\sqrt{3} \times \sqrt{3}$)R30° superstructure to the c(4 × 2) superstructure occurs, is characterized in the infrared by a very broad absorption feature that contains two absorption bands, as may be seen in Fig. 33. Finally, at $\theta = 0.5$, the band sharpens again at a frequency of 1936 cm^{-1}. The shift in the IR frequency as a function of coverage, shown explicitly in Fig. 34, demonstrates clearly the ordering of islands below $\theta = \frac{1}{3}$. Also evident in this figure is a large change in frequency associated with the transition from the ($\sqrt{3} \times \sqrt{3}$)R30° superstructure at $\theta = 0.33$ to the c(4 × 2) superstructure at $\theta = 0.5$. Between $\theta = 0.33$ and

[192] H. Conrad, G. Ertl, J. Koch, and E. E. Latta, *Surf. Sci.* **43**, 462 (1974).

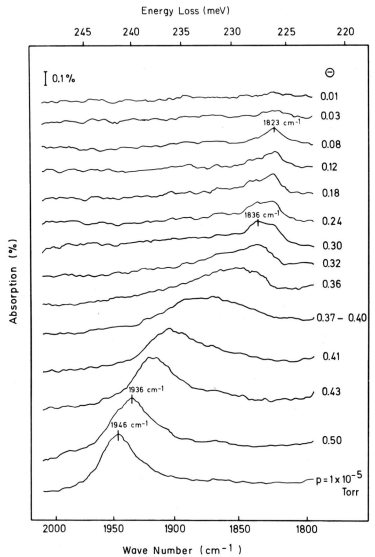

FIG. 33. IR absorption band for the C–O stretching frequency in the system CO/Pd(111) at 300 K as a function of increasing coverage. (From Bradshaw and Hoffmann.[54] Copyright North-Holland Physics Publishing, Amsterdam, 1978.)

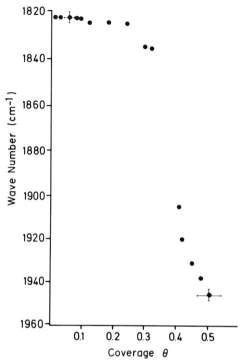

FIG. 34. The shift in IR frequency as a function of coverage for CO/Pd(111). (From Bradshaw and Hoffmann.[54] Copyright North-Holland Physics Publishing, Amsterdam, 1978.)

$\theta = 0.4$, very small changes in coverage cause a large shift in frequency, resulting finally in a frequency for $\theta > 0.4$ similar to that found for the bridge sites on Pd(100) (compare Fig. 31). This large frequency shift has been attributed to a change of the adsorption site from a three fold site in the $(\sqrt{3} \times \sqrt{3})R30°$ superstructure (Fig. 32a) to twofold bridged sites in the $c(4 \times 2)$ superstructure[53,54] (Fig. 32c). A similar behavior has been reported for the CO/Ni(111) system.[127] Increasing the coverage above $\theta = 0.5$ at lower temperatures leads to the appearance of several additional absorption bands,[163,183] the major one being at 2095 cm^{-1}. This band has been attributed to linearly bound CO and agrees well with earlier observations for the adsorption of CO on evaporated Pd films.[193,194]

2.3.4.3. CO on Pd(210). The (210) surface of Pd is particularly interesting because the surface layer does not contain close-packed atoms with a

[193] A. M. Bradshaw and F. M. Hoffmann, *Surf. Sci.* **52**, 449 (1975).
[194] F. M. Hoffmann and A. M. Bradshaw, *J. Catal.* **44**, 328 (1976).

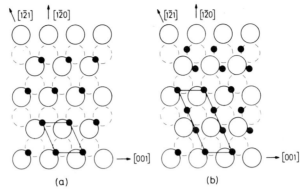

FIG. 35. For CO on Pd(210), (a) surface model of the assumed 1 × 1 structure and (b) surface model of the 1 × 2 structure. (From Bradshaw and Hoffmann.[54] Copyright North-Holland Physics Publishing, Amsterdam, 1978.)

nearest neighbor spacing of 2.73 Å [as are found on the (100) and (111) surfaces]. On Pd(210), the minimum distances are 3.88 and 4.75 Å in the [001] and [121] directions, respectively, as may be seen in Fig. 35. Therefore, one would expect the CO admolecules to populate either linear sites of type A in Fig. 36 (expected to cause a C–O stretching frequency above 2000 cm^{-1}) or bridge sites of type B–E between the top layer and the second layer of metal atoms. Occupation of the latter type of sites would result in a CO frequency in the range between 1900 and 2000 cm^{-1}, as observed for CO on Pd(100). Indeed, the (210) plane can be regarded as a stepped surface con-

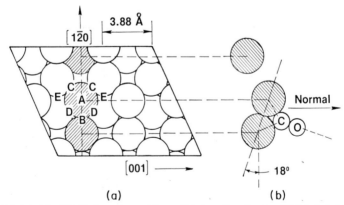

FIG. 36. Model of Pd(210) surface, with possible bonding sites for CO labeled: (a) top view and (b) cross-sectional view. The cross-sectional view illustrates how the CO molecule is "inclined" in the B sites. (From Madey et al.[195] Copyright North-Holland Physics Publishing, Amsterdam, 1979.)

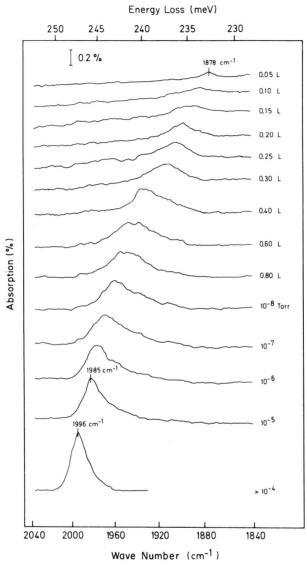

FIG. 37. IR absorption band for the C–O stretching frequency in the system CO/Pd(210) at 300 K as a function of increasing coverage. (From Bradshaw and Hoffmann.[54] Copyright North-Holland Physics Publishing, Amsterdam, 1978.)

sisting of terraces. Occupation of these terrace adsorption sites would result also in the inclination of the adsorbed CO with respect to the surface normal, as shown in Fig. 36.

The IR spectra of Fig. 37 and the frequency–coverage relationship of Fig. 38[54] show a behavior similar to that observed in the CO/Pd(100) system. This suggests, in conjunction with LEED observations,[54,192] that the structural model of Fig. 35a is appropriate. In this case, in which CO forms a 1 × 1 superstructure, adsorption in twofold bridge sites between Pd atoms of the first and second surface layers is suggested. However, on the basis of IR and LEED results alone, it is not possible to distinguish between the various sites B–E. Electron-stimulated desorption-ion angular distribution (ESDIAD) measurements have been used to verify this unusual bonding configuration.[195] With this method, it is possible to infer surface bond angles of an adsorbed molecule or atom from the angular distribution of the desorbed ions.[196,197] Indeed, the observed ESDIAD pattern confirms that for low cov-

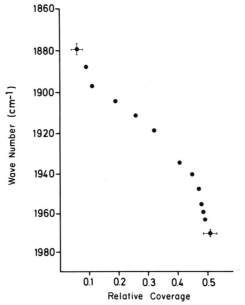

FIG. 38. The shift in IR frequency as a function of relative coverage for CO on Pd(210). (From Bradshaw and Hoffmann.[54] Copyright North-Holland Physics Publishing, Amsterdam, 1978.)

[195] T. E. Madey, J. T. Yates, Jr., A. M. Bradshaw, and F. M. Hoffmann, *Surf. Sci.* **89**, 370 (1979).
[196] T. E. Madey and J. T. Yates, Jr., *Chem. Phys. Lett.* **51**, 77 (1977).
[197] T. E. Madey and J. T. Yates, Jr., *Surf. Sci.* **76**, 397 (1978).

erages the CO molecules are inclined with respect to the surface normal as suggested in Fig. 36. Moreover, the symmetry of the angular distribution implies the existence of binding sites of type C, thus confirming the proposed structural model of Fig. 35a.

2.3.4.4. Synopsis: CO on Pd. To summarize, the combination of LEED and IR reflection–absorption spectroscopy provides a sensitive method for the determination of the geometry of the adsorption site. The band assignments made for the CO/Pd system yield for single coordination (linear bonding) a frequency from 2050 to 2120 cm^{-1}, for twofold bridge sites a frequency from 1880 to 2000 cm^{-1}, and for threefold coordination a frequency from 1800 to 1880 cm^{-1}. The observation of a constant frequency with increasing coverage indicates the formation of islands in the adphase. The observation of frequency shifts to higher wave numbers with increasing coverage can be attributed to intermolecular interactions and changes in the local adsorption site geometry.

2.4. Inelastic Electron Tunneling Spectroscopy

2.4.1. Introduction

A third experimental technique that can be utilized to probe the vibrational structure of molecular overlayers on surfaces of low area is inelastic electron tunneling spectroscopy (IETS). Although the nature of the phenomenon precludes an examination of adlayers on surfaces of single crystal metallic substrates, it is nevertheless appropriate to compare IETS with EELS and IRAS. The preponderance of past work with IETS has dealt with adsorption on oxide surfaces,[58–60] although in the past few years studies have been reported concerning the adsorption of CO on reduced clusters of metallic Rh[198–200] and Fe[201] "supported" on an oxide substrate and the adsorption of C_2H_5OH on similar Ag clusters.[202] It has been demonstrated that IETS can be used to describe the attachment of an inorganic cluster compound [$Zr(BH_4)_4$] on an oxide and to monitor the adsorption of C_2H_4 and C_3H_6 and the polymerization of C_2H_2 thereon.[203–205] Since there are several

---

[198] P. K. Hansma, W. C. Kaska, and R. M. Laine, *J. Am. Chem. Soc.* **98**, 6064 (1976).
[199] R. M. Kroeker, W. C. Kaska, and P. K. Hansma, *J. Catal.* **57**, 72 (1979).
[200] J. Klein, A. Léger, S. de Cheveigné, C. Guinet, M. Belin, and D. Defourneau, *Surf. Sci.* **82**, L288 (1979).
[201] R. M. Kroeker, W. C. Kaska, and P. K. Hansma, *J. Chem. Phys.* **72**, 4845 (1980).
[202] H. E. Evans, W. M. Bowser, and W. H. Weinberg, *Surf. Sci.* **85**, L497 (1979).
[203] H. E. Evans and W. H. Weinberg, *J. Am. Chem. Soc.* **102**, 872 (1980).
[204] H. E. Evans and W. H. Weinberg, *J. Am. Chem. Soc.* **102**, 2548 (1980).
[205] H. E. Evans and W. H. Weinberg, *J. Am. Chem. Soc.* **102**, 2554 (1980).

reviews available that discuss the concepts of IETS in detail, including its application to the study of chemisorption on oxides,[58-60] the discussion that follows will concentrate primarily on adsorption and catalytic reactions over reduced metal clusters and inorganic cluster compounds. Prior to this, a brief qualitative description of the relevant experimental and theoretical details will be presented.

2.4.2. Experimental Details

In IETS, a planar (diode) geometry is utilized consisting of a metallic film, a thin ($\sim 20-30$ Å) oxide layer and a top metallic film. When adsorbates are introduced onto the oxide surface prior to the evaporation of the top metal, the vibrational structure of the adspecies may be measured via the energy loss of electrons that tunnel inelastically through the barrier when a bias potential is applied across the two metal electrodes, as shown schematically in Fig. 39.[206,207] The second derivative of the tunneling current $[d^2I/d(eV)^2]$, which is easily measured electronically using a modulation and lock-in amplification detection scheme, is proportional to the oscillator strengths of the various vibrational modes of the molecular bonds that are detected. Thus, the preferred representation of the data is $d^2I/d(eV)^2$ as a function of applied voltage, as seen in Fig. 40, since this is the IETS analog of an EEL or IRA

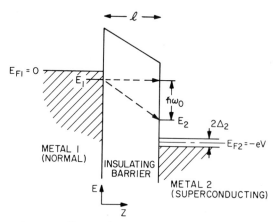

FIG. 39. Schematic energy diagram for both elastic (horizontal dashed line) and inelastic (oblique dashed line) electron tunneling through a barrier of width l. E_1 is the initial energy and E_2 the final energy, separated by $\hbar\omega_0$ (a vibrational excitation energy). E_{F1} and E_{F2} are the respective Fermi levels in the two metals. $2\Delta_2$ represents the superconducting gap in metal 2, and eV is the bias voltage.

[206] C. B. Duke, "Tunneling in Solids." Academic Press, New York, 1969.

[207] E. Burstein and S. Lundqvist, eds., "Tunneling Phenomena in Solids." Plenum, New York, 1969.

2.4. INELASTIC ELECTRON TUNNELING SPECTROSCOPY

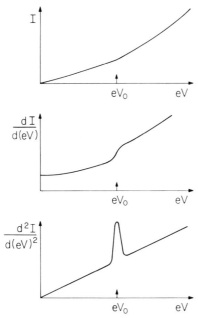

FIG. 40. Schematic representation showing the effect of a vibrational excitation at energy eV_0 on the tunneling current and its first $[dI/d(eV)]$ and second $[d^2I/d(eV)^2]$ derivatives as a function of applied voltage eV.

spectrum. Typical IET spectra extend from approximately 30 to 500 meV (i.e., from ~250 to 4000 cm^{-1}), although the upper limit can be extended to a few volts and is determined both by the breakdown voltage of the junction and the increased background present at the larger bias voltages. At voltages below approximately 30 meV, the phonon structure due to the metallic films dominates the IET spectra. Although numerous combinations of metal–oxide–metal junctions have been studied by using IETS,[207] emphasis has been placed on the Al–Al$_2$O$_3$–Pb junction for two reasons: (1) This junction is quite easy to prepare experimentally, and (2) the study of hydrocarbon reactions on alumina surfaces represents a technologically important problem.

The IETS effect was discovered by Jaklevic and Lambe and discussed first in 1966.[48,208] They not only discovered the effect, but, just as important, they gave a correct interpretation of the relevant physics. The impurity-assisted tunneling spectra of water (hydroxyl groups), methanol, ethanol, acetic acid, propionic acid, cyanoacetic acid, and hydrocarbon pump oil present on the

---

[208] R. C. Jaklevic and J. Lambe, *Phys. Rev. Lett.* **17**, 1139 (1966).

oxide surface in metal–oxide–metal tunnel junctions were measured for various combinations of junction materials. The majority of their work, however, was concerned with Al–Al$_2$O$_3$–Pb junctions, since oxidized Al forms a compact thin oxide rather easily. Also a Pb overlayer is convenient due to its chemical inertness vis-à-vis molecular adsorbates on the oxide, its large ionic diameter, which retards its diffusion into the oxide under the influence of an applied bias voltage, and its rather high superconducting transition temperature (7.2 K). It was also illustrated that (1) IETS is sensitive to both IR and Raman (permanent and induced dipolar) active modes, (2) a representation of the data as the second derivative of the tunneling current with respect to bias voltage as a function of bias voltage is the IETS analog of optical IR and Raman spectroscopy, (3) the inelastic tunneling current corresponds to an increase in the conductivity of the junction (the first derivative of the tunneling current with respect to bias voltage) of approximately 1%, (4) the sensitivity of the technique is at least 1% of a monolayer (which represents approximately 10^{11} molecules mm^{-2}, and a sample the area of which is 1 mm^2 is easily usable), and (5) the IETS lineshape is understood in the sense that the spectra are broadened at higher measurement temperatures according to the Fermi–Dirac distribution function of the electrons in the (normal) metallic films. In addition, it was shown that the resolution is improved further by having one or both of the metallic films superconducting, thereby creating a more sharply peaked electronic density-of-states function above the superconducting gap as compared to a normal metal.

A typical tunnel junction fabrication technique involves evaporation and plasma discharge oxidation. The first step in the fabrication of an Al–Al$_2$O$_3$–Pb junction is the evaporation of approximately 1000 Å of Al in a bell jar, the base pressure of which is below 10^{-6} Torr. Then the Al is oxidized in a plasma discharge (approximately 50 μ of O$_2$ or H$_2$O) to form a barrier the thickness of which is on the order of 25 Å. After *in situ* chemisorption or reaction studies, the final step in the junction preparation is the evaporation of Pb overlayer cross-strips of several thousand angstroms in thickness. This junction preparation scheme is shown schematically in Fig. 41. Such junctions may be heated resistively and their temperature measured via the known (measured) temperature coefficient of resistance of the Al film.[209]

Measurements in IETS invariably involve modulation and lock-in detection techniques (Fig. 42a). In the author's laboratory, the IET spectra are measured with a computer controlled analog detection system as shown in Fig. 42b. Typically, the current through the sample is modulated with a 50-kHz sine wave. A lock-in amplifier detects the resulting 100-kHz voltage

---

[209] W. M. Bowser and W. H. Weinberg, *Rev. Sci. Instrum.* **47**, 583 (1976).

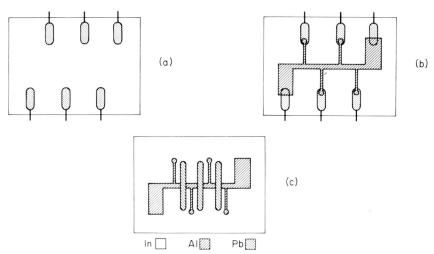

FIG. 41. Representation of IET sample preparation procedure showing (a) clean glass substrate with attached electrical leads, (b) evaporation of aluminum strip, followed by oxidation and exposure to reactants, and (c) deposition of top metal electrodes. Electrical connections are not shown in (c) for the sake of clarity.

across the sample, which is proportional to the second derivative of the current–voltage characteristic function of the junction. The steady-state voltage is generated by a ramp generator, which is controlled by the computer. A complete spectrum is obtained when the second derivative has been measured at several hundred different steady-state voltages. The computer-controlled analog detection scheme has three advantages over the more traditional analog schemes, namely, (1) that repeated fast scans and channel-by-channel sums reduce base-line drift (low-frequency noise), (2) that automatic multiplexing is possible (selected measurements of various windows of the spectrum), and (3) that the data are in digital form immediately, facilitating data analysis via background subtraction, peak-area integrations, and functional deconvolutions.

Now that the way in which tunnel junctions are prepared and the way in which IETS measurements are made have been discussed, it is necessary to clarify the detailed nature of the barrier in the junction on which chemisorption and chemical reactions occur. By comparing IET spectra of barriers prepared in an O_2 plasma at room temperature, and both an H_2O and a D_2O plasma at various temperatures from 300 to 525 K, Bowser and Weinberg[210]

[210] W. M. Bowser and W. H. Weinberg, *Surf. Sci.* **64**, 377 (1977).

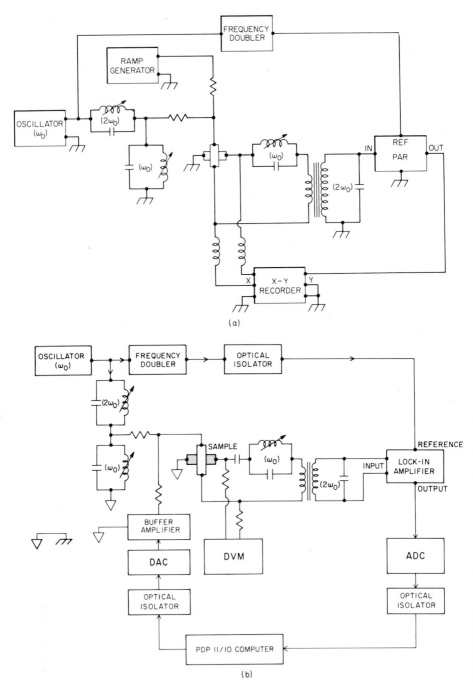

FIG. 42. Schematic of IETS computer-controlled measurement electronics (a) with standard modulation and lock-in techniques, and (b) with computer-controlled analog detection system added.

have shown that the barrier in IETS when prepared in an O_2 plasma is an oxide rather than a hydroxide of aluminum. In this study, all junctions were of the Al–barrier–Pb type. It was further shown that by using low substrate temperatures in conjunction with H_2O or D_2O plasmas, hydrates of aluminum may be prepared, but the dehydration reaction is complete after heating the barrier to 525 K. Thus, barriers formed by an O_2 plasma at room temperature are identical to those formed by an H_2O plasma at 525 K. Incidentally, the ability to heat the oxide surfaces, as demonstrated in the experiments by Bowser and Weinberg, is of obvious and crucial importance in the application of IETS both to surface chemistry and heterogeneous catalysis.[209,210] Subsequently, Evans et al.[211] have clarified further the nature of the insulating barrier formed on Al films as a result of various preparation schemes by using x-ray photoelectron spectroscopy.

Earlier, Geiger et al.[49] had suggested that the barrier in Al–barrier–metal tunnel junctions may be an aluminum hydroxide rather than an aluminum oxide, but results of Hansma et al.[212] are in agreement with those of Bowser and Weinberg[210] and Evans et al.;[211] namely, the barrier is in fact an aluminum oxide with OH groups present at the surface of the oxide, i.e., at the oxide–Pb interface in Al–oxide–Pb tunnel junctions. Hansma et al.[212] have also made the very important comparison of catalytic activity of air-oxidized aluminum versus the high-surface-area transition γ-alumina, which is a common industrial catalyst and catalyst support. It was found that the activity for 1-butene isomerization is actually somewhat greater for the air oxidized aluminum compared to γ-alumina after normalizing to specific surface area. Consequently, the film surfaces in tunnel junctions will be very representative models of high-surface-area transition aluminas used extensively on a commercial scale since air-oxidized and plasma-oxidized aluminum films are identical insofar as IETS measurements are concerned. In this same connection, Klein et al.[213] have investigated the MgO barrier in a Mg–MgO–Pb tunnel junction and have demonstrated that it is indeed an oxide, as opposed, for example, to a hydroxide.

2.4.3. Theoretical Concepts

Although a number of theories of IETS have appeared in the literature[129,214-219] the lack of a detailed understanding of the electron–

---

[211] H. E. Evans, W. M. Bowser, and W. H. Weinberg, *Appl. Surf. Sci.* **5**, 258 (1980).
[212] P. K. Hansma, D. A. Hickson, and J. A. Schwarz, *J. Catal.* **48**, 237 (1977).
[213] J. Klein, A. Léger, M. Belin, D. Défourneau, and M. J. L. Sangster, *Phys. Rev. B* **7**, 2336 (1973).

phonon coupling interaction represents a severe problem,[48,59,129,206-208,209,210,212-221] and it is certainly true that a completely satisfactory theory of IETS has yet to be formulated.

Originally, Scalapino and Marcus[214] applied the WKB approximation of tunneling through a barrier to IETS assuming charge-dipolar coupling between the tunneling electron and the adsorbate on the oxide barrier surface. The enhancement in the electron-tunneling current due to this inelastic scattering was calculated by using a "golden rule" formulation, and order-of-magnitude calculations for an $Al-Al_2O_3-Pb$ junction with OH groups present on the Al_2O_3 were found to be in qualitative agreement with experimental measurements.

More recently, Kirtley et al.[129] have used a transfer Hamiltonian formalism to describe the IETS intensity of vibrational modes of organic molecules chemisorbed on the oxide barrier in metal–oxide–metal tunnel junctions. The theory made correct qualitative predictions of the magnitude of integrated intensities in IETS, the magnitude of Raman-active versus IR-active modes, the observability of optically forbidden modes, and the orientation of the admolecules on the oxide surface. This theory has the advantage of being more realistic than the original formulation,[214] and, unlike many other previous theories,[215-218] it may be used to determine the dependence of peak intensity on adsorbate concentration. Herein lies a difficulty, however. It is assumed that the effect of one admolecule on the tunneling electron is independent of the other admolecules. This leads inevitably to the conclusion that the peak intensities vary linearly with adsorbate concentration. In the one experiment that has tested this hypothesis, which was performed by using tritiated benzoic acid (C_6H_5COOH) as the adsorbate in an $Al-Al_2O_3-Pb$ junction, the intensity rather was found to vary as $n^{4/3}$ for ranges of adsorbate concentration n between 5% and 90% of saturation coverage.[221]

Cunningham et al.[219] have used a "golden-rule" transfer-Hamiltonian formalism to calculate this inelastic tunneling probability that is proportional to the intensity measured in IETS. The potential that causes the inelastic transition is assumed to be due to a layer of dipoles at the adsorbate

---

[214] D. J. Scalapino and S. M. Marcus, *Phys. Rev. Lett.* **18**, 459 (1967).

[215] L. C. Davis, *Phys. Rev. B* **2**, 1714 (1970).

[216] C. Caroli, R. Combescot, D. Lederer, P. Nozieres, and D. Saint-James, *J. Phys. C* **4**, 2598 (1971).

[217] G. K. Birkner and W. Schattke, *Z. Phys.* **256**, 185 (1972).

[218] T. E. Feuchtwang, *Phys. Rev. B* **10**, 4135 (1974).

[219] S. L. Cunningham, W. H. Weinberg, and J. R. Hardy, *Appl. Surf. Sci.* **2**, 640 (1979); see also ref. 60, p. 125.

[220] N. O. Lipari, C. B. Duke, R. Bozio, A. Girlando, C. Pecile, and A. Padva, *Chem. Phys. Lett.* **44**, 236 (1976).

[221] J. D. Langan and P. K. Hansma, *Surf. Sci.* **52**, 211 (1975).

layer (as well as the infinite number of image dipoles that appear in both metal electrodes). When the cooperative effects of the adsorbate molecules are included, it was found that the model predictions agree qualitatively with the experimental results of Langan and Hansma[221] insofar as the dependence of IET spectral intensity as a function of surface concentration is concerned. The theory of Cunningham et al.[219] should prove extremely useful to the surface chemist interested either in chemisorption or heterogeneous catalysis, since the relation between IET spectral intensity and adsorbate concentration is of fundamental importance.

2.4.4. Experimental Results

2.4.4.1. Adsorption on Reduced Metallic Clusters.

Hansma and co-workers[198,199,222] have shown that it is possible to use IETS to investigate the chemisorption of CO on isolated atoms and/or reduced metallic clusters of Rh evaporated onto an oxide film, i.e., an $Al-Al_2O_3-Rh-CO-Pb$ tunneling junction was considered. This experimental configuration serves to model in a useful way industrially important supported metal catalysts. In particular, they have shown that at a statistical thickness of Rh of $\frac{1}{2}$ Å, CO chemisorbs with the appearance of a single Rh–CO mode at low frequency (408 cm^{-1}) and a CO stretching mode at a higher frequency (1935 cm^{-1}). This is suggestive of a single type of chemisorbed CO, perhaps on Rh monomers on the Al_2O_3 surface. At higher coverages of Rh (statistical thickness of 4 Å), five modes were observed, which occurred at 408, 454, 580, 1730, and 1935 cm^{-1}, as can be seen in Fig. 43. Based on the adsorption of $^{12}C^{16}O$, $^{13}C^{16}O$, and $^{12}C^{18}O$, Kroeker et al.[199] have concluded that the three lower-frequency modes have predominantly a bending character, whereas the two higher-frequency modes are, of course, two different CO stretching vibrations. Transmission electron microscopy revealed that a statistical thickness of 4 Å of Rh results in Rh clusters on the Al_2O_3, the average particle size of which is approximately 25 Å.[199]

Klein et al.[200] have repeated this measurement with $^{12}C^{16}O$ and $^{12}C^{18}O$. For a statistical thickness of Rh of 5 Å, five modes for $^{12}C^{16}O$ were observed, occurring at 408, 484, 588, 1736, and 1955 cm^{-1}, in rather good agreement with the results of Hansma et al.[198,199] However, at low coverages of Rh(< 3 Å) the modes at 484 and 1955 cm^{-1} appear first (compare with the modes at 408 and 1935 cm^{-1} in the work of Hansma et al.[198,199,222]), and the detailed assignment of the various modes differ also between the two groups. Klein et al.,[200] also on the basis of isotopic frequency shifts, conclude that (1) the mode at 408 cm^{-1} is due to a Rh–C stretching vibration from dissociatively adsorbed CO, (2) the mode at 484 cm^{-1} is due to the vibration of the

[222] P. K. Hansma, ref. 60, p. 186.

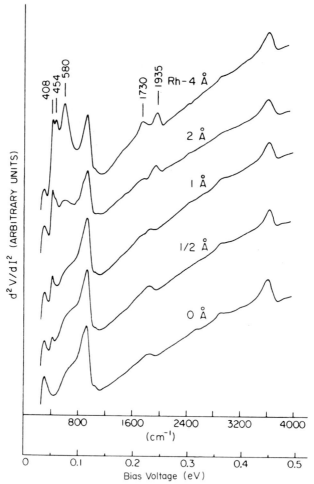

FIG. 43. Tunneling spectra of CO adsorbed on various thicknesses of Rh evaporated on alumina. Note that new peaks appear as the thickness of Rh increases, implying new types of bonding. For all these spectra, the CO exposure was 10^3 L. The modes near 950 and 3600 cm^{-1} are due to an aluminum–oxygen vibration of the Al_2O_3 film and an oxygen–hydrogen vibration of adsorbed OH groups present on the Al_2O_3 surface, respectively. (Reprinted with permission from Hansma et al.[198] Copyright 1976 American Chemical Society.)

CO molecule against a single Rh atom (linearly bound CO) with possible contributions also from a symmetric stretching vibration between more than one Rh atom and a CO molecule bound in a bridging site, (3) the mode at 588 cm^{-1} is due to the asymmetric stretching mode of bridge-bonded CO and (4) the modes at 1736 and 1955 cm^{-1} are due to CO stretching vibrations at bridged and linear sites, respectively.

2.4. INELASTIC ELECTRON TUNNELING SPECTROSCOPY

Since the selection rules governing the electron-vibrational coupling are not understood in detail in IETS, it is not possible to settle the controversy regarding the mode assignments for CO adsorbed on Rh. However, Yates et al.[223] have shown that at low pressures ($< 10^{-5}$ Torr, similar to that used in the IETS experiments), the probability of dissociation of CO on Rh(111) is quite low (below 10^{-4}) compared with the probability of molecular adsorption (near unity). Hence, the assignment of the 408-cm^{-1} mode as being due to a Rh–C vibration of dissociatively adsorbed CO should be viewed with some caution. The fact that the IETS results of both groups show that at low Rh coverages only the linear form of CO exists ($\hbar\omega_{CO} \sim 1935-1955$ cm^{-1}) whereas the bridged form occurs with higher Rh coverages is suggestive of a possible surface transition from Rh monomers at low coverages to large Rh clusters at higher coverages (e.g., an island size of approximately 25 Å at a statistical coverage of 4 Å).

Kroeker et al.[201] have considered CO adsorption on evaporated Fe particles just as in the CO/Rh experiments already described. In this case, the coverage of evaporated Fe was low (less than one monolayer) in order to avoid zero-bias anomalies due to magnetic impurities in a tunneling junction.[206] In this work, also, five modes were observed, occurring at 436, 519, 569, 1856, and 1900 cm^{-1}. On the basis of isotopic labeling experiments, the two lowest-frequency modes are predominantly bending in character, the mode at 569 cm a stretching vibration (either between the CO molecule and an Fe atom in a linear geometry or the asymmetric stretch between iron and carbon monoxide in a bridged geometry), and the modes at 1856 (intense) and 1900 (weak) cm^{-1} CO stretching vibrations.

For CO adsorbed both on Rh and on Fe, there is a downshift of the expected CO stretching frequencies of approximately 6% on the basis of transmission IR spectroscopy of supported Rh[224-227] and Fe[228-230] particles. In a model calculation, Kirtley and Hansma[231] have related this downshift in frequency to the dipole moment and the dipole derivative of the particular mode. They found that a large dipole derivative, which yields strong intensities in IR spectroscopy, can cause a large-frequency downshift in IETS. However, carbon–hydrogen stretching modes, which have also an appreciable dipole derivative, have been observed not to shift when hydrocarbons

---

[223] J. T. Yates, Jr., E. D. Williams, and W. H. Weinberg, *Surf. Sci.* **91**, 562 (1980).
[224] A. C. Yang and C. W. Garland, *J. Phys. Chem.* **61**, 1504 (1957).
[225] H. Aria and H. Tominaga, *J. Catal.* **43**, 131 (1976).
[226] H. C. Yao and W. G. Rothschild, *J. Chem. Phys.* **68**, 4774 (1978).
[227] J. T. Yates, Jr., M. T. Duncan, S. D. Worley, and R. W. Vaughan, *J. Chem. Phys.* **70**, 1219 (1979).
[228] R. P. Eischens and W. A. Pliskin, *Adv. Catal.* **10**, 1 (1958).
[229] G. Blyholder and L. D. Neff, *J. Phys. Chem.* **66**, 1464 (1962).
[230] H. Kölbel, M. Rálek, and P. Jíru, *Collect. Czech. Chem. Commun.* **36**, 512 (1971).
[231] J. Kirtley and P. K. Hansma, *Phys. Rev. B* **13**, 2910 (1976).

are adsorbed on Al_2O_3, when IETS and transmission IR spectroscopic results are compared.[232,233] Hence, there are two features of the CO stretching modes observed in IETS that are yet to be clarified fully: the large downshift in frequency and the low intensity relative to the low-frequency ($<$600-cm^{-1}) bending and stretching modes (Fig. 43). These issues point up clearly the need for a more sophisticated theory of IETS that addresses quantitatively the electron-vibrational excitation coupling mechanism.

Evans et al.[202] have monitored the adsorption of C_2H_5OH on Ag clusters evaporated onto the Al_2O_3 substrate in an $Al-Al_2O_3-Ag-C_2H_5OH-Pb$ tunneling junction. The statistically averaged thickness of the Ag was varied between 2.5 and 10 Å, which was related via transmission electron microscopy to an average cluster size between approximately 20 and 75 Å, as shown in Fig. 44. The solid lines in Fig. 44 indicate only a best fit to the available data points; the line showing average particle diameter was drawn to include also a point approximately representing the atomic diameter of Ag near zero-Ag coverage.

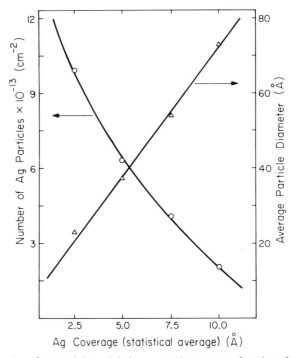

FIG. 44. Number of Ag particles and their average diameters as a function of Ag coverage on the alumina surface. (From Evans et al.[202] Copyright North-Holland Physics Publishing, Amsterdam, 1979.)

[232] H. E. Evans and W. H. Weinberg, *J. Chem. Phys.* **71**, 1537 (1979).
[233] H. E. Evans and W. H. Weinberg, *J. Chem. Phys.* **71**, 1537, 4789 (1979).

2.4. INELASTIC ELECTRON TUNNELING SPECTROSCOPY

At room temperature, ethanol chemisorbs dissociatively on alumina to form aluminum ethoxide.[232,234,235] Similar results have been noted on other surfaces,[236,237,238] and the same type of structure can be expected to form also on Ag. Even though ethoxide formation occurs on both Ag and alumina, slight differences between the two species are expected to be manifest in their vibrational spectra. Takezawa and Kobayashi[239] have studied alcohol adsorption on a number of surfaces and have established that hydrocarbon vibrational frequencies of surface ethoxides can be expected to increase with increasing electronegativity of the surface metal ion. The electronegativities of Ag and Al are 1.93 and 1.61, respectively, on the Pauling scale,[240] and thus one might expect slight shifts upward in frequency for the CH vibrations of silver ethoxide relative to the corresponding vibrational frequencies of the ethoxide formed on Al ions. Since even at the highest Ag coverages studied, much of the alumina surface is still available for ethanol adsorption, spectra resulting from the superposition of vibrations from both Ag and Al ethoxide would be anticipated.

Spectra at room temperature for saturation coverage of ethanol over a range of Ag coverages are shown in Figs. 45 and 46. The low intensities of the hydrocarbon peaks even at saturation coverage result from the blockage of additional adsorption sites below 400 to 500 K by H_2O or OH formed during ethanol adsorption.[232,234,235,237,238] Few noticeable changes occur in the spectra up to a statistical Ag coverage of 2.5 Å, probably due to the extremely large ratio of Al to Ag adsorption sites at these low coverages. As Ag coverage is increased beyond this point, however, features due to the Ag ethoxide species become more prominent. Figure 45 shows the CH_2 and CH_3 deformation bands (at 1472, 1440, 1380, and 1360 cm^{-1} on alumina). These data illustrate a decrease in intensity of the 1360-cm^{-1} bands as well as the growth of new structure between 1380 and 1440 cm^{-1} and at approximately 1470 cm^{-1} and above with increasing coverage of Ag. Similar effects are evident in Fig. 46, which shows the CH stretching bands. These bands appear at approximately 2863, 2915, and 2952 cm^{-1} for the Al ethoxide structure. Again, as Ag coverage is increased, the growth of new structure between 2865 and 2900 cm^{-1}, 2920 and 2950 cm^{-1}, and at approximately 2998 cm^{-1} can be observed.

As expected, due to the relative electronegativities of Ag and Al, ethanol adsorption on supported Ag produces an ethoxide with CH vibrational

---

[234] R. G. Greenler, *J. Chem. Phys.* **37**, 2094 (1962).
[235] H. Arai, Y. Saito, and Y. Yoneda, *Bull. Chem. Soc. Jpn.* **40**, 731 (1967).
[236] I. E. Wachs and R. J. Madix, *Appl. Surf. Sci.* 303 (1978).
[237] S. L. Parrott, J. W. Rogers, Jr., and J. M. White, *Appl. Surf. Sci.* **1**, 443 (1978).
[238] B. A. Morrow, L. W. Thompson, and R. W. Witmore, *J. Catal.* **28**, 332 (1973).
[239] N. Takezawa and H. Kobayashi, *J. Catal.* **25**, 179 (1972); **28**, 335 (1973).
[240] A. L. Allred, *J. Inorg. Nucl. Chem.* **17**, 215 (1961).

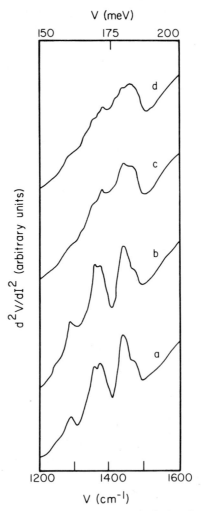

FIG. 45. IET spectra in the range 1200–1600 cm^{-1} of ethanol adsorbed on Ag/Al$_2$O for statistical Ag coverages of (a) 0 Å, (b) 2.5 Å, (c) 5.0 Å, and (d) 10.0 Å. (From Evans et al.[202] Copyright North-Holland Physics Publishing, Amsterdam, 1979.)

frequencies at slightly higher energies than the corresponding modes of ethoxide formed on the alumina support. The two different surface species are separated easily with IETS, demonstrating clearly the utility of this spectroscopy in the study of adsorption and catalytic reactions over supported metal clusters.

2.4.4.2. Attachment of Inorganic Cluster Compounds to Al$_2$O$_3$ Films. There is significant interest in newly developed supported-complex cata-

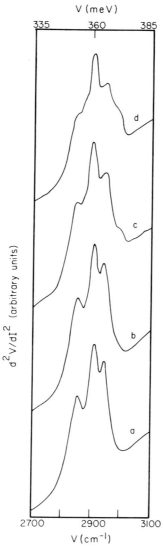

FIG. 46. IET spectra in the range 2700–3100 cm^{-1} of ethanol adsorbed on Ag/Al$_2$O$_3$ for statistical Ag coverages of (a) 0 Å, (b) 2.5 Å, (c) 5.0 Å, and (d) 10.0 Å. (From Evans et al.[202] Copyright North-Holland Physics Publishing, Amsterdam, 1979.)

lysts, which are formed by anchoring or grafting a homogeneous catalyst (a cluster compound) onto a high-surface-area support.[241,242] Such catalysts can combine the activity and selectivity found in homogeneous systems with the stability and ease of separation characteristic of heterogeneous catalysts, and they frequently exhibit activities an order of magnitude or more greater than the corresponding unsupported systems.[243] Progress in this area, however, has been hampered by a lack of detailed structural information for supported complexes. Characterization of supported complexes has been plagued by many of the same problems that arise in attempting to characterize traditional heterogeneous catalysts. The first structural determination of such a supported cluster compound was made by Evans and Weinberg[203-205,241], who used IETS. In particular, they reported the vibrational spectrum of the supported complex formed by the interaction of zirconium tetraborohydride, $Zr(BH_4)_4$, a known homogeneous polymerization catalyst, with an Al_2O_3 surface. An IET spectrum for a saturation coverage of $Zr(BH_4)_4$ [5 × 10^{-2} Torr of gaseous $Zr(BH_4)_4$ for 15 min] on Al_2O_3 at 300 K is shown in Fig. 47.[203]

Comparisons with IETS studies of "clean" Al_2O_3 indicate that the spectral features at 299, 945, and 1863 cm^{-1} can be assigned to a phonon in the underlying Al film, a bulk Al–O stretching mode and its harmonic overtone, respectively.[210] The 3675-cm^{-1} peak is the O–H stretching vibration of surface hydroxyl groups; the peak near 2930 cm^{-1} arises from the C–H stretching vibration of a small amount of adsorbed hydrocarbon contamination.[210] Contamination might also contribute to the intensity of features at 1030 cm^{-1} and in the 1300–1500-cm^{-1} region.

The boron atoms in $Zr(BH_4)_4$ are arranged tetrahedrally, each being bound to the central Zr atom in a tridentate manner with three bridging hydrogens.[244] During adsorption, one or more of the BH_4 ligands are lost as the Zr becomes either singly or multiply coordinated to oxygen atoms on the surface.[243] Since the surface becomes a virtual ligand, it might well affect bonding in the remaining BH_4 groups. For example, $(C_5H_5)_2Zr(BH_4)_2$ and $(C_5H_5)_2Zr(H)BH_4$ are both known to have bidentate bridging structures, and the surface could be expected to have a similar effect. Information concerning bonding can be obtained by examining the stretching vibrations of both terminal (H_t) and bridging (H_b) hydrogens. The B–H_t region shows at least three peaks near 2407, 2437, and 2490 cm^{-1}. For the tridentate structure, only one peak at 2560–2580 cm^{-1} is to be expected.[245] The ob-

[241] H. E. Evans and W. H. Weinberg, *J. Vac. Sci. Technol.* **17**, (1980).
[242] J. M. Basset and A. K. Smith, in "Fundamental Research in Homogeneous Catalysis" (M. Tsutsui and R. Ugo, eds.), p. 69. Plenum, New York, 1977.
[243] V. A. Zakharov and Y. I. Yermakov, *Catal. Rev.—Sci. Eng.* **19**, 67 (1979).
[244] P. H. Bird and M. R. Churchill, *Chem. Commun.* p. 403 (1967).
[245] N. Davis, M. G. H. Wallbridge, B. E. Smith, and B. D. James, *J.C.S. Dalton* p. 162 (1973).

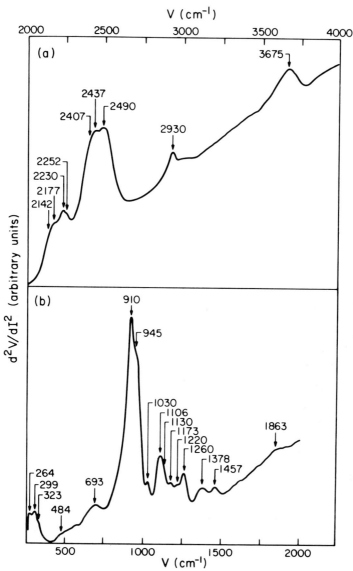

FIG. 47. IET spectrum for $Zr(BH_4)_4$ supported on Al_2O_3 at 300 K over the energy ranges (a) 2000–4000 cm^{-1} and (b) 240–2000 cm^{-1}. (Reprinted with permission from Evans and Weinberg.[203] Copyright 1980 American Chemical Society.)

served frequencies are more closely related to those reported for bidentate species (between 2375 and 2550 cm^{-1}),[246] yet the presence of more than two peaks, as well as the observation of four features in the B–H_b stretching region (at 2142, 2177, 2230, and 2252 cm^{-1}), indicate that there is more than one type of surface species. This is consistent with the assignments of features at 693 and 910 cm^{-1} to Zr–O stretching modes of multiply (bulklike) and singly (or, at least, less highly) coordinated Zr atoms.[247,248] Additionally, some of the BH_4 ligands displaced during adsorption might remain on the surface, forming

$$\begin{array}{ccc} HH & & HH \\ \diagdown\diagup & & \diagdown\diagup \\ B & \text{and} & B \\ \diagup\diagdown & & | \\ HH & & O \\ \diagdown\diagup & & \\ Al & & \end{array}$$

complexes. Similar surface species have been observed for diborane (B_2H_6) adsorption on Al_2O_3[249] and could also be expected to produce additional structure in the B–H stretching regions. Modes appearing at 1106, 1130, 1173, 1220, and 1260 cm^{-1} (and possibly 1030 cm^{-1}) can be assigned to deformations of the metal–BH_4 structure.[245] An AlH_3BH_2 deformation might also contribute to structure near 1457 cm^{-1}.[249] The 1378-cm^{-1} peak has been assigned previously to a B–O stretch, and a B–O vibration might also contribute to the intensity of the 1260-cm^{-1} peak.[249] Unresolved structure between 480 and 600 cm^{-1} can be assigned to Zr–B skeletal stretching modes,[250] with possible contributions from additional Zr–O vibrations.[247] Low-energy features near 264 and 323 cm^{-1} probably arise from BH_4–Zr–BH_4 bending and Zr–BH_4 torsional modes, respectively.[245]

To summarize, when $Zr(BH_4)_4$ is attached to Al_2O_3, at least two distinct types of surface zirconium complexes can be identified due to the appearance of both bulklike Zr–O stretching modes and modes indicative of Zr atoms less highly coordinated to surface oxygen. The migration of displaced ligands onto the surface to form Al–BH_4 and O–BH_2 groups was also observed. The nature of Zr–BH_4 bonds was determined to be bidentate for the supported complex as opposed to tridentate, which is observed for gaseous $Zr(BH_4)_4$.

After characterizing the supported complex, reactions were studied with ethylene, propylene, and acetylene.[203–205,241] Acetylene was observed to

[246] T. J. Marks, W. J. Kennelly, J. R. Kolb, and L. A. Shimp, *Inorg. Chem.* **11**, 2540 (1972).
[247] T. Maekawa and M. Terada, *Trans. Jpn. Inst. Met.* **4**, 57 (1963).
[248] G. A. Ozin, in "Vibrational Spectroscopy of Trapped Species" (H. E. Hallum, ed.), p. 237. Wiley, New York, 1973.
[249] T. Matsuda and H. Kawashima, *J. Catal.* **49**, 141 (1977).
[250] B. D. James, B. E. Smith, and H. F. Shurwell, *J. Mol. Struct.* **33**, 91 (1976).

polymerize on the surface as judged by its failure to attain saturation coverage, whereas ethylene and propylene both saturated the surface readily. Spectral features of adsorbed acetylene are consistent with the formation of polyacetylene, retaining some characteristics of unsaturated carbon–carbon bonds, whereas features for ethylene and propylene indicate no remaining unsaturation and provide no evidence of polymer formation. Thus, this study extended the use of IETS to include (for the first time) the study of supported complexes and contributed significant new structural information on an industrially important type of catalytic system. It appears almost certain that IETS will continue to be very important in these types of investigations.

2.5. Conclusions

2.5.1. Introduction

The advantages and disadvantages of the various vibrational spectroscopies have been discussed adequately in Chapters 2.2, 2.3, and 2.4 and were summarized in Table I. To conclude this chapter, in Section 2.5.2 below, a specific comparison will be made between EELS and IRAS insofar as the chemisorption of CO on the Ru(001) surface is concerned. Finally, in Section 2.5.3, tables are presented listing previous gas–surface systems investigated both by EELS and by IRAS on characterized surfaces. A table is also presented comparing the important features of each of the three techniques of vibrational spectroscopy.

2.5.2. CO Chemisorption on Ru(001)

2.5.2.1. *EELS Results.* Representative spectra for low and high coverages of CO on the Ru(001) surface at 100 K are shown in Fig. 48.[61,62] The 5-L exposure produces essentially saturation coverage. Two fundamental vibrations are observed in the spectra, a metal–carbon stretch at 445 cm^{-1} and a carbon–oxygen stretch, which varies from 1984 cm^{-1} for a 0.1-L exposure to 2080 cm^{-1} at saturation coverage as shown in Fig. 49. In addition, at high coverage the overtones and combination frequencies of these two bands are observable. The assignment of the frequency at 855 cm^{-1} (106 meV) as energy lost to excite a true harmonic ($0 \rightarrow 2$) vibration, rather than energy lost in multiple scattering events to excite two separate ($0 \rightarrow 1$) transitions, is borne out by the observation of this harmonic at a frequency below twice the frequency of the fundamental vibration, 888 cm^{-1} (110 meV), due to anharmonic effects. The observed frequencies are in good agreement with those of the corresponding modes in the $Ru_3(CO)_{12}$ complex in which CO is bonded

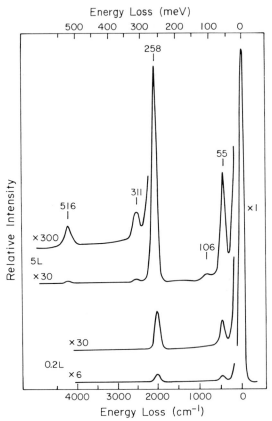

FIG. 48. Vibrational spectra of CO on the Ru(001) surface at low and high coverages measured by electron energy loss spectroscopy. The energies of the individual loss features are given in milli-electron volts. (From Thomas and Weinberg.[62])

linearly to the Ru atoms.[251] In this molecule, the axial carbonyl stretching frequency lies between 2028 and 2127 cm^{-1}, whereas the radial carbonyl stretching frequency lies between 1989 and 2011 cm^{-1}. The former would be a reasonable model for CO adsorbed on Ru(001) near saturation coverage, whereas the latter is the analog of CO chemisorbed at low coverages (based on CO–CO separation distances in the cluster). The mode in the cluster corresponding to the metal–carbon stretching vibration lies between 350 and 500 cm^{-1}, again in good agreement with that observed for CO chemisorbed on the extended Ru surface. Hence, it appears that CO is upright on the Ru(001) surface and that there is a single carbon–metal bond. Measure-

[251] C. O. Quicksall and T. G. Spiro, *Inorg. Chem.* **7**, 2365 (1968).

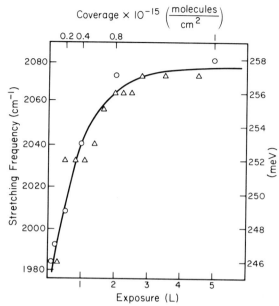

FIG. 49. Carbon–oxygen stretching frequency v_1 as a function of CO exposure on Ru(001) surface at 100 K. ○, obtained for separate exposures; △, obtained from measurements in a constant background of CO. (From Thomas and Weinberg.[62])

ments of the electron-stimulated desorption ion angular distributions of CO^+ and O^+ from this surface also indicate an upright geometry.[252] The two bending modes of this structure are not observed with EELS, presumably because they are parallel to the surface and have weak dipole-transition moments.

The most significant aspect of the vibrational spectra is that the carbon–oxygen stretching frequency is found exclusively in the range usually associated with an on-top CO at all coverages. The carbon–oxygen stretching region is shown in detail in Fig. 15; within the instrumental resolution (11–15-meV FWHM of the elastic peak), there is a single symmetric peak at all coverages. The carbon–metal frequency is constant at 445 ± 8 cm^{-1} for all coverages.

As reported by Madey and Menzel, at fractional coverages less than $\frac{1}{3}$ on this surface CO forms a $(\sqrt{3} \times \sqrt{3})R30°$ LEED structure.[143,253] The CO molecules may reside in on-top sites (Fig. 50a), as was discussed in Section 2.2.4. At higher coverages, the fractional order LEED beams split and move to

[252] T. E. Madey, *Surf. Sci.* **79**, 575 (1979).
[253] T. E. Madey and D. Menzel, *Proc. 2nd Int. Conf. Solid Surf., Kyoto*, 229 (1974).

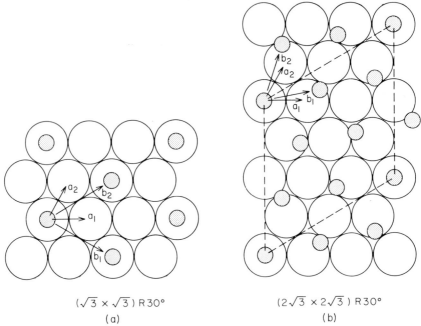

FIG. 50. Structures assigned to the low-temperature LEED patterns of CO on the Ru(001) surface at coverages of (a) $\frac{1}{3}$ and (b) $\frac{7}{12}$ of a monolayer. The large open circles represent substrate atoms and the shaded circles denote CO adsorption sites. (From Thomas and Weinberg.[62])

form a $(2\sqrt{3} \times 2\sqrt{3})R30°$ pattern.[143,253] A structure may be assigned to this pattern by assuming that the most intense nonsubstrate beams result from first-order diffraction from a CO superlattice. The primitive vectors of the superlattice are indicated in Fig. 50b as b_1 and b_2. The unit cell of the superlattice has been placed at an arbitrary origin on the substrate.

At saturation coverage at 100 K, a new LEED pattern appears, a $(5\sqrt{3} \times 5\sqrt{3})R30°$, which is due to a hexagonally close-packed layer of CO with a lattice spacing of 3.33 Å.[143] For this structure, the coverage corresponds to $\frac{49}{75}$ of a monolayer, or 1.04×10^{15} molecules cm^{-2}. Coverages, as determined from the areas under thermal desorption peaks, agree with these LEED structural observations. It is not possible to shift all the CO molecules to on-top sites at high coverage, unless an unreasonably small CO–CO spacing of 2.71 Å and an extremely complex ordering mechanism are assumed. Therefore, an inspection of the LEED patterns requires that the interpretation of the vibrational frequencies in terms of exclusively on-top CO be rejected.

To summarize, the vibrational spectra of CO on the Ru(001) surface indicate that CO is bound linearly to the surface at all coverages and that a bridged form is not present. Studies with LEED indicate that CO is found in a variety of asymmetric sites at high coverage. Consequently, the linear CO may not be assumed to be in on-top sites exclusively. Evidently, the carbon–oxygen stretching frequency is sensitive to the hybridization of the bonding orbitals of adsorbed CO but not necessarily to its position with respect to the substrate. This suggests that in regard to CO adsorption, the Ru(001) surface is electronically, as well as geometrically, homogeneous (i.e., there is only a small potential barrier for moving out of site registry) and that the adsorption site is not of primary chemical importance, at least at surface coverages greater than a third of a monolayer, at which point the $(\sqrt{3} \times \sqrt{3})R30°$ LEED structure begins to compress continuously. At a surface coverage of a third of a monolayer coadsorption of CO with NO has demonstrated that the CO occupies on-top sites.[146]

2.5.2.2. IRAS Results. In full agreement with EELS, the IR absorption band due to the CO stretching mode shifts from 1984 cm^{-1} at $\theta = 0.003$ to 2061 cm^{-1} at saturation coverage at 200 K, as may be seen in Fig. 51.[63] Concomitant with the ordering of the adlayer into the $(\sqrt{3} \times \sqrt{3})R30°$ superstructure, the band narrows markedly giving a half-width of 9 cm^{-1} at 2021 cm^{-1}. A similar, but less pronounced effect can be observed for the $(2\sqrt{3} \times 2\sqrt{3})R30°$ superstructure. The IR frequency–coverage relationship at three different temperatures is shown in Fig. 52, in which the coverage regions corresponding to ordering of the $(\sqrt{3} \times \sqrt{3})R30°$ and the $(2\sqrt{3} \times 2\sqrt{3})R30°$ superstructures in LEED are indicated by cross-hatching. At 80 K, the absorption band shifts almost linearly with coverage, whereas at 200 K two regions of constant wave number, or "plateaus," appear for $0.2 < \theta < 0.33$ and for $\theta \sim 0.5$. At 300 K, a similar behavior is observed, but the plateaus are shorter and less well defined. In the case for which no change in adsorption site takes place,[54] the overall shift of the band is due to molecular interactions, as indicated previously. Thus, the plateaus can be attributed to the growth of the adsorbed layer via island formation.[54,63] Since an incident molecule is adsorbed at an island boundary with the same intermolecular separation, the intermolecular interactions cause no further shift of the peak in these coverage regions. The simultaneous observation of a narrowing of the absorption band during ordering can be understood in a similar way: identical adsorption sites and intermolecular distances lead to a sharp common oscillatory mode.[124] This effect can be seen in Fig. 53, where the half-width of the absorption band as a function of coverage is shown. Again at 200 K, there is a marked decrease in half-width compared to 80 K when ordering in the $(\sqrt{3} \times \sqrt{3})R30°$ and $(2\sqrt{3} \times 2\sqrt{3})R30°$ superstructures is reached. At 300 K, only ordering in the $(\sqrt{3} \times \sqrt{3})R30°$ superstructure is observed, a result confirmed by Fig. 52 and by LEED results.

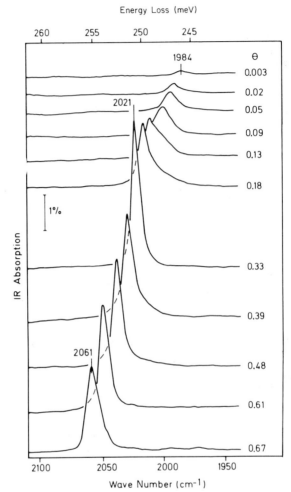

FIG. 51. IR absorption band due to the C–O stretching frequency as a function of increasing coverage for CO on Ru(001) at 200 K. (From Pfnür et al.[63] Copyright North-Holland Physics Publishing, Amsterdam, 1980.)

The data shown in Fig. 52 show also that the mobility of the admolecules at 80 K is insufficient to give rise to ordered structures. At 200 K, this is no longer the case, and an equilibrium prevails that appears to favor the formation of large ordered islands. At still higher temperatures, however, the equilibrium distribution appears to correspond to smaller islands, thus giving rise to shorter plateaus and less pronounced extra features in LEED.

The formation of ordered domains can be demonstrated by annealing an adlayer that is adsorbed at 80 K to room temperature, thus increasing the

2.5. CONCLUSIONS

mobility of the molecules to enable ordering. Recooling the surface to 80 K then results in "freezing" an equilibrium distribution of islands of different sizes. The results in Fig. 54 for ordering in the ($\sqrt{3} \times \sqrt{3}$)R30° superstructure for coverages between $\theta = 0.07$ and $\theta = 0.35$ show a splitting of the IR peak, the higher-frequency mode corresponding clearly to the ordered ($\sqrt{3} \times \sqrt{3}$)R30° superstructure as may be seen at $\theta = 0.33$. The peak at the lower frequency is attributed to molecules adsorbed at the boundary of the ordered domains. The reasoning is similar to that discussed in connection with the plateaus. Within the domains, the molecules feel an interaction from all the surrounding neighbors and therefore display a vibrational frequency typical

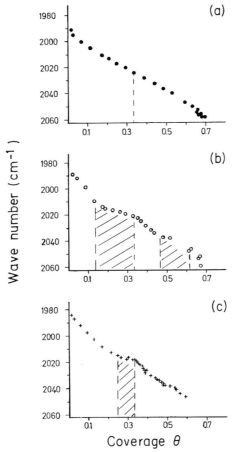

FIG. 52. Shift in IR frequency for CO on Ru(001) as a function of coverage for the temperatures (a) 80 K, (b) 200 K, and (c) 300 K. (From Pfnür et al.[63] Copyright North-Holland Physics Publishing, Amsterdam, 1980.)

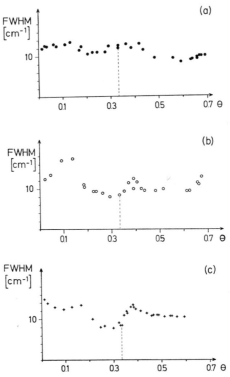

FIG. 53. Change in half-width of the IR absorption band for CO on Ru(001) as a function of increasing coverage for adsorption temperatures (a) 80 K, (b) 200 K, and (c) 300 K. (From Pfnür et al.[63] Copyright North-Holland Physics Publishing, Amsterdam, 1980.)

of $\theta = 0.33$. At the boundaries of the domains, only a part of this interaction is felt, resulting in a lower frequency. The ratio of high- to low-frequency peaks thus gives a measure of the domain size: as the high-frequency peak increases, the domains become larger and boundary effects decrease.

As the present study shows, the high sensitivity and intrinsically high resolution of the IR reflection–absorption method allows detailed studies of the position, shape, and intensity of surface vibrational bands, and hence of ordering processes in adsorbed layers. For the case of reversible order–disorder phase transitions this method permits a sensitive monitoring of these effects.

Although this application is not possible with EELS due to its lower resolution, the two techniques yielded identical results in this CO/Ru(001) system insofar as they could be compared. Furthermore, in EELS the observation of low-frequency modes makes it more likely that adsorbate site

2.5. CONCLUSIONS 121

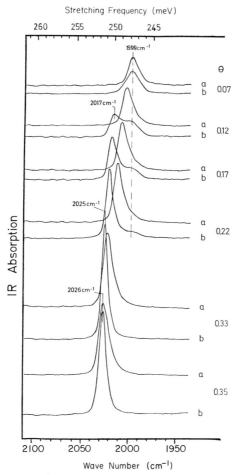

FIG. 54. Effect of annealing the adsorbed CO layer on the IR absorption band. (a) Spectrum directly after adsorption at 80 K and (b) spectrum after warming to 350 K and recooling to 80 K. (From Pfnür et al.[63] Copyright North-Holland Physics Publishing, Amsterdam, 1980.)

positions can be determined from a more complete normal-mode analysis and simple group theoretic arguments. Hence, the techniques of EELS and IRAS can be complementary to one another and at the same time individually valuable.

2.5.3. Tabulation of Previous EELS and IRAS Results

All previous investigations of adsorption using EELS and IRAS on characterized surfaces are presented in Tables IV and V, respectively. Since these

TABLE IV. Adsorbate–Substrate Systems Studied by Electron Energy Loss Spectroscopy

Gas	Substrate	References	Gas	Substrate	References
I. Hydrogen			X. Hydrocarbons		
	Si(111)	30	a. Cycloparaffins		
	Ni(100)	254, 255	C_6H_{12} (cyclohexane)	Pt(111)	273
	W(100)	27, 32, 33, 44, 67–71	b. Aromatics		
	W(110)	44	C_6H_6 (benzene)	Ni(100)	42, 274
	W(111)	43–45		Ni(111)	274–275
	Pt(111)	132, 256		Pt(111)	276
II. Carbon monoxide			c. Alkenes		
	Ni(100)	31, 117, 255	C_2H_4 (ethylene)	Ni(111)	273, 277
	Ni(111)	42, 257, 258		Pt(111)	273, 277
	Ni(7 9 11)	259		Ni(111)	79, 256, 273
	Cu(100)	260	C_6H_{10} (cyclohexene)	Ni(111)	278
	Cu(110)	261	d. Alkynes and substituted alkynes		
	Ru(001)	61, 62	C_2H_2 (acetylene)	Ni(100)	82
	Pd(100)	182		Ni(111)	275, 277, 279, 280
	W(100)	27, 32, 36, 47		Ni(7 9 11)	281
	W(110)	46		W(100)	45, 46
	W(111)	46		W(110)	46, 282
	Pt(111)	32, 36, 37, 40		W(111)	43, 46
	Pt(664)	37, 40		Pt(111)	79, 256
III. Oxygen			$CH_3C{\equiv}CCH_3$ (2-butyne)	Ni(100)	82
	Al(100)	38	$CH_3CH_2{\equiv}CCH_2CH_3$ (3-hexyne)	Ni(100)	82
	Si(111)	28, 29, 31	$CF_3C{\equiv}CCF_3$ (perfluoro-2-butyne)	Ni(100)	82
	Ni(100)	39, 81, 262	e. Nitriles and isocyanides		
	Cu(100)	263	$CH_3(CH_2)_3N{\equiv}C$	Ni(100)	82
	Ru(001)	61, 151	(n-butyl isocyanide)		
	W(100)	32, 34	$CH_3C{\equiv}N$ (acetonitrile)	Ni(100)	82
	Pt(111)	264, 265	$CF_3C{\equiv}N$ (trifluoroacetonitrile)	N(100)	82
IV. Sulfur	Ni(100)	81	XI. Coadsorption systems		
V. Nitrogen	W(100)	27, 82	Carbon monoxide and oxygen	Ni(100)	283
VI. Water	NiO(111)	117		Ru(001)	61
	Cu(100)	80		W(100)	35
	Ru(001)	266	Carbon monoxide and hydrogen	Ni(100)	254
	W(100)	27		Ni(111)	258
	Pt(100)	267			
	Pt(111)	265			

VII. Nitric oxide	Ru(001)	144, 145
	Pt(100)	268
	Pt(111)	269
VIII. Alcohols		
CH₃OH (methanol)	304 stainless steel	270
	Ni(111)	271
	Cu(100)	263
C₂H₅OH (ethanol)	304 stainless steel	270
	Cu(100)	263
IX. Carboxylic acids, ketones		
HCOOH (formic acid)	Cu(100)	263
CH₃COOH (acetic acid)	Cu(100)	272
CH₃COCH₃ (acetone)	304 stainless steel	270
Carbon monoxide and nitric oxide	W(100)	35
	Ru(001)	146
Carbon monoxide and hydrocarbons	Ni(111)	42, 284
	Pt(111)	284
Hydrogen and nitrogen	W(100)	27, 82
Hydrogen and nitric oxide	Ru(001)	266
Oxygen and water	304 stainless steel	270
	Ru(001)	266
	Pt(111)	265
Oxygen and ammonia	Ni(100)	117
Oxygen and alcohols	304 stainless steel	270
	Cu(100)	263

254 S. Andersson, *Chem. Phys. Lett.* **55**, 185 (1978).
255 S. Andersson, *Proc. 7th Int. Vac. Congr., Vienna*, 1019 (1977).
256 H. Ibach and S. Lehwald, *J. Vac. Sci. Technol.* **15**, 407 (1978).
257 W. Erley, H. Wagner, and H. Ibach, *Surf. Sci.* **80**, 612 (1979).
258 J. C. Bertolini and B. Imelik, *Surf. Sci.* **80**, 586 (1979).
259 W. Erley, H. Ibach, S. Lehwald, and H. Wagner, *Surf. Sci.* **83**, 585 (1979).
260 B. A. Sexton, *Chem. Phys. Lett.* **63**, 451 (1979).
261 J. F. Wendelken and M. V. K. Ulehla, *J. Vac. Sci. Technol.* **16**, 441 (1979).
262 G. Dalmai-Imelik, J. C. Bertolini, and J. Rousseau, *Surf. Sci.* **63**, 67 (1977).
263 B. A. Sexton, *Surf. Sci.* **88**, 299 (1979).
264 G. B. Fisher, B. A. Sexton, and J. L. Gland, *J. Vac. Sci. Technol.* **17**, 144 (1980).
265 B. A. Sexton, *Surf. Sci.* **94**, 435 (1980).
266 P. A. Thiel, F. M. Hoffmann, and W. H. Weinberg, *J. Chem. Phys.* **75**, 5556 (1981).
267 H. Ibach and S. Lehwald, *Surf. Sci.* **91**, 187 (1980).
268 G. Pirug, H. P. Bonzel, H. Hopster, and H. Ibach, *J. Chem. Phys.* **71**, 593 (1979).
269 H. Ibach and S. Lehwald, *Surf. Sci.* **76**, 1 (1978).
270 A. Adnot, ref. 130 p. 109.
271 J. E. Demuth and H. Ibach, *Chem. Phys. Lett.* **60**, 395 (1979).
272 B. A. Sexton, *Chem. Phys. Lett.* **65**, 469 (1979).
273 J. E. Demuth, H. Ibach, and S. Lehwald, *Phys. Rev. Lett.* **40**, 1044 (1978).
274 J. C. Bertolini, G. Dalmai-Imelik, and J. Rousseau, *Surf. Sci.* **67**, 478 (1977).
275 J. C. Bertolini, J. Massardier, and G. Dalmai-Imelik, *J.C.S. Faraday I* **74**, 1720 (1978).
276 S. Lehwald, H. Ibach, and J. E. Demuth, *Surf. Sci.* **78**, 577 (1978).
277 J. C. Bertolini and J. Rousseau, *Surf. Sci.* **83**, 531 (1979).
278 G. Dalmai-Imelik, J. C. Bertolini, J. Massardier, J. Rousseau, and B. Imelik, ref. 255, p. 1179.
279 J. E. Demuth and H. Ibach, *Surf. Sci.* **78**, L238 (1978).
280 J. E. Demuth and H. Ibach, *Surf. Sci.* **85**, 365 (1979).
281 S. Lehwald, W. Erley, H. Ibach, and H. Wagner, *Chem. Phys. Lett.* **62**, 360 (1979).
282 C. Backx and R. F. Willis, *Chem. Phys. Lett.* **53**, 471 (1978).
283 S. Andersson, ref. 255, p. 815.
284 H. Ibach and G. A. Somorjai, *Appl. Surf. Sci.* **3**, 293 (1979).

TABLE V. Adsorbate–Substrate Systems Studied by Reflection–Absorption Infrared Spectroscopy

Gas	Substrate	Reference(s)
I. Carbon monoxide	Ni(111)	127[a]
	Cu(100)	162, 285–287
	Cu(110)	285, 288
	Cu(111)	185[a], 285, 287, 289
	Cu(211), (311), (755)	285, 289
	Ru(001)	63, 64
	Pd(100), (111), (210)	53, 54, 163, 184[a]
	Ag(100)	162
	Pt(111) and Pt(111) ribbon	123, 141,[a] 166, 168, 290
	Pt(100)	291
II. Nitrogen	Ni(110)	162, 292
	Pt ribbon	290
III. Hydrocarbons		
C_2H_6 (ethane)	Cu(110)	160
	Pt(111)	170
CH_3OH (methanol)	Pt(111)	293
C_2H_5OH (ethanol)	Pt(111)	293

[a] Studied with isotopic mixtures.

[285] I. P. Batra and P. S. Bagus, *Solid State Commun.* **16**, 1097 (1975).
[286] K. Horn and J. Pritchard, *Surf. Sci.* **55**, 701 (1976).
[287] J. Pritchard, *Surf. Sci.* **79**, 231 (1979).
[288] K. Horn, M. Hussain, and J. Pritchard, *Surf. Sci.* **63**, 244 (1977).
[289] J. Pritchard, T. Catterick, and R. K. Grupta, *Surf. Sci.* **53**, 1 (1975).
[290] R. Shigeishi and D. A. King, *Surf. Sci.* **62**, 379 (1977).
[291] D. A. King, in "Vibrational Spectroscopies of Adsorbates" (R. F. Willis, ed.), Springer-Verlag, Berlin and New York (1980).
[292] M. Grunze, R. K. Driscoll, G. N. Burland, J. C. L. Cornish, and J. Pritchard, *Surf. Sci.* **89**, 381 (1979).
[293] H. J. Krebs and H. Lüth, ref. 130, p. 135.

are active research areas, these tables are likely to become dated rather quickly, but they do provide easy access to all relevant work published through the end of 1979. Finally, the most important characteristic of each of the three vibrational spectroscopic techniques discussed in this part are presented in Table VI.

2.5.4. Update

When this chapter was completed in late 1979, it represented a reasonably comprehensive review of electron energy loss, infrared reflection–absorption, and inelastic electron tunneling spectroscopies; and it still

2.5. CONCLUSIONS

TABLE VI. Comparison Among Vibration Spectroscopies

Characteristic	Technique		
	EELS	IRAS	IETS
State-of-the-art resolution (FWHM)	50–80 cm^{-1}	1–5 cm^{-1}	1–5 cm^{-1}
Approximate spectral range	5000–240 cm^{-1}	4000–1500 cm^{-1}	8000–240 cm^{-1}
Sensitivity	0.1% monolayer	0.1% monolayer	0.1% monolayer
Sample area	1 mm^2	10 mm^2	1 mm^2
Types of surfaces—applicability	Metallic single crystals, polycrystalline metals, semiconductors, insulators	Metallic single crystals, polycrystalline metals	Oxides; metallic particles on oxide supports; inorganic clusters on oxide supports
Ambient pressures under which experiments may be performed	≤10^{-6} Torr	≤1 atm	≤1 atmosphere for sample preparation

should provide a rather painless introduction to these topics. In the past four years, however, there has been much activity in these research areas, especially in EELS (probably due largely to the advent of commercially available instrumentation). Reviews that should prove particularly useful are the following:

1. H. Ibach and D. L. Mills, "Electron Energy Loss Spectroscopy and Surface Vibrations," Academic Press, New York, 1982.
2. W. H. Weinberg, *in* "Vibrational Spectra and Structure," (J. R. Durig, ed.), Chap. 11, 1 (1982).
3. F. M. Hoffmann, *Surf. Sci. Rep.* **3**, 107, 1983.
4. C. R. Brundle and H. Morawitz, eds., "Vibrations at Surfaces," Elsevier, Amsterdam, 1983.

Acknowledgments

The preparation of this chapter, as well as the research from the author's laboratory described therein, have been supported by the National Science Foundation.

It is also a pleasure to acknowledge the enthusiastic assistance of Howard Evans, Glenn Thomas, and especially Friedrich Hoffmann and Patricia Thiel. Without their help, this could not have been written.

3. PHOTOEMISSION SPECTROSCOPY OF VALENCE STATES

By

Giorgio Margaritondo

Department of Physics
University of Wisconsin
Madison, Wisconsin

and

John H. Weaver

Synchrotron Radiation Center
University of Wisconsin
Madison, Wisconsin and
Department of Chemical Engineering
and Materials Science
University of Minnesota
Minneapolis, Minnesota

3.1. Introduction: Three-Step Model, Escape Depth, Relevant Parameters

3.1.1. Photoemission: A Versatile Probe of the Valence States

Many of the important physical and chemical properties of solids and molecules are related to their valence electronic states. Since the development of quantum mechanics and the subsequent development of modern chemistry, a large number of techniques have been introduced to investigate valence electrons. For example, extensive optical spectroscopy experiments studied the details of the valence states near the Fermi level in metals and near the top of the valence band in semiconductors and insulators. In recent years, however, the conventional experimental probes have been overshadowed by the advent of photoemission spectroscopy and, in particular, by the photoemission techniques based on synchrotron radiation.[1,2] The success of

---

[1] See, e.g., M. Cardona and L. Ley eds., "Photoemission in Solids." Springer-Verlag, Berlin and New York, 1979; C. Kunz, "Synchrotron Radiation," Topics in Current Physics, Vol. 10.

photoemission is due to its extensive and flexible applicability. In particular, it is due to its capability to study the entire energy range occupied by valence electrons, and to probe both the energy distribution of the states and their wave functions. In this chapter we review the modern use of photoemission to study valence electrons. Particular emphasis will be given to those branches of photoemission that have given the most spectacular results, such as the novel solid-state photoemission experiments based on synchrotron radiation.[2] The discussion of these experiments will start in this section with a short review of the most relevant properties of the photoemission process.

In a photoemission process, a photon of energy $\hbar\omega$ is absorbed by an electron bound to an atom, a molecule, or a solid, and the electron is excited into a higher-energy state. This increase in energy may be sufficient to free the electron from the atom or solid so that it can be detected and analyzed in vacuum. Photoelectron spectroscopy is the spectroscopy of these free electrons. Its importance arises because the properties of the ejected electron depend on the state of the electron *inside* the system before the photoemission process took place. From the measured properties of the photoelectrons (energy, momentum, spin), one gains insight into the electronic states of the atom, molecule, or solid. This information is absolutely fundamental because of the many properties of a material that are directly related to the electrons and their quantum mechanical state.

The fundamental ideas which describe the photoemission process were introduced by Einstein[3] as early as 1905. Before 1905 it was known that free electrons were present in a metal. Even the highest in energy of these electrons were prevented from escaping into the vacuum by a potential barrier $e\Phi$ as shown in Fig. 1. The parameter Φ is termed the work function and it varies from material to material. When electromagnetic radiation is directed

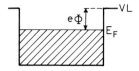

FIG. 1. Energy diagram of a metal surface. Φ is the work function, E_F the Fermi energy, and VL the vacuum level.

Springer-Verlag, Berlin and New York, 1979; B. Feuerbacher, B. Fitton, and R. F. Willis, "Photoemission and the Electronic Properties of Surfaces." Wiley, New York, 1978; H. Winick and S. Doniach, "Synchrotron Radiation Research." Plenum, New York, 1980.

[2] G. Margaritondo and J. E. Rowe, *J. Vac. Sci. Technol.* **17,** 561 (1980).

[3] A. Einstein, *Ann. Phys. (Leipzig)* **17,** 932 (1905).

onto the solid, some electrons escape as photoelectrons. Einstein's fundamental hypothesis was that photoelectrons escape after absorbing the energy of a photon $\hbar\omega$, so that the minimum photon energy for the process is

$$\hbar\omega_{min} = e\Phi \qquad (3.1.1)$$

The state of the electron in a solid is characterized in part by its energy. In this review the energy scale E for the bound electronic states will be referred to the Fermi energy E_F. The photoelectron outside the solid has only kinetic energy T such that

$$T = E + \hbar\omega - e\Phi \qquad (3.1.2)$$

The absolute threshold $\hbar\omega_{min}$ [Eq. (3.1.1)] corresponds to $T = 0$ and $E = 0$, i.e., to the ejection with zero velocity of an electron from a state at the Fermi level. Since $e\Phi$ and $\hbar\omega$ are known parameters in the experiment, a measurement of the kinetic energy T of the electron in vacuum gives information about the energy distribution of the electrons in the system.

3.1.2. The Three-Step Model

In the three-step model,[4] the one-step photoemission process is conceptually decomposed as three separate and independent steps leading from the electron lying in the ground state to the ejection of a photoelectron into free space. These steps are

(1) the excitation of the electron with the adsorption or annihilation of the photon,
(2) the transport of the excited electron through the solid to the surface, and
(3) the escape of the electron from the confines of the solid into vacuum.

The simple photoemission experiment illustrated in Fig. 2. provides an example of these steps. The electron energy analyzer in vacuum collects

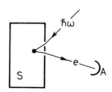

FIG. 2. Schematic diagram showing the three steps of the photoemission process. Incoming photon indicated by $\hbar\omega$, outgoing electron by e, the sample by S, and the analyzer by A. Refraction at the surface is indicated by bent paths.

[4] W. E. Spicer, *Phys. Rev.* **112**, 114 (1968).

the photoelectrons emitted from the surface and determines their energy distribution $N(T)$ defined in such a way that $N(T)\,dT$ = number of electrons collected in a unit time with kinetic energy in the interval $(T, T + dT)$. The three-step model[4] assumes that this energy distribution curve (EDC) is given by

$$N(T) = AN_0(T)L(T)S(T), \qquad (3.1.3)$$

where A is a constant, $N_0(T)$ a distribution in energy of electrons in the solid resulting from photoexcitation, the internal EDC, $L(T)$ the transport coefficient describing the propagation of the excited photoelectron to the surface, and $S(T)$ the threshold function describing the emission into vacuum.

The internal distribution curve $N_0(T)$ is related by the theory to the distribution in energy of the electrons before the excitation $N(E)$. According to Eq. (3.1.3), the relation between $N_0(T)$ and the measured EDC depends on both the second and third step of the photoemission process. Fortunately, however, the threshold function $S(T)$ is a smooth function of energy and it does not *per se* introduce structure in the measured spectra. The transport coefficient $L(T)$ is also a smooth function but plays a very important role in the photoemission process as we shall now see.

3.1.3. Escape Depth

Some elementary considerations are necessary to understand the crucial role of the transport process described by $L(T)$. An electron, after photoexcitation from an energy E to an energy $E + \hbar\omega$, may or may not reach the surface and escape. The probability of doing so depends on energy losses due to inelastic scattering involving other electrons or phonons. These inelastic scattering events give an electron mean free path $l(T) \sim 5-20$ Å at the kinetic energies of interest. This energy dependent mean free path is shown schematically in Fig. 3. Existing measurements of $l(T)$ indicate that it exhibits the same qualitative T dependence in a variety of different materials.[5]

A consequence of this escape depth is that there will be two kinds of photoelectrons observed *in vacuo*, primary electrons that have escaped from the solid prior to inelastic scattering and secondary electrons that have scattered at least once. The primaries are generally the electrons we are most concerned with because they carry information that is related to the initial state of the electron in the solid. The escape depth ensures an inherent surface sensitivity in photoemission experiments. Since interatomic distances in a solid are typically a few angstroms, photoemission with a mean free path of ~ 5 Å probes the outermost atomic planes of the solid. This leads

---

[5] See S. P. Kowalczyk, Ph.D. Thesis, Univ. of California, 1976 (Lawrence Berkeley Lab. Rep. LBL-4319).

3.1. THREE-STEP MODEL, ESCAPE DEPTH, RELEVANT PARAMETERS

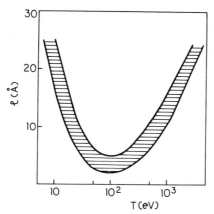

FIG. 3. Kinetic energy dependence of the mean free path of electrons in solids. The shaded area corresponds to the region occupied by the actual experimental points for most materials studied to date.

to comparable sensitivity for bulk states and localized electronic surface states and makes photoemission a fine tool to study surface chemical processes. In particular it makes it possible to monitor chemisorption and physisorption processes with sufficient sensitivity to observe changes induced by fractions of a monolayer of adatoms.

The surface sensitivity of photoemission is much more pronounced than that of optical experiments even though the first step in both experiments is the same, i.e., the photon absorption. If $\alpha(\hbar\omega)$ is the optical absorption coefficient then the photon beam falls to its $1/e$ value within a distance $1/\alpha$ of the surface. Typically, this depth is one or two orders of magnitude larger than the escape depth $l(T)$. Therefore only a fraction $\sim l(T)/[1/\alpha(\hbar\omega)]$ of the excited electrons escape into vacuum without scattering and the transport coefficient in Eq. (3.1.3) can be approximated by

$$L(T) = l(T)\alpha(\hbar\omega). \qquad (3.1.4)$$

Some electrons that lose energy in the solid after excitation can still reach the surface with sufficient energy to escape and be detected. However, their kinetic energy will not be given by Eq. (3.1.2). In an experiment, the two kinds of electrons are simultaneously detected and analyzed: primary electrons that are emitted without losing energy and secondary electrons, e.g., electrons that have lost part of their energy before reaching the vacuum. The measured EDCs are a superposition of the primary-electron and secondary-electron energy distribution curves, as shown in Fig. 4. Unfortunately, no satisfactory way exists to automatically separate contributions from these

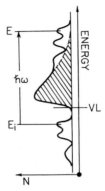

FIG. 4. The distribution in energy of the primary electrons corresponds to the density of occupied states N. The shaded region and the dashed line emphasize the contribution from secondary electrons. VL denotes the vacuum level, below which no electron can escape.

primary and secondary electrons. Corrections are usually attempted by drawing smooth lines more or less arbitrarily, as shown in Fig. 4. This procedure for background subtraction need not seriously affect the important information contained in the primary EDCs.

3.1.4. The Important Parameters

In the above discussion we have indicated that the aim of photoemission spectroscopy is to determine the state of photoelectrons outside the solid and from that to retrieve information about the original state of the electrons in the solid before the photoabsorption process took place. One measures, for example, the EDCs or $N(T)$ outside the solid and then one tries to retrieve the density of occupied states $N(E)$ via the internal EDC $N_0(T)$: $N(T) \rightarrow N_0(T) \rightarrow N(E)$.

The state of the electrons outside the solid is a free-electron state identified by its k vector **k** and by the spin orientation. The k vector is determined by measuring the kinetic energy T and the direction at which the photoelectron is emitted. This direction is defined by the polar and azimuthal angles as shown in Fig. 5.

The photoemission process depends on all the parameters that characterize the state of the electron outside the solid — which in turn are related to those of the original state of the electron inside the solid. Furthermore, it depends on the parameters that characterize the incident photon. The more parameters that are controlled and/or measured in a photoemission experiment, the more information is obtained from it.

3.2. INSTRUMENTATION

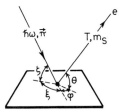

FIG. 5. The relevant parameters characterizing the photon and the photoelectron in a photoemission process: $\hbar\omega$ is the photon energy, $\boldsymbol{\pi}$ the photon-polarization vector, T the electron kinetic energy, and m_s the electron-spin polarization. The other parameters define the direction of the photon and that of the photoelectron. (Boldface indicates vector notation in text, arrows in illustration.)

In summary, the important parameters which define a photoemission experiment are the following:

(1) Photoelectron parameters: kinetic energy T (energy-resolved photoemission); direction of emission θ and ϕ (angle resolved photoemission) electron spin m_s (spin polarized photoemission)

(2) Photon parameters: photon energy $\hbar\omega$, incidence angles ξ and ζ and polarization vector $\boldsymbol{\pi}$.

This list of parameters hints at why photoemission spectroscopy is such a flexible and versatile technique with which to examine electronic properties. Different electronic properties can be studied by focusing attention on a group of these parameters, e.g., $\hbar\omega$-dependent, angle-integrated photoemission, to gather information about the bulk density of states or polarization-dependent, angle-resolved photoemission to examine adsorbate interactions on single-crystal surfaces. Recent progress reflects the increasing ability of the experimenter to control and measure more and more parameters.

3.2. Instrumentation

In this section we shall briefly examine the most important components of a photoemission system; the photon source and the electron analyzer. We shall not discuss the tools for sample preparation and characterization since these depend on the nature of the sample itself. Another important component of the photoemission system not discussed here is the ultrahigh-vacuum hardware required by the surface sensitivity of these measurements. All modern photoemission experiments are performed in an ultrahigh-vacuum environment at pressures below 10^{-9} Torr and preferentially in the 10^{-11}-Torr range.

3.2.1 Electron-Energy Analyzers

The function of the electron analyzer is to capture the emitted photoelectrons and to measure their energy and sometimes their take-off angle. (Some very specialized analyzers, such as the Mott detector, are able to measure the spin polarization of the electrons.) The needs of different experiments have lead to the development of a variety of analyzers, ranging from those that accept nearly all photoelectrons to those that are movable about the sample and select electrons emitted in a small solid angle.

The energy analyzers can be divided into two categories according to whether they detect photoelectrons only within a given energy window or detect all the electrons of energy greater than a given threshold. In the first case, the number of detected photoelectrons measured as a function of the energy of the window gives $N(T)$ or the EDC, as defined in Chapter 3.1. In the second, the number of detected photoelectrons measured as a function of the energy threshold gives the *integral* of the EDC $\int_T^\infty N(T')\, dT'$, from which $N(T)$ can be obtained by differentiation. This differentiation can be done numerically by computer or it can be done in an analog mode as, for example, by modulating the energy threshold and detecting the corresponding modulation in the number of collected photoelectrons. The energy window analyzers are the more commonly used today, but the retarding-field analyzers still offer some advantages.

3.2.1.1. Retarding-Field Analyzers. In its basic configuration, a retarding-field analyzer has a hemispherical retarding grid and hemispherical collector.[6] A negative voltage drop of magnitude V applied between the grid and the sample prevents electrons having energy greater than eV from reaching the collector. Those that are detected constitute a current flowing between this collector and ground. This basic configuration has been refined by examining the electric-field geometry, with the result that the energy resolution has improved. This analyzer can be used for photoemission experiments and also for LEED and Auger spectroscopy.

A typical retarding-grid analyzer is shown in Fig. 6. In that instrument, the resolution is improved over the single-grid analyzer by using a four-grid system. The first and fourth grids are at ground and a field-free region is created between the sample and the first grid. An electron-retarding voltage is applied between ground and *both* the second and third grids (the use of two retarding grids here instead of one enhances the energy resolution). A further retarding voltage is applied between the hemispherical collector and the outermost (fourth) grid to suppress stray low-energy electrons. Since the collector is at a negative voltage relative to ground, the photocurrent corresponds to a (small) voltage drop superimposed on the voltage bias across the

---

[6] See, e.g., P. W. Palmberg, *Appl. Phys. Lett.* **13**, 183 (1968).

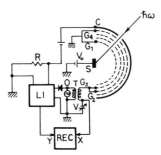

FIG. 6. Scheme of a four-grid retarding-field analyzer. The retarding voltage V is applied to grids G_2 and G_3. Photoelectrons of energy greater than eV are collected by the collector C and give a current in the load resistor R. The oscillator O can add a small modulation to the retarding voltage through the transformer T. This gives a modulation of the voltage drop across R, which is detected by the lock-in amplifier LI. The output of LI corresponds to the energy distribution curve and is plotted by the recorder REC.

load resistor R. This signal voltage should be directly proportional to $\int_T^\infty N(T')\,dT'$, where T corresponds to the cutoff energy for the analyzer. The retarding voltage V is provided by a variable voltage supply, typically an analog or a computer-controlled ramp. A small modulation of fixed amplitude of $\sim 0.3-0.4$ V is provided by an isolation transformer. The corresponding modulation in the photocurrent is detected as a modulation of the load-resistor voltage with a phase-sensitive (lock-in) amplifier. The first harmonic corresponds to the first derivative of the signal, which is the desired EDC $N(T)$. Thus a typical retarding-voltage photoemission experiment is carried out in a first-derivative mode. If the modulating voltage is not a perfectly sinusoidal function, then detection of higher-order-harmonic modulation becomes possible. For example, the second-harmonic modulation corresponds to $dN(T)/dT$ and thereby emphasizes small features in $N(T)$. This analog form of differentiation is rather simple, but it can have some practical difficulties, e.g., the spurious signal given by capacitive coupling.

An ultimate mode of operation of the retarding-grid analyzer uses numerical differentiation of $N(T)$. This approach is used, for example, in the modern display analyzers, discussed in Section 3.2.1.2.4., when they are operated as simple retarding-grid analyzers. The retarding-grid analyzer is not *per se* an angle-resolved analyzer capable of measuring θ and ϕ, but it can be transformed into such a device. For example, electron collection can be limited to a small solid area by using an electron multiplier mounted on a goniometer. Most of the early work with angle-resolved photoemission was based on analyzers of this kind.[7]

[7] N. V. Smith and M. M. Traum, *Phys. Rev. Lett.* **31**, 1247 (1973).

3.2.1.2. Energy-Window Analyzers.

3.2.1.2.1. THE CYLINDRICAL-MIRROR ANALYZER. The single-pass and double-pass cylindrical-mirror analyzers (CMAs) are probably the most widely used analyzers in electron spectroscopy.[8,9] In part this is because they have been commercially available since the early 1970s and because their versatility makes them a good choice for a large number of different experiments. Either can be used for both photoemission studies and Auger spectroscopy.

The basic scheme of a single-pass CMA is shown in Fig. 7a. The sample and the inner cylinder (IC) are kept at ground, and the outer cylinder (OC) is kept at a negative voltage V_p. In this way, a cylindrically symmetric field is created between IC and OC. Analysis of the electron optics of this device shows that electrons emitted by a point source at P are focused at P′, provided that their trajectories form an angle of 42.3° with respect to the axis of the CMA and their energy T is equal to ξeV_p, where $\xi = 1.7$. In practice, the geometry of a commercial CMA gives an acceptance cone of 42.3° ± 6° for

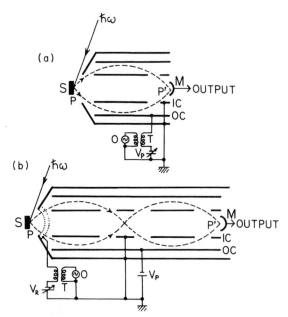

FIG. 7. (a) Single-pass and (b) double-pass CMA. M is the multiplier, and V_p is the voltage bias between the inner cylinder IC and the outer cylinder OC. A small modulation can be added to V_p by the oscillator O through the transformer T.

[8] P. W. Palmberg, G. K. Bohn, and J. C. Tracy, *Appl. Phys. Lett.* **15**, 254 (1969).
[9] P. W. Palmberg, *J. Electron Spectrosc. Relat. Phenom.* **5**, 691 (1974).

the trajectories and a corresponding energy window ΔT centered around $T = \xi e V_p$.

In a photoemission experiment the photon beam is incident on the sample at P and electrons of energy $\xi e V_p$ that are emitted into the 42.3° acceptance cone are filtered by the CMA and detected at P' by an electron multiplier. A single photoelectron reaching the front end of the multiplier gives rise to a much larger number of electrons at the rear end (gain $\sim 10^5$). Eventually there is a detectable pulse (voltage across the load resistor) that can be amplified for pulse counting or analog data handling. The number of pulses detected per unit time as a function of energy $T = \xi e V_p$ gives the energy distribution curves $N(T)$, except for a constant due to the difference in work function between sample and analyzer.

In principle, one could sweep the bias voltage V_p to obtain $N(T)$. This procedure is used, for example, in Auger spectroscopy (a small modulation is added to V_p so that derivatives of the EDCs can be plotted). In photoemission spectroscopy, however, this approach encounters two fundamental difficulties. First, for a constant acceptance cone the magnitude of the energy window ΔT is proportional to T, i.e., $\Delta T/T =$ const. Thus the response of the analyzer changes as the bias voltage is swept. The number of photoelectrons collected per unit time during the sweep gives $TN(T)$ rather than $N(T)$. Therefore, for $T \rightarrow 0$ the signal $TN(T)$ simply disappears. The second difficulty is related to the energy resolution—since $\Delta T/T$ is a constant, the resolution becomes very poor for high electron kinetic energies.

These difficulties are eliminated by the double-pass CMA[9] shown in Fig. 7b. This device differs from the single-pass analyzer in two essential aspects. First, the electrons are filtered in two equal stages, each of which is equivalent to a single-pass CMA. The first state acts as a lens for electrons to be sent into the second stage. This improves the flexibility of the analyzer in handling both large-area and small-area sources, and the double-pass CMA is then equally effective in electron-beam-stimulated experiments (Auger spectroscopy) and photon-beam-stimulated experiments (photoemission). Second, a retarding-grid system is added to the front end of the analyzer. The retarding grids can be used to keep the bias voltage V_p constant regardless of the electron kinetic energy. The distribution $N(T)$ can then be determined by sweeping the retarding (or accelerating) voltage between the grids. Since the first grid and the sample are at ground, the electrons travel in a field-free region until they enter the CMA. With V_p held constant, ΔT stays constant during the sweep, typically a few percent of V_p.

The main disadvantage of the CMA is that it is neither a completely angle-integrated nor a completely angle-resolved device. The acceptance cone provides a reasonable sampling of angles only if the normal-emission direction is included. In many applications, however, this disadvantage is minimal and is more than offset by the many advantages of the CMA. For example, the geometry of the CMA allows an electron gun to be added coaxially for Auger experiments. Photon- or electron-stimulated ion-desorption studies can be performed with a CMA by little more than a reversal of voltage polarities.[10] Furthermore, the retarding-grid system can be used to modulate the signal, and derivatives of the EDC can be taken to enhance weak structures. Angle-resolved measurements can be done with a CMA modification described by Smith et al.[11]

3.2.1.2.2. THE PLANE-MIRROR ANALYZER. The plane-mirror analyzer (PMA) is an energy-window angle-resolved device of particularly simple geometry.[12-14] It was developed by Green and Proca[13] and improved by Smith and co-workers.[14] The scheme is given in Fig. 8. For a given bias voltage V_p, only electrons of energy ξeV_p ($\xi \approx 1.8$) can pass through the analyzer. The energy resolution and the acceptance geometry are defined by the entrance and exit slits. As for the CMA, it is convenient to scan the photoelectron kinetic energies by using the retarding-grid system instead of sweeping the bias voltage. This fixes the pass energy for transmitted electrons regardless of their initial kinetic energy. A particularly delicate problem is related to the termination of the electric field, and this usually limits the

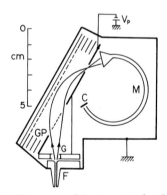

FIG. 8. Scheme of the PMA. The energy of electrons entering the multiplier M is determined by the voltage drop V_p between the plates. The field is terminated by the guard plates GP. The performance of the device is enhanced by front-end optics F, which include the grid system G (not shown in detail) to preretard/accelerate the electrons to enable one to scan the energy without changing V_p.

[10] M. M. Traum and D. P. Woodruff, J. Vac. Sci. Technol. **17**, 1202 (1980).
[11] N. V. Smith, M. M. Traum, J. A. Knapp, J. Anderson, and G. J. Lapeyre, Phys. Rev. **B 13**, 4462 (1976).
[12] N. G. Stoffel and G. Margaritondo, Rev. Sci. Instrum. **53**, 18 (1982), and references therein.
[13] T. S. Green and G. A. Proca, Rev. Sci. Instrum. **41**, 1409 (1970).
[14] N. V. Smith, P. K. Larsen, and M. M. Traum, Rev. Sci. Instrum. **48**, 454 (1977).

resolution (typically ~6% of the pass energy). Both guard plates and controlled-resistance walls have been used for field termination. The performance of the device can be improved by means of added electron optics at the front.[12]

The great advantages of the PMA are that it is simple and inexpensive to build and that it is compact. Its small size means that a PMA can be mounted on a light-weight goniometer for motion in vacuum. Indeed, the device is so inexpensive that two or more analyzers can be used for parallel measurements at several angles of emission. Smith and co-workers[14] mounted two PMAs on orthogonal axes to measure emission for the electric field vector of the light polarized parallel and perpendicular to the plane of incidence. Another promising aspect is that electrons of different energies are distributed along a line after the exit slit of the PMA and parallel detection of energies in a given range is possible with a position-sensitive detector.

3.2.1.2.3. THE SPHERICAL-SECTOR ANALYZER AND THE CYLINDRICAL-DEFLECTOR ANALYZER. Two widely used schemes for energy-window angle-resolved analyzers are the spherical-sector analyzer[15] and the cylindrical deflector analyzer.[16] The cylindrical deflector analyzer is shown in Fig. 9b. A voltage bias applied between the electrodes provides the energy filtering. Furthermore, it can be demonstrated[15] that this analyzer can be used as a

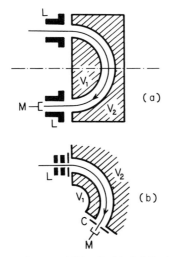

FIG. 9. (a) Spherical-sector analyzer and (b) cylindrical-deflector analyzer. The energies of the electrons entering the multiplier M are determined by the voltage bias $V_2 - V_1$. The performances of these devices are improved by the lenses L and by the collimator C.

[15] C. E. Kuyatt and J. A. Simpson, *Rev. Sci. Instrum.* **38**, 103 (1967), and references therein.
[16] H. Froitzheim and H. Ibach, *Z. Phys.* **269**, 17 (1974).

focusing device — the entrance slit corresponds to a focus at the exit slit. The best focusing conditions occur for an angle between front and rear plates of $\Omega = 127°$. Again, care must be taken to terminate the fields properly. This analyzer is very suitable for miniaturization, and its resolution can be as good as a few milli-electron-volts at energies of the order of a few electron volts. The construction is somewhat more complicated than a PMA.

The spherical-sector analyzer shown in Fig. 9a exhibits focusing that is best for an angle Ω of 180° between front and rear plates. The resolution of these sectors is superior to the PMA, but they are more difficult to fabricate. However, the PMA is not commercially available whereas the spherical-sector analyzer is available from several manufacturers.

3.2.1.2.4. DISPLAY ANALYZERS. Several angle-resolved spectrometers have been built that make it possible to simultaneously and independently detect photoelectrons emitted in many directions. These use a matrix of small electron-multiplying elements bound together to form a channel-plate multiplier. An electron reaching the front of the plate at a given position produces a cascade at the corresponding position on the opposite side, so that the array acts as a high-gain, position-sensitive detector. Electrons from the cascade can be detected with a fluorescent screen that may be monitored by a television camera coupled to a computer for fast data storage.

In the display spectrometer built by Weeks and co-workers,[17] the electrons are energy analyzed with a retarding-grid system and are then detected by the channel-plate/fluorescent screen system. The detection is actually made by a computer-controlled counting system, so that the EDCs can be obtained by numerical differentiation, as discussed for the retarding-grid analyzer. In the energy-window display analyzer built by Eastman and co-workers,[18] an ellipsoidal reflecting electrode focuses the image from the entrance slit to the exit slit, working as a low-pass energy filter, and a retarding-field system provides the high-pass energy filtering.

3.2.2. Photon Sources

The development of new and sophisticated photoemission techniques has gone hand in hand with the development of novel photon sources and in particular of synchrotron radiation sources. The early difficulties associated with finding suitable photon sources for photoemission experiments can be understood by considering the requirements imposed by the techniques on the source. We have seen that the minimum photon energy required to

---

[17] S. P. Weeks, J. E. E. Rowe, S. B. Christman, and E. E. Chaban, *Rev. Sci. Instrum.* **50**, 1249 (1979).

[18] D. E. Eastman, J. J. Donelon, N. C. Hien, and F. J. Himpsel, *Nucl. Instrum. Methods* **122**, 327 (1980).

produce photoelectrons is $\hbar\omega = e\Phi$. This limit should be exceeded by at least several electron volts in order that the photoemission feature in the EDC associated with the state of interest does not overlap the low-energy secondary-electron peak. For valence-band studies, we wish to examine initial states within ~5–15 eV of E_F. Since work functions typically range from 3 to 6 eV, photon energies larger than 10–20 eV are needed. Core levels have even higher binding energies and can only be probed with vacuum-ultraviolet photons ($6 \lesssim \hbar\omega \lesssim 100$ eV), soft x rays (100 eV $\lesssim \hbar\omega \lesssim 1000$ eV), or hard x rays ($\hbar\omega \gtrsim 1000$ eV), depending on the core level of interest. Before synchrotron radiation became available, there were very few sources that provided high-energy photons, and those that did had various disadvantages (noise, vacuum compatibility, lifetime, intensity, and others).

3.2.2.1. Conventional Photon Sources. The list of high-intensity photon sources in the vacuum ultraviolet and soft-x-ray ranges is short and disappointing. Below ~ 10 eV at the H_2 discharge continuum is available, but very few experiments can be effectively carried out at such low photon energies. Above ~ 10 eV we find good gas-discharge line sources: Ar at 11.2 eV, Ne I at 16.8 eV, He I at 21.2 eV, and He II at 40.8 eV. The next most suitable source is the Mg K_α x-ray emission line at 1256.7 eV, followed by the Al K_α line at 1486.5 eV. Other x-ray lines and gas discharge lines have been investigated but rarely used. It was this clear distinction of ultraviolet and x-ray line sources that lead to the now-fading separation of ultraviolet and x-ray photoelectron spectroscopy (UPS and XPS).

A problem with conventional gas-discharge sources is that they must be operated at pressures ~ 10^{-1} Torr. At the same time, the connection to the experimental vacuum chamber must be windowless, since any window would absorb the photon beam. To maintain a suitable pressure differential between the lamp and experimental chamber (where the pressure is in the 10^{-9}–10^{-11}-Torr range) sophisticated differential pumping systems and high-impedance capillaries are used.[19] Even capillaries cannot eliminate direct line-of-sight contamination from the high-pressure region. Fortunately, these high-pressure lamps use inert gases, which, by themselves, do not contaminate the sample surfaces.

3.2.2.2. Synchrotron-Radiation Sources. Synchrotron radiation is the electromagnetic radiation emitted by a relativistic charged particle when it is accelerated. As such, it comes as a natural consequence when electrons orbit a synchrotron or storage ring. Relativistic electrodynamics describes how the familiar Larmor radiation pattern of slowly moving centrifugally accelerated particles is modified at relativistic velocities and becomes sharply peaked in the forward direction of the electron.[20] It also describes how the spectral

---

[19] J. E. Rowe, S. B. Christman, and E. E. Chaban, *Rev. Sci. Instrum.* **44**, 1675 (1973).
[20] J. Schwinger, *Phys. Rev.* **75**, 1912 (1949).

distribution is made up of the orbital harmonic frequencies in such a way as to appear continuous.

A relativistic electron moving through the strong magnetic dipole fields of the binding magnets in a circular accelerator emit photons in the plane of the orbit, tangential to the orbit, with the electric field vector in the orbital plane. The emitted photons have a number of very interesting characteristics.[2]

3.2.2.3. The Spectral Output. In Fig. 10 we show the synchrotron radiation intensity as a function of the wavelength for several storage rings. The difference in intensity for the different facilities reflects the spectral dependence on the energy of the orbiting electrons and the bending radius of the machine. These intensities can be readily calculated through numerical treatments of the modified Bessel functions of fractional order.[20] The spectral distributions of Fig. 10 show that synchrotron sources provide photons over an extremely wide range of energies, including infrared, visible, ultraviolet, vacuum ultraviolet, and x rays. In photoemission we are most interested in the range $5 \lesssim \hbar\omega \lesssim 2000$ eV, and this source is nearly ideal for these experiments. Since the spectrum is continuous in energy, we can develop photoemission techniques that exploit the tunability of the photon energy, as we shall discuss in detail in subsequent sections.

The cutoff at high energy in the spectra of Fig. 10 is related to the energy of the electrons circulating in the storage ring, with high photon energies requiring high electron energies. For some applications, notably those that probe the deeper core levels or seek structural information through x-ray diffraction techniques, a high-energy facility is required, e.g., the 2-GeV ring

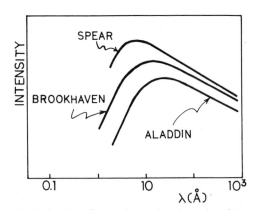

FIG. 10. Wavelength distribution of the typical emitted intensity of three synchrotron radiation sources: SPEAR at Stanford, Aladdin at the University of Wisconsin at Madison, and the new storage ring at the Brookhaven National Laboratory. Flux and wavelength are plotted in logarithmic scale.

at the Brookhaven National Synchrotron Light Source or the SPEAR facility at Stanford. However, the x-ray portion of the beam is not always desirable. Indeed, for electron spectroscopists working in the photon energy range below ~ 2 keV, it is actually undesirable because of the need for radiation shielding for personnel safety. In addition, the beams from the high-energy rings deliver copious amounts of x rays onto the first optical elements of each beamline, and special cooling devices are needed to prevent radiation damage to mirrors.

The intensity of the synchrotron radiation beam scales with the electron current circulating in the ring. It is therefore an advantage to have high beam currents, such as are expected for the advanced-design light sources at Brookhaven and Wisconsin. These new facilities represent storage rings that have been designed and constructed with the needs of the radiation users taken into account — these are accelerators built as light sources and should provide smaller beam profiles in the machine, high brightness, higher beam currents, and therefore higher count rates in the experiments.

3.2.2.4. Collimation. An important effect due to the relativistic transformation from the reference frame of the electron to the laboratory frame is the concentration of the synchrotron radiation close to the plane of the accelerator. The opening angle is proportional to $\gamma^{-1} = (1 - v^2/c^2)^{1/2}$ and this implies half-angles of the order of milliradians. This opening angle is also wavelength dependent so that the shorter the wavelength is the more confined to the orbital plane is the radiation. To gather virtually all of the harder radiation, one needs to collect only a very narrow beam in the orbital plane.

3.2.2.5. Polarization. For wavelengths close to λ_c, the critical wavelength of the radiation pattern,[1,20] the emitted radiation is linearly polarized with the electric vector lying in the orbital plane. This natural polarization is extremely useful in a number of photoemission techniques that explore the symmetry of the electronic states, as we shall see in Chapter 3.4. For wavelengths far from λ_c, the polarization is less complete, but the degree of polarization for photons reaching the sample can be enhanced if one ensures that all optical reflectances in the beam line deflect the beam up or down. These reflections preferentially reflect the component of the beam that has its electric field vector perpendicular to the optical plane of incidence.

3.2.2.6. Time Structure. The synchrotron radiation emitted by a storage ring has a time structure consisting of Gaussian pulses with duration of several nanoseconds separated by 100 – 500 ns. This structure changes from storage ring to storage ring, but, in practice, it is never noticed in photoemission experiments. In fact the source can be regarded as a cw emitter for photoemission purposes. We also emphasize that the intensity has a slow decay in time. The electron beam in a storage ring decays as electrons are scattered out of the bunch phase space by residual gas in the ring or because of electron-electron scattering (Touschek effect). The decay half-life is usually long enough that it can be neglected in a particular EDC, but if data are being handled by computer, it is straightforward to normalize to the instantaneous beam current.

3.2.2.7. Beam Lines. A photoemission experiment using synchrotron radiation requires a suitable beam line to transport the photons from the ring to the photoemission chamber. Along the line, the photons are reflected and focused with mirrors and are filtered in energy with a monochromator. The monochromator disperses the radiation, filtering out those photons outside a small energy range (bandpass) and provides the means for scanning the photon energy. There are at present a number of different monochromators that have been developed to operate in different parts of the photon energy range 2–1000 eV. Among these are the normal-incidence monochromators (2–35 eV), the Seya–Namioka monochromators (2–50 eV), the Brown–Lien grazing-incidence "grasshopper" and "extended-range grasshopper" monochromators (20–1000 eV), and the toroidal-grating monochromators (5–500 eV). Other instruments have been used, and still others are being developed at present.

Finally, it should be noted that synchrotron radiation sources have yet another advantage over other sources: they are inherently ultrahigh-vacuum in character. Since the environment of the storage ring is ultrahigh-vacuum, the beam lines connected to it are required to be under ultrahigh-vacuum. Therefore, no differential pumping system is required to couple a ultrahigh-vacuum photoemission experimental chamber to a synchrotron radiation source.

3.3. Conventional Photoemission

Conventional photoemission techniques are those that were developed when only line sources or low-photon-energy continuum sources were available. Because of the limitations of the source, measurements of the number of photoelectrons as a function of electron kinetic energy for a fixed photon energy were emphasized, i.e., the measurements of EDCs.

We have seen in Chapter 3.1 that the energy distribution curve $N(T)$ is related to the internal energy distribution curve $N_0(T)$ by Eq. (3.1.3) $N(T) = AN_0(T)L(T)S(T)$. The distribution of excited electrons in the solid $N_0(T)$ results from the optical excitation process. The quantity $N_0(T)$ can be related to the density of electronic states of the solid before photoemission, i.e., to the density of initial states $N(E)$.

3.3 CONVENTIONAL PHOTOEMISSION

3.3.1. Density of States, Joint Density of States, and Energy Distribution of the Joint Density of States

In relating the internal EDC $N_0(T)$ to the density of states (DOS) $N(E)$, the starting point is the concept of density of states as defined by the equation

$$N(E) \propto \int_{k\,\text{space}} \frac{dS_k}{|\nabla_k E(\mathbf{k})|} \tag{3.3.1}$$

where S_k is a surface in k space with constant energy E. To discuss photoabsorption processes in solids one must consider the energy distribution of both initial and final states. This leads[21] to the concept of *joint* density of states (JDOS)

$$\alpha(\hbar\omega) \propto \int_{k\,\text{space}} \frac{dS'_k}{|\nabla_k [E_f(\mathbf{k}) - E_i(\mathbf{k})]|}, \tag{3.3.2}$$

where α is the optical absorption coefficient, E_i and E_f are the initial-state and final-state energies, and S'_k is the surface defined by $E_f(\mathbf{k}) - E_i(\mathbf{k}) = \hbar\omega$. The integral in Eq. (3.3.2) is known as the joint density of states (JDOS) for optical adsorption. An approximation is hidden in Eq. (3.3.2), i.e., the assumption that the probability P_{if} of the transition $|i\rangle \rightarrow |f\rangle$ is independent of $\mathbf{k}$ ($|i\rangle, |f\rangle =$ initial, final states). This is a major approximation, which vastly simplifies the calculation of the absorption and works reasonably well over a limited range of photon energies.

The problem of relating $N_0(T)$ and $N(E)$ in photoemission experiments leads us to a third function somewhat related to the DOS and to the JDOS. This is the energy distribution of the joint density of states (EDJDOS). The photoemission effect requires not one but two different conditions:

(1) That the initial state $|i\rangle$ and the final state $|f\rangle$ have the same wave vector $\mathbf{k}$ and are separated in energy by $\hbar\omega$ such that

$$E_f(\mathbf{k}) - E(\mathbf{k}) = \hbar\omega, \tag{3.3.3}$$

which is equivalent to the condition found in the optical adsorption problem (E is now the initial-state energy);

(2) that the energy of the final state $E_f(\mathbf{k})$ must correspond to the energy T at which the electrons are collected by the analyzer

$$E_f(\mathbf{k}) - e\Phi = T. \tag{3.3.4}$$

Equations (3.3.3) and (3.3.4) define surfaces in k space, and their intercept

---

[21] See, e.g., M. Cardona, "Modulation Spectroscopy." Academic Press, New York, 1969.

defines a line L_k in k space. Hence, the formal analogy with the density of states calculation is limited. For the latter it was a matter of calculating the volume in k-space between the surfaces $E(\mathbf{k}) = E$ and $E(\mathbf{k}) = E + dE$. Now the problem is to calculate the volume confined by the four surfaces $E_f(\mathbf{k}) - E(\mathbf{k}) = \hbar\omega$, $E_f(\mathbf{k}) - E(\mathbf{k}) = \hbar\omega + d(\hbar\omega)$, $E_f(\mathbf{k}) = T + e\Phi$, and $E_f(\mathbf{k}) = T + dT + e\Phi$. Once again there is a geometric solution, so that

$$N_0(T) \propto \int_{k\,\text{space}} \frac{dL_k}{|\nabla_k[E_f(\mathbf{k}) - E(\mathbf{k})] \times \nabla_k[E_f(\mathbf{k})]|}, \quad (3.3.5)$$

where dL_k is the length element of the line L_k defined by the conditions of Eqs. (3.3.3) and (3.3.4). Since $\nabla_k E_f(\mathbf{k}) \times \nabla_k E_f(\mathbf{k}) = 0$, Eq. (3.3.5) can be simplified to a form symmetric in E and E_f

$$N_0(T) \propto \int_{k\,\text{space}} \frac{dL_k}{|\nabla_k E(\mathbf{k}) \times \nabla_k E_f(\mathbf{k})|}. \quad (3.3.6)$$

This expression is known as the energy distribution of the joint density of states (EDJDOS).

What the EDJDOS allows us to do in photoemission is to determine the energy of the initial state in the photon adsorption process. This cannot be done in an optical absorption study because optical spectroscopy must live within the constraints of the JDOS. Both spectroscopies share the first step, the photon absorption, which does not discriminate between the initial and final state pairs except to require that their energy separation be equal to the photon energy. Photoemission constrains the kinetic energy of the final state, since it is this that it measures. As long as we are within the one-electron-band picture and know $\hbar\omega$ we can identify E (final-state or many-body effects will complicate the interpretation).

It is worthwhile to emphasize that once again constant probabilities P_{if} were assumed for the $|i\rangle \rightarrow |f\rangle$ transitions in obtaining Eq. (3.3.6). There is no guarantee *a priori* that this assumption is valid, and, as a matter of fact, it is a rather poor approximation.

3.3.2. Transition Probabilities

The optical transition probability between the states $|i\rangle$ and $|f\rangle$ is defined by the Fermi "golden rule":

$$P_{if} \propto |\langle i|\nabla|f\rangle|^2. \quad (3.3.7)$$

Equation (3.3.7) illustrates the complexity of the problem of calculating P_{if} and therefore of estimating the errors due to the assumption that P_{if} is a constant. To do so, one must know the wave functions of the states $|i\rangle$ and

$|f\rangle$. In a few instances this calculation is now feasible. In most, however, it is not.

In an experiment, one can try to empirically estimate the importance of matrix-elements by scanning photon energies and measuring the EDCs. This effectively changes the final-state energy and $|f\rangle$, so that the extent to which the EDCs are modified reveals the validity of the constant-matrix-element approximation. When conventional sources are used, there are only a few photon energies that are available, and this is a limited option. With synchrotron radiation, the tunability of the source greatly facilitates such evaluations.

We shall see, however, that there are other factors that can cause the EDCs to change. One often reads in the literature that the changes shown in such-and-such a figure are due to matrix-element effects, but matrix elements are often blamed for a variety of other effects of different nature.

3.3.3. Interpreting the Energy-Distribution Curves

From Eqs. (3.3.1) and (3.3.6) it should be clear that there is no simple relation between $N_0(T)$ and $N(E)$. In particular, the initial and final states of $N(E)$ are heavily convolved in a complex way in the expression for $N_0(T)$ and it is unrealistic to expect every feature in the initial-state band structure to show up in $N_0(T)$ or, ultimately, the EDC in vacuum. However, one should observe from Eq. (3.3.1) that maxima in $N(E)$ occur when $\nabla_k E(\mathbf{k}) = 0$. Corresponding maxima will occur in $N_0(T)$ according to Eq. (3.3.6). These maxima will show up at final-state energies ($T = E_f$) equal to $E + \hbar\omega - e\Phi$. Therefore, although it is not true that $N_0(T)$ is simply $N(E)$ shifted by $\hbar\omega - e\Phi$, it is still true, in a first approximation, that maxima in $N(E)$ will produce structure in $N_0(T)$. To that extent, the EDCs can reflect features in the density of occupied states $N(E)$.

This is, of course, a somewhat simplistic picture and one must not automatically attribute all features in the EDC to DOS structure or be surprised if all DOS structures do not show up in the EDCs. For example, the presence of $\nabla_k E_f$ in Eq. (3.3.6) implies that the properties of the final state will also influence the spectral features by introducing new structure or removing some of the initial-state peaks. (These final-state features are interesting in their own right, and because of them photoemission is able to probe more than just the initial-state DOS.) To interpret features in the EDC then requires a study of both initial- and final-state structures, as can be done by changing the photon energy. As for matrix-element effects, one can learn about them much more easily by using synchrotron radiation than with conventional photon sources.

Processes other than direct photoemission can also contribute structure in

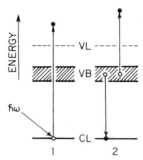

Fig. 11. Energy diagram of the Auger processes. CL is core level, VB valence band, VL vacuum level.

the EDCs. One of them is the Auger process shown schematically in Fig. 11. Creation of the core hole comes as a consequence of photoabsorption, so that there will be a direct photoemission structure associated with it. There will also be a feature that corresponds to the Auger decay of the core hole. One electron from a higher state will fall into the hole and a second electron will be ejected from a state with energy sufficient to conserve the total energy. The ejected electron will then appear in the EDC at a characteristic kinetic energy that is independent of the original photon energy. This constant kinetic energy is a fingerprint for an Auger process. These Auger derived emission features will appear in the measured EDCs. To identify them, one needs only scan $\hbar\omega$: a direct photoemission feature will track with $\hbar\omega$; an Auger feature will remain at fixed kinetic energy.

There is another kind of fixed kinetic process that is also a final-state effect, but that is not related to direct or primary-electron emission. Assume, for example, a distribution of the empty states of a solid such that there is a gap between two bands that falls above $E_F + e\Phi$. In this case, there can be primary electrons that scatter inelastically and ultimately decay into the upper band. Some of these may escape into vacuum, and, if this final-state band is flat and separated from the next lower band, there will probably be a feature in the EDC that corresponds to the escape of these secondaries. Again these will have fixed kinetic energy and they can be readily identified by scanning $\hbar\omega$.

3.3.4. Processes without k Conservation

Our discussion of the relationship between $N_0(T)$ and $N(E)$, including the EDJDOS, has been based on the hypothesis that excitation does not change **k** by an appreciable amount. Other mechanisms, however, could change **k** during the optical excitation. This possibility has been explored, for example, by Spicer[22] and Doniach.[23] They have shown that if **k** is not conserved, then the EDCs are proportional to $N(E)N(E + \hbar\omega)$, i.e., to the product of the density of occupied states and the density of unoccupied states at the final energy $E + \hbar\omega$. Most photoemission studies can be interpreted, however, assuming that one has **k** conservation. At very high photon energies, the photon wave vector is no longer negligibly small, and that assumption should be re-examined.

3.3.5. Processing the Photoemission EDCs

The discussion of the previous three sections gives rise to a number of practical rules to follow in acquiring and processing photoemission spectra taken in the conventional angle-integrated EDC mode. A few considerations that we suggest are the following:

(1) The data-taking process should include EDCs taken at as many photon energies as possible. For valence-band spectra it is advisable, in general, to go up to energies of at least 30–40 eV for reasons discussed below.

(2) For a sample, there must be a reference energy from which other energies are measured. The natural scale is the zero of kinetic energy. This energy scale, however, requires a precise knowledge of the work function and the photon energy. Comparison is made easier if the scale is referenced to the Fermi level E_F. To avoid problems due to the difference in work function between analyzer and sample, and hence the need to bias the sample, it is actually advisable to determine the position of the Fermi level from the spectra themselves. For metals with a strong signal near the Fermi-edge cutoff this is particularly easy. If the sample is not metallic or has a weak Fermi edge, then it may be convenient to determine the position of the zero *a posteriori* by evaporating a metal overlayer onto the sample.

For nonmetals, another frequently used zero for the energy scale is the top of the valence band. This also happens to be the most natural zero for calculations of the density of states. The top of the valence band E_v is usually found from the EDC itself by using, for example, a linear extrapolation of the leading edge. This must be done with care, however, since there may be surface-state emission above the top of the valence band, as we shall see below.

(3) The position of the spectral features should be compared in different spectra. One can identify three different kinds of spectral features:

(a) Peaks that remain in the same position regardless of photon energy (although they may be missing in some spectra due to matrix-element ef-

---

[22] W. E. Spicer, *Phys. Rev.* **154**, 385 (1967).
[23] S. Doniach, *Phys. Rev.* **B 2**, 3898 (1970).

fects). These can be associated with structure in the density of occupied states $N(E)$ provided the one-electron picture holds. Beyond the one-electron picture, one sometimes finds satellites of the DOS-like features due to collective excitation (plasmon) losses.

(b) Peaks that appear at a constant kinetic energy. These are most likely due to Auger emission or to final-state effects associated with secondaries, if their kinetic energy is low, as discussed previously.

(c) Peaks that change slightly with the photon energy. These most likely reflect features in the density of occupied states and the changes may be due to matrix-element and final-state modulation. In general, they stabilize in energy at high photon energies since the final states resemble more and more closely free-electron states, the nature and energy distribution of which have only slight dependence on the energy. For most materials, this so-called x-ray regime is reached for $\hbar\omega \sim 30-50$ eV.

Furthermore, artifacts in the spectra may arise from several kinds of spurious effects, including the second order of the monochromator, anomalies in the beam line output, etc. Once the spectra have been acquired, a reference energy established, and spurious features eliminated, one must turn to a calculation of the energy states to compare experiment with theory.

3.3.6. Experimental EDCs and Theoretical Curves: Bulk States

Let us now turn to some examples and show how theory can be compared to angle-integrated photoemission results. We shall consider a layered semiconductor, GaSe, and an intermetallic compound, FeTi.

Gallium selenide is one of the most widely studied layered compounds. These materials have played an important role in the development of photoemission techniques. Their quasi-two-dimensional character has been particularly valuable because with an entirely two-dimensional band structure, the bands $E(\mathbf{k}) = E(\mathbf{k}_\parallel, \mathbf{k}_\perp)$ depend only on the component of the crystal momentum parallel to the surface $\mathbf{k}_\parallel$. This simplification is particularly valuable, as will be discussed in Chapter 3.5, because $\mathbf{k}_\perp$ changes when the photoelectron crosses the surface. The measurements are simplified further by the relatively inert character of the surface and the ease with which layers separate by cleavage *in situ*.

A series of EDCs for GaSe are shown in Fig. 12.[24,25] The primary emission gives rise to structure superimposed in the smooth secondary-electron background on the low-energy side. The energy reference is the top of the valence band E_v. The various experimental features are present in EDCs taken at different photon energies, but the intensity, shape, and, to some extent,

[24] G. Margaritondo, J. E. Rowe, and H. Kasper, *Nuovo Cimento Soc. Ital. Fis. B* **38B**, 226 (1977).
[25] G. Margaritondo, J. E. Rowe, and S. B. Christman, *Phys. Rev.* **B 15**, 3844 (1977).

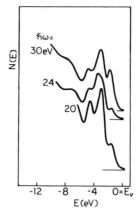

FIG. 12. Energy distribution curves taken at different photon energies with a double-pass CMA on the layered semiconductor GaSe. (After Margaritondo et al.[24,25])

position of these features depends on the photon energy. This modulation is due to matrix-element and final-state effects, which are quite dramatic in these materials because of the quasi-molecular character of the conduction-band states. On the other hand, the modulation vanishes at high photon energies, and the high-energy EDCs reveal structure that can be explained in terms of peaks in the theoretical density of states.[26]

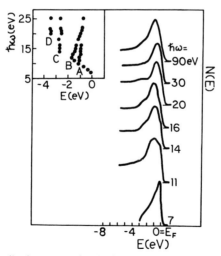

FIG. 13. Energy distribution curves for the intermetallic alloy FeTi.[27] The inset gives the positions of spectral features A, B, C, and D. These can be used as calibers of the band structure.

[26] M. Schluter and M. L. Cohen, *Phys. Rev.* **B 14**, 424 (1976).

The $\hbar\omega$ dependence of valence-band features can also be seen through examination of the results of Fig. 13, in which we show a series of EDCs for $7 \leq \hbar\omega \leq 90$ eV for the intermetallic alloy FeTi.[27] At low photon energy we are able to see only the feature near E_F. This structure disperses to greater binding energy as $\hbar\omega$ increases from 7 to 10 eV, and then two features appear at $\hbar\omega \simeq 11$ eV. Near $\hbar\omega = 14$ eV, another structure appears at -2.8 eV. The inset of Fig. 13 summarizes these initial-state energy positions as a function of photon energy. By examining the corresponding band calculations for FeTi, one can identify the origin of the various structures labeled A–D and can conclude that the calculations have overestimated by ~ 0.6 eV the energy positions of several important $(E, \mathbf{k})$ points in the band structure.

3.3.7. Surface-States Studies in the EDC Mode

The strong surface sensitivity of photoemission spectroscopy makes it possible to detect electronic states localized near the surface. These surface states can be either intrinsic and arise from the clean surfaces or extrinsic and arise from contamination. We shall review several examples of states of the two kinds.

One particularly important problem in dealing with surface states is to identify them and separate their contributions from bulk contributions. This can be done in different ways.

(1) *Surface-sensitivity–escape-depth methods.* By changing the photon energy, it is possible to change the surface sensitivity of the measurements. This comes through the energy dependence of the escape depth. As shown in Fig. 3, the thickness of the region from which primary electrons escape changes with their kinetic energy and thus with photon energy. The surface sensitivity is greatest for kinetic energies ranging from 20 to 150 eV. At lower energies the escape depth increases dramatically and bulk features are emphasized. Comparative measurements at low energies and at 20–150 eV can help one to identify the surface and bulk features. One must note, however, that the relative contribution of the surface features can also be modulated by the transition probability as discussed previously.

(2) *Surface state quenching.* Intrinsic surface states can be identified in part by their dependence on the quality of the surface. They can be removed by controlled surface contamination, which may or may not replace them with extrinsic adsorption states. To identify extrinsic states one can introduce controlled contamination and observe the appearance of characteristic spectral features. For a few common adsorbates, there are well-known signatures that can usually guide one in the assessment of surface cleanliness.

[27] J. H. Weaver and D. T. Peterson, *Phys. Rev.* **B 22**, 3624 (1980).

3.3. CONVENTIONAL PHOTOEMISSION

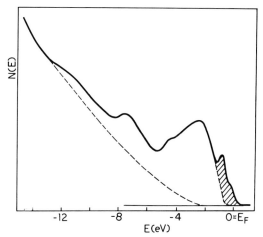

FIG. 14. Surface states detected in photoemission under 23-eV illumination. The energy distribution curve is of a cleaved Si(111) 12 × 1 surface. The surface-state contribution is emphasized by the shaded area. The dashed line shows the estimated secondary-electron background. (After Katnani et al.[28])

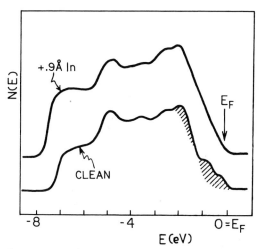

FIG. 15. The surface-state contributions to the energy distribution curves of an annealed Si(111) 7 × 7 surface (shaded area) are removed by a thin (~0.9 Å) indium overlayer. The top curve was shifted in energy to correct for the energy shift due to charge redistribution (see Fig. 17). The shift is emphasized by the different position of the Fermi level (see arrow). (After Margaritondo et al.[29])

Oxygen, for example, typically introduces a relatively broad bump at −6 eV; at heavy coverage (for which the character of the surface should not be in doubt), a double-structured oxygen 2p-derived feature appears. Hydrogen is generally observed as a relatively narrow chemisorption state at −5.0–5.5 eV. Carbon is more difficult to identify and may be better seen in Auger spectroscopy, although it often appears at −4 eV in photoemission spectra. Carbon monoxide is easily detected near −8 and −11 eV (or as a doublet with splitting of ~2.9 eV).

A classic example of intrinsic surface states are those found on the (111) surface of Si. In Fig. 14 we show EDCs taken on a Si(111) surface obtained by cleavage *in situ* at a pressure ~7 × 10^{-11} Torr.[28] The surface-state contribution to this spectrum is shown. These states are quenched very rapidly by any kind of contamination. Figure 15 shows, for example, how the intrinsic surface states of Si(111), as obtained by ion bombardment, annealing to 1300 K, and slow cooling, are removed by an indium overlayer with nominal thickness of 0.9 Å.[29]

3.3.8. Adsorption States

The study of the electronic states of adatoms on a surface represents a very powerful way of investigating surface chemical processes. Photoelectron spectroscopy is particularly effective when coupled to theory, either to simple chemical models or to sophisticated calculations of the local density of states.

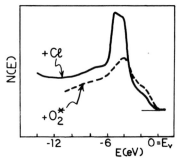

FIG. 16. Ge(111) surfaces with different kinds of adatom chemisorption give different energy distribution curves. Shown here are spectra taken after contamination with chlorine and with oxygen "activated" by a hot tungsten filament. (After Margaritondo *et al.*, Rowe *et al.*[29])

[28] A. D. Katnani, N. G. Stoffel, and G. Margaritondo (1979), unpublished work.

[29] G. Margaritondo, J. E. Rowe, and S. B. Christman, *Phys. Rev.* **B 14**, 5396 (1976); J. E. Rowe, G. Margaritondo, and S. B. Christman, *Phys. Rev.* **B 16**, 1581 (1977), and unpublished work.

3.3. CONVENTIONAL PHOTOEMISSION

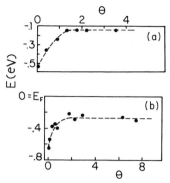

FIG. 17. The shift of the top of the valence band plotted as a function of the overlayer thickness Θ (in monolayers) for (a) indium-covered and (b) germanium-covered annealed Si(111) 7 × 7.

The ability to identify changes induced by chemisorbed atoms is demonstrated by the EDCs of Fig. 16 in which we show results for Ge(111) surfaces contaminated by oxygen and by chlorine.[29] The difference between these EDCs is quite evident. They reflect the details of the chemical bonds formed between the adatoms and atoms of the substrate.

Figure 17 shows some results of charge rearrangement at a semiconductor surface. These changes are caused by the modifications in the local electronic states induced by a thin overlayer evaporated on a clean silicon surface. The changes in the surface charge cause a change in the local electrostatic potential and therefore a rigid shift in energy of the electronic states with respect to the Fermi level. The shift is detected in practice as a rigid shift of the bulk features in the EDCs with respect to the Fermi level. Figure 17 shows two examples of this effect, for germanium adatoms and for aluminum adatoms on Si(111).[30] Measurements of these shifts are extremely important, since the final band bending corresponds to some very important interface parameters, e.g., the Schottky-barrier height for metal–semiconductor interfaces and the heterojunction discontinuities.[29,30]

Changes in the work function induced by chemisorption can easily be detected by measuring the change in energy separation between the low-energy cutoff of the EDC and the Fermi energy. Since the low-energy cutoff corresponds to the zero of the kinetic energy T, this distance, according to Eq. (3.1.2), corresponds to $e\Phi + \hbar\omega$ and Φ can be obtained by subtracting

[30] G. Margaritondo, N. G. Stoffel, A. D. Katnani, H. S. Edelman, and C. M. Bertoni, *J. Vac. Sci. Technol.* **18**, 784 (1981); A. D. Katnani, N. G. Stoffel, H. S. Edelman, and G. Margaritondo, *J. Vac. Sci. Technol.* **19**, 280 (1981).

$\hbar\omega$. Problems may occur, however, if the low-energy cutoff is due to the work function of the analyzer instead of that of the sample. This can be avoided trivially by biasing the sample with a few tenths of an electron volt relative to the analyzer, so that the measured EDC shifts rigidly in energy and the low-energy electrons are not artificially cut off by the analyzer work function.

In summary, photoemission spectroscopy can provide a great deal of valuable information about the electronic states of the bulk, the surface and the adatom even in the simple angle-integrated EDC mode. In the following chapters we shall see how this capability can be expanded by using other, nonconventional modes of photoemission.

3.4. Polarized-Photon Photoemission

Photoemission with polarized radiation makes it possible to enhance the selective process whereby some electronic states are examined while others are suppressed. This comes as a consequence of the matrix elements in photoabsorption, since there are well-defined symmetry requirements or selection rules for initial-to-final-state transitions involving the photon field. By controlling the polarization of the photon relative to the sample and in particular to its surface, we can exploit these symmetries and learn more about the electronic states. Polarization dependences are marked in photoemission and can be used rather easily. The ability to polarize the photon is independent of the nature of the photon source, and polarization-dependent studies can be done with any source — the natural polarization of the radiation from synchrotron sources only simplifies the experiments.

3.4.1. Photon-Polarization Selection Rules

The optical excitation probability P_{if} can be written

$$P_{if} \propto |\boldsymbol{\pi} \cdot \langle i|\nabla|f\rangle|^2, \quad (3.4.1)$$

where the polarization vector of the photon is given by $\boldsymbol{\pi}$. The matrix element in Eq. (3.4.1) is a vector, which we shall call ∇_{if}. Thus Eq. (3.4.1) becomes

$$P_{if} \propto |\boldsymbol{\pi} \cdot \nabla_{if}|^2. \quad (3.4.2)$$

The probability P_{if} therefore depends on the mutual orientation of the two vectors $\boldsymbol{\pi}$ and ∇_{if}. In turn, ∇_{if} depends on the character of the states $|i\rangle$ and $|f\rangle$. For example, if $\boldsymbol{\pi}$ is perpendicular to ∇_{if}, then $P_{if} = 0$ and there will be no optical excitation and, of course, no photoelectron.

Let us consider in detail an example in which $|i\rangle$ is a p-like atomic state and

3.4. POLARIZED-PHOTON PHOTOEMISSION

$|f\rangle$ is an s-like atomic state. In this case, $P_{if} \neq 0$ only for the p state that has the same symmetry as the electric field of the radiation. Therefore, for a polarized photon beam with electric field parallel to the z axis, only the $p_z \rightarrow$ s transition will have nonnegligible probability; the $p_x \rightarrow$ s and $p_y \rightarrow$ s transitions are forbidden by symmetry arguments.

A practical example of this rule is given by polarized photoemission studies that sought information about the chemisorption geometries of Cl on cleaved Si(111).[31] Two chemisorption geometries are possible, as illustrated in Fig. 18. In the first, the Cl atom sits above one of the Si atoms in a onefold symmetric geometry. In the second, it sits between three Si atoms of the substrate in a threefold geometry. In the first geometry, there are highly directional chemical bonds between Cl and Si atoms. Since the valence electrons in Cl are 3s and 3p, one expects the Cl 3p electrons to participate to the formation of these bonds. Of course, due to the directionality of the bonds that point along the z direction (perpendicular to the surface) one expects the p_z state to more involved than the p_x and p_y states. Since chemical bonding lowers the energy of the states participating in it, the hybridized Cl p_z states should fall at lower energies than the p_x and p_y states for this first, onefold-symmetric geometry. For the threefold symmetric geometry, no directional bonds are expected and, if anything, the p_x and p_y states, which point along the sides of the Cl atom, would be expected to contribute more to bonding than the p_z states. Therefore the p_z states should be at higher energies than the p_x and p_y states. Hence, polarized photoemission studies that would reveal the position in energy of the Cl p_z states with respect to the Cl p_x and p_z states would provide a fingerprint of the chemisorption geometry by allowing structural information to be inferred from electronic energies.

In the experiment, the p_z states can be identified by changing the polarization of the radiation relative to the surface by using the matrix element to emphasize the p_z states. The results, shown in Fig. 19 demonstrate that for p-polarized light there is a sharp peak at the low-energy side of the main Cl-induced EDC feature, and that it nearly vanishes when the polarization is

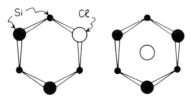

FIG. 18. Top view of two possible chemisorption geometries for chlorine on Si(111). The large Si atoms are in the first atomic plane and the small Si atoms are in the second atomic plane.

[31] J. E. Rowe, G. Margaritondo, and S. B. Christman, *Phys. Rev.* **B 16**, 1581 (1977).

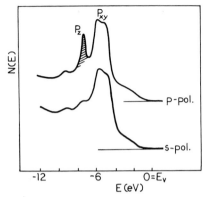

FIG. 19. The change from s polarization with the polarization vector π parallel to the surface to p polarization in Cl adsorbed on Si(111) enhances the p_z-like peaks in the energy distribution curves.

changed to de-emphasize p_z states. This indicates that the narrow, polarization-dependent peak is p_z-like, whereas the main feature is due to p_x and p_y states. Since the p_z states are below the p_x and p_y states for Cl on Si(111), the onefold or on-top geometry can be identified as the correct one.[31]

The validity of this approach has now been demonstrated on a variety of interesting chemisorption problems. Since the p-like states are frequently involved in the formation of chemisorption bonds and the final states can be often viewed as symmetric states (similar to the s-states), the Si(111)–Cl example has a widespread transferability to other cases in which only a few simple chemisorption geometries are conceivable *a priori*. The limitation is, of course, that more complicated chemisorption geometries were left out of the analysis.

3.4.2. Experimental Problems in the Photon-Polarization Techniques

These polarized-photon photoemission techniques appear quite straightforward and require only the measurement of EDCs for different orientations of the polarization vector of the photons relative to the sample surface. Changing the mutual orientation, however, may present some serious technical problems. One could, in principle, change either the orientation of the sample or the polarization vector of the photon beam. For conventional photon sources, this second option corresponds to rotating the mirror system used to polarize the beam without changing the optical path. For synchrotron radiation it would correspond to a rotation—of the entire storage ring. Not surprisingly, these polarization studies are generally done by changing the orientation of the sample.

3.4. POLARIZED-PHOTON PHOTOEMISSION

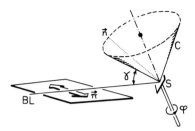

FIG. 20. A simple geometry to change the orientation of the photon polarization vector π with respect to the sample without changing the angle-integrated photoelectron collection geometry. The sample S is rotated around the axis of the CMA so that its normal n sweeps the conical acceptance of the CMA C.

Changing the orientation or position of the sample can easily introduce spurious effects that obscure the process of interest. As we shall see in the next section, the photoemission current depends on the direction along which the photoelectrons are emitted, and a change in the collection geometry will alter the EDCs. This problem can be easily solved in angle-resolved photoemission because the change in the sample position can be compensated by a corresponding change in the position of the analyzer. When we are dealing with angle-integrated techniques, however, we must deal with the problem of changing the sample position (and therefore the polarization) without changing the collection geometry.

This geometric problem has been solved in a number of ways. For example, a useful geometry for polarization-dependent work is shown in Fig. 20. Several versions of this geometry have been used by Rowe,[32] by Lapeyre and Anderson,[33] and by Margaritondo et al.[34] The invariance of the collection geometry is based on the geometry of a CMA. By definition, this analyzer has cylindrical symmetry around its axis and any rotation of the sample around that axis will not change the collection geometry, except for changes due to the intrinsic asymmetries in the transmission efficiency of the CMA. This rotation does produce the desired changes in the mutual orientation of the sample surface and the polarization vector.

3.4.3. Photon-Polarization Effects in Angle-Resolved Photoemission: The Hermanson Rule

The polarization effects we have discussed so far were observed in angle-integrated photoelectron collection geometries. As we shall see in Chapter 3.5, much progress has been made in angle-resolved photoemission. We

---

[32] J. E. Rowe, *Phys. Rev. Lett.* **34**, 398 (1975).
[33] G. J. Lapeyre and J. Anderson, *Phys. Rev. Lett.* **35**, 117 (1975).
[34] G. Margaritondo, J. H. Weaver, and N. G. Stoffel, *J. Phys. E* **12**, 662 (1979).

shall anticipate in this section some of the concepts developed there in the following brief examination of photon polarization effects in angle-resolved photoemission. The interplay of angular and photon polarization effects can be very effective in obtaining detailed information about the initial, intermediate, and final states in the photoemission process. This interplay, however, can be extremely complicated and it is especially important to plan *a priori* the experiments to be carried out. It is trivially easy to obtain interesting sets of EDCs that are full of angle and polarization effects; most of such spectra are so complex that they cannot be easily interpreted.

In the planning of a polarization-dependent angle-resolved photoemission experiment, there is one particularly important and simple polarization rule that should be kept in mind. This rule, as formulated by Hermanson,[35] is applied to photoelectrons emitted by a single crystal (or an ordered system in general) in a direction normal to the surface, or confined to a plane of mirror symmetry of the crystal that contains the normal. For these electrons, the initial state of the process must have the same symmetry (with respect to the mirror plane) as the electric field of the photon, as we shall demonstrate now.

The final state of the photoemission process has — under the above assumptions — even parity. This parity is preserved during the second and third step of the photoemission process. Thus the final state of the optical excitation *inside* the solid also has even parity under reflection at the mirror plane. Since this identifies the final-state symmetry for the optical excitation process, one must ask how the final-state symmetry is related to the initial-state symmetry, i.e., how the excitation process changes the symmetry of the state. The answer is that a photon polarized parallel to the mirror plane does not change the parity of the initial state, but all other photons do change its parity. Therefore since the final states have even parity, the initial state must have

(1) even parity if the photon is polarized parallel to the mirror plane or
(2) odd parity otherwise.

Thus, by changing the polarization of the photon one can identify the symmetry of the initial state. For example, spectral features that correspond to states of even parity will be observed in EDCs for electrons emitted in one of the directions of the mirror plane if the photons are polarized parallel to the mirror plane. When the polarization is changed, the even-initial-state features should disappear and be replaced by odd-initial-state features. This rule can be somewhat weakened in real cases, but it is still strong enough to be applied for practical purposes. Other more stringent rules can be obtained for emission along the normal direction.

---

[35] J. Hermanson, *Solid State Commun.* **22**, 9 (1977).

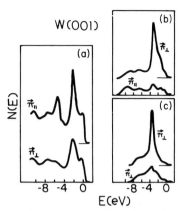

FIG. 21. An application of the Hermanson rule.[35] Angle-resolved energy distribution curves are taken on W(001) along the mirror planes and under illumination of photon energy as given. (a) $\hbar\omega = 20$ eV, $\phi = (100)$; (b) $\hbar\omega = 16$ eV, $\phi = (100)$; (c) $\hbar\omega = 16$ eV, $\phi = (110)$. (Data from Anderson et al.[36])

An elegant experimental example of the Hermanson rule is given by the EDCs of Fig. 21. These curves were reported by Anderson et al.[36] for tungsten(001) with photoelectron collection geometry in the mirror planes defined by the sample normal and its (100) and (110) directions. One sees that the strong peak is almost entirely removed when the polarization of the photon is in the mirror plane. Thus this peak corresponds to an initial state with odd parity, in agreement with calculations of the electronic states of tungsten.

3.5. Angle-Resolved Photoemission: The Band-Mapping Technique

The Hermanson rule introduced in the previous section serves as a nice example of the power of angle-resolved photoemission experiments in examining specific electronic states. In any angle-resolved measurement, the goal is to determine both the energy and the k vector for a photoelectron outside the solid and to relate these properties to those of the corresponding state in the solid. In this section, we shall distinguish between these two k vectors, identifying the k vector outside the solid as $\mathbf{k}^o$ and that inside the solid as $\mathbf{k}$. Each of these will have components parallel and perpendicular to the surface, labelled $k_\parallel$, $k_\parallel^o$, and $k_\perp$, $k_\perp^o$ respectively.

[36] J. Anderson, G. J. Lapeyre, and R. J. Smith, *Phys. Rev.* **B 17**, 2436 (1978), and references therein.

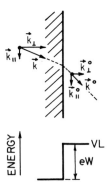

FIG. 22. The changes in the perpendicular component of the k vector due to the surface barrier eW encountered by the photoelectrons.

Consider first the problem of an electron at a potential barrier corresponding to the surface of the sample. For simplicity, this barrier is usually taken to be a step, as shown in Fig. 22. Encountering this barrier will change the magnitude and direction of the k vector·

$$(k_\perp^{o2} + k_\parallel^{o2}) - (2meW/\hbar^2) = (k_\perp^2 + k_\parallel^2) \tag{3.5.1}$$

and

$$k_\parallel^o = k_\parallel. \tag{3.5.2}$$

Superficially, Eqs. (3.5.1) and (3.5.2) relate the k vectors inside and outside the solid in a simple way. Unfortunately, however, we have no *a priori* knowledge of the value of the depth of the potential well W which in general is energy dependent. The relation between **k** and **k**^{o} is further complicated by other factors, as we shall discuss now.

3.5.1. The Surface Reciprocal-Lattice Vectors

The k vectors of electrons in a periodic potential can be changed — with no change in the energy — by a reciprocal-lattice vector **G**

$$\mathbf{k'} - \mathbf{k} = \mathbf{G}.$$

The parallel k-conservation rule of Eq. (3.5.2) must then be reformulated to take into account the possibility that the k-vector changes by a G vector as the photoelectron escapes:

$$\mathbf{k}_\parallel^o = \mathbf{k}_\parallel + \mathbf{G}_\parallel^{\text{surf}}, \tag{3.5.3}$$

where $\mathbf{G}^{\text{surf}}$ is any of the G vectors parallel to the surface. Since the periodicity of the crystal at the surface can be different from that of the bulk, the surface

G vector appearing in Eq. (3.5.3) need not coincide with the parallel component of a *bulk* G vector. For example, a reconstructed surface will exhibit a new periodicity, as may an adsorbed overlayer on the surface.

Hence, the parallel component of the k vector can be changed by a surface G vector as the electron leaves the sample to become a photoelectron. Thus the same final state of the optical excitation can give rise to photoelectrons emitted in different directions. One can distinguish between the emission of a photoelectron with no contribution of surface G vectors and that with a G vector contribution. For free-electron metals or similar systems, these two kinds of emission are related to the so-called primary and secondary Mahan cones.[37]

3.5.2. Mahan Cones

The concepts developed in the previous section must be extended if we are to consider all possible contributions of the G vectors to the photoemission process. Although we have discussed the changes in the k vectors as the electron escapes, we have not considered G-vector contributions to the optical excitation step of photoemission. Let $\mathbf{k}_i$ be the k vector of the electron in the initial state, and let the final state correspond to the k vector $\mathbf{k}$ considered in the preceding section. Without taking the G vector into account, these two k vectors would be so very similar that they can be taken to be nearly equal, $\mathbf{k}_i = \mathbf{k}$. The equivalence of $\mathbf{k}$ and $\mathbf{k}_i$ does not, however, include the possibility that the lattice as a whole participates in the optical excitation process, changing the balance of the k vectors by a G vector. Therefore we must write the more general expression

$$\mathbf{k}_i + \mathbf{G} = \mathbf{k}. \tag{3.5.4}$$

Let us now assume that the electrons can be treated as nearly free particles in the sample so that $E = p^2/2m$. Conservation of energy requires

$$\hbar^2 k_i^2/2m + \hbar\omega = \hbar^2 k^2/2m. \tag{3.5.5}$$

Substitution from Eq. (3.5.4) gives

$$\hbar(\mathbf{k} - \mathbf{G})^2/2m + \omega = \hbar k^2/2m. \tag{3.5.6}$$

Since $(\mathbf{k} - \mathbf{G})^2 = k^2 + G^2 - 2kG\cos\mu$, we have $G^2 + (2m\omega/\hbar^2) = 2kG\cos\mu$, where μ is the angle between $\mathbf{k}$ and $\mathbf{G}$, or

$$\cos\mu = [G^2 + (2m\omega/\hbar^2)]/2kG. \tag{3.5.7}$$

This tells us that if the electron in some final state of a free-electron metal has a given energy, then the magnitude of its k vector is fixed by the value of the

---

[37] G. D. Mahan, *Phys. Rev.* **B 2**, 4334 (1970).

energy as already discussed. Thus, for a given G vector participating in the optical excitation, the value of μ is determined by Eq. (3.5.7). For a given G vector, all the excited electrons of a given energy travel along trajectories that have the same angle with respect to **G**. These trajectories therefore define a cone, called the Mahan cone. Each G vector has associated with it a Mahan cone.[37]

At the surface, the electron k vector is changed such that $\mathbf{k}_\perp$ is not conserved and the parallel component of the k vector can be modified according to $\mathbf{k}_\parallel^0 = \mathbf{k}_\parallel + \mathbf{G}_\parallel^{surf}$. Hence the direction of each Mahan cone will be modified accordingly.

Photoemission processes can be distinguished according to whether there is or is not a contribution from a surface G vector. If $\mathbf{G}_\parallel^{surf} = 0$, only $\mathbf{k}_\perp$ will change, and each Mahan cone will be modified as the electron leaves the sample. The set of directions in the vacuum are termed the primary Mahan cones. If $\mathbf{G}_\parallel^{surf} \neq 0$, both components of **k** are modified and each Mahan cone will give rise to several subsets of external directions, which correspond to different $\mathbf{G}_\parallel^{surf}$'s. These subsets are called secondary Mahan cones.

The emission involving primary and secondary cones will not, of course, occur with equal probability. One can demonstrate that secondary-cone emission is less probable than primary-cone emission, but it cannot be neglected in a complete accounting of directional emission of photoelectrons.

The above considerations are rigorous only within the confines of the free-electron model for metals. The terminology, however, is much more widely used.

3.5.3. Physical Information from the Angular Distribution of Photoelectrons

We have seen in the previous section that the direction at which a photoelectron is emitted is related to the **k** vector **k** of the excited electron, although the relation may be quite complicated. In turn, this k vector is related to that of the initial state $\mathbf{k}_i$. Hence, one can measure the energy distribution of photoelectrons as a function of their direction relative to the sample and extract from it information about a specific state of the solid. Our analysis will start at an elementary level with a number of simplifying assumptions.

Consider the first step of a photoemission process described by $|i\rangle \rightarrow |f\rangle$. The probability for these optical transitions is given by Eq. (3.3.7), namely,

$$P_{if} \propto |\langle i|\nabla|f\rangle|^2.$$

Now assume that the final-state wave function can be approximated by a free-electron wave such that

$$|f\rangle \sim |\exp(i\mathbf{k} \cdot \mathbf{r})\rangle, \quad (3.5.8)$$

3.5. ANGLE-RESOLVED PHOTOEMISSION

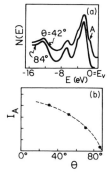

FIG. 23. (a) Dependence of the GaSe angle-resolved EDCs on the angle θ. (b) The change in intensity of peak A. The dependence of this intensity on θ is consistent with a p_z character of the corresponding states. (After Williams et al.[38])

which, when substituted in Eq. (3.3.7), gives

$$P_{if} \propto |\langle i|\nabla|\exp(i\mathbf{k}\cdot\mathbf{r})\rangle|^2 \propto k^2|\langle i|\exp(i\mathbf{k}\cdot\mathbf{r})\rangle|^2. \quad (3.5.9)$$

The term in brackets is just the Fourier transform of the wave function of the state $|i\rangle$. Thus, by measuring the photoemission intensity for a given k vector, it is possible to study the corresponding Fourier transform of the wave function—or at least its square modulus. This approach is complicated by the complex relation between the k vector $\mathbf{k}$ of the electron inside the solid and that of the detected photoelectron $\mathbf{k}^o$, as discussed earlier. Hence, a simple mapping of the Fourier coefficient is not possible unless some correction is carried out to relate $\mathbf{k}^o$ and $\mathbf{k}$.

In Fig. 23, we give an example of an experimental result obtained by this angle-resolved technique.[38] The energy accepted by the analyzer is selected to emphasize a single initial-state feature and the analyzer is moved to scan the angle θ (ordinarily without changing the angle ϕ). The intensity of the EDC feature of interest is then measured as a function of θ, and the resulting profile gives a fingerprint of the atomic symmetry of the initial state. For the example shown in Fig. 23, we see that the dependence on θ of the intensity of peak A is consistent with that of p_z states. Thus we conclude that the states at the top of the valence band of GaSe have p_z character.

3.5.4. Band-Structure Mapping for Two-Dimensional Crystals

A complete knowledge of the state of the electrons in a solid is equivalent to knowing their wave functions. Different photoemission approaches give different levels of knowledge of the electronic states. The first level gives the distribution in energy of the electrons, as we discussed in Chapter 3.3. The

[38] R. H. Williams, P. C. Kemeny, and L. Ley, *Solid State Commun.* **19**, 495 (1976).

second reveals the properties of the wave function itself. We have seen how wave functions can be explored to gain information about their spatial symmetry. The third level provides direct information about the band structure, i.e., the relation between E and **k**. This is explored by the band-mapping technique[39] discussed here.

The band-structure mapping technique is among the most spectacular developments of photoemission spectroscopy. It gives us a way of directly visualizing the band structure of crystalline solids. For two-dimensional ordered systems, the technique is relatively straightforward.[39-42] Recent developments have extended the band-mapping technique to three-dimensional samples.[43-45]

Let us consider an EDC taken by an infinitely narrow detector. Since we do not average over different directions, this EDC does not relate directly to the density of states as did the angle-integrated spectra discussed in Chapter 3.3. Instead, we can assume that each peak corresponds to a given electronic state of well-defined energy and k vector. The relation between the energy and k vector for the initial and final states is that discussed earlier. The final-state energy is, of course, measured by the analyzer and the position of the analyzer defines **k**.

An experimental feature at a kinetic energy T taken along a direction defined by the angles θ, ϕ, with a photon energy $\hbar\omega$, corresponds to a final state of energy T and k vector of magnitude $k^\circ = \sqrt{2mT}/\hbar$ directed along the same direction. The free-electron final state is entirely determined in this way. On the other hand the relation between $\mathbf{k}^\circ$ and $\mathbf{k}_i$ via **k** can be somewhat complicated. For simplicity, assume $\mathbf{k} = \mathbf{k}_i$ so that the optical excitation does not change the k vector (this corresponds in particular to describing the excitation process in the first Brillouin zone).

Two-dimensional systems exhibit spatial periodicity only in directions parallel to the crystal surface. In these directions there is a simple relationship between $\mathbf{k}^\circ$ and **k**, and much of the pioneering angle-resolved work was done with them. The band structure for such a system is two-dimensional, i.e., $E = E(\mathbf{k}_\parallel)$. The measured parallel component of the k vector $\mathbf{k}_\parallel^\circ$ directly gives the corresponding $\mathbf{k}_\parallel$ for the leading primary-cone emission,

---

[39] N. V. Smith and M. M. Traum, *Phys. Rev.* **B 11**, 2087 (1975).
[40] P. K. Larsen, G. Margaritondo, J. E. Rowe, M. Schluter, and N. V. Smith, *Phys. Lett. A* **58A**, 623 (1976).
[41] P. K. Larsen, N. V. Smith, M. Schluter, H. H. Farrel, K. M. Ho, and M. L. Cohen, *Phys. Rev.* **B 17**, 2612 (1978).
[42] M. Schluter, K. M. Ho, and M. L. Cohen, personal communication; see also ref. 41.
[43] T. Grandke, L. Ley, and M. Cardona, *Phys. Rev. Lett.* **33**, 1033 (1977).
[44] T. -C. Chiang and D. E. Eastman, *Phys. Rev.* **B 22**, 2940 (1980).
[45] E. Dietz and D. E. Eastman, *Phys. Rev. Lett.* **41**, 1674 (1978).

3.5. ANGLE-RESOLVED PHOTOEMISSION

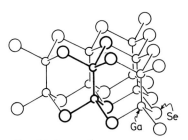

FIG. 24. Structure of a layer of GaSe.

that is $\mathbf{k}_\parallel = \mathbf{k}_\parallel^o$. Given the measured values of T, θ, and ϕ, we can immediately retrieve the initial-state energy and parallel k vector by using

$$|\mathbf{k}_\parallel^o| = |\mathbf{k}^o \sin\theta| = (\sqrt{2mT}/\hbar)\sin\theta \quad \text{and} \quad E = T - \hbar\omega + e\Phi. \quad (3.5.10)$$

We can then actually plot the band structure $E(\mathbf{k}_\parallel)$ by measuring the energy position of peaks in the angle-resolved EDCs taken in different directions θ, ϕ.

The layer compounds are nearly ideal two-dimensional systems. We have already dealt with a classic example in earlier discussions, namely, the layer compound GaSe (GaS and InSe are of the same family). The crystal structure

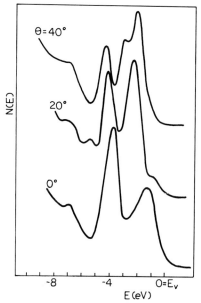

FIG. 25. Angle-resolved EDCs of GaSe with photons of energy 19 eV. The ϕ angle was selected to correspond to the $\Gamma M\Gamma$ direction in k space. (After Larson et al.[40])

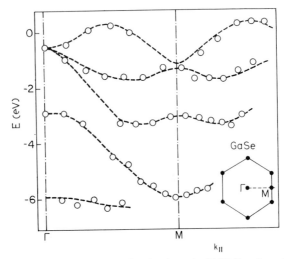

FIG. 26. Experimental band structure of GaSe along the ΓMΓ direction of the two-dimensional Brillouin zone (bottom right).

is shown in Fig. 24 where one can see the chemical bonds within each layer—which are much stronger than those between layers. To a reasonable first approximation, the interlayer interactions can be neglected and each layer can be considered as an isolated system.

A series of angle-resolved EDCs for GaSe[40] are shown in Fig. 25. In these measurements, the angle θ was scanned while the angle ϕ was kept constant. The value of ϕ was selected *a priori* to correspond to one of the high-symmetry directions of the crystal corresponding to the ΓM line of the two-dimensional first Brillouin zone, shown in Fig. 26. The EDCs show energy dispersion of features as θ is varied and these reflect the dependence of E on **k**. A plot of $(E, k_\parallel)$ obtained from these spectra constitutes an empirical plot of the band structure, as shown by the open circles in Fig. 26. The smooth dashed lines emphasize the mirror symmetry of the experimental band structure around the M point. The good correspondence between theoretical and experimental band structures[40] demonstrates the validity of this band-mapping approach.

3.5.5. Band Structure of Adsorbed Overlayers

The preceding section established the practicality of mapping a band structure for two-dimensional systems. We need not restrict our attention to layered compounds, however, when we consider two-dimensional systems. Indeed, very interesting band-mapping results have been reported for ordered adatom overlayers. For such an overlayer, the localized electronic states are related to a two-dimensional band structure, and the same approach used for two-dimensional layered compounds can be used for overlayers.

The first successful attempt to map the band structure of a chemisorbed overlayer was reported by Larsen and co-workers.[41] Their experimental band structure for Si(111)–Cl was in good agreement with the theoretical band structure calculated by Schluter et al.[42] The theoretical curves were obtained for a given chemisorption geometry, and different geometries would have given quite different results. Since the details of the band structure are so sensitive to the chemisorption geometry, band structure mapping is a sensitive probe of the properties of the chemisorption process. One should mention that in the particular case of Si(111)–Cl the deduced properties of the chemisorption properties, and in particular the chemisorption geometry, agree with and are a complement to the results obtained by polarization-dependent photoemission results discussed in the previous section.[31]

The two-dimensional band-structure-mapping approach described here basically consists of measuring the dispersion of the energy of a photoemission peak as a function of $k_\parallel^0$. Usually, the value of ϕ is fixed and the magnitude of $k_\parallel^0$ is changed so to sweep along a high-symmetry direction of k space. In the examples discussed so far, this has been done by changing θ. An alternate approach is to change T by changing the photon energy $\hbar\omega$. In principle, these two approaches could appear equivalent. In practice they could be complementary rather than equivalent since matrix-element and final-state effects will act in different ways when $\hbar\omega$ is scanned and when θ is scanned. Therefore, a comparative use of both approaches makes it easier to identify and eliminate spurious effects.

Although it may be advisable to scan both $\hbar\omega$ and θ, it may not be practical in many cases. For example, scanning $\hbar\omega$ is impossible without a synchrotron radiation source. Conversely, scanning θ at a fixed ϕ is extremely difficult with some kinds of analyzers, e.g., the modified CMA. Clearly, it is very important to decide *a priori* what kind of techniques will be of interest and to design the photoemission system in such a way as to leave open as many options as possible.

3.5.6. Three-Dimensional Band-Structure Mapping

The nonconservation of the perpendicular component of the k vector is the main difficulty to be solved in extending the band structure mapping from two to three dimensions. For three-dimensional crystals, data processing schemes must be found to overcome this problem and to identify the position in k space of the observed electronic states. Some of these data-processing schemes are based on a parallel use of theory and experiment. Therefore, they may require a sophisticated theoretical treatment not only to interpret the data but also to process them.[43-45] The results of these schemes are quite spectacular, and they are being extended to more and more three-dimensional systems. In this section we shall discuss two examples of band structure mapping in three-dimensional crystals.

A first approach to band mapping of three-dimensional systems attempts to identify the k vector in spite of the nonconservation of $\mathbf{k}_\perp$. This approach is based on the study of the photon energy dependence of angle-resolved spectra taken in a given direction, as was used, for example, by Dietz and Eastman[45] in mapping the band structure of copper. Let us assume that angle-resolved photoemission spectra are taken at different photon energies for a given direction, θ and ϕ. The dependence of these spectra on $\hbar\omega$ is generally smooth except for dramatic changes observed at certain critical photon energies. To understand the origin of these changes, first note that as the photon energy is scanned, the magnitude of the k vector of the photoelectrons corresponding to a given feature varies, but its direction remains unchanged as required by the invariance in analyzer position. However, both magnitude and direction of the corresponding excited-electron k vector $\mathbf{k}$ inside the crystal do vary. Scanning $\hbar\omega$ will cause a given spectral feature to scan the k vector $\mathbf{k}$. Dramatic changes in the photoemission spectra are likely to correspond to k vectors at the boundary between two Brillouin zones. The analysis of the EDCs, therefore, enables one to identify exactly the photoelectron k vectors corresponding to the zone boundaries. The procedure can be repeated for different directions in k space. In this way one can relate the internal k vectors corresponding to zone boundaries to external k vectors. After establishing this relation between external and internal k vectors it becomes possible to convert the external energy–wave-vector plots into internal energy–wave-vector plots, i.e., into band structure plots.[45]

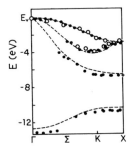

FIG. 27. Experimental band structure map obtained by Chiang and Eastman[44] for GaAs. The dashed lines are pseudopotential theoretical bands.

Another example of three-dimensional band mapping is the experimental band structure determination of III–V semiconductors. The experimental bands of GaAs shown in Fig. 27 were obtained by Chiang and Eastman[44] with a rather simple approach based on Eqs. (3.5.1) and (3.5.2) and on the knowledge of the parameter W in that equation. In order to scan in k space along the special symmetry direction shown in Fig. 27, it was necessary to adjust the photon energy and direction of collection very carefully in a way determined by the geometry of the experiment. The accuracy in determining $E(k)$ may be limited by the lifetime of the excited state, but the agreement of theory and experiment appears excellent.

3.6. Synchrotron Radiation Techniques

The extreme utility of synchrotron radiation for photoemission has been directly or indirectly illustrated in many of the techniques discussed in the previous sections. However, we have not as yet discussed any techniques that strictly require a synchrotron radiation source. The techniques examined so far all represent a measurement of the energy distribution curve in one form or another, including angle integrated or angle resolved with or without polarized photons. These are all techniques that can be exploited with a conventional source, although, of course, they profit from a synchrotron radiation source. For example, only the wide-range tunability of the source distinguishes an EDC taken with a conventional source from one (or many) taken with a synchrotron radiation source—but this is a major distinction since it makes possible a better examination of electronic structures.

When synchrotron radiation sources became available, it was a question of how to exploit their unique properties. To fully realize the potential of synchrotron radiation photoemission, spectroscopy, it is necessary to reevaluate the photoemission process itself. The best starting point is the reconsideration of the relevant parameters in photoemission as introduced in Chapter 3.1. The important parameters emerging from that discussion were T, θ, ϕ, and m_s for the electron and $\hbar\omega$, ξ, ζ, π for the photon. In its most general definition, photoemission spectroscopy is the study of the distribution of photoelectrons as a function of one or a combination of several of these parameters. However, only a limited number of these are of practical interest. In an EDC experiment, electrons are detected by averaging over θ and ϕ with the photon energy $\hbar\omega$ held constant and the photoelectron kinetic energy T scanned. This is not necessarily the best way to do the experiment. We have already seen that the measured EDC is related to the density of states through a rather complicated convolution of the initial density of

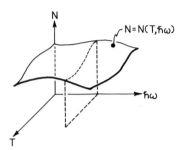

FIG. 28. The function $N(T, \hbar\omega)$ of kinetic energy T, where N is the number of collected photoelectrons of kinetic energy T, a surface in the space $(N, T, \hbar\omega)$.

states and final density of states, the EDJDOS. This suggests that the condition $\hbar\omega = $ const may not be the best possible choice.

Let us confine our attention to the energy variables in the above list, $\hbar\omega$ and T. We can regard the number of calculated photoelectrons N as a function of these two variables

$$N = N(T, \hbar\omega). \qquad (3.6.1)$$

In a three-dimensional space defined by the coordinates $(T, \hbar\omega, N)$, Eq. (3.6.1) corresponds to a surface as shown in Fig. 28. Energy distribution curves obtained by sweeping this surface under the condition

$$\hbar\omega = \text{const}, \qquad (3.6.2)$$

which defines a plane in the space $(T, \hbar\omega, N)$. Of course Eq. (3.6.2) is not the only condition under which the surface $N = N(T, \hbar\omega)$ can be swept. We can imagine a number of ways to sweep that surface, and, in principle, each would correspond to a mode of photoemission spectroscopy. Some of these can be extremely useful and they will be discussed in detail in this chapter. We shall consider in particular two modes introduced by Lapeyre and co-workers,[46] namely,

(1) the mode in which T is kept constant while $\hbar\omega$ is swept, called constant-final-state-(energy) (CFS) mode;

(2) the mode in which $T - \hbar\omega$ is kept constant while T and $\hbar\omega$ are synchronously swept, called constant-initial-state-(energy) (CIS) mode.

[46] G. L. Lapeyre, A. D. Baer, J. Hermanson, J. Anderson, J. A. Knapp, and P. L. Gobby, *Solid State Commun.* **15**, 1601 (1974); G. J. Lapeyre, J. Anderson, P. L. Gobby, and J. A. Knapp, *Phys. Rev. Lett.* **33**, 1290 (1977).

3.6.1. Constant-Final-State-Energy Spectroscopy and Partial-Yield Spectroscopy

The CFS mode has some fundamental advantages over the EDC mode in a study of the density of occupied states. In particular, it eliminates the convolution of initial and final states that is intrinsic in the EDC mode. The CFS mode is defined by the condition

$$T = \text{const}, \quad \hbar\omega \text{ varies.} \quad (3.6.3)$$

This is illustrated in Fig. 29, from which it should be clear that all collected photoelectrons come from excited states of the sample having the same energy—hence the acronym. Since the final-state energy is kept constant, a CFS curve reflects the distribution in energy of the initial states. The same conclusion can be reached on a slightly more formal ground. We have seen in Chapter 3.3 that the number of electrons emitted at an energy T on excitation by a photon of energy $\hbar\omega$ is given by the EDJDOS

$$N_0 \propto \int_{k\text{ space}} \frac{dL_k}{|\nabla_k E(\mathbf{k}) \times \nabla_k E_f(\mathbf{k})|}.$$

In the EDC mode we regarded this formula as a function of T. Only by assuming that $\nabla_k E_f(\mathbf{k})$ has no structure can we relate peaks in the EDC $N_0(T)$ to the energies $E(=E_f - \hbar\omega)$ for which $\nabla_k E = 0$, i.e., to maxima in the DOS at an energy $\hbar\omega$ below $E_f = T$. We can also examine the same formula as a function of $\hbar\omega$ while keeping T, and therefore E_f, constant. The gradient $\nabla_k E_f$ is fixed for all k-space points of the $E_f(\mathbf{k}) = T$ surface; peaks in the CFS cannot be influenced by strong energy variations in $\nabla_k E_f(\mathbf{k})$. From this we conclude that the relation between CFS peaks and density-of-states peaks is more direct than that found in the EDCs.

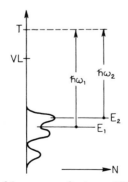

FIG. 29. Schematic diagram of the constant-final-state (CFS) mode. T is kept constant while $\hbar\omega$ is swept. (After Lapeyre et al.[46])

A few words of caution are necessary regarding possible spurious features in the CFS. First, the experimental spectrum is a superposition of primary and secondary electrons. In the CFS mode, the secondaries give an $\hbar\omega$-dependent background. This can be understood as follows. When the energy of the collected electron is set to a value T, we accept primary electrons created by the photon $\hbar\omega$ at that energy and the secondary electrons as schematically illustrated by Fig. 30. These secondary electrons escape at a final-state energy T after excitation to some higher state — either directly or through an Auger process — and possible subsequent inelastic scatterings. The total number of secondaries at a given energy will be proportional to the probability that the photon of energy $\hbar\omega$ directly or indirectly creates electrons of energy greater than T. This probability corresponds to the optical absorption coefficient for photons of energy $\hbar\omega$ and reintroduces the JDOS ideas.[47] Therefore, the secondary-electron contribution to the CFS *reflects the optical absorption coefficient*.[47] Indeed, CFS measurements are frequently used to determine the absorption coefficient α. The association of a CFS feature with density-of-states features is safe only if it is known that the optical absorption coefficient has no structure in the same region.

The use of the CFS mode to measure absorption coefficients represents a very interesting technique. In final-state-energy regions in which the primary signal has a "flat" dependence on $\hbar\omega$, it can be used to determine optical absorption coefficients in a nonoptical way — that is, without photon detection.[47] This use of the CFS technique is so important that a special name was introduced for it, partial yield spectroscopy (PYS). The term *partial yield* implies that the optical absorption coefficient is monitored by detecting

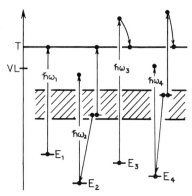

FIG. 30. Direct and indirect processes involving Auger recombination and/or energy losses that give rise to secondary electrons of energy T.

[47] W. Gudat and C. Kunz, *Phys. Rev. Lett.* **29**, 169 (1972).

3.6. SYNCHROTRON RADIATION TECHNIQUES

secondary electrons. Thus one measures the yield of secondary electrons or, better, the partial yield in a limited kinetic energy window. This window can be selected and we shall see that this option further increases the practical capabilities of partial yield spectroscopy.

Figure 31 provides an example of a PYS curve corresponding to the $L_{2,3}$ absorption edge for Si.[48] The inset of Fig. 31 shows the transitions corresponding to this absorption edge. One very remarkable thing is that the initial state in this case is a level rather than a state in a band, i.e., the 2p core level of Si. Thus the optical absorption coefficient is a picture of the density of those final states than can be reached from that core level according to the selection rules for optical transitions. Partial yield spectroscopy curves are a complement to the ordinary CFS and EDC spectra that study the density of occupied states. The use of a series of core levels rather than just one core level for the initial state is the key to overcoming the limitations due to the selection rules, i.e., to the transition probability P_{if}. The core-level PYS technique is one of the two ways in which modern photoemission spectroscopy studies unoccupied states.

A reasonable question at this point might be, "Why should we use the PYS

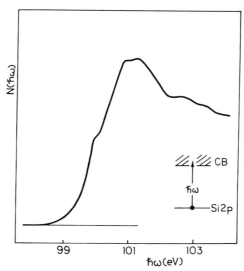

FIG. 31. Partial yield spectrum showing the optical absorption coefficient of Si(111) 2 × 1 in the spectral region corresponding to transitions from the Si 2p core level to states near the bottom of the conduction band CB (see inset). Photoelectron kinetic energy $T = 25$ eV. (After Margaritondo et al.[48])

---

[48] G. Margaritondo, A. Franciosi, N. G. Stoffel, and H. S. Edelman, *Solid State Commun.* **36**, 297 (1980).

technique to measure optical absorption coefficients when more direct or conventional approaches are available?" In a conventional measurement the absorption coefficient $\alpha(\hbar\omega)$ is determined by measuring

$$I(X) = I_0 \exp[-\alpha(\hbar\omega)X], \tag{3.6.4}$$

where I_0 is the intensity of the photon beam at a point X_0 in space and $I(X)$ the intensity at a point in space at a distance X from X_0 along the direction of the propagation of the beam. The intensities I and I_0 are not measured inside the absorbing medium but outside, in vacuum or air. These measurements must therefore be corrected for the reflection and scattering of light at the two boundaries between the absorbing sample and vacuum (or air). Furthermore in the photon energy region of interest to the core-level excitations, i.e., ultraviolet and soft x rays, the absorption coefficient is usually so strong ($\sim 10^5 - 10^6$ cm^{-1}) that very thin samples must be used ($X \sim 10^{-5} - 10^{-6}$ cm or 100 – 1000 Å). There are considerable difficulties in preparing thin samples of that kind that must be either self-supporting or placed on optically neutral substrates. Measurements of crystalline materials make these difficulties extremely severe, since the films are typically polycrystalline rather than single crystals. All these problems are so hard that an alternative approach is welcome. It should be noted that these measurements can be performed without modifications in an existing photoemission system so that one can measure both photoemission spectra and absorption coefficients. This obviously enhances the effective use of the system.

An even more important advantage gained by using the PYS technique instead of conventional measurements of the optical absorption coefficient arises from the tunable surface sensitivity of the PYS technique. As with other electron spectroscopies, the surface sensitivity comes from the transport step in photoemission and is reflected by the escape depth of the photoelectron. In PYS the emphasis is usually on the low-energy secondary electron. As was shown in Chapter 3.1, low-energy electrons have the greater mean free path and are therefore the most effective in giving bulk information. Somewhat higher energies are much more surface sensitive. Hence, if we wish to emphasize surface or bulk phenomena, we can do so by changing the kinetic energy of the collected electrons.

A spurious effect that is common to both the CFS and PYS technique is the spectral structure arising from the photon beam itself. We have seen in Chapter 3.2 that the spectral output of a synchrotron radiation source is structureless. This output, however, is filtered by a monochromator that will have its own spectral response and by mirror or grating reflections that introduce structure related to the absorption coefficient of the mirror or its coating (e.g., an Au-coated mirror with carbon surface contamination will introduce structure at the carbon k edge and at the Au 4f, 5p edges). A CFS or

PYS experiment requires a careful measurement of the spectral intensity versus $\hbar\omega$ at the end of the beam line. These spectral intensity curves and in particular the sharp spurious spectral structure in them are time dependent since the metal coating of a mirror or of a grating will change quite dramatically with time because of radiation-induced damage.

3.6.2. Constant-Initial-State Spectroscopy

The CIS mode of photoemission consists of sweeping the surface $N(\hbar\omega, T)$ along a line defined by

$$T - \hbar\omega = \text{const.} \tag{3.6.5}$$

The reasons for this are emphasized in Fig. 32. The quantity $T - \hbar\omega$, which is the final-state energy minus the photon energy, gives the energy (measured from the vacuum level) of the initial state for primary photoelectrons of the photoemission process. All of these photoelectrons have the same initial-state energy, regardless of $\hbar\omega$, and it is reasonable to assume that structure in N versus T will correspond to features in the final-state density of states.

The relation between the CIS curves (or CISs) and the final-state density of states can be understood by again using the EDJDOS:

$$N_0 \propto \int_{k\,\text{space}} \frac{dL_K}{|\nabla_k E(\mathbf{k}) \times \nabla_k E_f(\mathbf{k})|}.$$

Since $E = T - \hbar\omega + e\Phi = \text{const}$ in the CIS mode, structure in the CIS should rise primarily from the condition $\nabla_k E_f(\mathbf{k}) = 0$. This, of course, corresponds to peaks in the density of final states. As with the CFS mode,

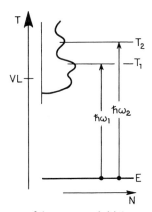

FIG. 32. Schematic diagram of the constant-initial-state mode of photoemission.[46]

178   3. PHOTOEMISSION SPECTROSCOPY OF VALENCE STATES

however, the EDJDOS can also diverge for a constant $\nabla_k E(\mathbf{k})$ if $\nabla_k E$ is parallel to $\nabla_k E_f$. This condition could, in principle, introduce spurious structure, but, in practice, more serious artifacts are introduced by the secondary electrons and by the spectral output of the monochromator.

The condition expressed by Eq. (3.6.5) implies that T and $\hbar\omega$ are to be scanned synchronously. One way of doing this is shown in Fig. 33. Here the photon energy is scanned by the monochromator and an analog output for the wavelength is provided. This is transformed into an analog signal proportional to the photon energy $\hbar\omega$. A dc supply provides the constant required by Eq. (3.6.5), which, when added to the $\hbar\omega$ analog signal, gives the T signal to drive the electron analyzer. This method is simple but it has drawbacks, among them that the wavelength-to-energy conversion can be unstable and difficult to calibrate. An alternate, more reliable approach is also shown in Fig. 33, in which both $\hbar\omega$ and T are shown to be controlled by a minicomputer. With appropriate interface units, the minicomputer provides the pulses to drive the stepping motor of the monochromator and the

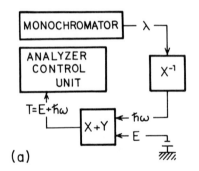

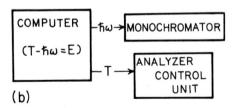

FIG. 33. Two practical ways to implement the CIS mode. In (a) the monochromator scans the photon wavelength and provides an analog signal proportional to the wavelength λ, which is converted into a photon energy signal, added to a constant voltage, and used to drive the analyzer control unit. In (b) both the monochromator and the analyzer control unit are driven by a minicomputer that keeps constant the difference $T - \hbar\omega$.

3.6. SYNCHROTRON RADIATION TECHNIQUES

analog voltage to drive the analyzer. The condition of Eq. (3.6.5) is satisfied by software.

In general, the secondary-electron background is not flat because the optical absorption coefficient is a function of photon energy. More absorption corresponds in general to more final products, i.e., more secondary electrons at any energy of the secondary-electron energy distribution curve. Therefore, a peak in the absorption coefficient would produce a corresponding CIS peak even if we scan T at the same time. Moreover, the relation between the CIS secondary-background and the optical absorption coefficient is not as straightforward as it was in the CFS mode. For example, when we scan T the surface sensitivity of the measurement changes because the escape depth depends on T. Thus the secondary-electron background corresponds to an optical absorption coefficient measured with continuously varying surface sensitivity, with prominent features of the absorption coefficient being visible in the CIS. This convolution makes it impractical to use the CIS mode as an alternate way of measuring partial yield.

Needless to say, the spectral output of the monochromator and beamline mirror system will affect the CIS curves. The structure of the spectral intensity curve will appear as spurious structure in the CIS curves, and all the words of caution given above for the CFSs are equally valid for the CIS mode.

After drawing attention to so many possible spurious effects—and not including other corrections such as matrix-element effects similar to those present for the EDCs—one might suspect that a typical CIS would actually bear little resemblance to the density of final states. Experimental evidence provides the necessary reassurance and demonstrates the utility of this technique. One of the first examples of CIS curves, obtained by Lapeyre and co-workers,[46] is shown in Fig. 34. As they discussed, the peaks present in this spectrum could be related to peaks in the density of final states of KCl. The low-energy cutoff of these curves arises because no electron can be emitted at energies below the vacuum level. The peak present in these CIS curves immediately above the cutoff may be spurious and related to the cutoff itself.

In Chapters 3.3, 3.4, and 3.5, we showed how important it is to take photoemission spectra with different choices of parameters. For the EDCs it was necessary to take spectra at different photon energies and polarizations to estimate the matrix-element and final-state effects and to correct possibly spurious features such as the Auger peaks. For the CFS curves it was similarly necessary to take spectra for different final-state energies, again to estimate and correct matrix-element effects, and to separate partial yield (or absorption coefficient) features from true CFS features. A similar requirement was found in partial yield spectroscopy, in which selection rules must be considered when examining core levels. The requirement for CIS curves is that

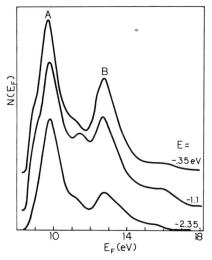

FIG. 34. CIS curves taken on KCl reveal two features in the conduction-band density of states, peaks A and B. The curves were taken for different values of the initial-state energy E (measured from the valence-band edge). (After Lapeyre et al.[46])

several different initial-state energies — and photon polarizations — must be used if the final-state analysis is to be reliable.

The results shown in Fig. 35 for the layer compound GaSe[49] demonstrate the need for different photon polarizations in taking CIS curves. We see that the two CIS curves have some features in common, but that other features do not appear in one of the curves. The reason is again linked to matrix-element effects. Initial states in a given energy range quite often have well-defined symmetry reflecting their angular momentum character.

Our discussion of the CIS mode in terms of the EDJDOS does not take into account the matrix-element factor $|\langle i|\nabla|f\rangle|^2$. There are two separate things to be considered. First, the matrix-element factor was taken out of the k-space integral defining the EDJDOS, implying that this factor is independent of $\mathbf{k}$. Second, the matrix element can be zero or nearly zero for certain configurations of initial and final states and for certain polarizations. These forbidden transitions account for the disappearance of certain features in Fig. 35. A complete picture of the unoccupied states cannot be given by a single CIS curve. A collection of different CIS curves taken for different values of the constant E and for different polarizations can help solve the problem. In particular, it is useful to use theoretical input (e.g., symmetry) to

[49] G. Margaritondo, J. E. Rowe, and H. Kasper, *Nuovo Cimento Soc. Ital. Fis. B* **38B**, 234 (1977).

3.6. SYNCHROTRON RADIATION TECHNIQUES

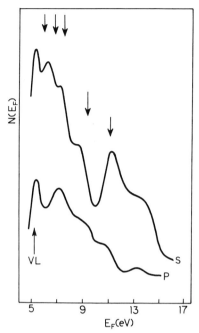

FIG. 35. CIS curves taken on GaSe for both s- and p-polarized photons. These curves reveal a series of peaks in the conduction band that are correlated to structure in the theoretical density of states (vertical arrows). (After Margaritondo et al.[49])

pick initial-state energies that are to be used to determine the unoccupied states.

The above matrix-element effects can be a source of valuable information about the properties of the electronic states. For example, the appearance or disappearance of CIS features can be linked to the symmetry of the initial and final states, but, in general, something must be known about each initial state either from theory or from other experiments.

As a final example, we offer the results of a CIS study of chemisorption geometries. In Fig. 36 we show the results for the Si(111)–Cl system previously discussed in terms of photon-polarized EDCs and two-dimensional band structure mapping. To try to identify the chemisorption geometry, we can compare the CIS curves to the calculated density of unoccupied states determined for different hypothetical chemisorption geometries. As shown, reasonable agreement between theory and experiment can be found for the chemisorption geometry already identified in Chapter 3.4, namely Cl on top of the substrate Si atoms. This demonstrates that the CIS curves are sensitive to the chemisorption geometry, and therefore they can be used to help

182   3. PHOTOEMISSION SPECTROSCOPY OF VALENCE STATES

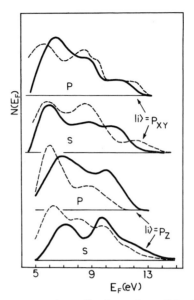

FIG. 36. Correspondence between theory[50] and experiment[31] for the CIS curves of Cl chemisorbed on cleaved Si(111). ——, experiment; ---, theory.

identify unknown chemisorption geometries. The use of the CIS as a surface-structure probe is very powerful when the chemisorption-induced CIS features are in an energy region in which the substrate introduces no CIS structure — as is the case for many covalent semiconductors such as silicon.

3.7. Conclusions and Future Prospects

A review of modern photoemission techniques (1) should emphasize the large number of powerful and sophisticated techniques that have been introduced during the development of photoemission spectroscopy and (2) should point out that new techniques are still being invented and developed. Our review has dealt primarily with the former — but the new developments should not be neglected by the reader who wishes to make full use of photoemission. Many major developments in photoemission can be related to advances in instrumentation and in theory. In particular, the widespread use of synchrotron radiation has had a profound impact on photoemission research. Other fundamental instrumentation advances have concerned the electron analyzers, surface-analysis techniques, and ultrahigh-vacuum tech-

[50] M. Schluter and M. L. Cohen, *Phys. Rev.* **B 17**, 716 (1978).

3.7. CONCLUSIONS AND FUTURE PROSPECTS

nology in general. As for the theory, a number of powerful computational schemes have been developed that consider bulk electronic states, surface states, adatom interactions, and scattering effects. These computational schemes are continually being challenged to interpret the flow of advanced photoemission results.

The different sections of this review have illustrated a number of photoemission approaches capable of studying the properties of solids and solid surfaces from many points of view. We shall now briefly summarize what physical and chemical properties can be investigated by the different techniques.

(1) The density of occupied electronic states can be investigated by the EDC technique (see Chapter 3.2) and by the CFS technique (see Chapter 3.6).

(2) The density of unoccupied electronic states can be investigated by the core-level partial yield spectroscopy and — above the vacuum level — by the CIS technique (see Chapter 3.6).

(3) The surface and bulk states can be distinguished from each other using the photon tunability of synchrotron radiation (see Chapter 3.2).

(4) The parity and character of the electronic states can be explored by the EDC, CIS, and partial yield techniques, either directly by means of a theoretical feedback (see Chapters 3.2 and 3.6) or by investigating photon-polarization effects (see Chapter 3.4). This leads in particular to the identification of surface bonding geometries.

(5) The electronic energy bands of two- and three-dimensional systems can be experimentally mapped using the band mapping technique illustrated in Chapter 3.4.

We emphasize that it was not the purpose of this chapter to review the great mass of results obtained by these techniques. The few results discussed in the text are only practical examples of their use. They were selected because of the particular familiarity of the authors with them. A list (not to mention a description) of all the excellent results produced by photoemission spectroscopists in the last two decades would require much more than a chapter. Such a list would, however, demonstrate the fundamental role that photoemission techniques are playing in the development of condensed matter sciences.

3.7.1. Future Prospects

The techniques of photoemission have matured a great deal in the last two decades, starting with the work by Spicer[4] and ultimately including contributions from a large body of outstanding scientists. In the future, photoelectron spectroscopy can be expected to evolve by expanding the number of practitioners who use it as one of their research tools and by increasing still further the flexibility of the technique. A great many surface science laboratories have x-ray or gaseous-discharge sources for conventional photoemission. In addition to these sources there is an increasing number of synchrotron radiation facilities both in the U.S.A. and abroad. Two major laboratories in the U.S.A. have new, dedicated synchrotron radiation sources: the l-GeV source in Wisconsin and the 750-MeV and 2-GeV sources at Brookhaven. Together with the previously existing sources (Stanford, National Bureau of Standards, and Cornell), the new sources will assure access to synchrotron radiation for the U.S. scientific community. These synchrotron radiation sources make accessible, in particular, the spectral range 50 eV to ~ 2 keV, in which there are many core levels of fundamental importance to chemistry, including the ls levels of C, N, and O. Furthermore, they allow greater flexibility in examining valence-band states and in tuning surface versus bulk sensitivities.

The increased flexibility of photoemission will undoubtly come from the use of synchrotron radiation coupled to other instrumentation advances. The new photon sources increase by several orders of magnitude the flux reaching the sample because of improved optical systems, new monochromators and better source characteristics. Such increases will permit real-time studies of dynamic events and high-resolution studies that have been impossible in the past. For example, high resolution and high flux in the spectral region of the C, N, and O ls levels will make possible studies of surface chemistry through dynamic core-level shifts. With the development of monochromators in the range in which traditional grating and crystal monochromators are inadequate, new and exciting structural and electronic studies of solids, surfaces, and interfaces will become feasible.

An extremely interesting area of development is the use of high-field magnetic bumps to enhance emission in the short-wavelength portion of the synchrotron radiation spectrum. These "wigglers" or "undulators"[1] have the potential of increasing the flux by orders of magnitude and of extending the emitted spectral range by several octaves. Another futuristic development could be the use of free-electron lasers in the ultraviolet.

Major advances are also expected in other areas of instrumentation. Computer-controlled analyzers based on channelplate detection will enhance the use of parallel detection of electrons at different angles or at different energies. The pioneering work done in this field by Eastman and co-workers[18] and Rowe and co-workers[19] (see Chapter 3.2) has already produced very spectacular results. In particular, it will become a routine matter to take specialized kinds of photoemission spectra, e.g., those in selected regions in k space. New kinds of detectors are also improving the efficiency of spin

3.7. CONCLUSIONS AND FUTURE PROSPECTS

polarization analysis of photoelectrons. With the increased flexibility in controlling the photoemission parameters new techniques could be introduced along the lines that have given, for example, the CIS and CFS techniques.

The development of computational schemes for electronic properties of great efficiency and reliability is a very active area of research at the present time. This shows promise of continuing the close collaboration between theory and experiment that has been one of the major factors of the success of photoemission spectroscopy. It is likely that the present theoretical advances will meet the demanding requirements of the present and future photoemission experiments. Ultimately, however, the limits in photoemission experimentation and in its theoretical feedback are the creativity and skill of the involved scientist. Indeed, their creativity and skill have been the major factor of the successful development of photoemission spectroscopy in recent years.

Acknowledgment

It is a pleasure to thank Ms. Mary Arttus for her expert help and patient handling of the manuscript with a computer text-editing system.

4. CORE-LEVEL SPECTROSCOPIES

By Robert L. Park

Department of Physics and Astronomy
University of Maryland
College Park, Maryland

4.1. Introduction

Prior to 1913, nickel was assigned the position of element 27 in the periodic table and cobalt the position of element 28. These assignments were based on measurements of the nuclear mass of the naturally occurring elements. In that year, however, Moseley undertook his famous systematic study of the x-ray spectra of the elements.[1] This remarkable study, which used a potassium ferrocyanide crystal as a diffraction grating, followed by only a single year von Laue's discovery of the diffraction of x-rays. As predicted by the Bohr theory of the atom, Moseley found a simple linear relationship between the square root of the frequency of K-shell x-ray lines and the atomic number of the emitting element. It was apparent from his results, however, that the positions of cobalt and nickel in the periodic table were reversed, thus establishing core-level spectroscopy as the unambiguous means of elemental identification.

The elemental analysis of materials remains the principal application of core-level spectroscopy. There is, however, much more to be learned from a careful study of core-level spectra. Although not involved directly in chemical bonding, the positions of the core levels are measurably shifted in response to changes in the distribution of the valence electrons. These "chemical shifts" are small and do not seriously interfere with elemental identification; they can, however, serve as an important indicator of the chemical states of the atoms. Moreover, the electronic transitions involved in the excitation or decay of a core state may involve states within a few electron volts of the Fermi energy that are involved in chemical bonding. The line shapes associated with these transitions carry information concerning the distribution of valence and conduction states. This indirect view of the electronic states, however, may be very different from the view provided

---

[1] H. G. J. Moseley, *Philos. Mag.* **26**, 1024 (1913); **27**, 703 (1914).

by the direct spectroscopies discussed in the previous chapter. The core state wave function overlaps only a small region of the valence and conduction bands and thus provides a window through which the local electronic structure in the vicinity of the excited atom is viewed. In a system such as an alloy or compound, this makes it possible to study the electronic states of the constitutents separately.

There is also fine structure associated with the excitation of a core state, resulting from the interference of the outgoing wave of an ejected core electron with backscattered components from the neighboring atoms. This structure, which may extend for hundreds of electron volts above an excitation edge, provides a unique local view of short-range order in the vicinity of the excited atom. Thus the core-level spectroscopies provide information not only on the elemental composition of a material but also on the chemical state of the atoms and even their structural arrangement.

There are three distinct classes of experiment by which the core-level structure of matter can be studied: (1) measurements of the threshold energies for the creation of excited core states, (2) measurements of the kinetic energies of ejected core electrons that have absorbed a known amount of energy, and (3) measurements of the energies of electrons or photons emitted to conserve energy in the decay back to the ground state.

Before separately discussing each of the surface core-level spectroscopic techniques, it may be useful to briefly summarize some of what is known in general about the core-level structure of the atoms and to discuss the electron–solid interactions that limit these spectroscopies to the near-surface region.

4.2. The Core-Level Structure of Atoms

The energy levels of a system are never viewed directly. They can only be inferred from the study of transitions between levels. In principle, any incident particle of energy greater than the binding energy of an inner-shell electron can excite that electron into an unoccupied state above the Fermi level. The core vacancy left behind will be filled from a higher level as the atom, in a series of transitions, convulses its way back to the ground state. Energy is conserved in these decay transitions by the emission of x-ray photons or Auger electrons.

Since the decay time of a core hole is long compared to the excitation time, the decay of the hole is generally independent of the mode of excitation. The decay time is, however, sufficiently finite so as to produce a measurable uncertainty or "width" in the energy of the deep level.

4.2.1. Binding Energies

Experimentally, the binding energies are determined either from the threshold energies for excitation of the core state, as in x-ray absorption and appearance-potential measurements, or from the kinetic energies of ejected core electrons that have absorbed a known amount of energy, as in x-ray photoelectron spectroscopy and characteristic energy loss measurements.

The first extensive table of core-electron binding energies, published by M. Siegbahn,[2] relied on x-ray absorption measurements to establish a reference level for each element. For $Z > 51$ Siegbahn used the L_3 ($2p_{3/2}$) edge as a reference because of its intrinsic sharpness. For lower-Z elements, however, he used the K edge. The energies of the remaining levels were calculated from x-ray emission wavelengths, which give a measure of the separation between levels. Siegbahn's table was revised and expanded in 1952 by Hill et al.[3] who used more accurate values of the physical constants.

Subsequent tabulations of binding energies by Bearden and Burr[4] and by Siegbahn et al.[5] have relied on x-ray photoelectron measurements, rather than absorption edges, to establish the reference scale wherever possible. As a consequence of the relatively short mean free path for inelastic scattering of electrons, however, it is clear that x-ray photoelectron spectroscopy (XPS) samples a comparatively shallow region near the surface of a solid. Unfortunately, the XPS measurements on which these tabulations relied were not taken under the ultraclean conditions generally regarded as necessary for a surface-sensitive technique. Shirley et al.[6] have therefore redetermined many of these binding energies on clean surfaces under ultrahigh-vacuum conditions.

However, a comparison of XPS binding energies measured by Shirley et al. for the L_3 levels of the 3d transition metal series, with absolute measurements of the threshold energies for inelastic scattering of electrons from these levels,[7] reveals serious differences. Similar discrepancies are reported for x-ray photoelectron measurements by different laboratories on presumably identical samples.[8] We shall defer a full discussion of this problem to Chapter

---

[2] M. Siegbahn, "Spectroskopie der Rontgenstrahlen." Springer-Verlag, Berlin, and New York, 1931.

[3] R. H. Hill, E. L. Church, and J. W. Mihelich, *Rev. Sci. Instrum.* **23**, 523 (1952).

[4] J. A. Bearden and A. F. Burr, *Rev. Mod. Phys.* **39**, 125 (1967).

[5] K. Siegbahn et al., "ESCA, Atomic Molecular and Solid State Structure Studied by Means of Electron Spectroscopy." Almqvist & Wiksell, Stockholm, 1967.

[6] D. A. Shirley, R. L. Martin, S. P. Kowalcyzk, F. R. McFeely, and L. Ley, *Phys. Rev. B* **15**, 544 (1977).

[7] Y. Fukuda, W. T. Elam, and R. L. Park, *Phys. Rev. B* **16**, 3322 (1977).

[8] C. J. Powell, N. E. Erickson, and T. E. Madey, *J. Electron Spectrosc. Relat. Phenom.* **17**, 361 (1979).

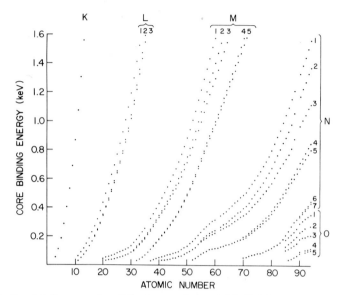

FIG. 1. Core binding energies of the elements below 1.6 keV. More precise values can be obtained from various tabulations, but the figure serves to indicate which levels are generally involved in surface experiments.

4.6, by which time we shall have had a chance to examine these techniques in greater detail.

The energy levels of greatest interest for surface studies are those with binding energies less than about 1.5 keV. This is the approximate energy of an aluminum K_α x ray, which is the most frequently used excitation radiation in XPS. It also corresponds to an electron mean free path for inelastic scattering of perhaps 10–30 Å.[9] For much higher-energy electrons, the surface contribution to the scattering would be slight.

The energy levels of the elements below 1.6 keV are shown in Fig. 1. To obtain more accurate values of the binding energies, reference should be made to the tabulations discussed above. The diagrams in Fig. 1, however, serve to indicate what levels are generally involved in surface studies.

4.2.2. Auger Yields and Lifetime Broadening

In the energy range of interest for surface studies, a core hole is overwhelmingly likely to decay by a radiationless or Auger transition. This is illustrated in Fig. 2, which shows the relative Auger yield for a number of levels as a function of Z. Data on fluorescence and radiation loss yields have been

[9] C. J. Powell, *Surf. Sci.* **44**, 29 (1974).

4.2. THE CORE-LEVEL STRUCTURE OF ATOMS

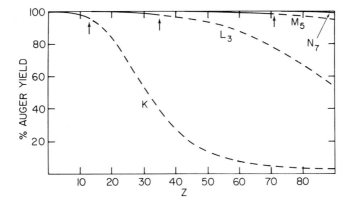

FIG. 2. Relative Auger yields as a function of Z. The yield for a given level decreases with Z, but above about 1.5 keV the levels are not important for surface studies. This "limit" is indicated on the curves by arrows. Thus for levels of most interest in surface science the yield is overwhelmingly Auger.

collected by Krause.[10] For a given level the Auger yield drops with increasing Z, but at sufficiently high Z the binding energies are too great for excitation to take place. We have therefore used a dashed curve to represent the portions of the plots for which the binding energy exceeds the 1.5-keV "limit" discussed in the previous section. This has the effect of limiting our interest in the K shell to $Z \lesssim 13$ and in the L shell to $Z \lesssim 35$. Thus, for those levels of most interest in surface research, the Auger yield is greater than 95%.

The core levels have a natural width governed solely by the lifetime of the core hole that must be created to examine it. The width is related to the lifetime by

$$\Gamma\tau = \hbar, \qquad (4.2.1)$$

where Γ is the width and τ the lifetime. This is no different from the uncertainty in localizing charge that produces a valence band, and in some cases a core level may be broader than the valence band. The natural widths serve not only to limit the precision with which chemical shifts in binding energy can be measured; they also obscure the view of the local electronic structure. In extracting information from line shapes or level positions, therefore, it is desirable to study transitions involving core levels with relatively small natural widths.

For a given principal quantum number, the sharpest level will always correspond to the least tightly bound subshell. A hole in a more tightly bound subshell can always be filled by a transition from another level with the same

[10] M. O. Krause, *J. Phys. Chem. Ref. Data* **8**, 307 (1979).

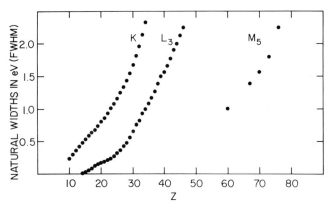

FIG. 3. Calculated Auger lifetime broadening of core levels as a function of Z. (After McGuire.[11])

principal quantum number. Such transitions, which are called Coster–Kronig transitions, have very high rates and thus result in relatively short lifetimes for the core hole. Thus, for example, the L_2 ($2p_{1/2}$) levels are measurably broader than the L_3 ($2p_{3/2}$), because of the strong L_2 (hole) → L_3 (hole) Coster–Kronig transition. For the low energy levels of interest in surface studies the lifetimes of the K (1s) and L_3 ($2p_{3/2}$) levels are dominated by the Auger process. Calculated lifetimes[11] for these levels are shown in Fig. 3. Semiempirical widths for K and L shells have been collected by Krause and Oliver.[12]

For most purposes a simple one-electron model is adequate to describe core-state transitions. On this basis, lifetime broadening produces a Lorentzian distribution for the core level. Therefore, the width of a core level is just the algebraic sum of the widths resulting from each transition that contributes to recombination of the hole, i.e.,

$$\Gamma = \Gamma_R + \Gamma_A + \Gamma_C, \qquad (4.2.2)$$

where Γ_R is the radiative rate of decay, Γ_A the Auger rate, and Γ_C the Coster–Kronig rate.

4.3. The Interaction of Electrons with a Solid

The sensitivity of the core-level spectroscopies to the surface region of a solid is in each case dependent on the mean free path for inelastic scattering of electrons. It is a perverse fact, however, that the very inelastic damping

---

[11] E. J. McGuire, *Phys. Rev. A* **2**, 273 (1970); **3**, 587 (1971); **5**, 1043 (1972).
[12] M. O. Krause and J. H. Oliver, *J. Phys. Chem. Ref. Data* **8**, 329 (1979).

4.3. THE INTERACTION OF ELECTRONS WITH A SOLID

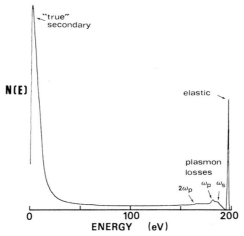

FIG. 4. Secondary-electron spectrum of aluminum. The obvious features include quasi-elastic peak at the incident electron energy, a region of characteristic loss features, and a large true secondary maximum. Core excitation losses and Auger emission features are too weak to be seen in the spectrum.

that enables us to restrict our view to the surface, distorts that view in a way that cannot be entirely corrected for. This problem was recognized as early as 1923 by Robinson,[13] who used the term "electron straggling" to describe the distortion of the XPS spectrum by inelastic scattering. Thus, a single sharp spectral line is detected as a complete secondary-emission spectrum. We may therefore regard the secondary-electron energy distribution as a sort of Green's function that describes the response of the solid to monoenergetic electrons. To understand this response function, it may be useful to consider a particular case in detail. For this purpose, the secondary-electron spectrum of aluminum has been studied more thoroughly, both by theory and by experiment, than any other material (Fig. 4).

4.3.1. The Secondary-Electron Energy Distribution

It is instructive to separate secondary-electron emission into two parts: (1) those electrons that have undergone some inelastic scattering event and then escaped from the surface, and (2) those electrons that are emitted in the decay of the excited states created in step (1). Many electrons will, of course, undergo more than one inelastic scattering event sequentially, just as those electrons emitted in the nonradiative decay of an excited state may subsequently suffer an inelastic scattering event.

The various scattering events that must be considered and the secondary

---

[13] H. H. Robinson, *Proc. R. Soc. London, Ser. A* **104**, 455 (1923).

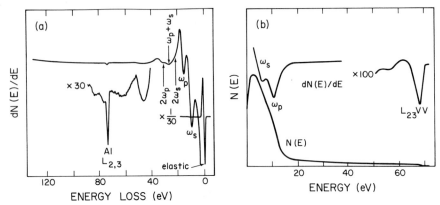

FIG. 5. Separation of the secondary-electron spectrum of Al into (a) loss (with incident electron energy 150 eV) and (b) emission (with incident electron energy 200 eV) parts. The separation was based on whether the energy of a spectral feature was correlated with the incident electron energy (loss features), or independent of the incident electron energy (emission features).

processes they produce are summarized in Table I. The secondary electron spectrum has been divided into two parts. The first, under the column labelled "loss spectrum," represents primary electrons that have undergone some sort of inelastic scattering. The energies of electrons in this column are therefore always referenced to the primary-electron energy. The column labelled "electron emission spectrum" represents "true" secondary events in the sense that these electrons result from the decay of the excited states produced by scattering of the primary electrons, although with some probability these excited states may instead decay by photon emission. It must be emphasized that both the inelastically scattered primary and emitted true secondary electrons will continue to scatter until they escape from the solid or fall below the scattering thresholds.

The separation of the spectrum into emission and loss features was until recently effected in an arbitrary manner, with authors defining the true secondary or emission portion as those features below $\frac{1}{2}E_0$.[14] Experimentally, however, the separation of emission and loss spectra is quite straightforward.[15] A small oscillation is superimposed on the energy of the incident electrons. The loss features, which are correlated with the incident electron energy, exhibit the same modulation, whereas the emission features remain essentially unaffected.

To obtain the emission spectrum, the filament potential is modulated relative to the sample and analyzer. The emission spectrum then corre-

---

[14] K. G. McKay, *Adv. Electron.* **1**, 66 (1948).
[15] R. L. Gerlach, J. E. Houston, and R. L. Park, *Appl. Phys. Lett.* **16**, 179 (1970).

4.3. THE INTERACTION OF ELECTRONS WITH A SOLID

TABLE I. Excitations and Emissions Produced by Electron Bombardment

Scattering event	Loss spectrum[a]	Emission spectrum	
		Secondary electrons[b]	Photons
Collective excitation—bulk	Bulk plasmon loss[c] $E = E_0 - E_P$	Electron–hole pair $E \leq E_P$	Radiative decay $hv = E_P$
Collective excitation—surface	Surface plasmon loss[d] $E = E_0 - E_S$	Electron–hole pair $E \leq E_S$	Radiative decay (rough surface $hv = E_S$
Radiative capture	Continuum		Bremsstrahlung $hv \leq E_0$
Electron–electron scattering	Joint density of states	Continuum electron emission	Transition radiat
Core excitation	Core level loss[e] $E = E_0 - E_B$	Auger emission	Characteristic x-r

[a] Energies referenced to incident electron energy E_0.
[b] Energies referenced to Fermi level of sample.
[c] E_P = bulk plasmon energy.
[d] E_S = surface plasmon energy.
[e] E_B = core binding energies.

sponds to the modulated portion of the analyzer current. If, on the other hand, the modulation is placed on the sample potential, the analyzer window moves synchronously with the incident electron energy, and the loss features, which are fixed relative to the primary energy, exhibit no modulation. The result of this separation for the aluminum spectrum (Fig. 4) is shown in Fig. 5.

4.3.2. The Loss Spectrum

It is now necessary to specify the relative importance of the excitation and emission processes shown in Table I and Fig. 5. The excitation processes are, of course, strongly energy dependent. For electrons with energies of the order of 100 eV, however, the dominant energy loss process in aluminum is the creation of plasmons. The mean free path for plasmon creation in Al, calculated by Quinn,[16] is shown in Fig. 6.

Features corresponding to plasmon losses are evident in Figs. 4 and 5a, but by plotting the derivative of the secondary-electron energy distribution these features are seen even more clearly.[17] In addition to structures corresponding

[16] J. J. Quinn, *Phys. Rev.* **126**, 1453 (962).
[17] R. L. Park, M. denBoer, and Y. Fukuda *in* "Characterization of Metal and Polymer Surfaces" (L.-H. Lee, ed.), Vol. 1. Academic Press, New York, 1977.

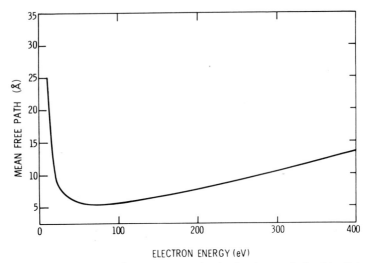

FIG. 6. The mean free path for plasmon creation in aluminum calculated by Quinn.[16]

to the creation of single-bulk (15.4-eV) and surface (10.9-eV) plasmons, additional features can be associated with multiple plasmon creation.

The energy loss spectrum is determined by the same excitations that give rise to the optical properties of solids.[18] The principal difference lies in the extreme sensitivity of the electron energy loss spectrum to the surface region.

In the derivative spectrum of Fig. 5a, a weak characteristic loss feature is observed 73 eV below the elastic peak, corresponding to the excitation of the

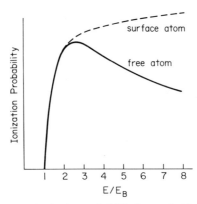

FIG. 7. Electron-impact core ionization probability for an s level of a free atom (solid curve), and for an atom on the surface of an electron-bombarded solid. The enhancement for the surface atom results from backscattering.

[18] S. Ohtani, K. Terada, and Y. Murato, *Phys. Rev. Lett.* **32**, 415 (1974).

4.3. THE INTERACTION OF ELECTRONS WITH A SOLID

2p core state. A Coulomb–Born theory for the excitation of an s core state by electron impact[19] predicts a dependence on electron energy of the form shown in Fig. 7. A variety of other classical and quantum mechanical models predicts a similar form. The cross section rises to a maximum at about 2.5 times the binding energy of the core electron, and then declines slowly. Comparison of these theories with scattering from free atoms is straightforward, but in the bombardment of surfaces a correction must be made for electrons that have already been scattered. Thus the effective excitation as a function of the energy of electrons incident on the surface will frequently continue to rise with energy, as indicated by the dashed curve in Fig. 8.

4.3.3. The Emission Spectrum

It is evident from Fig. 5 that the emission spectrum must be closely linked to the excitation processes revealed in the loss spectrum. Indeed, since the principal mode of energy loss in aluminum is plasmon creation, it is not surprising that the decay of plasmons accounts for most of the true secondary electrons, as proposed by Pillon et al.,[20] and by Everhart et al.[21] They based this proposal on structure in the true secondary maximum that could be related to the maximum electron energies for decay of plasmons into electron–hole pairs.[22] This is seen clearly in the derivative of the emission spectrum shown in Fig. 5b, taken from the thesis of denBoer.[23] Even more convincingly, denBoer has shown that the true secondary maximum is extinguished if the incident electron energy is reduced below the thresholds for plasmon creation.

As indicated in Fig. 5, bulk plasmons may also decay by photon emission. Surface plasmons, however, have phase velocities less than that of light in vacuum and cannot couple with the radiation field except in the presence of surface irregularities.[24] This has been confirmed for aluminum.[25]

In addition to the plasmon decay features in Fig. 5b, a weak feature at about 68 eV corresponds to the Auger recombination of the 2p core hole. This feature is some two orders of magnitude weaker than the peak produced by plasmon decay, as we would expect from the strength of the 2p loss feature relative to the plasmon losses.

---

[19] M. R. H. Rudge and S. B. Schwartz, *Proc. Phys. Soc., London* **88**, 563 (1966).
[20] J. Pillon, D. Roptin, and M. Cailler, *Surf. Sci.* **59**, 741 (1976).
[21] T. E. Everhart, N. Saski, R. Shimizu, and T. Koshihawa, *J. Appl. Phys.* **47**, 2941 (1976).
[22] N. S. Chung and T. E. Everhart, *Phys. Rev. B* **15**, 4699 (1976).
[23] M. L. denBoer, Ph.D. Thesis, Univ. of Maryland, College Park, 1979.
[24] E. A. Stern, in "Optical Properties and Electronic Structure of Metals and Alloys" (F. Abeles, ed.). North-Holland Publ., Amsterdam, 1966.
[25] A. J. Braundmeier, M. W. Williams, E. T. Arakawa, and R. H. Ritchie, *Phys. Rev. B* **5**, 2754 (1972).

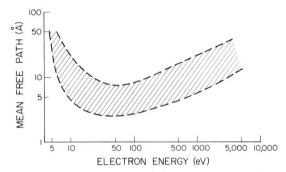

Fig. 8. Sampling depth for electron spectroscopies. Measured mean free paths for inelastic scattering of electrons in solids fall within the shaded region. The general trend resembles the mean free path for plasmon creation (Fig. 6).

4.3.4. The Inelastic Scattering Mean Free Path

The cross sections of the various loss processes listed in Table I combine to determine a mean free path for inelastic scattering. It is this attenuation length that determines the region sampled by the electron spectroscopies. For a simple metal such as aluminum, it is possible to calculate the mean free path on the basis of the observed fact that it is controlled almost entirely by plasmon excitation. This gives a minimum mean free path of about 5 Å at 50 eV (Fig. 6).

There have been a number of experimental determinations of the attenuation length, based on Auger and x-ray photoelectron measurements on solid overlayers, the widths of Bragg maxima in LEED, and the dependence of x-ray photoelectron yields on the angle of incidence of the x rays. The results of many of these determinations have been compiled by Seah and Dench.[26] There are, of course, substantial experimental uncertainties in most of these determinations. However, for a rather wide range of materials, mostly metals, the results lie in the shaded region of Fig. 8. The general trend follows the mean free path shown in Fig. 6.

4.4. Appearance-Potential Spectroscopy

In 1911 Franck and Hertz[27] provided one of the first and most direct proofs of the existence of discrete electronic energy states in atoms, when they discovered the threshold potential for inelastic scattering of electrons from atoms in a metal vapor, and correlated this threshold with the appear-

[26] M. P. Seah and W. A. Dench, *Surf. Interface Anal.* **1**, 2 (1979).
[27] J. Franck and G. Hertz, *Verh. Dtsch. Phys. Ges.* **16**, 12 (1911).

4.4. APPEARANCE-POTENTIAL SPECTROSCOPY

ance of characteristic light emission. This concept was extended during the 1920s to the core levels of atoms in a solid and was used to construct x-ray energy-level diagrams of the elements. This early work has been briefly reviewed by Park and Houston.[28] The method consisted of detecting abrupt, albeit small, changes in the total x-ray yield of an anode as a function of the applied potential. It was not a particularly sensitive method because of the large bremsstrahlung background, which tends to obscure the subtle changes in total yield that result from the excitation of characteristic x rays. Indeed, it may seem remarkable that these thresholds can be detected at all at the energies of interest in surface studies in view of the low fluorescence yields.

Since excited core states at these energies are almost certain to decay by nonradiative transitions, it might seem that it would be easier to detect the excitation thresholds in the secondary-electron emission. In fact, small inflections in plots of the total yield of secondary electrons versus primary-electron energy were reported more than fifty years ago and attributed by some researchers to the excitation of core states.[29,30] However, the existence of these inflections was discounted by later researchers.[14] The difficulty is that signal detection is more frequently limited by unwanted background emission than by noise, and whereas the bremsstrahlung background in the x-ray case is well behaved, the secondary-electron yield is not.

Potential modulation differentiation has made the extraction of these thresholds in either the x-ray[28] or secondary-electron yield[31] commonplace. It has also been shown that they can be detected in the elastic scattering yield,[32] and, in the case of adsorbed atoms or molecules, even in the photon-stimulated desorption yield.[33]

4.4.1. Core-Hole Excitation

It is instructive to contrast the electron excitation probability with x-ray absorption. An incident photon can be absorbed by a core electron if its energy $h\nu$ exceeds the core-state binding energy E_B. The ejected core electron will have an energy $E = h\nu - E_B$ relative to the Fermi level of the sample as shown in Fig. 9. To lowest order, and if dipole selection rules are satisfied, the excitation probability is proportional to the integral of the product of the density $N(E)$ of unfilled states at E and the initial-state distribution $\rho(E)$:

$$N\gamma(E) = \int_0^\infty N(E)\rho(E + E_B - E') \, dE'. \tag{4.4.1}$$

---

[28] R. L. Park and J. E. Houston, *J. Vac. Sci. Technol.* **11**, 1 (1974).
[29] R. L. Petry, *Phys. Rev.* **28**, 362 (1926).
[30] H. Krefft, *Ann. Phys. (Leipzig)* **84**, 639 (1927).
[31] R. L. Park, *Appl. Surf. Sci.* **4**, 250 (1980).
[32] J. Kirschner and P. Staib, *Phys. Lett. A.* **42A**, 335 (1973).
[33] M. L. Knotek, V. O. Jones, and V. Rehn, *Surf. Sci.* **102**, 566 (1981).

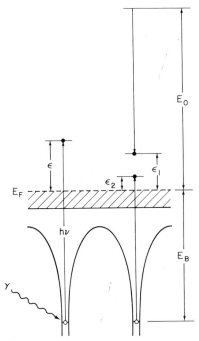

FIG. 9. Energy-level diagram contrasting x-ray absorption with electron bombardment excitation of a core state. In electron excitation, the final-state energy is shared between two electrons.

The core-state distribution $\rho(E)$ is a Lorentzian of the form

$$\rho(E) = \frac{1}{2\pi} \frac{\Gamma}{E^2 + \Gamma^2} \tag{4.4.2}$$

corresponding to a core-hole lifetime of

$$\tau = \hbar/\Gamma. \tag{4.4.3}$$

This assumes the oscillator strength of the transition is a slowly varying function of E. The identification of $N(E)$ with the one-electron plane-wave density of unfilled states is probably not a bad approximation near the threshold, where the wavelength of the electron is very long. In fact, however, the localization of the core hole on a specific atom requires that the ejected electron be treated as a spherical wave, and the "local" density of states can be regarded as arising from the interference properties of the spherical waves. We shall return to this when we discuss extended fine structure.

There is also a threshold energy for inelastic electron scattering from a core state, when the incident electron energy E_0 is equal to E_B. Above the threshold, however, the incident electron energy E_0 is equal to E_B. Above the threshold, however, the incident electron need not give up all its energy to the core electron, and the excitation probability $N_\beta(E)$ depends on the states available to two electrons. To lowest order, this probability will vary as the integral product of the density of unfilled states $N(E)$ with the one-electron transition density for the x-ray absorption case $N_\gamma(E)$, i.e.,

$$N_\beta(E) = \int_0^E N(E')N_\gamma(E-E')\,dE'. \tag{4.4.4}$$

This integral has the effect of obscuring much of the detail in $N_\gamma(E)$. For metals, however, in which the density of unfilled states rises abruptly at the Fermi energy, this detail is recovered by examining the derivative of $N_\beta(E)$

$$\frac{dN_\beta(E)}{dE} = \int_0^E \frac{dn(E')}{dE'} N_\gamma(E-E')\,dE' \tag{4.4.5}$$

where the lower limit of integration is taken just below the edge such that $N(0) = 0$. To the extent that $dN(E)/dE$ is dominated by the Fermi discontinuity, it can be approximated by a delta function:

$$dN(E)/dE \simeq N(E_F)\delta(E), \tag{4.4.6}$$

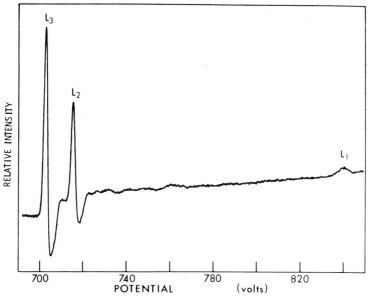

FIG. 10. L-shell appearance-potential spectrum of iron. The large peak at threshold corresponds to the unfilled portion of the 3d band. Extended fine structure can be observed above the $L_{2,3}$ edges.

where $N(E_F)$ is a constant determined by the density of states at the Fermi level. With this substitution, Eq. (4.4.5) can be integrated to yield

$$dN_\beta(E)/dE \simeq N(E_F)N_\gamma(E). \quad (4.4.7)$$

Thus, the derivative of the electron excitation function should resemble the x-ray excitation function. This approximation is best for free-electron-like metals, in which case $N(E)$ exhibits a steplike increase at the Fermi energy. For transition metals, in which $N(E)$ contains a strong peak at E_F, Eq. (4.4.7) is less satisfactory and the derivative spectrum exhibits a pronounced undershoot following the peak as shown in Fig. 10.

4.4.2. Background Suppression

To observe the core-level excitation probability, represented by Eq. (4.4.4), it must first be distinguished from the background of unrelated emissions that have been stimulated by the incident electrons (Table I). In the case of soft-x-ray appearance-potential spectroscopy, in which the total soft-x-ray yield is detected, this background is primarily bremsstrahlung. In Auger electron appearance-potential spectroscopy, in which the total secondary-electron yield is detected, the background is mostly due to true secondary-electron emission.

The suppression of the background is based on the fact that it is a relatively slowly varying function of incident electron energy, and it is achieved by differentiating the yield. This has the effect of weighting the Fourier components of the spectrum by their frequency. To more fully suppress the background, it may be desirable to go to the second derivative, in which case the Fourier components of the spectrum are weighted by the square of their frequency. The extent to which differentiation assists in extracting the core excitation edges from a smoothly varying background is evident from the comparison of the total soft-x-ray yield of stainless steel with its second derivative in Fig. 11.

It might be supposed that the relatively simple functional dependence of the bremsstrahlung background would allow it to be subtracted. To understand why this is not so, we must consider the sources of noise that ultimately limit the sensitivity. These can be divided into statistical or "white" noise sources, such as shot effect and thermal or "Johnson" noise in the measurement circuit, and low-frequency "flicker" noise, which can be regarded as a measure of the stability of the entire measurement system. An example of flicker noise would be fluctuations in the primary-electron-beam current due to reactions at the surface of the emitter. Flicker noise is usually represented by a $1/f$ spectrum, which has the unpleasant characteristic that its contribution increases in direct proportion to the time required to take the

4.4. APPEARANCE-POTENTIAL SPECTROSCOPY

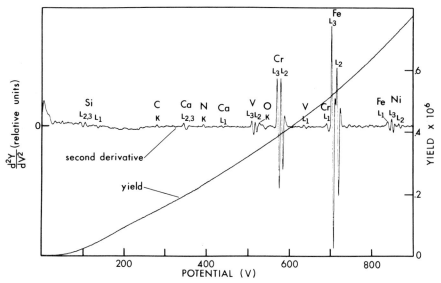

FIG. 11. Second derivative of the soft-x-ray yield of an electron-bombarded stainless steel surface. The yield was measured photoelectrically.

measurement. By contrast the statistical white noise can be reduced arbitrarily by integrating for a sufficient period. Every measurement therefore represents a compromise between these two sources of noise. Differentiation of the spectrum reduces the problem of flicker noise by suppressing the low-frequency Fourier components of the spectrum. At the same time, however, it emphasizes high-frequency noise. Indeed, it would be catastrophic to actually record the derivative of the measured yield. This catastrophe is averted by the inability of the instrument to respond to high frequencies. Thus, associated with differentiation there must be an instrument response function that suppresses the high-frequency Fourier components.

The most versatile means of differentiating the spectrum is by the potential modulation technique.[34] If the potential is modulated about some value E such that

$$E(t) = E + e_0 \cos \omega t, \quad (4.4.8)$$

the yield will also be modulated, as shown in Fig. 12. If the yield curve $I(E)$ is linear over the region of modulation, the output signal will also be a sinusoid whose amplitude is proportional to the slope of $I(E)$. If $I(E)$ is nonlinear, however, the output wave form will be distorted. The spectrum of harmonic

[34] J. E. Houston and R. L. Park, *Rev. Sci. Instrum.* **43**, 1437 (1972).

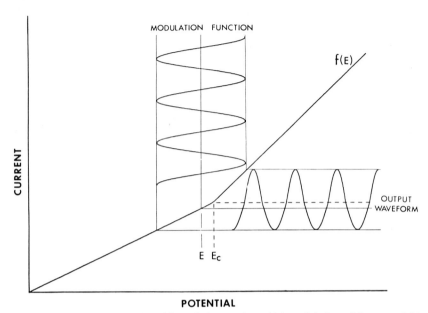

FIG. 12. Potential modulation differentiation. A sinusoidal modulation of the potential is mapped into a variation of the signal. Nonlinearities in the function $f(E)$ produce harmonic distortion in the output wave form. In the limit of small oscillations, a derivative of order n is given by the amplitude of the corresponding harmonic.

frequencies in this distorted wave form can be obtained from the Fourier cosine transform of the functional $I[E(t)]$:

$$F\{I[E(t)]\} = \frac{2}{\pi} \int_0^\pi I[E(t)] \cos n\omega t \, d\omega t. \quad (4.4.9)$$

The spectral components derived from Eq. (4.4.9) represent the broadened derivatives of $I(E)$. To obtain an instrument response function we consider Eq. (4.4.9) for the hypothetical case of a yield function in the form of a unit impulse. Equation (4.4.9) can then be integrated to give

$$F\{\delta(E + e_0 \cos \omega t)\} = \frac{2}{e_0 \pi} \frac{\cos n\theta}{\cos \theta}, \quad (4.4.10)$$

where

$$\cos \theta = -E/e_0. \quad (4.4.11)$$

Equation (4.4.10) represents a Green's function that operates on the yield to give its nth derivative, broadened by an amount related to the modulation amplitude e_0. A more useful Green's function is the response function

$T_n(E, e_0)$ that would smooth the "true" nth derivative of the yield to give the measured spectrum, such that

$$\left(\frac{d^n I(E)}{dE^n}\right)_{\text{measured}} = \int_0^E T_n(E', e_0) \frac{d^n I(E')}{dE'^n} dE'. \quad (4.4.12)$$

The response function $T_n(E, e_0)$ is obtained by simply integrating Eq. (4.4.10) n times.[28] For the first and second derivatives, this gives

$$T_1(E, e_0) = (2/\pi)[1 - (E/e_0)^2]^{1/2} \quad (4.4.13)$$

and

$$T_2(E, e_0) = (2/3\pi)[1 - (E/e_0)^2]^{3/2}. \quad (4.4.14)$$

The function $T_1(E, e_0)$ is a semielliptical broadening function. It should be stressed that it is absolutely essential to have some smoothing associated with differentiation of a measured spectrum. How much smoothing depends on the amount of high-frequency noise in the measured spectrum.

4.4.3. Resolution

Appearance-potential spectroscopy is the highest-resolution core-level spectroscopy available. In contrast to the other electron spectroscopies, it requires no dispersive analyzer. It is only the energy of the incident electron that is measured, and this is determined by the potential between the electron source and the sample. Moreover, dispersive analyzers actually select on the basis of momentum rather than energy and are thus restricted to electrons emerging from a well-defined point. The appearance-potential technique has no such spatial limitations. Aside from the broadening introduced by differentiation, therefore, the instrumental resolution is determined solely by the spread in incident electron energies.

Most appearance-potential spectra have been taken using simple thermionic emission electron sources. Since thermionic emission represents just the tail of the Fermi–Dirac distribution that extends above the work function barrier, it is quite independent of the band structure of the emitter. In fact, as Richardson demonstrated,[35] the velocity distribution of emitted electrons is identical to that predicted for a Maxwell–Boltzmann distribution in the emitter.

If the emitted electrons are subsequently accelerated to energies large compared to thermal energies, we need consider only the distribution of velocities away from the emitting surface, in which case the distribution of

---

[35] O. W. Richardson, "Emission of Electricity from Hot Bodies." Longmans, New York, 1921.

electron energies at the target can be treated by a distribution function of the form

$$J_t(E, T, \phi) \sim \sqrt{E - e\phi} \exp[(-E + e\phi)/kT]. \qquad (4.4.15)$$

The distribution has its peak at $kT/2$ above the work function and the average energy above E is kT. The width of the distribution at half-maximum is about $1.5kT$. Even for a pure tungsten filament operating at 2700 K, this is only 0.34 eV.

In fact, some additional broadening will occur if the work function is not uniform over the emitter or if there is a potential variation over the emitter. For a pure tungsten emitter, however, the work function is a fairly constant 4.52 eV, and the potential drop along the emitting portion of a directly heated filament can be kept negligible.

Although generally much less convenient, a field emission electron source is far superior in terms of resolution. Rather than surmounting the work function barrier, field emitted electrons tunnel through it. Although electrons can tunnel into the vacuum from any state in the valence band, the tunnel current is a strong function of the barrier width, with the result that emission drops off rapidly for states below the Fermi level. Young[36] has shown that for a free-electron metal the energy distribution of field-emitted electrons is given by

$$J_f(E, d, T) \sim (1/d) \exp(E/d)[1 + \exp(E/kT)]^{-1/2}, \qquad (4.4.16)$$

where d is a parameter determined by the work function and the applied field.

Tungsten, which is the most frequently used field-emitter material, is, of course, not a free-electron metal. For the purpose of an instrument response function, however, the difference is slight since the transmission coefficient for d band tunneling is reduced from that for s band tunneling by several orders of magnitude.[37]

The width of $J_f(E, T, d)$ at half-maximum is typically <0.2 eV, which is somewhat better than $J_t(E, T, \phi)$, but this is by no means the only advantage of field emission. The energy distribution cuts off abruptly at $E = 0$, with only the thermal spread of the Fermi distribution producing a slight smearing of the high-energy edge. This can be further reduced by cooling the emitter. As a result, the identification of the threshold for inelastic scattering from a core state is not significantly impaired by the energy distribution of field-emitted electrons, in contrast to the thermionic case in which the maximum electron energy is not well defined.

[36] R. D. Young, *Phys. Rev.* **113**, 110 (1959).
[37] J. W. Gadzuk, *Phys. Rev.* **182**, 416 (1969).

The most significant advantage of the field emission source, however, is that it does not require a knowledge of the work function. Thus, if a field emission source is used, threshold potentials for inelastic scattering from a core state provide an absolute measure of the core binding energy.

What limits the application of field emitters to appearance-potential spectroscopy is the comparatively low sustained currents they can supply. It is not practical to operate a single tip for sustained periods at currents exceeding a few microamperes. Recently, however, arrays of field emitters produced by microcircuit techniques have been made available.[38] These arrays consist of 5×10^3 arrays mm^2 with the extractor built in. They are capable of supplying many tens of milliamperes for indefinite periods. Unfortunately, they are subject to catastrophic failure and require many hours to turn on, even under ideal conditions. If these devices can be made more reliable, they will be of enormous benefit not only to appearance-potential spectroscopy but to many other projects in science and technology.

The total instrument response function for appearance-potential spectroscopy is given by the convolution product of either Eq. (4.4.15) or (4.4.16) with the derivative broadening function, Eq. (4.4.13).

4.4.4. Soft-X-Ray Appearance-Potential Spectroscopy

In view of the low fluorescence yields for core-level excitations in the energy range below 1.5 keV, it may seem remarkable that the appearance potentials were first detected in the soft-x-ray yield. The reason is that the background in the soft-x-ray case consists almost entirely of bremsstrahlung radiation, and the low fluorescence yield is largely compensated for by the small probability that an incident electron will give up its energy by the direct emission of photons. Even more important, as we shall see, is the fact that the bremsstrahlung yield is a smoothly increasing function of the incident electron energy.

Bremsstrahlung is produced by the radiative capture of an incident electron in states above the Fermi level. The short-wavelength limit of the spectrum will therefore correspond to the case in which the incident electron is captured at the Fermi level with the emission of a single photon of energy $h\nu = E_0$. Near the short-wavelength limit, the bremsstrahlung spectrum will approximately reflect the density of states available to the incident electron. At lower energies, however, the spectrum will exhibit an approximately $1/E$ dependence, which results in a so-called infrared catastrophe at long wavelengths.

This infrared catastrophe can be understood in terms of a simple model in which the incident electron is equally likely to be captured at any energy

---

[38] I. Brodie and C. A. Spindt, *Appl. Surf. Sci.* **2**, 149 (1979).

above the Fermi level. We can imagine, then, dividing the density of states into discrete levels separated by an energy Δ. The photon flux at the short-wavelength limit is therefore

$$\rho(E_0) \sim \Delta/E_0. \tag{4.4.17}$$

The photon flux at $E_0 - \Delta$ will include Δ/E_0 from the initial decay plus a contribution from the $E_0 - \Delta$ level, i.e.,

$$\rho(E_0 - \Delta) \sim \frac{\Delta}{E} + \frac{\Delta}{E}\left(\frac{\Delta}{E - \Delta}\right) = \frac{\Delta}{E_0 - \Delta}. \tag{4.4.18}$$

Similarly,

$$\rho(E_0 - n) = \frac{\Delta}{E_0 - n}. \tag{4.4.19}$$

If we identify $E_0 - n\Delta$ with the photon energy, then

$$\rho(E) \sim dE/E \quad \text{for} \quad E \geq E_0. \tag{4.4.20}$$

In the appearance-potential experiment, the contribution of characteristic soft x rays must be detected above this bremsstrahlung radiation. It is quite clear then that the experiment must discriminate against the long wavelengths.

Most appearance-potential spectra have been measured using a simple photoelectric detector. Such a detector is unresponsive to photons with energies below the work function of the photocathode. A schematic diagram of a soft-x-ray appearance-potential spectrometer using photoelectric detection is shown in Fig. 13. The sample is bombarded by electrons from a bare filament, which is usually tungsten. Although the relatively high temperature of the tungsten emitter is a disadvantage from the standpoint of resolution ($kT \simeq 0.25$ eV), its work function is accurately known (4.52 eV)[39] and very stable in ultrahigh vacuum. Photons produced by electron impact pass through a grid biased to reject electrons and strike a cylindrical photocathode. Electrons from the photocathode are collected on a positively biased coaxial collector wire. The work function of the photocathode should be sufficiently high to discriminate not only against low-energy bremsstrahlung but also against most of the filament incandescence. The short-wavelength tail of the filament radiation does contribute some shot noise, however, and a cooler filament offers an advantage from this standpoint. Additional shot noise results from ions desorbed from the surface by the incident electrons, which neutralize at the photocathode surface by the emission of an Auger electron. To obtain the derivative spectrum of the collector current, a small

[39] W. B. Nottingham, *Phys. Rev.* **47**, 806 (1935).

4.4. APPEARANCE-POTENTIAL SPECTROSCOPY

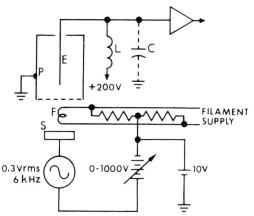

FIG. 13. Schematic diagram of a soft-x-ray appearance-potential spectrometer with photoelectric detection. The sample S is bombarded with electrons from a tungsten filament F. X rays from the sample pass through the grid and impinge on the walls of the photocathode, producing photoelectrons that are collected on the electrode E. That portion of the current that varies at the modulation frequency is selected by a tuned circuit.

sinusoidal oscillation is superimposed on the potential of the sample, as is discussed in Section 4.4.2. That portion of the collector current that varies at the frequency of the oscillation or one of its higher harmonics is selected by a high-Q resonant LC circuit and further filtered and detected by a phase-lock amplifier. By using the distributed capacitance of the collector cable in a resonant tank circuit, high input impedance can be achieved without a preamplifier, since the impedance at resonance is given by $Q/\omega C$.

If resolution is not critical, the greater background suppression that can be achieved in the second derivative makes it possible to use larger oscillation amplitudes and hence achieve greater sensitivity.

The most serious limitation of the simple photoelectric spectrometer shown in Fig. 13 is the high primary-electron current required to obtain spectra in a reasonable period of time. Typically, currents of a milliampere or greater are required for high-quality spectra. This is the consequence not only of small fluorescence yields but also of the poor quantum efficiency for the detection of soft x rays by photoemission. Typically, only about 1% of the x-ray photons striking the photocathode produce a photoelectron. At high primary-electron currents, sample heating can be substantial and in this form the technique is not well suited to the study of chemisorption. Electron-induced desorption, it should be noted, is no more serious than in techniques such as Auger electron spectroscopy, since the primary current can be spread over a large area of the surface with no degradation of resolution.

The problem of low quantum efficiencies was overcome by Andersson et al.[40] who employed a surface-barrier detector with quantum efficiency near unity for the detection of soft x rays. With this detector Andersson and Nyberg[41] were able to study a variety of chemisorbed systems without appreciable sample heating of electon-induced desorption.

As Lee[42] has pointed out, solid-state detectors are ideally suited to appearance-potential spectroscopy, since the output is weighted by the energy of the photons. This provides optimum discrimination against the $1/E$ bremsstrahlung spectrum. In their spectrometer, Andersson et al. found it necessary to use a thin aluminum window in front of the detector. As Lee points out in his analysis, however, such a window does not assist in discriminating against the breamsstrahlung and was presumably necessitated by the incandescence of the electron emitter. This problem was overcome by Morar,[43] who employed a field emission array as a dark electron source, coupled with a nude solid-state detector.

Single field-emitter electron sources have previously been used in Auger electron appearance-potential spectroscopy,[44] but the current that can be drawn from a single tip is generally below a microampere, which is inadequate for most purposes. Generally speaking, soft-x-ray appearance-potential spectroscopy using solid-state detection seems to require about $100\,\mu A$ or primary current. Experience with the new field emission arrays, which are produced by microcircuit techniques,[38] is limited, but they represent a potentially important step forward in the technology of electron-beam analysis.

4.4.5. Auger Electron Appearance-Potential Spectroscopy

As we pointed out in Section 4.2.2, in the soft-x-ray region, excited core states are much more likely to decay by an Auger process than by radiative recombination. It is not surprising, therefore, that appearance-potential spectra can also be obtained from changes in the total secondary-electron yield, in which case the technique is termed Auger electron appearance-potential spectroscopy (AEAPS).[45]

The apparatus required to measure the secondary-electron yield is quite simple. As shown schematically in Fig. 14, it may consist only of a thermionic emitter and an anode with a pinhole. Electrons passing through the pinhole are allowed to impinge on the sample. If the anode potential V_0 is

---

[40] S. Andersson, H. Hammarqvist, and C. Nyberg, *Rev. Sci. Instrum.* **45**, 877 (1972).
[41] S. Andersson and C. Nyberg, *Surf. Sci.* **52**, 489 (1975).
[42] R. N. Lee, *Rev. Sci. Instrum.* **48**, 1603 (1977).
[43] J. Morar, Ph.D. Thesis, Univ. of Maryland, College Park, 1981.
[44] Y. Fukuda, W. T. Elam, and R. L. Park, *Phys. Rev. B* **16**, 3322 (1977).
[45] J. E. Houston and R. L. Park, *Phys. Rev. B* **5**, 3808 (1972).

4.4. APPEARANCE-POTENTIAL SPECTROSCOPY

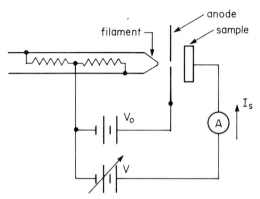

FIG. 14. Schematic diagram of experiment to measure the secondary-electron yield. Secondary electrons are collected on the anode, which is fixed to a potential V_0 greater than the sample potential V. The primary current is essentially constant, therefore changes in secondary yield are accurately reflected by changes in the sample current.

held fixed, the current striking the sample is independent of the emitter-sample potential V. If $V_0 > V$, secondary electrons from the sample will be collected on the anode. Thus the secondary yield $Y(E)$ is just

$$Y(E) = 1 - I_s(E)/I_p, \qquad (4.4.21)$$

where $I_s(E)$ is the net sample current flowing through the measurement impedance Z, and I_p is the constant primary current passing through the aperture. The primary-electron energy E relative to the Fermi energy of the sample is given by

$$E = eV + e\phi_c + kT, \qquad (4.4.22)$$

where $e\phi_c$ is the work function of the emitter. The general features of $Y(E)$ have been known for many years and are remarkably similar for metals, semiconductors, and insulators. The yield rises smoothly at low energies and reaches a maximum at several hundred electron volts, after which it slowly decreases. For most materials, the yield crosses unity at an energy of 50–200 eV and again between 1 and 2 keV. The similarity of such yield plots for very diverse materials was emphasized by Baroody[46] who plotted so-called reduced yield curves in which the yield is normalized to its maximum value and the energy is divided by the energy at which the maximum occurs. Such plots are remarkably similar for quite different materials, which would seem to argue that the total yield spectrum is insensitive as a means of characterizing the surface.

[46] E. M. Baroody, *Phys. Rev.* **78**, 780 (1950).

The derivative of the yield, however, reveals a multitude of fine structure, an example of which is shown in Fig. 15. The structure can be divided into two types: (1) At low energies there are very strong oscillations in the derivative of the secondary-electron yield, resulting from diffraction of the incident electrons. In effect, if there are a large number of states available in the solid at the momentum of the incident electron, the electron tends to be transmitted. If there are fewer states available, it increases the elastic reflection coefficient. Depending on the crystallinity of the sample and its Debye temperature, these diffraction features usually damp out by 400–500 eV. (2) At energies where the diffraction features are sufficiently damped, the appearance-potential features due to the excitation of core states are clearly evident.

If one neglects the obscuring effect of diffraction at low energies, the sensitivity of the Auger electron appearance-potential technique is roughly comparable to the soft-x-ray technique when a solid-state detector is used. The advantage of a high Auger yield is largely offset by the very large background of unrelated secondary-emission events.

The principal disadvantage of the Auger electron appearance-potential technique, however, is the diffraction effect at low energies, which on crystalline samples tends to obscure the K-shell edges of the light elements that are so important in chemisorption studies. The obscuring effects of diffraction are even more disastrous when one attempts to exploit the extended fine

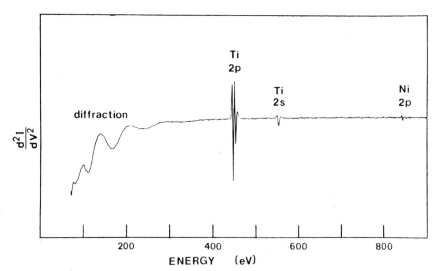

FIG. 15. Second derivative of the secondary-electron yield of a titanium–nickel alloy. The structure at low energies is a consequence of diffraction of the incident electron beam. In more crystalline materials the diffraction structure is even more pronounced.

4.4. APPEARANCE-POTENTIAL SPECTROSCOPY

structure above the excitation edges, as we shall see in the section on extended fine structure analysis.

Because of its great sensitivity, however, it is advantageous to use the Auger electron appearance-potential technique for certain classes of problems: in particular, studies of surfaces at high temperature, in which case the Debye–Waller effect suppresses the diffraction structure; studies of highly disordered or amorphous surfaces; and studies involving core levels with relatively high binding energies. Indeed, it is by no means clear that there is any practical limitation in the energies to which this technique can be extended. The sensitivity to the surface region, of course, diminishes with the energy and the technique becomes essentially a bulk probe.

4.4.6. Disappearance-Potential Spectroscopy

In all cases examined so far, the total secondary-electron yield increases at the threshold for core excitation. In 1973, however, Kirschner and Staib[32] observed that the elastic yield decreased above the critical potentials, which led them to refer to their technique as disappearance-potential spectroscopy (DAPS). Thus even as a new channel for secondary emission is opened by the excitation of a core level, so also a new channel is created for inelastic scattering. The observed fact that the total yield increases above the critical potentials is not therefore a result that could have been predicted with certainty. Indeed, it may prove not to be the case for all levels of all materials.

The elastic yield can be obtained by simply measuring the current to the fluorescent screen of a conventional spherical grid low-energy electron diffraction (LEED) system, in which case a retarding grid is biased as a high-pass filter to pass only quasi-elastically scattered electrons. The DAPS spectrum and the AEAPS spectrum have almost identical shapes but are inverted with respect to one another. By changing the potential on the retarding grid of the analyzer, more and more secondary electrons can be included in the measurement until finally all secondary electrons are collected, which is just the AEAPS spectrum. Clearly there will be some setting of the retarding grid at which the decrease in the elastic yield due to excitation of a core level is just offset by the increase in secondary emission as the core hole recombines. This experiment has been carried out for the 2p spectrum of titanium.[47] It was found that most of the secondary electrons contributing to the 2p AEAPS spectrum had energies below 30 eV; thus they are not contributed directly by the Auger recombination of the core hole but are mostly produced by secondary processes in stopping the Auger electrons. One conclusion that can be drawn from this is that the disappearance-potential spectrum should be somewhat more sensitive to the surface region than the

---

[47] M. L. denBoer, P. I. Cohen, and R. L. Park, *Surf. Sci.* **70**, 643 (1978).

Auger electron appearance-potential spectrum. An elegant confirmation of this occurs for the 2p spectrum from the basal plane of a titanium single crystal. As Feibelman et al.[48] have predicted, this surface exhibits a strong band of surface states lying at the Fermi energy. The unifilled portion of this band of surface states is clearly evident in the 2p appearance-potential spectrum. Calculations show that the surface states are contributed only by the surface layer. Comparing the DAPS and AEAPS spectra from this surface the relative contribution of the surface states in DAPS is clearly greater.[49]

Unfortunately, disappearance-potential spectroscopy suffers essentially the same limitation as AEAPS; that is, the spectrum is obscured up to energies of several hundred electron volts by structure arising from the diffraction of the incident electron beam. Indeed, it appears that the diffraction effects are carried almost entirely by the elastic secondaries.[50]

Disappearance-potential spectroscopy is also the only one of the appearance-potential techniques that requires the use of a dispersive analyzer. It should be emphasized, however, that the spectral resolution, as with the other appearance-potential techniques, is limited solely by the energy spread of the incident electrons. The only purpose of the dispersive analyzer is to suppress inelastic secondaries.

4.5. X-Ray Photoelectron Spectroscopy

The appearance-potential technique discussed in the previous section consists, in effect, of measuring the excitation probability of a core state as a function of the energy of a beam of incident electrons. It is, of course, possible to make a similar measurement using a monochromatic beam of soft x rays. Such measurements have been carried out for the surface region of solids, as we shall discuss in greater detail in Chapter 4.8 on extended fine structure analysis. However, there are two principal obstacles to this type of measurement: (1) the attenuation of the incident beam is much too slight to confine the measurement to the surface region, and (2) tunable sources of x-ray photons of sufficient intensity are both scarce and expensive. In contrast to ionization by electron impact, however, a photon absorbed by a core electron gives up all its energy, as is shown in Fig. 12. It is possible, therefore, to study the core-electron structure from the energy distribution of electrons ejected by soft x rays of a fixed energy.

Such measurements were carried out as early as 1914 by Robinson and

---

[48] P. J. Feibelman, J. A. Appelbaum, and D. R. Hamann, *Phys. Rev. B* **20**, 1433 (1977).
[49] B. T. Jonker, J. F. Morar, and R. L. Park, *Phys. Rev. B* **24**, 2951 (1981).
[50] M. L. denBoer, P. I. Cohen, and R. L. Park, *J. Vac. Sci. Technol.* **15**, 502 (1978).

Rawlinson.[51] Primary x rays of a single energy were found to produce what might be called a line spectrum of secondary electrons from the target, this line spectrum being made up of electrons from the different levels of atoms in the solid. This technique was revived by Siegbahn and co-workers[5] at the University of Uppsala, who developed sophisticated electron spectrometers for this purpose and demonstrated the utility of the technique for chemical analysis. Analysis is based primarily on an accurate determination of the core-electron binding energies, which are measurably shifted by changes in the distribution of valence electrons. Siegbahn shared the 1981 Nobel Prize in Physics for his contributions to this development.

4.5.1. Core-Hole Excitation

The x-ray photoelectron spectroscopy (XPS) technique is illustrated by the energy-level diagram of Fig. 16. X-ray photons of known wavelength are allowed to impinge on the sample surface. If the photon energy $h\nu$ exceeds

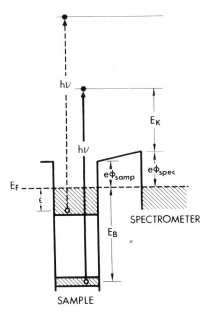

FIG. 16. Energy-level diagram of the x-ray photoelectron experiment. The absorption of an x-ray photon of energy $h\nu$ by a core electron with binding energy E_B relative to the Fermi level results in a photoelectron of energy $E_K = h\nu - E_B - e\phi_{spec}$. The work function of the spectrometer is not a well defined quantity, and spectrometers are generally calibrated with respect to lines of known energy.

[51] H. Robinson and W. F. Rawlinson, *Philos. Mag.* **28**, 277 (1914).

the binding energy E_B of a core electron, it may be absorbed by the electron, which is then excited into an available state above the Fermi energy. In contrast to the case of excitation by electron bombardment, incoherent scattering of the incident photons is negligible, that is, the full energy of the photon is imparted to the electron. If the energy of the ejected core electron exceeds the work function of the sample, it may be emitted into the vacuum. The XPS experiment consists of determining the kinetic energies of these photoelectrons.

The overwhelming majority of XPS measurements are obtained using unmonochromatized K_α radiation from magnesium or aluminum targets. The peak of the unresolved $K_{\alpha_{1,2}}$ doublet of magnesium occurs at 1254.6 eV and has a full width at half-maximum of about 0.8 eV. The aluminum $K_{\alpha_{1,2}}$ peak lies at 1486.6 eV and has a natural width at half-maximum of about 0.9 eV. The reasons for the almost exclusive use of these two x-ray anode materials are quite simple. The width of the line is determined in part from the lifetime broadening of the 1s hole and the unresolved 2p doublet from which the hole is filled. For higher-Z elements, the spin–orbit splitting of the 2p doublet increases, as does the lifetime broadening of the 2p and 1s levels. Of the elements with lower Z, most cannot be used because the 2p levels from which the 1s hole fills merge with the valence band. Sodium, at $Z = 11$, would provide a suitable x-ray line, but it is simply not practical to fabricate as an x-ray anode.

The x rays from the anode, of course, are not limited to the $K_{\alpha_{1,2}}$ doublet, which results from filling a 1s hole from the 2p levels. There is in addition a relatively broad K_β line corresponding to the filling of the 1s hole from the valence band. For aluminum the K_β emission band is about 80 V above the $K_{\alpha_{1,2}}$ doublet. The peak intensity of the K_β band, however, is only about 1% that of the $K_{\alpha_{1,2}}$. The satellite lines resulting from multiple ionization of the atoms of the anode are a more serious problem. The K_{α_3} and K_{α_4} satellites are particularly troubling (see Fig. 17). Depending on the electron bombardment energy of the x-ray anode, their peak intensities can be more than 15% of the intensity of the $K_{\alpha_{1,2}}$ doublet. All of these lines are, of course, superimposed on a bremsstrahlung background, which should exhibit the $1/h\nu$ dependence described in the previous section.

In most commercial x-ray photoelectron spectrometers, the x rays are unfiltered, and care must always be taken that photoelectrons excited by satellite lines are not mistaken for weak photoelectron peaks produced by the primary $K_{\alpha_{1,2}}$ radiation. In some spectrometers, however, this problem is eliminated by the use of a bent crystal monochrometer.[52] The use of mono-

---

[52] K. Siegbahn, D. Hammond, H. Fellner-Feldagg, and E. F. Barnett, *Science (Washington, D.C.)* **176**, 245 (1972).

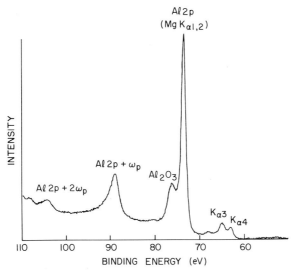

FIG. 17. The Al 2p x-ray photoelectron spectrum obtained with unmonochromatized Mg K_α radiation. A chemically shifted 2p line reveals the presence of Al_2O_3 on the surface. Additional satellites on the high-binding-energy side of the main peak are plasmon loss replicas. On the low-binding-energy side, peaks resulting from Mg $K_{\alpha_{3,4}}$ radiation are evident. (The spectrum was taken by C. R. Anderson of the Naval Surface Weapons Center.)

chromatized x-rays also results in an improvement of the achievable resolution. When coupled with a dispersion-compensated electrostatic analyzer, this is capable of reducing the effective line width to about 0.25 eV, although at this resolution the sensitivity is greatly reduced.

In addition to the photoelectron peaks, the electron spectrum from a surface will exhibit characteristic lines due to Auger recombination of the core holes produced by the incident radiation (see Fig. 18). The Auger spectrum will be discussed in greater detail in Chapter 4.6. On the high-binding-energy side of each photoelectron and Auger feature, the "background" rises as a result of secondary-electron processes (Chapter 4.3).

Synchrotron radiation, especially from the large electron–positron storage rings, provides an intense broadband source of radiation from which a narrow line can be extracted by a monochromater. Such sources have a number of inherent advantages in addition to high intensity, not the least of which is that the experimentalist is not confined to work with photons of just one or two energies. In addition, synchrotron radiation is polarized, which is an advantage for certain types of measurement. The use of such sources, however, is by no means routine and for the foreseeable future most XPS measurements will continue to be made with commercial spectrometers utilizing magnesium and aluminum x-ray sources.

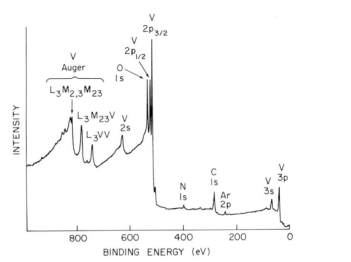

FIG. 18. X-ray photoelectron spectrum from a vanadium surface taken over a wide range. In addition to the photoelectron peaks of vanadium and various surface contaminants, the L Auger spectrum is evident. The Auger spectrum may in some cases interfere with the direct photoelectron spectrum. (The spectrum was taken by C. R. Anderson of the Naval Surface Weapons Center.)

For most purposes, therefore, the resolution in x-ray photoelectron spectroscopy is limited by the width of the magnesium K_α doublet to about 0.8 eV.

4.5.2. Electron Spectroscopy

The kinetic energies of the electrons ejected as a result of the absorption of the incident x rays are determined by means of a dispersive analyzer in which the electron trajectories are deflected by electrostatic or magnetic fields, the strength of the field required to produce a given deflection being a measure of the initial kinetic energy of the electrons. Properly speaking, of course, such dispersive analyzers select on the basis of momentum rather than energy. This is not a trivial consideration. It means, among other things, that the resolution of the spectrometer is dependent on the spatial extent of the electron source and that the accuracy of the technique is dependent on the precise positioning of the source, the absence of stray electric or magnetic fields, including the fields produced by nonuniform work functions of the spectrometer surfaces, dimensional inaccuracies, etc.

The earliest x-ray photoelectron spectrometer, used by Robinson and Rawlinson[51] prior to 1914, used a magnetic field to deflect electrons ejected through a slit. The entire spectrum was recorded simultaneously on a photographic film. The modern revival of x-ray photoelectron spectroscopy by Siegbahn *et al.* at Uppsala also relied initially on magnetic spectrometers originally developed for nuclear spectroscopy.[5] Magnetic spectrometers have been analyzed in detail by Siegbahn[53] but are rarely used today because of the difficulties of operation at relatively low energies. These difficulties result primarily from stray magnetic fields.

The simplest form of electrostatic analyzer is the spherical-grid retarding-potential analyzer shown schematically in Fig. 19, in which the electron source is at the center of curvature of a set of concentric spherical-grid segments. The first grid is at the same potential as the electron source and simply provides a field-free region in which the electrons follow straight trajectories. The second, or retarding, grid is at some potential V_R below the sample potential, which we shall take to be ground. The retarding grid forms a high-pass filter, in which electrons of energy greater than eV_R will pass through the grid if they approach it at normal incidence. Slower electrons will be rejected. Often two concentric grids, electrically connected, are used as the retarding grid to avoid field penetration. If the electrons passing through the retarding grid are collected by a spherical collector, the collector current represents the integral under the electron energy distribution curve for energies greater than eV_R. A derivative of the collector current as a function of V_R therefore yields the true secondary-electron energy distribu-

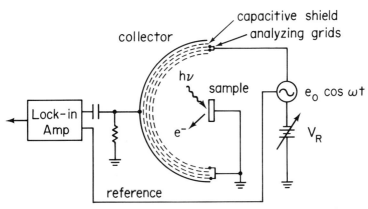

FIG. 19. Spherical-grid retarding-potential analyzer. This spectrometer is a high-pass filter with the pass energy set by the analyzing grids. The energy distribution is given by the derivative of the collector current as a function of pass energy. This is accomplished by potential-modulation differentiation. Such spectrometers are widely used in electron-excited Auger spectroscopy, but infrequently in photoelectron spectroscopy.

[53] K. Siegbahn, "Alpha-Beta and Gamma-Ray Spectroscopy." North-Holland Publ., Amsterdam, 1965.

tion.[54] The derivative is generally taken by the potential modulation technique described in Section 4.4.2, with the potential modulation superimposed on the retarding grid potential V_R. An additional grid between the retarding grid and the collector serves as a capacitive shield to prevent direct capacitive coupling of the retarding grid modulation to the collector.

The principal advantage of the spherical-grid retarding-potential analyzer is the very high luminosity that can be obtained, since the photoelectrons can be collected over the entire half-solid angle. This would represent a considerable advantage over other types of deflection analyzers if the spectrum truly consisted of discrete lines corresponding to electrons from the various levels of the atoms. In fact, however, associated with each discrete line, there is a complete distribution of secondary electrons resulting from photoelectrons that failed to escape the solid without an inelastic collision. Generally speaking, the number of these secondary electrons vastly exceeds the number contained in the sharp lines, as discussed in Section 4.3.1. This has a profound effect on the signal-to-noise ratio, since the signal current at a given setting of the retarding potential consists of electrons with energies just above or just below the pass energy, whereas the current contributing to the shot noise includes all the electrons with energies above the pass energy. For this reason, retarding-potential analyzers have found relatively little use in x-ray photoelectron spectroscopy, although, as we will see, they have found widespread use in electron-excited Auger electron spectroscopy.

To overcome the noise problems inherent in a high-pass filter, an electrostatic deflection analyzer can be used, and today virtually all x-ray photoelectron spectrometers are electrostatic deflection capacitors of one sort or another. The principle of the capacitor analyzers is most simply described with reference to the parallel-plate analyzer introduced by Harrower.[55] Electrons leaving a source at an angle of $\pi/4$ pass through an aperture in the positive plate of a capacitor. These electrons travel in parabolic paths and are refocused upon returning to the lower plate. The horizontal distance each electron travels is determined by its initial kinetic energy and the applied field. Consequently, those electrons able to pass through the second slit and reach the collector will have been selected according to their initial velocity. There is a linear relation between the voltage applied between the plates and the velocity of an electron able to pass through both apertures. If the voltage between the plates is varied linearly, the current reaching the collector will represent the energy distribution of electrons from the source. As with all capacitor-type analyzers, the resolution of the parallel-plate analyzer can be

---

[54] D. A. Huchital and J. D. Rigden, "Electron Spectroscopy." North-Holland Publ., Amsterdam, 1972.

[55] G. A. Harrower, *Rev. Sci. Instrum.* **26**, 850 (1955).

given as a fixed percentage of the electron energy. Thus, as Harrower demonstrated, the resolution for the case in which all electrons entered the first aperture at $\pi/4$ is given by

$$\frac{\Delta E}{E} = \frac{\Delta X_1 - \Delta X_2}{X_0} \tag{4.5.1}$$

It has subsequently been shown by Green and Proca[56] that for an entrance angle of $\pi/6$ the addition of a field-free section to the normal parallel-plate analyzer will yield second-order focusing (Fig. 20a).

Despite its simplicity, the simple parallel-plate analyzer has not been widely used because of the small solid angle that it accepts. Two modifications of the geometry can be easily imagined to increase this acceptance. In the fountain analyzer, both the entrance and exit apertures are extended into concentric annular slits, centered about a normal passing through the source. This geometry, first introduced by Edelman and Ulmer,[57] is quite simple to construct and accepts a relatively large solid angle. Although it probably deserves to be more widely used, its principal drawback is that it requires a large ring detector, which makes single particle detection impractical.

A more frequently used modification of the parallel-plate analyzer is the cylindrical-mirror spectrometer (Fig. 20b) first used by Blauth[58] in 1957. Its focusing properties, which were described by Sar-El,[59] include second-order focusing. As with the fountain analyzer, the cylindrical-mirror analyzer makes use of the full 2π azimuthal angle and thus has a relatively high luminosity. Its great advantage over the fountain analyzer is that the electrons are focused to a point, thus simplifying the problem of electron detection.

Preretardation of the electrons can be applied to the cylindrical-mirror spectrometer and indeed most instruments today are operated in this fashion. As with all dispersive analyzers, the cylindrical-mirror analyzer is susceptible to stray magnetic fields. Its geometry, however, makes it relatively simple to eliminate stray magnetic fields by mumetal shielding rather than the necessity of using compensating coils. To increase the resolution, it is possible to simply add another section to the cylindrical-mirror analyzer, and commercial x-ray photoelectron spectrometers utilizing a double-pass cylindrical-mirror analyzer are available. As with the parallel-plate analyzer, the problem of fabrication of the cylindrical-mirror analyzer is not severe.

---

[56] T. S. Green and G. A. Proca, *Rev. Sci. Instrum.* **41**, 1409 (1970).
[57] F. Edelman and K. Ulmer, *Z. Angew. Phys.* **18**, 308 (1965).
[58] E. Blauth, *Z. Phys.* **147**, 228 (1957).
[59] H. Z. Sar-El, *Rev. Sci. Instrum.* **38**, 1210 (1967).

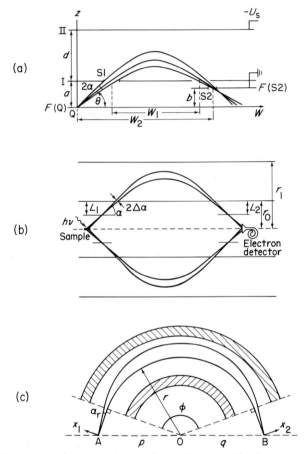

FIG. 20. Three types of electrostatic capacitor analyzers used in electron spectroscopy: (a) The parallel-plate analyzer. For an entrance angle of $\pi/6$, the addition of a field-free section yields second-order focusing. (b) The cylindrical-mirror analyzer, which makes use of the full 2π azimuthal angle. It is frequently operated with a retarding stage at the input. (c) The spherical-capacitor analyzer. It also can be constructed to accept a full 2π azimuth, but is more frequently constructed as a spherical segment.

One complication results from the problem of stray fields at the ends of the cylinders. At the source end in particular the cylinders must be truncated to enable the sample to be illuminated conveniently. This problem is generally corrected by the use of resistive end plates, which give a properly graded electric field. Problems of imperfect geometry are usually minimized by using a retarding input stage and keeping the pass energy of the analyzer fixed. Thus, in a sense, it is really a retarding analyzer in which the deflection stage serves only to reduce the shot noise associated with the high-pass filter.

4.5. X-RAY PHOTOELECTRON SPECTROSCOPY

The other widely used and commercially available electrostatic analyzer is the spherical-capacitor analyzer, first used by Purcell[60] in 1938. As with the cylindrical-mirror analyzer and the fountain analyzer, the spherical-capacitor analyzer can be constructed to take advantage of the entire 2π azimuthal angle. As shown in Fig. 20c, the electron source, center of curvature and image in the spherical capacitor lie in a straight line. Therefore, as with the cylindrical mirror, the entire structure can be rotated around this line, thus utilizing the entire azimuthal angle. It is, however, more frequently used as a spherical segment.

4.5.3. Chemical Shifts

The principal application of x-ray photoelectron spectroscopy is to identify the chemical environment of an element by comparison of its core-level binding energies with those of a set of reference compounds involving the same element. The shift in core-level binding energy from one oxidation state to another is generally of the order of a few electron volts (see Fig. 17). The factors producing this shift are quite simple, although as we shall see, its quantitative interpretation may not be. The energy of an electron in a core state is determined by the attractive potential of the nuclei and the repulsive Coulomb interaction with all of the other electrons of the system. A change in the chemical environment of a particular atom results in a redistribution of the valence charges. The resulting binding energy difference of a core-level of an atom in two different compounds, designated a and b, was described by Gelius[61] by a simple two term equation:

$$\Delta E = k \left(\frac{q_a}{r_a} - \frac{q_b}{r_b} \right) + \left(\sum_{i \neq a} \frac{q_a}{r_{ai}} - \sum_{i \neq b} \frac{q_b}{r_{ab}} \right). \quad (4.5.2)$$

The first term in the equation describes the difference in the electron–electron interaction between the core orbital and the valence charge. The coupling constant k is the two-electron integral between core and valence electrons. The second term represents a Madelung potential resulting from the other ions of the material. Although Eq. (4.5.2) is helpful conceptually, it often fails even to predict the sign of the change.

The difficulty is not that the model represents an incorrect description of the core electron energies. Rather, it is that it is a ground state model, and as we pointed out in the Introduction, the ground state of a system cannot be viewed directly. Therefore to properly determine the binding energy of the core electron, the very considerable energy involved in the rearrangement of the conduction electrons to screen the suddenly created core hole must be

---

[60] E. M. Purcell, *Phys. Rev.* **54**, 818 (1938).
[61] U. Gelius, *Phys. Scr.* **9**, 133 (1974).

included[62]. The screening lowers the energy of the core-hole state and therefore lowers the measured binding energy as well. Since this relaxation energy E_R will also depend on the chemical environment of the atom, an additional term representing the change in relaxation energy must be added to Eq. (4.5.2). As with the ground state energy of the core electron, the relaxation energy can be separated into two parts: the intra-atomic relaxation energy, which represents the rearrangement of the atom's own electrons, and the extra-atomic relaxation energy, corresponding to the movement of charge in neighboring atoms.

It is apparent, therefore, that to calculate the chemical shift, it is necessary at a minimum to know the lengths and angles of bonds to neighboring atoms. For the most interesting surface cases, this is information that is often unavailable. In other cases, the application of x-ray photoelectron spectroscopy is quite straightforward. The 2p binding energy of aluminum, for example, is increased by about 2.7 eV in going from the pure metal to Al_2O_3 (Fig. 17). This shift is easily resolved in XPS, with the result that the presence of Al_2O_3 on an aluminum surface can be readily detected.

In other cases, however, the chemical shift may be much more subtle. Atoms in the outermost layer of a metal reside in a different chemical environment than their bulk counterparts simply by virtue of their reduced coordination. This produces a "surface chemical shift" between the energies required to excite a core electron in an atom at the surface and one in the bulk. As with other chemical shifts, there are two contributions: a change in the position of the core-level eigenvalue, and a change in the screening energy when the valence electrons relax around the resulting core hole. Laramore and Camp[63] predict that a core hole at the surface is screened more effectively than one in the bulk. This is due to the two-dimensional nature of the charge fluctuation produced by the surface plasmons.

Experimental verification of a surface chemical shift is difficult for two reasons: (a) The predicted shift is of the same order as the widths of experimentally accessible core levels. Thus the surface contribution to the core-level spectra may not be resolved from the bulk contribution. (b) The predicted shift is of the same order as the uncertainty in the determination of absolute binding energies. Thus to ascertain the shift by comparison of bulk-sensitive spectra with surface spectroscopies requires that they be carried out *in situ* using the same reference standard.

This latter approach was followed by Houston *et al.*,[64] who compared electron-excited soft x-ray appearance potentials, which are surface sensitive, with photon-excited Auger electron appearance potentials, which have

[62] A. Barrie, *Chem. Phys. Lett.* **19**, 109 (1973).
[63] G. E. Laramore and W. J. Camp, *Phys. Rev. B* **9**, 3270 (1974).
[64] J. E. Houston, R. L. Park, and G. E. Laramore, *Phys. Rev. Lett.* **30**, 846 (1973).

little surface contribution. A single tungsten thermionic electron source served as a reference for both measurements.

It should, of course, also be possible to determine the surface chemical shift from x-ray photoelectron spectroscopy. If, however, the contribution from the surface atoms is unresolved from that of the bulk, it will result only in an asymmetry of the photoelectron peak, and as we shall see in the next section, there are a number of effects producing asymmetries in the photoelectron peaks.

By detecting photoelectrons at different takeoff angles, however, Citrin et al.[65] were able to demonstrate the presence of a surface-shifted component in the gold $4f_{7/2}$ photoelectron spectrum. This observation was made using monochromatized aluminum K_α radiation for the excitation.

The higher resolution and greater sensitivity possible with synchrotron radiation enabled fully resolved, surface-shifted components to be observed in a number of elements.[66] In particular, the $4f_{7/2}$ levels of the 5d transition metals are narrow and readily accessible to synchrotron-radiation-excited photoemission spectroscopy. Rosengren and Johansson[67] relate the surface shifts to surface energies of the 5d elements. Their calculated shifts show a strong dependence on the surface structure as observed by van der Veen et al.,[68] and accounts in a simple way for the change of the sign of the shift through the 5d series.

4.6. Comparison of Binding Energy Measurements

The widespread use of x-ray photoelectron spectroscopy as an analytical tool stems from its ability to detect chemical shifts of a fraction of an electron volt in core electron binding energies of up to 1500 eV. The reproducibility of binding energy measurements on a given instrument is generally within 0.1 eV, which is sufficient for most chemical shift measurements. On an absolute scale, however, discrepancies between binding energy measurements reported by different laboratories for presumably identical samples are far greater, as demonstrated by the Round Robin Study organized under the auspices of the American Society for Testing and Materials. The results of this round robin study, which were reported by Powell et al.,[69] reveals a surprisingly large spread in binding energy determinations by different labo-

---

[65] P. H. Citrin, G. K. Wertheim, and Y. Baer, *Phys. Rev. Lett.* **41**, 1425 (1978).
[66] D. E. Eastman, F. J. Himpsel, and J. F. van der Veen, *J. Vac. Sci. Technol.* **20**, 609 (1982).
[67] A. Rosengren and J. Johansson, *Phys. Rev. B* **22**, 3706 (1980).
[68] J. F. van der Veen, F. J. Himpsel, and D. E. Eastman, *Phys. Rev. Lett.* **44**, 189 (1980).
[69] C. J. Powell, N. E. Erickson, and T. E. Madey, *J. Electron Spectrosc. Relat. Phenom.* **17**, 361 (1979).

ratories and different instruments. These results have generally been regarded as demonstrating a need for standard calibration techniques. Since x-ray photoelectron spectroscopy depends on a dispersive analyzer, it is actually the momentum of the ejected core electron that is measured rather than kinetic energy, and thus measurements are generally referenced to some standard photoelectron line. A second calibration point is provided by the Fermi level of the sample. For a metal, the most energetic electron in the photoelectron spectrum is presumably ejected from a level lying at the Fermi energy. Since the energy of the incident photons is accurately known, it should be possible to measure binding energies relative to this point in the spectrum. This assumes that errors due to imperfect geometry or stray fields can be corrected for by a linear function. Lee[70] has concluded, however, after an analysis of the round robin results, that a linear response function is not an adequate representation of most instruments and that indeed the nonlinearities often appear most severe at low binding energies, that is, in the vicinity of the Fermi level. If the problem is nonlinearities, it is clear that the size of the error depends on the energy separation between the measured photoelectron peak and the peak used as a calibration reference.

This difficulty can be overcome by introducing a reference electron source into the XPS instrument, as Powell and Jach[71] and Anderson et al.[72] have demonstrated. The method of Anderson et al., which they term FRESCA, for field-emission-referenced electron spectroscopy for chemical analysis, has the additional refinement that the electron source is a field emission tip, thus eliminating any uncertainty over the work function of the electron source, since the most energetic electrons tunnel directly from the Fermi level of the field emitter.

4.7. Electron-Excited Auger Electron Spectroscopy

Electron-excited Auger electron spectroscopy is probably the most widely used of all surface analytical techniques. It has become the accepted standard for establishing surface cleanliness as a starting point for almost every surface experiment.

In principle, any incident particle of energy greater than the binding energy of an inner shell electron can excite that electron into an unoccupied

---

[70] R. N. Lee, to be published.
[71] C. J. Powell and T. Jach, *J. Vac. Sci. Technol.* **20**, 625 (1982).
[72] C. R. Anderson, R. N. Lee, J. F. Morar, and R. L. Park, *J. Vac. Sci. Technol.* **20**, 617 (1982).

4.7. ELECTRON-EXCITED AUGER ELECTRON SPECTROSCOPY

state above the Fermi level. The core vacancy left behind will be filled by an electron from a higher level, as the atom, in a series of transitions, convulses its way back to the ground state. This reorganization is generally independent of the mode of excitation because the decay time is long compared to the excitation time. Energy is conserved in the decay transitions by the emission of x-ray photons or Auger electrons. As discussed in Section 4.2.2, for energy levels of most interest in surface research, the decay of a core hole is overwhelmingly likely to be by the Auger process.

Auger electrons were first identified in secondary-electron energy distributions of electron-bombarded surfaces by Lander[73] in 1953, but only the most intense Auger transitions could be detected above the background of secondary electrons resulting from the interaction of the electron beam with the valence-electron fluid of the solid. Electron-excited Auger electron spectroscopy thus did not appear to offer a sensitive means of surface analysis until Harris[74] demonstrated that the Auger emission features could be greatly enhanced by simply taking the derivative of the secondary-emission spectrum, as shown in Fig. 5b.

The acceptance of Auger spectroscopy was made easier because many surface physicists already had spherical-grid low-energy-electron diffraction systems that could be converted to retarding-potential energy analyzers by the addition of external electronics.[75] Because of the obvious advantages of coupling LEED with Auger spectroscopy, the retarding analyzer remains an important instrument. As pointed out in Section 4.5.2, however, the spherical-grid retarding-potential analyzer has an inherently high shot noise level, and today most Auger electron spectra are obtained with cylindrical-mirror analyzers, which are commercially available. Such analyzers frequently have an integral electron gun contained within the inner cylinder. These analyzers permit the spectrum to be scanned in a relatively short period of time.

4.7.1. The Auger Transition Energies

The acquisition of the Auger spectrum may be far simpler than its interpretation. The complexity results from the fact that an Auger line represents term differences between three levels. For example, a transition labeled KL_1L_2 refers to an initial vacancy in the K-shell (1s) that undergoes a transition to holes in the L_1 (2s) and L_2 ($2p_{1/2}$) shells plus an Auger electron. For a heavy element, the L-shell Auger spectrum alone consists of hundreds of

---

[73] J. J. Lander, *Phys. Rev.* **91**, 1382 (1953).
[74] L. A. Harris, *J. Appl. Phys.* **39**, 1419 (1968).
[75] R. E. Weber and W. T. Peria, *J. Appl. Phys.* **38**, 4355 (1967).

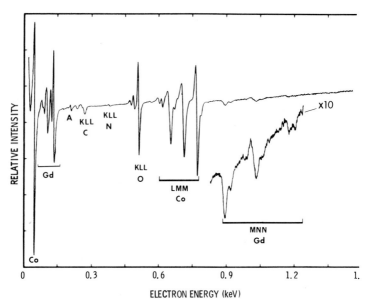

FIG. 21. Electron-excited Auger-electron spectrum of a GdCo$_5$ alloy surface taken with a cylindrical-mirror analyzer. Groups of Auger lines can sometimes be identified, but individual transitions often cannot, except in the case of very low-Z elements. (The spectrum was taken by R. N. Musket of Sandia Laboratories.)

lines, the energies of which are not susceptible to precise first-principle calculations. As a result, it it generally possible to identify spectral groups, as shown in Fig. 21 but not individual features within those groups. Serious ambiguities therefore arise in the identification of Auger transitions for all but the lightest elements, and elemental analysis is generally based on matching spectra against "standard" plots taken from samples of known composition.

The theory of the KLL transitions has been treated more extensively than that of any other series. In electron-excited Auger spectroscopy of surfaces, however, the incident electrons usually have an energy of 3 keV or less. This is less than the K-shell binding energy of elements above $Z = 17$. For these light elements, the coupling can be taken as pure $l-s$, which results in only five lines, as compared to the nine lines predicted by intermediate coupling. This has been confirmed experimentally for gases. The L Auger spectrum is far more complex than the K spectrum, and theoretical calculations taking into account relativistic and intermediate coupling effects have not been carried out. The result is that serious ambiguities arise in the identification of L Auger peaks.

4.7. ELECTRON-EXCITED AUGER ELECTRON SPECTROSCOPY

The energy of the Auger electron emitted as a result of the transition ABC can be written in terms of the binding energies as

$$E(ABC) = E(A) - E(B) - E(C) + U(BC), \quad (4.7.1)$$

where $U(BC)$ is the effective interaction energy of the final two-hole state. Matthews[76] has discussed approximate methods for calculating $U(BC)$ for metals, which involve a relaxation energy due to screening by the valence-electron fluid, in addition to atomic relaxation.[77] Since the relaxation energies reduce the total interaction energy, the Auger energies for atoms condensed in a solid are less than for free atoms.

For purposes of elemental identification, there is a simplification that is often helpful. For a given initial hole, the most energetic Auger electrons will be those for which the final-state holes are in the valence band (a core–valence–valence transition). In this case, for a free-electron-like metal, delocalization of the final-state charge reduces the interaction energy nearly to zero. Thus, the high-energy limit in a group of Auger transitions can often be identified with the binding energy of the level that was initially excited.

Characteristic loss replicas of the principal transitions displaced by the plasmon energy are an obvious feature of every core-level spectroscopy (Fig. 17). In Auger electron spectroscopy, they may result simply from Auger electrons that suffer a characteristic loss before escaping from the solid, in which case they are referred to as extrinsic Auger satellites. There is, however, also the possibility of generating intrinsic plasmons directly from the screening response of the valence electrons to the suddenly altered core potential. This direct coupling to the recombination of the core hole is physically indistinguishable from the two-step process, leading to the possibility of interference effects.

We have, of course, been assuming that the recombination of the core hole is independent of the mode of excitation, since the decay time is long compared to the collision time. Evidence for coupling between the excitation and recombination of a core state is the existence of plasmon gain satellites. Such satellites have been reported from time to time for many years but were generally found to be associated with Auger transitions involving a doubly ionized core level. The most likely candidates for plasmon gains would be free-electron metals in which the plasmon lifetimes approach the lifetimes of the core hole, and Fuggle et al.[78] have reported unambiguous evidence for weak plasmon gain satellites in KLL transitions of sodium and magnesium.

---

[76] J. A. D. Matthews, *Surf. Sci.* **89**, 596 (1979).
[77] D. A. Shirley, *Phys. Rev. A* **7**, 1520 (1973).
[78] J. C. Fuggle, R. Lasser, O. Gunnarson, and K. Schonhammer, *Phys. Rev. Lett.* **44**, 1090 (1980).

4.7.2. The Auger Line Shape

If one or both of the final-state holes in an Auger transition lies in the valence band, the shape of the spectral line corresponding to that transition should contain information concerning the local density of states in the region of the excited atom. In his pioneering paper on Auger spectroscopy, Lander[73] discussed the case in which both final-state holes lie in the valence band. In this case the energy distribution of Auger electrons must take into account all possible combinations of hole energies allowed by the conservation of energy. The transition density $T(E)$ is therefore given in this simple picture by the self-convolution of the valence-band density of states $N_v(E)$ broadened by the width of the core level, i.e.,

$$T(E) = \int_0^E N_{2v}(E')N_i(E + E_B - E')\,dE', \quad (4.7.2)$$

where

$$N_{2v}(E) = \int_0^E N_v(E')N_v(E - E')\,dE'. \quad (4.7.3)$$

Unfortunately there seem to be no cases for which this simple picture holds. For aluminum and silicon,[79] the width of the L–valence–valence (LVV) Auger line is about right, but the shape of the line is very different from that predicted by Eq. (4.7.2). In other cases, such as silver and copper,[80] the line shapes, as contrasted to their position, resemble more nearly that which would be expected from free atoms rather than from a solid. The reasons for the failure of the simple self-convolution theory are quite different in these two cases. Self-convolution fails for simple metals such as aluminum because of transition rates and screening problems.[81] For d band metals such as silver, it fails because the two holes are localized on a single site.[82]

The role of transition matrix elements in determining the shape of Auger lines is convincingly demonstrated for silicon, where the LVV Auger line shows the correct total width but whose shape bears little relation to the self-convolution of the total density of states. Although there are no rigid dipole selection rules in the case of Auger transitions, where angular momentum can always be conserved, explicit calculation shows that certain decay channels are strongly preferred. In particular, an initial 2p core hole is overwhelmingly likely to decay by a transition from a p-like state in the

---

[79] J. E. Houston, *J. Vac. Sci. Technol.* **12,** 255 (1975).
[80] C. J. Powell, *Phys. Rev. Lett.* **30,** 1169 (1973); L. Yin, I. Adler, T. Tsang, M. H. Chen, D. A. Ringers, and B. Craseman, *Phys. Rev. A* **9,** 1070 (1974).
[81] J. W. Gadzuk, *Phys. Rev. B* **9,** 1978 (1974).
[82] M. Cini, *Solid State Commun.* **24,** 682 (1977).

4.7. ELECTRON-EXCITED AUGER ELECTRON SPECTROSCOPY

valence band, and the self-convolution of the p-partial density of states gives a satisfactory fit to the LVV silicon Auger line.[83]

The situation is very different for a d band metal such as copper, where the shape of the Auger line is dominated by final-state correlation effects resulting from the localization of the two final-state holes on a single site. Cini[82] has shown that this results in a quasi-atomic behavior.

Some asymmetry in the Auger line shape will also result from the dynamic screening of the core hole by the valence electrons.[84] The effect is quite small compared to density-of-states effects, however, with the result, as Weissmann and Muller[85] have pointed out, that Auger electron spectroscopy provides a local probe of the density of states, however distorted that view may be by the effects discussed above.

4.7.3. AES and Surface Composition

Initially scientists were content to let Auger spectroscopy serve as a monitor of the cleanliness of experimental surfaces, and it is this application that has had the greatest impact on surface science. Clearly, however, if Auger spectroscopy could be used to quantitatively measure elemental abundances in the surface region rather than simply identify contaminants, it would open up whole new fields of inquiry.

The difficulty is that unless the structure of the surface is known, its composition cannot be defined. We cannot simply attach a percentage to each element as we do for homogeneous bulk samples, since for any interesting case the surface is necessarily inhomogeneous along its normal. Thus, except in cases for which the surface structure is known, as in certain examples of monolayer formation, Auger spectroscopy can provide at best a qualitative notion of elemental abundance. The same limitation of course applies to the other core-level spectroscopies we have discussed. Nevertheless, references to quantitative Auger analysis are fairly common and depend on the use of so-called elemental sensitivity factors.[86] The accuracy of this approach was tested on clean magnesium oxide samples that were believed on other grounds to be homogeneous in composition out to the surface.[87] Applying elemental sensitivity factors to the Auger spectrum, it was verified that the composition was made up of roughly equal numbers of magnesium and oxygen atoms. Unfortunately, this demonstrates only that, if the com-

---

[83] P. J. Feibelman, E. J. McGuire, and K. C. Pandey, *Phys. Rev. B* **15**, 2202 (1977).
[84] P. H. Citrin, G. K. Wertheim, and Y. Baer, *Phys. Rev. B* **16**, 4256 (1977).
[85] R. Weissmann and K. Muller, *Surf. Sci. Rep.* **1**, 251 (1981).
[86] C. C. Chang, *in* "Characterization of Solid Surfaces" (P. F. Kane and G. B. Larrabee, eds.), Plenum, New York, 1974.
[87] P. W. Palmberg, *J. Vac. Sci. Technol.* **13**, 214 (1976).

position of a sample is completely homogeneous, Auger electron spectroscopy can be used to perform a bulk analysis.

In some cases, of course, the structure of the surface may be quite well known, as in the study of certain chemisorbed monolayers. In such cases, Auger intensities may provide the most accurate means of determining surface coverage. If, for example, the maximum development of a c(2 × 2) structure on a square substrate can be associated with the formation of one-half of a monolayer, smaller amounts of the overlayer can be determined by simply scaling the Auger intensity. This assumes, of course, that at lower coverages, the adsorbed atoms are occupying the same kinds of sites.

4.8. Extended Fine Structure Analysis of Surfaces

Extended fine structure refers to a modulation in the excitation cross section of atomic core levels resulting from interference of the ejected core electron with backscattering from neighboring atoms. If the backscattered wave is in phase, excitation is aided. If it is out of phase, excitation is inhibited. Apart from corrections for the scattering phase shifts, therefore, periodicities in the excitation cross section as a function of the momentum of the ejected core electron are the reciprocal interatomic spacings. This fine structure, which may extend for hundreds of electron volts above an excitation threshold, has been observed for many years above x-ray absorption edges and has been used to obtain the interatomic spacings of bulk materials without the necessity of accurately modeling the structure.[88] It can be employed even for noncrystalline materials and, moreover, it has the unique advantage that the structure is associated with a single elemental constituent.

It is this latter characteristic that has motivated attempts to adapt extended fine structure analysis techniques to the study of the surface region, since it provides a means of sorting out the bond lengths of adsorbed atoms from substrate–substrate bond distances. It is the contribution of substrate scattering that ultimately limits the accuracy that can be obtained by low-energy electron diffraction.

There have been two approaches to adapting extended fine structure analysis to the surface region. Both approaches employ the short mean free path for inelastic scattering of low-energy electrons as a means of restricting observations to the near surface region. In one approach, referred to as surface-extended x-ray-absorption fine structure (SEXAFS), photoelectron yield, rather than a direct observation of x-ray absorption, is used as a measure of the core-excitation probability. In the second approach, called extended appearance-potential fine structure (EAPFS), electrons rather than

[88] E. A. Stern, D. E. Sayers, and F. W. Lytle, *Phys. Rev. B* **11**, 4836 (1975).

photons are used to create the core excitation. In this section we shall contrast these two approaches and describe in some detail the experimental problems involved.

4.8.1. Surface-Extended X-Ray-Absorption Fine Structure

If dipole selection rules are satisfied, the probability that an x ray will be absorbed by a core electron depends on the states available to the ejected electron. This is frequently taken to be just the one-electron density of states of the material, modulated by the oscillator strength of the transition. In fact, the ejected core electron should be represented as a spherical wave. Thus the plane-wave density of states is a good approximation only near the edge, where the wavelength is very long. If the wavelength of the ejected core electron is shorter than the interatomic separation, the inclusion of the backscattered components in the final-state wave function results in a sinusoidal modulation of the absorption coefficient as a function of k, where k is given by

$$k = [2m(\epsilon + E_\mathrm{I})]^{1/2}/h. \quad (4.8.1)$$

Here, E_I is the inner potential. The fractional change χ in the absorption coefficient as a function of k is given by

$$\chi(k) = \sum_j \frac{N_j f_j(k, \pi)}{kR_j^2} \exp(-2\sigma_j^2 k^2 - 2R_j/\lambda) \sin[2kR_j + \theta_j(k)]. \quad (4.8.2)$$

In this equation N_j is the number of neighboring atoms at a distance R_j from the absorbing atom; $|f_j(k, \pi)|$ is the amplitude for backscattering from one of these atoms; $\exp(-2\sigma_j^2 k^2)$ is a Debye–Waller-like factor accounting for thermal vibrations with a root-mean-square pairwise fluctuation σ_j; and $\exp(-2R_j/\lambda)$ represents the loss of ejected core electrons by inelastic scattering with a mean free path of λ. The factor $\sin[2kR_j + \theta_j(k)]$ is the interference term producing the modulations in the absorption cross section, where $\theta_j(k)$ is the energy-dependent phase shift.

Since the phase shifts are essentially an atomic property, they can, to a good approximation, be transferred from one system to another. If no suitable system is available for comparison, they must be determined by calculation. The inner potential, which affects the value of k, may also be determined by direct calculation, since the spacing determined by Fourier inversion of Eq. (4.8.2) is not very sensitive to the choice of the inner potential. Lee and Beni[89] suggest that the inner potential can be determined by requiring spacings to agree in both the absolute value and the imaginary part of the optical Fourier transform.

[89] P. A. Lee and J. Beni, *Phys. Rev. B* **15**, 2862 (1977).

The inversion of the extended fine structure is not as straightforward as it might seem from published accounts. Multiple scattering effects strongly influence spectra near the threshold and the inclusion of information in the first 50–100 eV of the spectrum is a questionable practice. In addition, the rapid decrease in the absorption coefficient with energy makes it difficult to obtain a data range extending more than 500 V above the threshold. It is, in addition, generally necessary to filter low-frequency components from the spectrum resulting from instrumental and other factors, and the choice of filter will have some influence on the measured spacings. In spite of these problems, however, the technique is capable of providing high-precision determinations of interatomic spacings in a bulk solid.

Except in very special cases of high-surface-area materials, however, the surface contribution to x-ray absorption is too slight to be utilized, and a different measure of core excitation probability must be found. One such measure is the yield of Auger electrons emitted in the recombination of the core hole. By using this approach Citrin *et al.*[90] were able to measure the silver–iodine bond length for iodine absorbed on the (111) surface of silver to an accuracy of a few hundredths of an angstrom. This was possible only through the use of the intense synchrotron x-ray source at SPEAR with the aid of a focusing mirror.

The selection of the iodine–silver adsorption system was by no means random. A principal limitation in EXAFS studies is imposed by overlapping spectra of the constituents. Thus extended fine structure above an absorption edge of one constituent may be prematurely terminated by an absorption edge of a second constituent. This limits the combinations of elements that can be successfully studied by this technique.

It was expected that utilizing the Auger yield would circumvent the problem of overlapping edges. Unfortunately, in addition to the characteristic Auger electron emission peaks in the photoelectron spectrum, there are the generally more intense direct x-ray photoelectron peaks whose energies are fixed not with respect to the Fermi level of the solid but rather with respect to the energy of the incident photons. Thus as the energy of the x rays incident on the sample is varied across the range of interest, photoelectron peaks may be swept through the fixed window of the electron energy analyzer, which is set to accept a particular characteristic Auger electron. This can produce an artifact that completely overwhelms the fine structure variations in the Auger signal. Measurements are therefore confined to systems for which there are no interfering photoelectron lines. In particular, this has made it difficult to use the technique for the study of low-Z adsorbates, such as oxygen and carbon, which are of particularly great practical interest.

To avoid the problem of interfering photoelectron lines, most SEXAFS

---

[90] P. H. Citrin, P. Eisenberger, and R. C. Hewitt, *Phys. Rev. Lett.* **41**, 309 (1978).

4.8. EXTENDED FINE STRUCTURE ANALYSIS OF SURFACES

measurements use the yield of true secondary electrons, rather than the yield of characteristic Auger electrons, to monitor the core excitation probability.[91] Much of the surface sensitivity is lost in taking this approach, since the characteristic Auger electrons must be inelastically scattered in order to contribute to the signal. Moreover, although it avoids the problem of interfering photoelectron lines, this approach does not avoid the problem of overlapping edges.

One very important advantage of using the synchrotron light source for excitation is that the synchrotron light is polarized. By comparing the amplitude of the interference terms for various polarization directions, information is provided on the adsorption site rather than just the bond length. This has been demonstrated by Brennan et al.[92] who used polarization-dependent SEXAFS to study the c(2 × 2) structure of sulfur on Ni(100). Their results are in good agreement with LEED determinations.

4.8.2. Extended Appearance-Potential Fine Structure

It is possible to utilize extended fine structure above appearance-potential thresholds in a manner analagous to extended x-ray-absorption fine structure. Indeed, as shown in Section 4.4.1, the derivative of the electron excitation function of a core state should resemble the x-ray excitation function. However, as Laramore has discussed in detail,[93] there are certain fundamental differences between the electronic excitation case and photoexcitation case that must be dealt with if the extended appearance-potential fine-structure (EAPFS) technique is to be utilized. In the photoexcitation case, dipole selection rules determine the partial-wave character of the outgoing electron. By contrast, in the case of electronic excitation, angular momentum can be conserved in a variety of ways, thus creating the prospect of a mixture of partial waves, each with its own k-dependent phase shift. Explicit calculations, however, indicate that a single partial wave will generally be dominant.

There are, of course, a variety of methods by which the electron bombardment excitation of a core level can be monitored. These methods were discussed in Chapter 4.4. It was pointed out that if the core excitation is monitored by the total, or elastic, secondary yield, the spectrum is obscured at low energies by structure resulting from the diffraction of the incident electrons. This diffraction structure damps out at higher energies, due to the Debye–Waller effect. Thus, if the sample temperature is sufficiently high, or if the structure is sufficiently disordered, the diffraction structure may be sufficiently suppressed to permit analysis of extended fine structure variations above core thresholds at a few hundred electron volts. The interaction

---

[91] J. Stohr, L. Johannson, I. Lindau, and P. Pianetta, *Phys. Rev. B* **20**, 664 (1979).
[92] S. Brennan, J. Stohr, and R. Jaeger, *Phys. Rev. B* **24**, 4871 (1981).
[93] G. E. Laramore, *Phys. Rev. B* **18**, 5254 (1978).

of oxygen with the (100) surface of aluminum, for example, produces a very noncrystalline layer at the surface, which has been analyzed by studying the extended fine structure variations in the elastic scattering yield above the oxygen K threshold.[94] The results yield an oxygen–aluminum spacing that agrees well with the larger of the two oxygen–aluminum spacings for bulk Al_2O_3. Based on the structure of the bulk oxide, this has been interpreted as evidence that the oxygen has penetrated below the aluminum surface. An attempt to make a similar study of silicon dioxide films thermally grown on the (100) surface of silicon, however, was unsuccessful. Although such films are generally characterized as noncrystalline, diffraction effects were nevertheless sufficiently pronounced as to obscure the extended fine structure above the oxygen k edge.

Diffraction of the incident electron beam, however, has essentially no effect on the soft-x-ray yield. It is thus possible to use the soft-x-ray appearance-potential technique even for well-ordered single crystal surfaces, the problem in this case being largely one of sensitivity. By using a nude solid-state detector together with a field-emission-array electron source, as discussed in Section 4.4.4, it was possible to obtain not only the oxygen–silicon spacing in a thin SiO_2 film thermally grown on a single crystal silicon surface but also, for the first time, the oxygen–oxygen spacing.[95]

Although the EAPFS technique, by utilizing a simple electron source in contrast to a synchrotron storage ring, is more convenient than SEXAFS, it is also generally more destructive and cannot be used on surfaces that are damaged by prolonged electron bombardment. Moreover it retains the principal limitation of SEXAFS, which is the problem of overlapping edges restricting the range over which data can be analyzed.

Very recent experiments, however, demonstrate that this limitation can be overcome in the electron excitation case by monitoring the Auger electron yield.[96] As we pointed out in the last section, attempts to use the Auger yield in photon excitation experiments were frustrated in most cases by photoelectron peaks being swept through the analyzer window. In the electron excitation case the corresponding features, which would be the core-level loss edges, are far weaker and can be almost completely eliminated by the modulation techniques described in Chapter 4.3. It seems likely that the Auger-monitored extended appearance-potential fine structure measurements, which can be carried out with a conventional cylindrical-mirror Auger spectrometer, will become widespread in the coming years.

[94] M. L. den Boer, T. L. Einstein, W. T. Elam, R. L. Park, L. D. Roelofs, and G. E. Laramore, *Phys. Rev. Lett.* **44**, 496 (1980).

[95] T. L. Einstein, M. L. den Boer, J. F. Morar, R. L. Park, and G. E. Laramore, *J. Vac. Sci. Technol.* **18**, 490 (1981).

[96] J. F. Morar and R. L. Park, to be published.

5. DIFFRACTION TECHNIQUES

By Max G. Lagally

Department of Metallurgical and Mineral Engineering
and Materials Science Center
University of Wisconsin
Madison, Wisconsin

5.1. Introduction

Probably the most important of all surface analysis techniques are those that give information about the surface structure. Apart from secondary and thermionic emission measurements, they are also the oldest. In December 1927, these introductory paragraphs appeared in an article in *Physical Review*:

"The investigation reported in this paper was begun as the result of an accident that occurred in this laboratory in April 1927. At that time, we were continuing an investigation . . . of the distribution in angle of electrons scattered by a target of polycrystalline nickel. During the course of this work a liquid-air bottle exploded at a time when the target was at a high temperature; the experimental tube was broken, and the target heavily oxidized by the inrushing air. The oxide was eventually reduced and a layer of the target removed by vaporization, but only after prolonged heating at various high temperatures in hydrogen and in vacuum. When the experiments were continued it was found that the distribution-in-angle of the scattered electrons had completely changed. . . . The most striking characteristic of [the observed strong beams in particular directions] is the one-to-one correspondence [that they] bear to the same beams that would be found issuing from the same crystal if the incident beam were a beam of x-rays."[1]

Thus was the diffraction of electrons discovered. It may seem that little has changed since that first low-energy electron diffraction (LEED) experiment. Vacuum accidents still happen (perhaps with less serendipitous returns) and Ni has continued to be a favorite material for LEED studies. More remarkable, however, is that most of the important features of the diffraction of

---

[1] C. J. Davisson and L. H. Germer, *Phys. Rev.* **30**, 705 (1927).

electrons were recognized within a few years of the Davisson–Germer paper.[1] Since then, there have been considerable refinements in the basic ideas (brought about especially by the advent of ultrahigh vacuum), many applications, and better methods of analysis and quantification, as well as a periodic rediscovery of the fundamental ideas. There has been an expansion to other energy ranges, leading to the acronyms MEED (medium-energy electron diffraction) and RHEED (reflection high-energy electron diffraction). There have been considerable instrumental developments,[2-12] as it became apparent that the resolving power and sensitivity of existing instruments were insufficient to meet the goals of the experiments being undertaken. There have been theoretical developments, particularly in the construction of dynamical theories of low-energy electron diffraction,[13-17] but also in the analysis of structural defects.[18-27] And finally, there has been the growing recognition that crystallographic information is necessary to interpret the output of other surface spectroscopies, with the consequence that a LEED or RHEED capability has become standard equipment on more or less all experimental systems for surface research.

[2] W. T. Sproull, *Rev. Sci. Instrum.* **4**, 193 (1933).
[3] W. Ehrenberg, *Philos. Mag.* **18**, 878 (1934).
[4] J. J. Lander, J. Morrison, and F. Unterwald, *Rev. Sci. Instrum.* **33**, 782 (1962).
[5] C. W. Caldwell, *Rev. Sci. Instrum.* **36**, 1500 (1965).
[6] R. L. Park and H. E. Farnsworth, *Rev. Sci. Instrum.* **35**, 1592 (1964).
[7] P. Heilman, E. Lang, K. Heinz, and K. Müller, *Appl. Phys.* **9**, 247 (1976); E. Lang, P. Heilman, G. Hanke, and K. Müller, *Appl. Phys.* **19**, 287 (1979).
[8] M. D. Chinn and S. C. Fain, *J. Vac. Sci. Technol.* **14**, 314 (1977).
[9] D. G. Welkie and M. G. Lagally, *Appl. Surf. Sci.* **3**, 272 (1979).
[10] P. C. Stair, *Rev. Sci. Instrum.* **51**, 132 (1980).
[11] K. D. Gronwald and M. Henzler, *Surf. Sci.* **117**, 180 (1982).
[12] J. A. Martin and M. G. Lagally, *J. Vac. Sci. Technol., A* **1**, 1210 (1983).
[13] E. G. McRae, *J. Chem. Phys.* **45**, 3258 (1966).
[14] C. B. Duke and C. W. Tucker, Jr., *Surf. Sci.* **15**, 231 (1969).
[15] D. W. Jepsen, P. M. Marcus, and F. Jona, *Phys. Rev. B* **5**, 3933 (1972); **6**, 3684 (1972).
[16] J. B. Pendry, "Low-Energy Electron Diffraction." Academic Press, New York, 1974.
[17] M. A. Van Hove and S. Y. Tong, "Surface Crystallography by LEED." Springer-Verlag, Berlin and New York, 1979.
[18] R. L. Park, *J. Appl. Phys.* **37**, 295 (1966).
[19] P. J. Estrup and J. Anderson. *Surf. Sci.* **8**, 101 (1967).
[20] W. P. Ellis and R. L. Schwoebel, *Surf. Sci.* **11**, 82 (1968).
[21] J. C. Tracy and J. M. Blakely, *Surf. Sci.* **15**, 257 (1969); J. C. Tracy, Ph.D. Thesis, Cornell Univ., Ithaca, New York, 1968.
[22] J. E. Houston and R. L. Park, *Surf. Sci.* **21**, 209 (1970); **26**, 269 (1971).
[23] C. S. McKee, D. L. Perry, and M. W. Roberts, *Surf. Sci.* **39**, 176 (1973).
[24] M. Henzler, *Surf. Sci.* **22**, 12 (1970).
[25] M. Henzler, *Top. Curr. Phys.* **4**, 117 (1977).
[26] T.-M. Lu and M. G. Lagally. *Surf. Sci.* **120**, 42 (1982).
[27] T.-M. Lu, L.-H. Zhao, M. G. Lagally, G.-C. Wang, and J. E. Houston, *Surf. Sci.* **122**, 519 (1982).

5.1. INTRODUCTION

This chapter is an attempt to summarize in a tutorial fashion the most important experimental aspects of surface-sensitive diffraction. The desired result is that the reader will understand enough of the essentials to set up a diffraction experiment and interpret the results. It is not a compendium of recent, or even the most important, results of surface crystallography. Thus no attempt has been made to develop a comprehensive list of references; only those most germane to the goal stated above are included. A number of reviews exist on various aspects of surface crystallography.[16,17,25-46] Especially the older ones are well worth reading to obtain a clear overview of diffraction techniques.

We begin with the motivation for doing diffraction, namely, to learn about the atomic structure of surfaces, by discussing the general types of surface crystallographic information that can be obtained with a surface-sensitive diffraction technique. This is done in the context of simple diffraction theory. After the establishment of this framework in Chapter 5.2, the rest of this chapter will concentrate on experimental aspects of diffraction. Two major techniques, LEED and RHEED, will be discussed. The emphasis will

---

[28] J. J. Lander, *Prog. Solid State Chem.* **2**, 26 (1965).

[29] J. J. Lander and J. Morrison, *J. Appl. Phys.* **34**, 3517 (1963).

[30] P. J. Estrup, *in* "Modern Diffraction and Imaging Techniques in Materials Science" (S. Amelinckx, R. Gevers, G. Remaut, and J. Van Landuyt, eds.), p. 377. North-Holland Publ., Amsterdam, 1970.

[31] J. W. May, *Adv. Catal.* **21**, 152 (1970).

[32] P. J. Estrup and E. G. McRae, *Surf. Sci.* **25**, 1 (1971).

[33] G. A. Somorjai and H. H. Farrell, *Adv. Chem. Phys.* **20**, 215 (1972); G. A. Somorjai, *Surf. Sci.* **34**, 156 (1973).

[34] M. B. Webb and M. G. Lagally, *Solid State Phys.* **28**, 301 (1973).

[35] G. Ertl and J. Küppers, "Low-Energy Electrons and Surface Chemistry," Chaps. 9 and 10. Verlag Chemie, Weinheim, 1974.

[36] J. A. Strozier, Jr., D. W. Jepsen, and F. Jona, *in* "Surface Physics of Materials" (J. M. Blakely, ed.), p. 2. Academic Press, New York, 1975.

[37] M. G. Lagally, *in* "Surface Physics of Materials" (J. M. Blakely, ed.), p. 419. Academic Press, New York, 1975.

[38] F. Jona, *J. Phys. C* **11**, 4271 (1978).

[39] M. A. Van Hove, *in* "The Nature of the Surface Chemical Bond" (T. N. Rhodin and G. Ertl, eds.) p. 275. North-Holland Publ., Amsterdam, 1979.

[40] M. G. Lagally, G.-C.Wang, and T.-M. Lu, "Chemistry and Physics of Solid Surfaces" (R. Vanselow, ed.), Vol. 2, p. 153. CRC Press, Boca Raton, Florida, 1979.

[41] M. G. Lagally, T.-M. Lu, and D. G. Welkie, *J. Vac. Sci. Technol.* **17**, 233 (1980).

[42] J. E. Cromwell, R. J. Koestnev, L. H. Dubois, M. A. Van Hove, and G. A. Somorjai, *in* "Recent Advances in Analytical Spectroscopy" (K. Fuwa, ed.), p. 211. Pergamon, Oxford, 1982.

[43] M. Henzler, *Appl. Surf. Sci.* **11/12**, 450 (1982).

[44] M. G. Lagally, *Appl. Surf. Sci.* **13**, 260 (1982).

[45] M. G. Lagally, *Springer Ser. Chem. Phys.* **20**, 281 (1982).

[46] M. G. Lagally and J. A. Martin, *Rev. Sci. Instrum.* **54**, 1273 (1983).

be on LEED because of the much greater activity historically and presently in LEED. However, much of the description of LEED applies as well to RHEED. Additional discussion of RHEED will focus on those experimental aspects that are unique, different, or of particular value in certain experiments. Chapter 5.3 will deal with measurements of diffracted intensity distributions in general terms, including a discussion of resolving power. In Chapter 5.4 specific measurements are discussed and in Chapter 5.5 the instrumentation required to carry out these measurements is described. Some representative results are given in Chapter 5.6 and a brief conclusion follows in Chapter 5.7.

5.2. Elements of Diffraction Theory

In almost all interpretations of surface electronic, chemical, or transport properties, the positions of atoms, their separations, their deviations from perfect periodicity, or the positions of their nearest or next-nearest neighbors are necessary inputs. These properties are ideally probed with a diffraction technique. If one considers the elastic scattering of radiation with momentum $\mathbf{k}_0$ from a rigid cubic crystal with lattice points

$$\mathbf{r}_j = m_1\mathbf{a} + m_2\mathbf{b} + m_3\mathbf{c}, \tag{5.2.1}$$

where m_1, m_2, and m_3 are integers, with atoms located within each unit cell at positions

$$\boldsymbol{\rho}_n = u_n\mathbf{a} + v_n\mathbf{b} + w_n\mathbf{c}, \tag{5.2.2}$$

where u_n, v_n, and w_n are fractions, then the amplitude at a given momentum transfer $\mathbf{S} = \mathbf{k} - \mathbf{k}_0$ is given by

$$A(\mathbf{S}) = \sum_{j,n} f_n(\theta, E) \exp[i\mathbf{S} \cdot (\mathbf{r}_j + \boldsymbol{\rho}_n)]. \tag{5.2.3}$$

The sum is over lattice sites j and the atoms n within a unit cell. The function $f_n(\theta,E)$ is the atomic scattering factor of the nth atom, where θ is half the scattering angle and E is the energy of the radiation. After separating the sums, one finds

$$A(\mathbf{S}) = F(\theta, E) \sum_j \exp[i(\mathbf{S} \cdot \mathbf{r}_j)], \tag{5.2.4}$$

where

$$F(\theta,E) = \sum_n f_n(\theta,E) \exp[i\mathbf{S} \cdot \boldsymbol{\rho}_n] \tag{5.2.5}$$

5.2. ELEMENTS OF DIFFRACTION THEORY

is the structure factor. The intensity is

$$I(\mathbf{S}) = A(\mathbf{S})A^*(\mathbf{S}) = |F(\theta,E)|^2 \mathcal{J}(\mathbf{S}), \tag{5.2.6}$$

where $\mathcal{J}(\mathbf{S})$ is called the interference function. The function $\mathcal{J}(\mathbf{S})$ can be visualized by using the concept of the reciprocal lattice. For a three-dimensional infinite crystal, the reciprocal lattice is a three-dimensional array of points whose positions are given by the reciprocal-lattice vectors $\mathbf{G}_{hkl}$, where $|\mathbf{G}_{hkl}| = n2\pi/d_{hkl}$ and d_{hkl} is the distance between (hkl) planes. The interference function is periodic with $\mathbf{G}_{hkl}$ and, for an arbitrary momentum transfer $\mathbf{S}$, can be written in terms of $\mathbf{G}_{hkl}$ and the deviation parameter $\mathbf{s} = \mathbf{S} - \mathbf{G}_{hkl}$. For a crystal with dimensions $N_1 a$, $N_2 b$, and $N_3 c$, where a, b, and c, are the lattice constants,

$$\mathcal{J}(\mathbf{G}_{hkl} + \mathbf{s}) = \frac{\sin^2[\tfrac{1}{2}N_1(\mathbf{G}_{hkl} + \mathbf{s}) \cdot \mathbf{a}]}{\sin^2[\tfrac{1}{2}(\mathbf{G}_{hkl} + \mathbf{s}) \cdot \mathbf{a}]} \frac{\sin^2[\tfrac{1}{2}N_2(\mathbf{G}_{hkl} + \mathbf{s}) \cdot \mathbf{b}]}{\sin^2[\tfrac{1}{2}(\mathbf{G}_{hkl} + \mathbf{s}) \cdot \mathbf{b}]}$$
$$\times \frac{\sin^2[\tfrac{1}{2}N_3(\mathbf{G}_{hkl} + \mathbf{s}) \cdot \mathbf{c}]}{\sin^2[\tfrac{1}{2}(\mathbf{G}_{hkl} + \mathbf{s}) \cdot \mathbf{c}]}. \tag{5.2.7}$$

The value of $\mathcal{J}(\mathbf{G}_{hkl} + \mathbf{s})$ is at its maximum, $\mathcal{J}(\mathbf{G}_{hkl})$, when $\mathbf{s} = 0$, i.e., when $\mathbf{S}$ satisfies the Laue conditions

$$\mathbf{S} = \mathbf{G}_{hkl} \quad \text{or} \quad \mathbf{S} \cdot \mathbf{a} = 2\pi h, \quad \mathbf{S} \cdot \mathbf{b} = 2\pi k, \quad \mathbf{S} \cdot \mathbf{c} = 2\pi l, \tag{5.2.8}$$

where h, k, and l are integers. The maxima in the interference function have heights proportional to $(N_1 N_2 N_3)^2$ and widths in three orthogonal directions in $\mathbf{S}$ space proportional to $1/N_1 a$, $1/N_2 b$, and $1/N_3 c$.

A useful representation of diffraction from a lattice is in terms of its reciprocal lattice and the Ewald construction. The Ewald sphere gives simply the conservation of energy for elastic scattering, i.e., $\lambda_{in} = \lambda_{out}$ or $|\mathbf{k}_0| = |\mathbf{k}|$, where λ and $\mathbf{k}$ are respectively the electron wavelength and momentum. The superposition of the Ewald sphere onto the reciprocal lattice shows conservation of momentum as well as energy in the form of the Laue conditions $\mathbf{S} = \mathbf{G}_{hkl}$. The diffracted-intensity distribution in angle at constant energy is then given by the intersection of the Ewald sphere with the reciprocal lattice. The Ewald construction for penetrating radiation is shown in Fig. 1a for the momentum and diffraction geometry appropriate for LEED. An incident beam with a perfectly defined momentum is assumed, and as a result the Ewald sphere is perfectly sharp. Figure 1b shows the situation for penetrating radiation for the momentum and diffraction geometry appropriate for RHEED or grazing-angle x-ray diffraction. As the energy or angle is varied, diffraction spots will appear or disappear, but only one or several diffracted beams will be excited and hence visible in the diffraction pattern at one set of diffraction conditions.

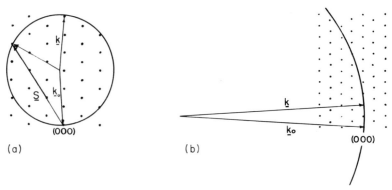

FIG. 1. Reciprocal lattice and Ewald construction for a crystal illuminated by penetrating radiation. (a) Diffraction geometry and momentum appropriate for LEED. (b) Diffraction geometry and momentum appropriate for RHEED or glancing angle x-ray diffraction. The relationship between the separation of reciprocal-lattice points and the length of the $\mathbf{k}$ vectors corresponds to a lattice with a 3-Å lattice constant and x-ray energies of ~ 12 keV and 70 keV, respectively.

5.2.1. Diffraction from Surfaces

For an absorbed monolayer (or in any case in which a phase is only one atomic layer thick) it is easy to deomonstrate that the reciprocal lattice becomes a set of rods normal to the plane of the layer. In Eq. (5.2.7) $N_3 = 1$ and the third term equals one, implying that the interference function has a constant value for all values of $(\mathbf{G}_{hkl} + \mathbf{s}) \cdot \mathbf{c}$. Figure 2 shows the Ewald constructions corresponding to LEED and RHEED for this case. It is evident that now many more diffracted beams are excited for any diffraction geometry and that they will remain excited as the energy or diffraction geometry is varied.

Intermediate between the limit of an infinite three-dimensional crystal and an infinite two-dimensional crystal is one that has a finite dimension in the third direction. This situation is approximated for any infinite three-dimensional crystal if the radiation used for the diffraction experiment does not penetrate the sample to very large depths. This must clearly be the case for all techniques that are surface sensitive, but the limited penetration is achieved in different ways. Under optimum conditions only one or two atomic planes make a significant contribution to the diffracted intensity.

The interference function for a crystal that is artificially limited in the z direction by a scattering density of exponentially decreasing magnitude can be described by defining[47] an attenuation coefficient α such that

$$\alpha = A_{n+1}/A_n, \quad (5.2.9)$$

[47] J. J. Lander and J. Morrison, *J. Appl. Phys.* **34**, 3517 (1963); E. R. Jones, J. T. McKinney, and M. B. Webb, *Phys. Rev.* **151**, 476 (1966); see also ref. 34.

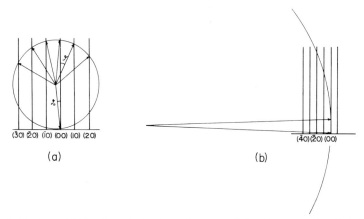

FIG. 2. Reciprocal lattice for a single plane of atoms and Ewald construction for (a) LEED and (b) RHEED geometries. The relationship between the separation of rods and the lengths of the **k** vectors corresponds to a lattice with a 3-Å row spacing and electron energies of ~ 150 eV and 12,000 eV, respectively. The intersection of the Ewald sphere and reciprocal-lattice rod in (b) is misleading. The widths of the lines for both the Ewald sphere and the reciprocal-lattice rod vastly exceed physically realistic values.

where A_n is the amplitude scattered by the nth atomic plane. The total scattered amplitude is then

$$A(\mathbf{S}) = \sum_{j,n} F_j(\theta,E)\alpha^{n_j} \exp(i\mathbf{S} \cdot \mathbf{r}_j). \quad (5.2.10)$$

where n_j specifies the plane containing the jth atom. The interference function becomes

$$\mathcal{J}(\mathbf{G}_{hkl} + \mathbf{s}) = |F(\theta,E)|^2 \frac{\sin^2[\tfrac{1}{2}N_1(\mathbf{G}_{hkl} + \mathbf{s}) \cdot \mathbf{a}]}{\sin^2[\tfrac{1}{2}(\mathbf{G}_{hkl} + \mathbf{s}) \cdot \mathbf{a}]} \frac{\sin^2[\tfrac{1}{2}N_2(\mathbf{G}_{hkl} + \mathbf{s}) \cdot \mathbf{b}]}{\sin^2[\tfrac{1}{2}(\mathbf{G}_{hkl} + \mathbf{s}) \cdot \mathbf{b}]}$$
$$\times \frac{1}{1 + \alpha^2 - 2\alpha \cos[(\mathbf{G}_{hkl} + \mathbf{s}) \cdot \mathbf{c}]}. \quad (5.2.11)$$

Unlike a grating with a finite number of lines N_3, which produces $N_3 - 2$ intermediate interference maxima, such an attenuated grating produces only broad maxima that are smoothly connected. This implies that the reciprocal lattice consists of neither points (3-D) nor "rods" (2-D) but rather of "elongated points" or "cigars" in the $\mathbf{c}^*$ direction. The interference function thus is modulated. This modulation can be expressed in terms of the mean free path for inelastic scattering λ_{inel} by[34,47]

$$\alpha = \exp\left[\frac{c}{2\lambda_{\text{inel}}}\left(\frac{1}{\cos \vartheta_0} - \frac{1}{\cos \vartheta}\right)\right], \quad (5.2.12)$$

where ϑ_0 and ϑ are, respectively, the angles the incident and diffracted beams make with the surface normal. Because the last term in Eq. (5.2.11) never goes to zero for attenuations typical of surface-sensitive diffraction techniques, it is customary to describe the reciprocal lattice as a set of rods, as for a single layer, keeping in mind, however, that the intensity along these rods is modulated with a period that reflects the interlayer distance perpendicular to the surface. This is illustrated in Fig. 3.

The limited penetration of the radiation in surface-sensitive diffraction techniques is achieved in different ways. At the energies used in LEED, 10–1000 eV, the cross sections for both elastic and inelastic scattering are so large (of the order of several square angstroms) that the penetration of the beam is very small. The inelastic and elastic scattering cross sections have approximately the same magnitude. The sensitivity of LEED to the surface is in large part due to the strong inelastic scattering. The inelastic mean free path of electrons, i.e., the distance an electron beam travels in a crystal before inelastic collisions reduce the beam intensity by a factor of $1/e$, is plotted as a

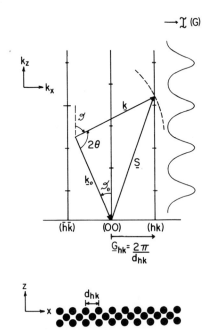

FIG. 3. Reciprocal lattice and Ewald construction for a crystal illuminated by weakly penetrating radiation. The reciprocal lattice is generally drawn as a set of rods with the positions of the third Laue condition indicated. The Ewald construction is drawn appropriate for LEED. The variation of the interference function with $G_\perp$ along a rod is indicated on the right. The real lattice to which this reciprocal lattice corresponds is shown on the bottom.

5.2. ELEMENTS OF DIFFRACTION THEORY

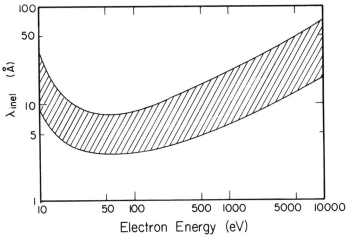

FIG. 4. Dependence of the mean free path of electrons for inelastic scattering λ_{inel} on the electron kinetic energy. The hatched area indicates the range of values of λ_{inel} that have been measured for different materials.

function of energy in Fig. 4. In the energy range of 100 eV, the inelastic mean free path may be as little as 4 Å. Because of the strong elastic scattering, the "extinction distance" of the beam is even less. For a reflection diffraction experiment, in which a monoenergetic beam must enter the crystal, and beams with the same energy must exit again, the mean sampling depth is half the value of the extinction distance.

For high-energy electrons or x rays, the mean free path is much larger. However, as Eq. (5.2.12) shows, the limited penetration necessary for surface sensitivity can be achieved by making ϑ_0 and ϑ large, i.e., by using grazing incidence and exit. Equation (5.2.11) therefore also gives the interference function for RHEED and grazing-angle x-ray diffraction. One can again think of this reciprocal lattice in terms of the approximation of rods.

5.2.2. Surface Defects

So far it has been assumed that the crystal surface is infinite, perfect, and rigid. In this case, the reciprocal-lattice rods will have zero width and the diffraction spots will be sharp. In the presence of defects the reciprocal-lattice rods have a finite width. This can be readily seen by letting N_1 or N_2 be finite in Eq. (5.2.7), in which case the corresponding term is no longer a delta function, but rather a function of the form $\sin^2 Nx / \sin^2 x$ (where $x = \frac{1}{2} \mathbf{S} \cdot \mathbf{a}$), which has a nearly Gaussian shape in its center, a full width at half-maximum (FWHM) proportional to $1/N$, and wings higher than those of a Gaussian profile, with $N - 2$ side bands that are distinct if N is small and unresolvable if N is large. Limitations in the order or the size of a crystal in the dimension parallel to the surface thus cause a behavior different from limitation of the crystal in the direction perpendicular to the surface caused by finite penetration. In the latter case subsidiary maxima do not occur because the amplitude from each plane is different. In the former, each row of atoms scatters the same amplitude, and therefore, if the finite-size effects are regular, distinct subsidiary maxima can be observed. In many cases, of course, the disorder is random (e.g., there are many small ordered regions of different sizes), and the subsidiary maxima wash out to leave only a broadened Bragg peak and a diffuse background.

5.2.2.1. Clean Surfaces. An example that illustrates generally the broadening of reciprocal-lattice rods caused by defects can be given by considering a distorted crystal, where the position of the jth lattice site (or unit cell) is now

$$\mathbf{r}'_j = \mathbf{r}_j + \mathbf{R}_j, \tag{5.2.13}$$

where $\mathbf{R}_j$ is the vector giving the displacement of the unit cell from its proper position $\mathbf{r}_j$. To describe this density function, more Fourier components are necessary than for a perfect crystal. Around a particular reflection $\mathbf{G}_{hkl}$, Eq. (5.2.4) becomes

$$A(\mathbf{G}_{hkl} + \mathbf{s}) = F(\theta, E) \sum_j \exp[i(\mathbf{G}_{hkl} + \mathbf{s}) \cdot (\mathbf{r}_j + \mathbf{R}_j)]. \tag{5.2.14}$$

The presence of the imperfection thus introduces an added phase factor $\exp[i(\mathbf{G}_{hkl} + \mathbf{s}) \cdot \mathbf{R}_j]$. Consider $\mathbf{R}_j$ to be a sinusoidal variation in spacing in the **a** direction, i.e.,

$$\mathbf{R}_j = \delta \cos(2\pi x_j/\Lambda), \tag{5.2.15}$$

where δ is very small compared to **a** and where Λ is the wavelength of the modulation. Then for the reflection $\mathbf{G}_{hkl}$

$$A(\mathbf{G}_{hkl} + \mathbf{s}) = F(\theta, E) \sum_j \exp[i(\mathbf{G}_{hkl} + \mathbf{s}) \cdot \mathbf{r}_j]$$
$$\times \exp[i\mathbf{G}_{hkl} \cdot \delta \cos(2\pi x_j/\Lambda)], \tag{5.2.16}$$

where the term $(\mathbf{s} \cdot \mathbf{R}_j)$ is neglected. Because δ is small, the second exponential can be expanded to give

$$A(\mathbf{G}_{hkl} + \mathbf{s}) = F(\theta, E) \sum_j \{\exp[i(\mathbf{G}_{hkl} + s) \cdot \mathbf{r}_j]$$
$$+ i\mathbf{G}_{hkl} \cdot \delta \cos(2\pi x_j/\Lambda) \exp[i(\mathbf{G}_{hkl} + \mathbf{s}) \cdot \mathbf{r}_j]\}. \tag{5.2.17}$$

The first term is just the Fourier transform of the perfect crystal, whereas the second leads to sidebands at positions $s_x = \pm 2\pi/\Lambda$ away from the main peak. If the crystal has a dimension $N_1 a$ in the **a** direction, then the rods will be broadened, with a FWHM, $\Delta s_x = 2\pi/N_1 a$, of the rods in the **a** direction. If $\Lambda \geq N_1 a$, then the sidebands merge with the main rod and cause further broadening that increases with increasing h because of the factor $\mathbf{G}_{hkl} \cdot \boldsymbol{\delta}$. If $\Lambda \ll N_1 a$, the extra intensity goes into a background and leaves the main rod unchanged in width but reduced in intensity.

The effect on the reciprocal lattice of several types of defects can be illustrated by considering their wavelength Λ. For example, most of the Fourier components of displacements due to thermal vibrations have Λ of the order of atomic dimensions, which is much smaller than the dimension of a typical ordered region on a crystal surface. Hence thermal vibrations cause a diffuse background intensity that reduces the intensity at the (hk) rod but causes no broadening. Similarly, point defects have Fourier components that have mostly small Λ, and they thus produce only a diffuse background. On the other hand, random lattice strain can have a much longer wavelength, i.e., $\Lambda \simeq N_1 a$, the size of the crystal. Hence random strain causes broadening of reciprocal-lattice rods that increases with increasing $\mathbf{G}_\parallel$, the parallel component of a reciprocal-lattice vector. The dependence of the broadening on $\mathbf{G}_\parallel$ due to random lattice strain at the surface[48] is shown schematically in Fig. 5.

Other defects that are not as easily explained in the above context can also

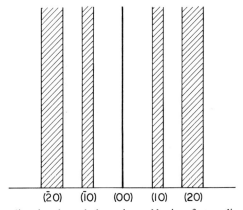

FIG. 5. Cut in the **a** direction through the reciprocal lattice of a two-dimensional crystal that contains random lattice strain in the **a** direction. The reciprocal-lattice rods broaden with increasing order h of reflection. The hatched regions represent the full width at half-maximum of the intensity distribution. The (00) rod is not sensitive to strain broadening. The widths are greatly exaggerated.

[48] D. G. Welkie, Ph.D. Thesis, Univ. of Wisconsin, Madison, 1981; D. G. Welkie and M. G. Lagally, *Thin Solid Films* **93**, 21 (1982).

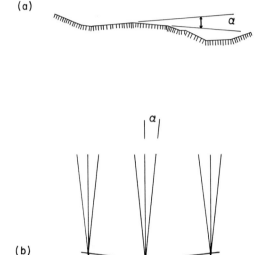

Fig. 6. Schematic illustration of mosaic structure in a crystal and the reciprocal lattice corresponding to the surface. (a) Mosaic structure, with an average misorientation angle α. The misorientation is vastly exaggerated. (b) Cut in the **a** direction through the corresponding reciprocal lattice, which consists of many reciprocal lattices all with a common origin but misoriented with respect to each other by the angle α. The crystallite size is assumed large enough to produce narrow rods as shown.

exists at a surface. For example, subgrain (mosaic) structure in a crystal, shown in Fig. 6a, manifests itself at the surface as finite-size domains that have small misorientations with respect to each other. Typical mosaic dimensions in well-grown crystals are of the order of 1 μm or larger, and misorientations are of the order of a tenth to several tenths of a degree. This misorientation causes a broadening of all reflections with $G_\perp$, the normal component of a reciprocal-lattice vector, as is illustrated in Fig. 6b. The broadening is easily understood by recognizing that each crystallite has its own reciprocal lattice normal to its surface and that these must have a common origin. Misorientations of 0.1° are readily observable by making angular profile measurements at various energies (i.e., $G_\perp$).[49,50]

Surface steps represent an entirely different type of defect. Many arrangements of steps are possible on a surface, with distinctive effects on the

[49] D. G. Welkie, M. G. Lagally, and R. L. Palmer, *J. Vac. Sci. Technol.* **17**, 453 (1980).

[50] H. M. Clearfield, D. G. Welkie, T.-M. Lu, and M. G. Lagally, *J. Vac. Sci. Technol.* **19**, 323 (1981).

5.2. ELEMENTS OF DIFFRACTION THEORY

reciprocal lattice. The step arrangement for which the reciprocal lattice is visualized most easily is a monotonically increasing or decreasing step array with constant terrace size. As shown in Fig. 7, the reciprocal lattice can be considered as the product of the terrace structure factor and the reciprocal lattice associated with the average surface. This is equivalent to saying that the surface is the convolution of the single terrace unit with the step "lattice". The lattice points for this lattice represent the repeat units for the average surface, i,e., there is one lattice point associated with each terrace. Because the average surface consists of many such lattice points, the reciprocal-lattice rods corresponding to the average surface will be sharp. Because the lattice points are far apart, the reciprocal-lattice rods are close together. Their separation is inversely related to the cosine of the angle of cut: the greater the

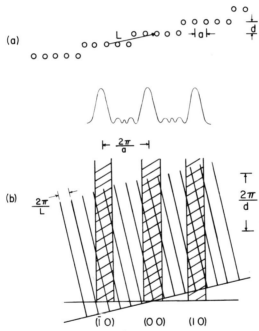

FIG. 7. Cut in the **a** direction through the reciprocal lattice of a vicinal surface with monotonically increasing steps and a constant terrace size. (a) Surface: a is the row spacing, d the plane spacing, and L the separation of terraces. (b) Reciprocal lattice: the narrow rods represent the reciprocal lattice of a lattice whose unit vector is L. The separation of rods and their inclination depend on the terrace size L. The hatched rods represent the reciprocal lattice of a finite lattice (each terrace) whose unit vector is **a**. The complete reciprocal lattice for each terrace, called the terrace-structure factor, is shown above the figure. Diffraction features occur when the product of the two reciprocal lattices is nonzero. The period in the appearance of diffraction features gives the inverse of the layer spacing d. The figure is drawn for a terrace size of 5 atoms.

deviation from singular, the farther apart these rods will be. The terrace structure factor is just the reciprocal lattice associated with a single terrace. Because the terrace has finite dimensions $N_1 a$ and $N_2 b$, the reciprocal-lattice rods corresponding to a single terrace will be broad, as discussed earlier. The greater the deviation of the average surface from singular, the smaller will be the terrace size and hence the greater will be the broadening. It should be evident that as the distribution of intensity due to the terrace structure factor gets broader (i.e., smaller terraces), the reciprocal-lattice rods of the average surface get farther apart (i.e., the repeat unit for the average lattice gets smaller). The product of these two factors, analogous to Eq. (5.2.6), is observed in reciprocal space.

Other step distributions produce different reciprocal lattices.[25] For example, if the terrace size remains uniform but the steps are alternately up and down, the reciprocal lattice consists of rods modulated in $G_\perp$ that are not inclined, because the average surface is flat. This is illustrated in Fig. 8 and can be explained simply in the following manner. As shown in Fig. 8a, the periodicity of the step structure is $2Na$, where N is the effective number of scatterers in one terrace. The interference function can be described as the product of three structure factors: one that describes the periodicity with $2Na$, one that reflects the scattering from one terrace of dimension Na, and one that gives the interference between terrace and trough. The first, $|F_1|^2$, is a reciprocal lattice consisting of rods that are delta functions (because this structure is infinite) spaced $2\pi/2Na$ apart and oriented normal to the surface. The separation of adjacent rods, i.e., $n = 0$ and $n = 1$, reflects the size of one terrace plus one trough. The structure factor $|F_2|^2$ of one terrace of dimension Na consists of an intensity function of the form $\sin^2(N\mathbf{S} \cdot \mathbf{a}/2)/\sin^2(\mathbf{S} \cdot \mathbf{a}/2)$ with maxima spaced $2\pi/a$ apart (because a is the repeat unit in a terrace) and with $N - 2$ subsidiary maxima. The rods are again oriented normal to the surface. The main maxima have a FWHM of $2\pi/Na$, and overlap three of the delta function rods. The minima of the terrace structure factor occur on alternate delta function rods, as do the subsidiary maxima. The structure factor $|F_3|^2$ takes the difference in z spacing between terraces and troughs into account. This can easily be shown to be a function of the form $|F_3|^2 = 2[1 + \cos(h\pi)\cos(G_\perp d)]$, where $h = 0, 1, 2, \ldots$ is the order of the reflection, and d is the height difference between terraces and troughs. If this function is evaluated at different values of $G_\perp$, it is found that its zeros occur alternately at the $n = $ even and $n = $ odd rods. The product of all three structure factors gives the reciprocal lattice shown in Fig. 8c, with the rods with $n = 0, 2N, \ldots$ having maximum intensity, the adjacent ones having about half as much intensity, and the subsidiary rods having a few percent of the intensity of the main rods, depending on the size of a terrace. They are all, in principle, visible, although the subsidiary maxima may be quite weak.

5.2. ELEMENTS OF DIFFRACTION THEORY

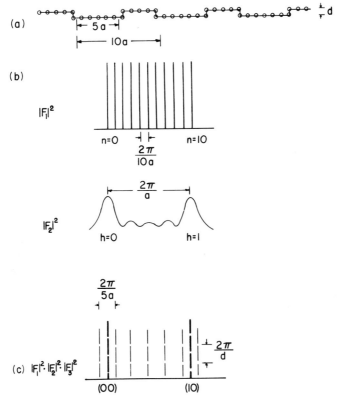

FIG. 8. Cut in the **a** direction through the reciprocal lattice of a stepped surface with constant terrace size and steps alternately up and down. (a) Surface, (b) two structure factor components, (c) complete reciprocal lattice. $|F_1|^2$ is the structure factor of the (infinite) lattice whose unit mesh vector is $10a$. $|F_2|^2$ is the structure factor of a single terrace (five rows wide) with unit mesh vector a. The complete reciprocal lattice consists of delta function rods that have zero intensity at periodic positions in $G_\perp$. The blackness of the lines is meant to illustrate the intensity. The periodicity reflects the inverse of the layer spacing. The periods in panel (c) are not in scale with the lattice periodicity shown in panel (a).

The periodicity of the oscillation in $G_\perp$ is quite evidently related to the step height, as can be seen from the third structure factor. Physically this can be interpreted as constructive and destructive interference between terraces and troughs. When the interference is constructive, the diffraction does not recognize the existence of steps, and only the reflections corresponding to an infinite lattice with lattice constant a appear, at $2\pi h/a$. At other $G_\perp$'s, all the other rods appear, and at characteristic $G_\perp$'s the rods at $2\pi h/a$ disappear while all the others are present. Each rod displays the same periodicity in $G_\perp$, but with minima displaced in $G_\perp$ because the phase shift due to $\mathbf{G}_\parallel$ is proportional to h. This periodicity in $G_\perp$ is reflected in all stepped structures that have a step height that is a unique value or a multiple thereof. It is always possible in such cases to extract the step height, simply from the periodicity of the oscillation or modulation in width of the reciprocal-lattice rods.

Introducing a distribution of terrace sizes causes broadening of reflections rather than a set of sharp delta functions. The most general case, shown in Fig. 9, is for random up and down step edges occurring at random intervals.[25] This situation is approximated by most surfaces that are nominally flat but contain steps. It is similar to the case of alternate up and down terraces of uniform size, except that, because of the randomness, the original delta function term $|F_1|^2$ is absent and $|F_2|^2$ does not have subsidiary maxima but only a diffuse monotonically changing intensity between the maxima. The reciprocal-lattice rods alternately broaden and narrow as $G_\perp$ is changed. The periodicity of the broadening is again related easily to the step height. At the conditions for which all the terraces scatter in phase, the diffraction spots will be sharp. At other $G_\perp$'s there will be partial destructive interference and the spots will broaden. They will be broadest (for step distributions that contain only monatomic steps or a predominance of them) halfway between the sharp spots.

If there is a predominance of steps of a height different from monatomic,

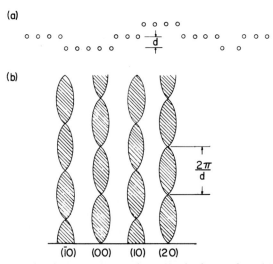

FIG. 9. Cut in the **a** direction through the reciprocal lattice for a surface with AB stacking that has random up and down step edges occurring at random intervals. (a) Surface, (b) reciprocal lattice. The hatched regions represent the FWHM of the intensity distribution. This width is related to the average terrace size. The periodicity in $G_\perp$ reflects the inverse of the step height.

5.2. ELEMENTS OF DIFFRACTION THEORY

the broadening will be different.[26] This is simply illustrated by considering all the steps to be double height, causing the periodicity in $G_\perp$ to halve. As the step height multiplicity increases, the period of oscillation decreases. A superposition of steps with a range of multiple-step heights leads to a distribution in reciprocal space as shown in Fig. 10. The broadening becomes flat over most of the range of $G_\perp$, with very sharp minima in width occurring at the positions of constructive interference, which do not change. A physical analog of multiatomic steps is slip planes emanating at a surface. Slip in crystals can result, for example, during cleavage, and may result in step heights of 20 to 500 lattice constants.

5.2.2.2. *Overlayers.* Most of the phenomena mentioned above for clean surfaces have analogs in adsorbed overlayers. The most commonly observed (if not the most common) form of adsorbed monolayer is one that is commensurate with the substrate (i.e., adsorbed in regular lattice sites of the substrate) and with a unit mesh larger than that of the substrate (i.e., with a superlattice). Such layers are identified with a standard notation indexed to the substrate unit mesh, such as W(110)p(2 × 1)–0, which indicates an overlayer of O on the W(110) face that has a unit mesh that is twice the size of the substrate unit mesh dimension in the **a** direction and the same size as the substrate unit mesh in the **b** direction. The "p" indicates that the overlayer unit mesh is primitive, i.e., it contains only one atom. Similarly W(100)c(2 × 2)–H indicates an overlayer of H on the W(100) face that has unit mesh vectors twice that of the substrate in both **a** and **b** directions, but additionally includes an atom in the center of the mesh. It is thus not primitive. An equivalent notation for this overlayer that does give a primitive mesh is W(100)p($\sqrt{2} \times \sqrt{2}$)R45°–H, where R45° indicates that the overlayer unit mesh is rotated by 45° relative to that of the substrate. Overlayers with a periodicity different from that of the substrate produce addi-

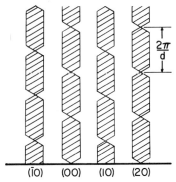

FIG. 10. Cut in the **a** direction through the reciprocal lattice for a surface with AB stacking that contains a wide distribution of step heights that are multiples of the monatomic-step height.

tional, or superlattice, reciprocal-lattice rods. The positions of these superlattice rods can be established by substitution of the proper superlattice unit mesh vectors for **a** and **b** into Eq. (5.2.7), or else by an argument similar to that given by Eqs. (5.2.13)–(5.2.17), which will demonstrate that the new Fourier components produce a sideband at the proper reciprocal-lattice positions. The reciprocal lattice for a complete, infinite, perfectly ordered monolayer with a double periodicity in the **a** direction [e.g., p(2 × 1), p(2 × 2), c(2 × 2)] is shown in Fig. 11a. The intensity distribution consists of delta function rods.

Another form of adsorbed layer is an "out-of-registry" layer, where the spacing of the overlayer atoms is slightly different from that of the substrate. If one considers an atom at an origin lining up with a substrate atom, then the $(N + 1)$st overlayer atom will again line up, but with the Nth substrate atom, leading to a superlattice with a large wavelength. The Fourier transform of this spacing gives sidebands, close to the main maximum, called statellite reciprocal-lattice rods. The formation of satellite lines can be caused by an overlayer that may be rotationally or translationally out of registry, by a periodic lattice distortion of the substrate caused by overlayer adsorption, or by similar effects.[51,52] If the distortion or displacement has a definite period, the satellite reflections will be sharp. Because the wavelength Λ of the distortion is generally large relative to a lattice constant, the satellite reflections will lie close to the position of the reflections for the undistorted or undisplaced lattice. There should be a multiplicity of sidebands, at values of $s_x = 2h\pi/\Lambda$, with the one nearest the main peak being the most intense.

Any commensurately adsorbed monolayer that forms a superlattice will have translational and possibly also rotational (depending on the symmetry of the unit mesh) antiphase domains. A monolayer may exist in the form of two-dimensional antiphase islands[21,17] at some finite temperature if there is a net attractive interaction between the adsorbate atoms. There are a number of reasons why a submonolayer might exist as a distribution of islands, the most important being kinetic limitations and substrate point or line defects that act as nucleation sites. The reciprocal lattice appropriate for a submonolayer that forms antiphase islands depends on the coverage and on the type of antiphase boundaries that can exist for the structure. A reciprocal lattice that is generally applicable at low coverages for overlayers that form a superlattice is shown in Fig. 11b. A distribution of overlayer island sizes is assumed. The superlattice reciprocal-lattice rods broaden, whereas the fundamental reflections consist of a broadened contribution, due to the over-

---

[51] S. C. Fain and M. D. Chinn, *J. Phys. Colloq. (Orsay, Fr.)* **38**, Suppl 10, C4-99 (1977).
[52] S. Semancik and P. J. Estrup, *J. Vac. Sci. Technol.* **18**, 541 (1981); R. A. Barker and P. J. Estrup, *J. Chem. Phys.* **74**, 1442 (1981); see also ref. 32.

5.2. ELEMENTS OF DIFFRACTION THEORY

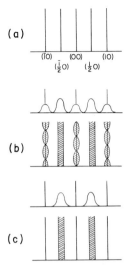

FIG. 11. Cuts in the **a** direction through reciprocal lattices for a overlayer with double periodicity in the **a** direction adsorbed on an infinite, perfectly ordered substrate. The overlayer atoms are assumed to sit in bridge sites, so that the layer and substrate appear to form an AB stacking sequence. (a) Complete, infinite, and perfectly ordered overlayer. Sharp half-order rods appear. (b) Low-coverage overlayer broken up into finite-size islands. Upper panel: cut at a particular $G_\perp$. Lower panel: behavior as a function of $G_\perp$. The hatched areas represent the FWHM. The half-order rods are broad and the fundamental reflections contain a diffuse-intensity halo. Its behavior with $G_\perp$ depends on the scattering phase-shift differences between overlayer and substrate atoms and on the overlayer–substrate distance. The width and shape of the superlattice rods are related to the average island size and the size distribution. (c) Saturation-coverage overlayer with antiphase domain boundaries. Upper and lower panels as in (b). The superlattice reflections are broad as in (b). Their width and shape reflect the average domain size. The fundamental reflections do not have a diffuse-intensity halo.

layer islands, and a delta function resulting from the periodicity imposed by the substrate. This same reciprocal lattice results at any coverage for overlayer structures in which rotational antiphase domains are allowed [e.g., p(2 × 1) and p(1 × 2)]. This is easy to understand. At low coverages, islands will be separated by a considerable amount of "sea." Because of the translational antiphase boundaries that always exist in overlayers with a superperiod, these widely separated ordered overlayer regions will be uncorrelated in phase (except at the precise Laue conditions). The diffracted intensity from the overlayer consists of the sum of intensities from the individual islands.[21,27] There is interference, however, between the amplitudes scattered from the substrate and from the randomly arranged islands, which makes the relative strengths of the delta function and the scattering from the overlayer dependent on diffraction conditions.[53] The reason that an overlayer with rotational antiphase domains acts in the same way, independent of coverage, is that these domains cannot interfere with each other, and thus one always acts as sea for the other.

For an overlayer that forms *only* translational antiphase boundaries [e.g., $p(2 \times 2)$] the reciprocal lattice is identical that shown in Fig. 11b at low coverages. At *saturation* coverage there is no sea, and because the fundamental reflections are not sensitive to antiphase boundaries that occur at integral multiples of the fundamental spacing, these rods will not have the diffuse wings, but will be sharp, as shown in Fig. 11c. The superlattice rods are broadened, because they are sensitive to antiphase boundaries.

If the overlayer adsorbs commensurately but without a superlattice [$p(1 \times 1)$ structure], there obviously will be no superlattice reciprocal-lattice rods. If two-dimensional islands form, the reciprocal-lattice rods at low coverage will be identical to the fundamental reflections shown in Fig. 11b. For a $p(1 \times 1)$ layer there can be no rotational antiphase domains. There can also be no translational antiphase domains in the sense that we have so far described, i.e., occupation of the same type of site but translationally displaced by a multiple of the substrate lattice constant. At saturation coverage the rods will therefore ordinarily be sharp, as shown in Fig. 11c for the fundamental reflections. However, a translational antiphase boundary of a different type can occur. This requires the occupation of two types of sites, e.g., hexagonal close-packed (hcp) and face-centered cubic (fcc) sites on an fcc (111) surface. The resulting antiphase boundaries (called twin or stacking-fault boundaries in bulk films or crystals) cause selective broadening of some of the reciprocal-lattice rods. Figure 12 shows a schematic diagram of an overlayer with a twin boundary and the resulting reciprocal lattice in one direction. The selective broadening of some rods can easily be understood physically by recognizing that the translational mistake at a twin boundary is less than one lattice constant, thus causing broadening of all those rods for which the phase does not sum up to one, and no broadening of those for which it does.

Finally, overlayers may form as completely incommensurate layers that, in a sense, form two-dimensional ordered "rafts" that have no definite phase relationship to the substrate on which they are adsorbed. The overlayer has its own reciprocal lattice. At any coverage below a monolayer, this system behaves like a two-dimensional mosaic, i.e., there is a random phase relationship between ordered islands, but, if the substrate is flat, there is no out-of-plane misorientation of the overlayer islands. If only translational

---

[53] D. A. Saloner and M. G. Lagally, *Bull. Am. Phys. Soc.* **28**, 457 (1983); D. Saloner and M. G. Lagally, *J. Vac. Sci. Technol., A* **2**, 935 (1984), and references therein.

5.2. ELEMENTS OF DIFFRACTION THEORY

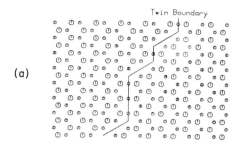

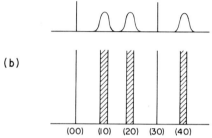

FIG. 12. Schematic diagram of an overlayer containing a twin boundary and the resulting reciprocal lattice. (a) Surface: Overlayer atoms on one side of the twin boundary occupy sites that are displaced by $\frac{1}{3}$ of the substrate lattice constant. Large symbols: top-layer atoms; small symbols: second-layer atoms. (b) Cut in the **a** direction through the corresponding reciprocal lattice. The hatched areas represent the FWHM. Every third rod is not sensitive to the twin boundary, because all atoms scatter in phase there. Other rods broaden and reflect the twin size distribution. Relatively thick overlayers are assumed so that there is no interference with the substrate. The upper panel shows a cut at a particular $G_\perp$. (From Welkie.[48])

randomness exists, (i.e., all the islands are oriented in the same manner) the reciprocal lattice is as shown in Fig. 13. The (00) rod is a delta function, because the specular reflection is not sensitive to lateral phase shifts. All the other rods will be uniformly broadened. Because completely incommensurate layers will generally also involve rotational randomness, all but the (00) rod turn into rings of finite width.

Islands adsorbed on a substrate cause a modulation of the shape of the fundamental reflections with $G_\perp$ for the same reasons that steps do, namely, there occurs an interference between substrate and adsorbate layer. For self-adsorbed layers, which form p(1 × 1) structures, this is obvious, because they are directly analogous to terraces. The same reciprocal-lattice-rod shapes described for step structures are possible for such systems. For sys-

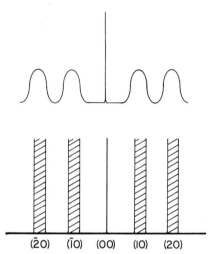

Fig. 13. Cut in the **a** direction through the reciprocal lattice corresponding to a layer that consists of ordered regions that are translationally random in the **a** direction. The layer is assumed thick enough so that no interference with the substrate occurs. The upper panel shows a cut at a particular $G_\perp$.

tems in which the overlayer atom is different from the substrate atom a weak modulation of the shape of fundamental reflections should also exist, but it may no longer be simply oscillatory with a period that is proportional to the inverse of the layer spacing. Because the scattering factors are different for substrate and overlayer atoms, there may be a phase difference upon scattering from the overlayer and substrate. This phase adds to that due to the path difference and may shift the minima or distort the period of modulation.

In this section we have summarized basic elements of diffraction theory in the kinematic limit. We have shown how surface-sensitive diffraction techniques can be discussed in terms of the Ewald construction and the relevant reciprocal lattice. A number of reciprocal lattices that correspond to various types of surface disorder have been presented. A perfect instrument has been assumed throughout this section. In the next section we take the opposite approach. We assume that the crystal is perfect but that the instrument, like all real instruments, has a finite sensitivity and a finite resolving power. We discuss the influence of these limitations on the measurement of diffracted intensities and on the accuracy of surface crystallographic determinations.

5.3. The Measurement of Diffracted-Intensity Distributions

A diffraction experiment requires the creation of a beam of radiation to use as a probe and the detection, at specific diffraction geometries, of the

5.3. MEASUREMENT OF DIFFRACTED-INTENSITY DISTRIBUTIONS

radiation scattered elastically by the sample. For accurate structural analysis, the intensities must be precisely measurable, and intensities scattered into different directions or at different energies must be distinguishable. The performance of a diffractometer in these respects can be described in terms of its sensitivity and its resolving power. An instrument cannot be optimized in both; maximum sensitivity can only be achieved at the expense of resolution and vice versa.

5.3.1. Sensitivity

In a diffraction experiment one is, in essence, counting particles arriving in a particular direction, and therefore the best obtainable ratio of the intensity of the true signal to the noise current can be described by the well-known relationship between signal and shot noise,[54]

$$\frac{J_{\text{true}}}{J_{\text{shot noise}}} \propto (\sigma i t)^{1/2}, \qquad (5.3.1)$$

where i is the incident beam current, t the time of measurement, and σ the probability of measuring a diffracted particle for each incident particle. The probability σ can therefore be identified with the *sensitivity* in a measurement. It includes both physical factors, such as scattering powers and the inelastic scattering cross section, and instrumental factors, such as detector size and sensitivity. It is clear from Eq. (5.3.1) that the simplest way to increase the signal-to-shot-noise ratio is to increase the dose, i.e., to raise the incident current i or measure for a longer time t. In surface diffraction experiments it is frequently not possible to increase i arbitrarily because structural or chemical changes are introduced as a result of surface heating. It is also not possible to measure for arbitrarily long times, because the surface becomes contaminated. Surface contamination can, in fact, be considered the $1/f$ or flicker noise in the measurement. In addition, electron-beam damage is dose dependent, and many surfaces and overlayers are quite sensitive to electron beams in the applicable energy ranges.

In LEED and RHEED, low sensitivity manifests itself in the inability to measure the intensity of small peaks accurately and reproducibly or to obtain an intensity map with sufficient data for reliable structural analysis in the time available before the surface structure begins to change. The magnitude of the background intensity is also significant, because the shot noise is proportional to the sum of the true and background intensities. A large

---

[54] See, e.g., K. R. Spangenberg, "Fundamentals of Electron Devices." McGraw-Hill, New York, 1957; R. L. Park, *in* "Experimental Methods in Catalytic Research" (M. Dawson and C. A. Anderson, eds.), Vol. 3, p. 1. Academic Press, New York, 1976.

background can therefore significantly degrade the reliability of intensity data.

The most effective approach to improving the signal quality is to increase the sensitivity of the instrument. This will be addressed briefly later.

5.3.2. Resolving Power

Instrumental factors also limit the *resolving power* of the instrument, i.e., they cause a loss in attainable accuracy in measuring the shapes of reciprocal-lattice rods. It is possible to quantify this contribution.[55] If $T(\vartheta,E)$ represents the broadening, or instrument response, function,[56] the measured intensity is given by the convolution

$$J(\vartheta,E) = I(\vartheta,E) * T(\vartheta,E), \quad (5.3.2)$$

or equivalently

$$J(\mathbf{S}) = I(\mathbf{S}) * T(\mathbf{S}). \quad (5.3.3)$$

The intensity $I(\vartheta,E)$, the true signal, results from the incoherent sum of diffraction patterns of individual electrons all with the same momentum. It is a delta function if the surface is perfect and has some angular spread or physical width if the surface is not perfect. The function $T(\vartheta,E)$ can be thought of as a shape function, whose integral is unity, that distributes the true signal $I(\vartheta,E)$ over a range of angles and energies in reciprocal space. The major sources of instrumental broadening in diffractometers are the incident beam divergence or "source extension" γ of the electron gun, the energy uncertainty ΔE in the incident beam, the incident beam diameter D, and the detector aperture width d. As discussed by Park et al.,[56] the instrument response function, i.e., the distribution in momentum of all the electrons arriving at the detector if the sample is perfectly periodic, will be

$$T(\vartheta,E) = T(\vartheta,E)_\gamma * T(\vartheta,E)_E * T(\vartheta,E)_D * T(\vartheta,E)_d. \quad (5.3.4)$$

The total instrument response measured for a typical commercially available LEED instrument has a Gaussian profile near its center, with wings that are more Lorentzian. Response functions have not been measured as accurately for RHEED systems.

The different uncertainties that make up $T(\vartheta,E)$ influence the accurate determination of the reciprocal lattice in different ways, some causing inaccuracies in $G_\parallel$, others in $G_\perp$, but most in both. It is instructive to consider the contributions to $T(\vartheta,E)$ separately using the reciprocal lattice of a two-di-

---

[55] T. M. Lu and M. G. Lagally, *Surf. Sci.* **99**, 695 (1980).
[56] R. L. Park, J. E. Houston, and D. G. Schreiner, *Rev. Sci. Instrum.* **42**, 60 (1971).

5.3. MEASUREMENT OF DIFFRACTED-INTENSITY DISTRIBUTIONS

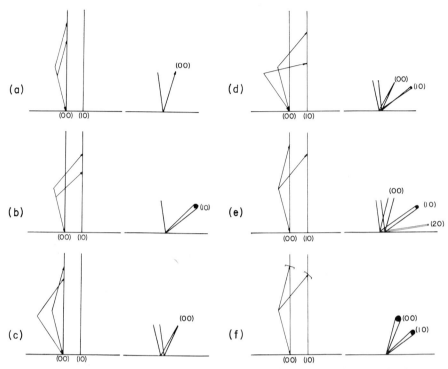

FIG. 14. Schematic diagrams in real and reciprocal space illustrating various contributions to instrumental broadening. (a) Energy spread (either in the incident beam or due to detector resolution), (00) reflection. Integration over a range in $G_\perp$ occurs but no spread in $\mathbf{G}_\parallel$. (b) Energy spread, nonspecular reflections. An integration over a range in $G_\perp$ and a spreading in $\mathbf{G}_\parallel$ result. (c) Angular divergence, for a beam focused on the detector. An integration over a range in $G_\perp$ results, but, because the particular beam [here chosen to be (00)] was focused on the detector, there is no uncertainty in $\mathbf{G}_\parallel$. (d) Angular divergence, general case. Both an integration over a range in $G_\perp$ and an uncertainty in $\mathbf{G}_\parallel$ result. (e) Finite beam diameter. An uncertainty in $\mathbf{G}_\parallel$ results that differs for different beams. (f) Finite detector aperture. An uncertainty in $G_\parallel$ results that is the same for all beams. In all cases the contributions are greatly exaggerated for illustration.

mensional perfect crystal. Figure 14 illustrates some of these contributions in both real and reciprocal space. Consider first an uncertainty in the energy of the incident beam. In order to isolate just this one uncertainty, assume an infinitely narrow parallel beam, i.e., two coaxial rays, with slightly different $|\mathbf{k}_0|$, falling onto an infinite perfectly periodic two-dimensional crystal. The reciprocal lattice and diffraction geometry for the (00) reflection are shown in Fig. 14a. Because the wave vectors $\mathbf{k}$ must all end on a rod, it is clear that an averaging in $G_\perp$ results, i.e., all values of intensity along the rod between the extremal values of the **k** vectors are sampled. No broadening of the specular [(00)] beam results from an energy uncertainty in the incident beam because the **k** vectors are all parallel. Thus if the only uncertainty is one in the energy of the incident beam, the (00) diffraction peak forms an indefinite perfect two-dimensional crystal that will be a delta function in angle whose intensity is some integral of the intensity along the rod over the range in $G_\perp$ that is sampled.

For nonspecular beams, an energy uncertainty in the incident beam causes uncertainty in $\mathbf{G}_\parallel$ as well as an integration over a range in $G_\perp$, as shown in Fig. 14b. The **k** vectors must still fall onto the rod, which gives the average in $G_\perp$, but in order to do so, they emanate from the crystal at various angles of incidence. For an infinitely narrow parallel incident beam, this would produce a diffracted beam that has a finite width at the detector. As indicated below, such a finite width represents an inability to determine precisely the value of $\mathbf{G}_\parallel$, because it will give a reflection broadened in angle. Thus for nonspecular reflections, an energy uncertainty gives measured profiles that are some sample over $G_\perp$ in which, as well, the value of $\mathbf{G}_\parallel$ is imprecise.

A similar situation obtains if there is an angular divergence in the incident beam. It is, however, somewhat more complicated to visualize. The condition of minimum influence of angular divergence occurs if the beam is focused such that the particular diffracted beam that is being measured has minimum size at the detector. Even the best focusing effort will result in a beam that has finite size and divergence at the detector. Finite beam size and finite divergence can be treated separately by considering each point in the beam area to be the focal point for rays. Then the complete beam can be described by the convolution, as in Eq. (5.3.4), of the divergence contribution and the finite-beam-size contribution. Thus, for the purposes of describing the influence of the beam divergence, one should assume that it is possible to focus perfectly, i.e., to make an infinitely narrow beam at the detector that does, however, have some divergence. The corresponding real- and reciprocal-space diagrams for the (00) rod are shown in Fig. 14c. As for an energy uncertainty, an integration over a range in $G_\perp$ results. The **k** vectors are also not parallel. However, if it is assumed that this beam is focused at the detector, there is no broadening in $\mathbf{G}_\parallel$ due to the beam divergence.

If the focusing conditions are set to focus the (00) beam on the detector, none of the other diffracted beams will be in focus. This can be seen from simple geometry by recognizing that the other reflections come from "mirrors" oriented differently with respect to the incident beam. Thus for any beam other than the one that is chosen for best focus, beam divergence will lead also to an uncertainty in $\mathbf{G}_\parallel$, as shown in Fig. 14d. If the incident beam is focused on the sample, all diffracted beams will be broadened in angle.

5.3. MEASUREMENT OF DIFFRACTED-INTENSITY DISTRIBUTIONS

The important quantity in considering the contribution of finite beam width to the resolving power is the beam width at the detector. After removing the divergence, the finite-beam-width contribution can be considered separately as resulting from a parallel beam that has a size given by the size of the beam at the detector. Each ray in this beam will be incident at the origin of reciprocal space, hence no averaging over $G_\perp$ results from a finite beam diameter. Uncertainty in $G_\parallel$ results because a finite beam diameter translates into an angular uncertainty. The angular width differs for different beams, as shown in Fig. 14e. It will be most severe for beams near normal exit. A lens before the detector may be used to focus a parallel beam to reduce the effect of beam diameter.[56]

The effect of finite detector aperture size is the same as that of finite beam diameter, i.e., it causes a broadening in $G_\parallel$ because of the angle subtended by the detector at the sample. Unlike that of the finite beam diameter, the angular width of the finite detector aperture is not dependent on exit angle of the beam, as is shown in Fig. 14f.

Several of these contributions influence the resolving power differently in different directions, causing asymmetric-spot-shape effects. To illustrate this, consider again the energy uncertainty in the incident beam. For the (00) beam, it causes integration only in $G_\perp$. For nonspecular reflections uncertainty in $G_\parallel$ also results, but for some configurations, e.g., when $\mathbf{k}_0$ and $\mathbf{k}$ are coplanar, broadening occurs in only one direction in $G_\parallel$. This can be readily visualized with the help of the reciprocal-lattice constructions in Fig. 14.

The above discussion has illustrated the various contributions to $T(\vartheta,E)$ and their effect on a measurement. It is evident that the instrument response differs at different diffraction conditions. The resolving power of an instrument at any diffraction condition can be quantified by defining a minimum angle of resolution. If one represents the instrument response function $T(\vartheta,E)$ by a Gaussian with a full width at half-maximum b_T and the accuracy to which this width is known and to which a measurement $J(\vartheta,E)$ has been made as $X\%$, then the smallest value of the width of a signal $I(\vartheta,E)$ that can be resolved by the instrument is[55]

$$\vartheta_{\min} = [(b_T + X\% \times b_T)^2 - (b_T - X\% \times b_T)^2]^{1/2} = 2b_T(X\%)^{1/2}. \quad (5.3.5)$$

Equation (5.3.5) represents the worst possible case for the resolving power of a given instrument, because it is based on extremal values of signal allowed by the error bars. The uncertainty X in the measurement is clearly related to the system sensitivity and thus is a function of the incident beam current, the detector efficiency, and the measurement time. Methods to increase the resolving power by improving electron gun characteristics to give a better instrument response and by improving detector efficiency to give greater sensitivity are briefly addressed in a later section.

It has been mentioned earlier that different diffraction techniques observe the reciprocal lattice in different ways. The limiting cases are those appropriate for LEED and for RHEED. In the former, the incident beam is generally nearly normal to the surface, and the most frequently observed diffracted beams are those that emanate nearly normal to the surface. In the latter, the incident beam is very near grazing incidence (fractions of one degree to several degrees), and the diffracted beams likewise exit near grazing angles. The two techniques therefore give quite different cuts through the reciprocal lattice with a consequently significant influence on the resolving power. In LEED, the beams emanating near backward directions represent cuts nearly perpendicular to a reciprocal-lattice rod. For a cut that is strictly perpendicular to the rod, no averaging in $G_\perp$ results except that introduced by the instrument response $T(\vartheta,E)$. Any other cut through a rod always gives an integral over $G_\perp$ in addition to that introduced by $T(\vartheta,E)$. This is an important disadvantage in many experiments, especially on surfaces with defects, as is discussed in the next section, where several of the ways of measuring diffraction spots are more fully explored. Cuts not perpendicular to a rod also have an important positive consequence, however, in that they increase the resolving power in the plane of the cut.[57] This can easily be seen by considering a rod with finite width $\Delta G_\parallel$. A cut normal to the rod will give an angular width

$$\Delta \vartheta = \Delta G_\parallel / k. \quad (5.3.6)$$

The intensity function with this width is, of course, convoluted with the instrument function. Whether it is resolvable depends on the minimum angle of resolution ϑ_{min}. A cut at an angle ϑ will give an angular width

$$\Delta \vartheta = \Delta G_\parallel / (k \cos \vartheta), \quad (5.3.7)$$

where ϑ is the angle the exiting beam makes with the surface normal. Thus, as ϑ increases, $\Delta \vartheta$ increases until, at an exit angle of 1°, the angular width measured (in the plane of **k** and the surface normal) for a rod of a given width $\Delta G_\parallel$ is about 50 times that near normal exit. As the minimum angle of resolution is not significantly affected by exit angle (because of the dominance of the beam-size contribution to the instrument response in typical systems[58]), it becomes possible to distinguish much smaller $\Delta G_\parallel$, and hence the resolving power increases. A cut at 45° at LEED energies increases the angular width of a reflection by 1.4; a cut at 70° increases it by a factor of 3. In RHEED, the length of the **k** vector is greater, which to some degree negates the effect of the small angle. Nevertheless, for a 1° exit angle and a 10-keV

---

[57] J. M. Van Hove, P. Pukite, P. I. Cohen, and C. S. Lent, *J. Vac. Sci. Technol., A* **1**, 609 (1983).

[58] D. E. Savage, unpublished work.

5.3. MEASUREMENT OF DIFFRACTED-INTENSITY DISTRIBUTIONS

beam, the angular width ("streak length") of a reflection should be about seven times that observed for LEED at 150 eV and normal exit, and hence, the resolving power in the plane defined by **k** and the surface normal is more than seven times as great as in LEED. In the plane perpendicular to this one the resolving power is, in fact, smaller than that of LEED by the ratio of the lengths of the **k** vectors.

The resolving power of the instrument is meaningful only in terms of the real-space distances that can be resolved. Some simple examples will illustrate that for a given instrument the resolving power depends on the type of defect that is present on the surface. We assume an instrument with a minimum angle of resolution $\vartheta_{min} = 0.5°$, typical for a conventional LEED system at 50 eV. We consider three cases:[55]

(1) *Mosaic structure with translationally random phases only.* In this model the surface is flat, but has domains of mosaic structure that are separated by random phases in a direction parallel to the surface. An example might be a saturation-coverage-incommensurate overlayer consisting of many domains that have random translational phase relationships. Within each domain, the structure is assumed perfectly periodic with a lattice constant **a**. For simplicity, all domains are assumed to have the same size $N_1 a$. The interference function is given by Eq. (5.2.7),

$$\mathcal{J}(\mathbf{G}_{hkl} + \mathbf{s}) = M \frac{\sin^2 \tfrac{1}{2} N_1 (\mathbf{G}_{hkl} + \mathbf{s}) \cdot \mathbf{a}}{\sin^2 \tfrac{1}{2} (\mathbf{G}_{hkl} + \mathbf{s}) \cdot \mathbf{a})}, \tag{5.3.8}$$

where N_1 is the total number of atoms in each domain, and M is the number of domains within the diameter of the incident beam. For a normally incident beam, the FWHM b_I of the intensity function is related to the domain size by

$$b_I = 0.888 \lambda a / [(N_1 a)(a^2 - h^2 \lambda^2)^{1/2}], \tag{5.3.9}$$

where λ is the wavelength of the electron and h is the order of the diffracted beam from a set of surface rows, i.e., (00), (10), (20), etc. For the first-order ($h = 1$) diffraction maximum for diffraction from a surface with a lattice constant $|\mathbf{a}| = 3$ Å the maximum size of flat mosaic domain that the instrument can resolve at 50-eV incident-beam energy is

$$(N_1 a)_{max} = 0.888 \lambda a / \vartheta_{min} (a^2 - \lambda^2) \cong 215 \text{ Å}. \tag{5.3.10}$$

If ϑ_{min} is improved to $0.2°$, either by increasing the accuracy of the measurement or by improving the instrument response, the maximum domain size that can be resolved is $(N_1 a)_{max} \cong 535$ Å.

As can be seen from Eq. (5.3.10), the maximum observable domain size depends on the energy of the incident beam, the order of the diffracted beam (hence the incident and exit angles), the lattice constant, and the minimum angle of resolution.

(2) *Flat continuous surface layer with translational antiphase boundaries and equal-size domains.* If the surface consists of a continuous phase containing translational antiphase boundaries, instead of a continuous phase with random phase boundaries as in Model (1), a definite phase relationship exists between the domains. This is the approximate situation for many commensurate overlayers with high-symmetry superlattice structures [e.g., p(2 × 2)] at their saturation coverage. If there are a roughly equal number of each type of antiphase domain, all of size $N_1 a$, the superlattice Bragg reflections will be split into two peaks separated by an angle[55]

$$b_1 = \lambda a / [N_1 a (a_2 - h^2 \lambda^2)^{1/2}], \tag{5.3.11}$$

where the angular separation b_I reflects the periodicity introduced by the repetition of antiphase boundaries at regular intervals given by $N_1 a$, the domain size. The minimum angle of resolution is then interpreted as the smallest angular separation of the two spots that the instrument can resolve. For $\vartheta_{\min} = 0.5°$, $E = 50$ eV, $a = 3$ Å, and $h = 1$, the largest antiphase domain of this type that the instrument can resolve is

$$(N_1 a)_{\max} = \lambda a / \vartheta_{\min} (a^2 - \lambda^2)^{1/2} \cong 240 \text{ Å}. \tag{5.3.12}$$

(3) *Flat continuous surface layer with translational antiphase boundaries and random-size domains.* If the domains in Model (2) are not the same size, but have a random distribution of sizes, with only translational antiphase boundaries between them, then instead of splitting, the superlattice Bragg reflections will be broadened. The FWHM b_I is related to the average domain size $\langle N_1 a \rangle$ by[55]

$$b_I = \frac{\lambda}{\pi (a^2 - h^2 \lambda^2)^{1/2}} \cos^{-1} \alpha, \tag{5.3.13}$$

where $\alpha \equiv (4f - f^2 - 1)/2f$ and $f \equiv 1 - 2a/\langle N_1 a \rangle$. Again for $\vartheta_{\min} = 0.5°$, $E = 50$ eV, $a = 3$ Å, and $h = 1$ the maximum average domain size that the instrument can resolve is $(N_1 a)_{\max} \cong 155$ Å.

It is evident from this discussion that for a given instrumental response the resolving power differs for different types of defects that limit the order on a surface or in an overlayer. In addition, the resolving power may be different in different directions and for different techniques. Similarly, the sensitivity may differ for different techniques and different instruments. Some diffraction measurements require high sensitivity, others high resolving power. In

5.4. Surface Crystallography Measurements

the next two sections we discuss the major types of diffraction measurements and the optimization of diffractometers in terms of sensitivity and resolving power.

5.4. Surface Crystallography Measurements

In this section, the major types of surface crystallography measurements are described and the limitations of these measurements explored with reference to the last section.

5.4.1 Diffraction Patterns

The most common diffraction measurement is, of course, the observation of the diffraction pattern. From it the size and shape of the surface unit mesh and the existence of any superlattice can be obtained by inspection. Simply the presence or absence of reflections, their position, and their behavior with incident-beam energy can be used in investigations of the overlayer symmetry, the nature of the defects present, and the existence of possible phase transformations in an overlayer. Figure 15a illustrates the corresponding LEED geometry in real and in reciprocal space. Figure 16a illustrates the equivalent RHEED geometry. Generally, the sensitivity of the diffractometer is of greater concern that the resolving power in overall visual observations, particularly in the search for weak diffraction features that may indicate the presence of an overlayer phase.

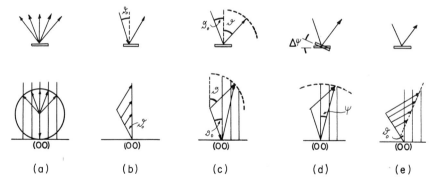

FIG. 15. Schematic diagrams in real and reciprocal space of the major types of LEED measurements. ϑ_0 is the angle of incidence measured from the surface normal, ϑ the angle the scattered beam makes with the surface normal, and ψ the tilt angle of the crystal. (a) Spot pattern, (b) intensity profile, (c), (d), (e) beam angular profiles J versus ϑ, J versus ψ, and J versus E, respectively.

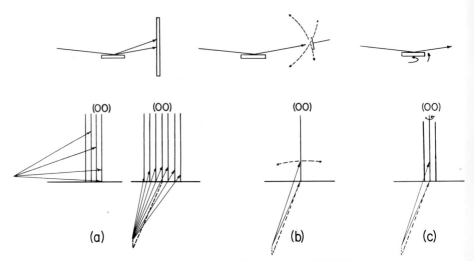

FIG. 16. Schematic diagrams in real and reciprocal space of RHEED measurements. (a) Pattern, (b) angular profile, (c) rotation plots. The dashed line in (b) indicates the path of the detector or the deflection of the diffracted beam across a fixed detector.

5.4.2. Equilibrium Position Determinations

Much of the activity in LEED has been in the determination of the equilibrium position of surface or overlayer atoms. Other surface-sensitive diffraction techniques have, on the other hand, essentially not been used for this purpose. Although the required data may be obtained in a variety of ways, the most common form is a measurement of the integrated intensity-versus-energy profile [commonly called an "I versus E" curve but more properly identified as $\int$ detector $J_{hk}(\vartheta,E)\, d\Omega$ versus E curve]. In this measurement, the intensity in a particular reflection $J_{hk}(\vartheta,E)$, integrated over the solid angle of the detector $d\Omega = d\vartheta\, d\phi$, is determined as a function of incident-beam energy so that one obtains effectively a scan of the reciprocal lattice in $G_\perp$ at fixed $G_\parallel$. This is illustrated for LEED in real and reciprocal space in Fig. 16b. the sensitivity of the diffractometer is of primary importance in this measurement. Frequently, this is simply because some peaks in the diffracted intensity (e.g., from fractional monolayers) are small. In other cases electron-beam damage to the overlayer or surface structure requires use of as low a total dose as possible. Frequently, however, the need for sensitivity implies simply a need for speed in data acquisition, because a large data base is required for accurate analysis of the equilibrium positions, and because the LEED intensity is usually quite sensitive to surface contamination.

5.4. SURFACE CRYSTALLOGRAPHY MEASUREMENTS

The integrated-intensity-versus-energy measurement represents a determination of the inteference among the several layers that are illuminated by the incident beam (see Fig. 3). Because an integral is usually taken over the width $\Delta G_\parallel = k \cos \vartheta \, \Delta \vartheta$ of a particular reflection, finite-size effects are not resolved and generally will not affect the equilibrium position determination. This need not be true in all cases, however.[59] The possible influence of finite-size effects on intensity-versus-energy profiles can be illustrated with Fig. 17, which show a LEED measurement at various energies for a surface that contains random steps. An integrated-intensity measurement is obtained by using a detector (Faraday cup or telephotometer/screen) with a fixed aperture size chosen to be "large enough to collect the whole diffracted beam." This means in practice that the detector is approximately the size of the full width at half-maximum of the beam at some energy and the measurement is $\int_{\text{detector}} J_{hk}(\vartheta,E) \, d\Omega$ versus E. However, a fixed-aperture detector, because it subtends a constant solid angle at the crystal, collects an increasingly large fraction of the intensity in the Brillouin zone (including thermal-diffuse-scattering intensity) as the energy is increased. The effect is in all cases a more or less monotonically increasing background in I versus E curves, which causes a worse signal-to-noise ratio and a consequently larger uncertainty for peaks at high energies than for peaks at low energies. Additionally,

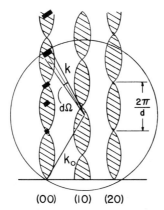

FIG. 17. Schematic diagram of a fixed-aperture measurement at various energies (an intensity-versus-energy profile) for a surface that contains randomly distributed steps, illustrating the effect on the fraction of the Bragg intensity that is collected. The solid bar represents the detector aperture. It always has the same solid angle $d\Omega$. At an in-phase condition, all of the Bragg intensity is collected. At an out-of-phase condition, a fraction that depends on energy and aperture size is collected. (From Lagally.[44] Copyright North-Holland Physics Publishing Amsterdam, 1981.)

[59] T.-M. Lu, M. G. Lagally, and G.-C. Wang, *Surf. Sci.* **104**, L229 (1981).

for a surface containing randomly distributed steps, the intensity distribution oscillates in width along the reciprocal-lattice rods (Fig. 9), and as a result, as Fig. 17 illustrates, the fraction of the Bragg intensity that is collected can vary rapidly even within a few electron volts. This phenomenon can distort peak shapes or shift peaks in integrated-intensity-versus-energy measurements.

Although a high resolving power is not a priori necessary for integrated-intensity measurements, the instrument response function must at least be well known in order to extract reliable data. This can be illustrated as follows. Because LEED instruments in different laboratories commonly have different beam parameters and detector widths, and because intensity-versus-energy data are usually collected with the detector centered on the maximum intensity rather than by scanning through the reflection, the measured integrated intensity $\int_{\text{detector}} J_{hk}(\vartheta, E) \, d\Omega$ can differ markedly for the same $I_{hk}(\vartheta,E)$ (i.e., for surfaces with identical structures and degree of order). Thus, for reliable equilibrium position determinations, a knowledge of $T(\vartheta,E)$ is a necessity to allow an accurate evaluation of what the measured intensity represents.[59]

Intensity-versus-energy measurements are much more difficult with RHEED because it is not possible to obtain the intensity integrated over $\Delta G_\parallel$ at any point $G_\perp$ on a rod. Any finite-size effect gives a broadened rod. Because the Ewald sphere cuts rods at a grazing angle in RHEED, a streak results for any broadened rod. In principle, it is possible to use a very small detector to obtain the intensity along this streak to get an intensity-versus-energy profile. This measurement is, however, complicated by instrument response (leading to some integration in $G_\perp$ at every point in the measurement) and by the distribution of intensity in $\mathbf{G}_\parallel$ on the cut across the rod that is made by the Ewald sphere. Hence such measurements may not reflect very reliably the integrated intensity as a function of $G_\perp$.

It is customary in LEED to calculate directly the intensity distribution expected from a given model structure and to compare this with the measured intensity. A number of calculations for different structures are required to search for the most probable structure. In x-ray diffraction, where the interaction of the radiation with the material is weak, a single-scattering or kinematic calculation of the diffracted intensity is usually sufficient for structural analysis. Because of the strong interaction of electrons with the material, there are usually strong multiple-scattering events in LEED and RHEED. The energy positions of intensity maxima in a diffracted beam depend on the relative phases of the electron waves diffracted from the atoms in the surface and thus depend both on the positions of the atoms and the phase shifts on scattering from the atoms. The intensities are influenced by a number of effects, including thermal vibrations, structural order, energy loss

5.4. SURFACE CRYSTALLOGRAPHY MEASUREMENTS

mechanisms, and experimental factors. Dynamic LEED theories, which include multiple scattering, have been developed and used for most surface-structure determinations.[13-17] Because of the sensitivity of the energy positions of peaks in I versus E curves to geometric positions of atoms, the structural analysis emphasizes a fit between peak positions in calculations and experiments. For simple structural problems (relaxation of the outer layer of a clean metal, overlayers with small unit mesh) visual comparison has been used successfully for structure determination, because the eye acts as an excellent discriminator and/or noise filter. For comparisons of large amounts of data, as is required with larger unit meshes, this becomes cumbersome. As a result, automated criteria, called reliability or R-factors, have been developed.[60] The R-factors are single numbers summarizing the level of agreement between sets of curves, a small R-factor indicating better agreement. In order to be useful in discriminating one structural model from another, R-factors should be sensitive to surface atom equilibrium positions but insensitive to nonstructural parameters such as the scattering potential or even to other structural parameters such as surface defects or thermal disorder. Unfortunately, because an R-factor can be sensitive to any of a number of features in an I versus E curve, e.g., peak positions, slopes, relative peak heights, small peaks, large peaks, etc., it is difficult to design one that objectively measures the best fit between theory and experiment. Attempts to develop a globally sensitive R-factor have so far failed.[60] No single R-factor presently in use is totally satisfactory for reliable structure analysis. The best structure analysis results from an average of all R-factors. The spread in R-factor values for any structure may be significant for structure determination: the smaller this spread, the more likely the structure is correct, independent of the absolute values of the R-factors.[60]

The existence of multiple scattering in principle makes electron diffraction very sensitive to small structural differences. Unfortunately, the resulting theoretical problem also becomes quickly unmanageable in terms of the computing time needed to perform a range of structural searches. One approach that alleviates this difficulty is to limit the required range of search by extracting the single-scattering intensity in I versus E curves and using it to provide likely bounds to some of the structural parameters. Because it depends only on the momentum transfer vector S and not on k_0 and k, the single-scattering intensity can be determined by making a number of measurements at constant momentum transfer S and averaging them.[61,62] The single-scattering intensity can then be interpreted with simple modifications

---

[60] For a summary, see M. A. Van Hove, in "Surface Structure by LEED" (P. M. Marcus, ed.). Plenum, New York, 1985, to be published.
[61] M. G. Lagally, T. C. Ngoc, and M. B. Webb, Phys. Rev. Lett. **26**, 1557 (1971).
[62] T. C. Ngoc, M. G. Lagally, and M. B. Webb, Surf. Sci. **35**, 117 (1973).

of methods that were developed for x-ray diffraction to provide approximate values of surface or overlayer atom structural parameters. Two ways of making I versus E measurements that keep **S** fixed but change **k** and $\mathbf{k}_0$ so that the multiple scattering changes are shown in Fig. 18.

5.4.3. Structural Defects

A major and increasingly popular application of surface-sensitive diffraction techniques is in the determination of surface defects. The required measurement is the angular distribution of intensity in individual diffracted beams $\int J_{hk}(\vartheta,E)\, d\Omega$ versus ϑ. When this measurement is made at various diffraction conditions (e.g., energy, angle of incidence, and for various diffracted beams), different surface defects can be identified because of their differing but characteristic influence on the beam profile. The precision of such measurements depends on the resolving power of the diffractometer, as has already been indicated. This implies a need for highly accurate measurements and good electron beam characteristics, including low divergence and energy spread and a small beam size, as well as a small detector aperture. Although much work has been done on increasing both sensitivity and resolution, especially in LEED diffractometers, very high resolution can be obtained only at the expense of sensitivity. Thus most high-resolution surface defect studies are so far made on static defect distributions on surfaces stable against beam damage or rapid contamination, for which sufficient time is available to signal average the intensities.

Angular profile measurements can be made in a number of different ways. These are shown schematically in real and reciprocal space in Figs. 15c, d, e and 16b, c. In the most commonly used method, shown in Figs. 15c and 16b, the detector is scanned through a diffracted beam. In the second method (Fig. 15d) the detector is fixed and the crystal rocked so that the desired

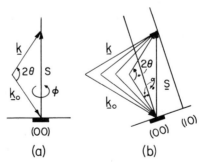

FIG. 18. Schematic diagrams of integrated-intensity-versus-energy measurements at constant momentum transfer. (a) Constant colatitude angle ϑ_0, varying azimuthal angle ϕ. (b) Constant azimuthal angle ϕ, varying colatitude angle ϑ_0.

5.4. SURFACE CRYSTALLOGRAPHY MEASUREMENTS

reflection moves across the detector. There are two advantages of this method for LEED. The cut across a rod is flatter than if the detector is moved, because it has a radius of $|S|$ rather than of $|k|$ and, because the triangle k_0, k, S stays fixed, the scattering angle 2θ is constant. As a result, there are no variations in the scattering factor $f(\theta,E)$ across a profile with this method.[63] Asymmetries in angular profiles caused by a changing $f(\theta,E)$ can be significant over the width of a Brillouin zone, especially at low energies at which $f(\theta,E)$ can vary rapidly with angle and the width of a zone represents a large angular change (i.e., the diffracted-beam separation is large at low energies). The disadvantage is that a mechanical motion, i.e., tilting the sample, is required. This generally cannot be done as accurately as scanning the detector, especially if electronic detection is used. In addition, the angle of incidence changes, introducing possible multiple-scattering effects involving other diffracted beams. A third method for angular profile measurement (Fig. 15e), advanced originally for LEED intensity-versus-energy-profile measurements,[6,18] consists of scanning the energy of the incident beam sufficiently so that the diffracted beam moves across a fixed detector. The cut across the rod is elongated for diffracted beams near normal incidence, which, as has been discussed, can be both an advantage and a disadvantage. The advantage is high resolving power in the plane containing the surface normal and the diffracted beam. A disadvantage is the large range of $G_\perp$ that is included, causing an uncertain interpretation of the beam shape if the rod itself has structure. A second disadvantage is the possible introduction of multiple-scattering effects that are dependent on energy. An advantage is that no mechanical motion is required. A fourth method of angular profile measurement (Fig. 16c) is rotation of the sample.

There are a number of inherent difficulties in making LEED angular profile measurements with high precision. The most important of these is suppression of inelastically scattered electrons. The angular distribution of these electrons will in general differ from that of the elastically scattered ones and will therefore distort a measurement of the latter, especially in the wings of the angular profile, which are the most sensitive to the defect density and defect distribution. It is not a simple task to remove all inelastically scattered electrons without disturbing the angular distribution of the elastically scattered ones. The usual method of suppression of inelastically scattered electrons is with a high-pass filter consisting of a retarding-potential grid in front of the detector. It has been shown[9] that, in order not to disturb measurably the angular distribution of the diffracted electrons passing through a typical LEED grid structure, the bias on the retarding grids can be no more than

---

[63] J. T. McKinney, Ph.D. Thesis, Univ. of Wisconsin, Madison, 1966; J. T. McKinney, E. R. Jones, Jr., and M. B. Webb, *Phys. Rev.* **160**, 523 (1967).

about 80–90% of the energy of the diffracted beam. This implies that all those inelastically scattered electrons that have energies within 10–20% of the elastically scattered electrons will contribute to the measured profile. In order to remove this contribution, it is necessary to know the angular distribution of those inelastically scattered electrons that pass through the grids. It is usually assumed to be a constant, in which case a constant background can be subtracted from the angular profile. More precise methods for measuring the elastically scattered electrons while suppressing the inelastically scattered electrons involve the use of magnetic fields or special geometries of a Faraday cup.[46] The most accurate measurement of the elastically scattered electron current can be made with an electron energy analyzer in front of a Faraday cup collector. Detectors will be discussed in a later section.

Although a major fraction of the inelastically scattered electrons can always be removed without noticeably affecting the angular profile, the remaining inelastic scattering intensity can cause significant noise in the wings of the profile. Statistical noise goes up as the square root of the signal plus background; thus even after background subtraction, the elastically scattered intensity may have large uncertainties.

A different approach to removal of the inelastic scattering is to consider it as a part of the instrument response. If it is assumed that the distribution in angle and the relative magnitude of the inelastic scattering do not differ for a "perfect" surface and one containing defects, the perfect-surface profile including the inelastic scattering can be used as the appropriate response function for deconvolution for all other profiles. The advantage of this approach is that no analysis of the inelastic scattering is required; disadvantages are the need for assumptions and large noise in the wings of the profile.

An even more serious problem is the thermal-diffuse or phonon scattering.[63-65] Because the energy losses (or gains) associated with phonons are only of the order of tens of milli-electron-volts, they are too small to be resolved with the energy resolution typically achievable at the energies of LEED or RHEED experiments.[66] Thus phonon scattering is present to some degree in all angular profiles of diffracted intensity. The magnitude of the phonon contribution to the total intensity can be roughly estimated from a knowledge of the Debye temperature of the surface under investigation or from a measurement of the Debye–Waller factor of this surface. The

---

[64] R. F. Barnes, M. G. Lagally, and M. B. Webb, *Phys. Rev.* **171**, 627 (1968).

[65] M. G. Lagally and M. B. Webb, *in* "The Structure and Chemistry of Solid Surfaces" (G. A. Somorjai, ed.) Chap. 20. Wiley, New York, 1969.

[66] Phonon dispersion relations have been measured with high-resolution electron energy loss spectroscopy; see S. Lehwald, J. M. Szeftel, H. Ibach, T. S. Rahman, and D. L. Mills, *Phys. Rev. Lett.* **50**, 518 (1983).

Debye–Waller factor, which is obtained by measuring the decay of the Bragg peak intensity with increasing temperature at any energy,

$$J(\mathbf{S}, T_2) = J(\mathbf{S}, T_1) \exp(-2M) \tag{5.4.1}$$

gives the value of

$$2M = \langle \mathbf{S} \cdot \mathbf{u} \rangle^2 = 4\pi (\sin^2 \theta) \hbar^2 T / \lambda^2 m k_B \theta_D^2, \tag{5.4.2}$$

where $\langle u^2 \rangle$ is the mean square vibrational amplitude of surface and near-surface atoms, θ the scattering angle, λ the wavelength of the radiation, T the temperature, m the mass of the surface atoms, k_B Boltzmann's constant, and θ_D the effective surface Debye temperature of the material. Most materials have a value of $2M$ of the order of 2 or 3 at room temperature and 100-eV incident-beam energy. Figure 19 shows a plot of the dependence on $2M$ of the thermal-diffuse-scattering intensity integrated over a Brillouin zone. The phonon scattering can be considered to consist of one-phonon and multiphonon contributions. The multiphonon scattering is uniformly distributed[64,65] over the Brillouin zone, whereas the one-phonon scattering is peaked near the diffraction maximum and falls off roughly as $1/\mathbf{s}_\parallel$, where $\mathbf{s}_\parallel$ is the deviation parameter (see Eq. (5.2.7)] parallel to the crystal surface.[64] The multiphonon scattering can thus be removed by subtracting a constant background. The one-phonon scattering must be modeled[67] on the basis of the value of $2M$ and the resulting integrated intensities from Fig. 19.

The simplest and most reassuring experiment that serves at least approximately to estimate the phonon scattering is to compare measurements of angular profiles of a high-quality surface taken at low and high temperatures. If no differences are observed in the two profiles, the phonon scattering is not a significant factor in the angular distribution, to the accuracy that it is measured. The phonon scattering can be reduced by proper choice of the diffraction parameters (i.e., measurements at low momentum transfer). With the assumption that the phonon scattering does not depend significantly on surface or overlayer order or defect density, the profile measured on the high-quality surface will serve as an instrument response function that also accounts for the thermal-diffuse-scattering contribution to the angular profile taken on the surface containing defects. This is the same approach used above in deconvoluting the inelastic scattering.

Angular profile measurements in RHEED are illustrated in Fig. 16b. These could be made with a movable detector or by scanning the patterns with a vidicon. However, experimental considerations (discussed below)

---

[67] R. L. Dennis and M. B. Webb, *J. Vac. Sci. Technol.* **10**, 192 (1973).

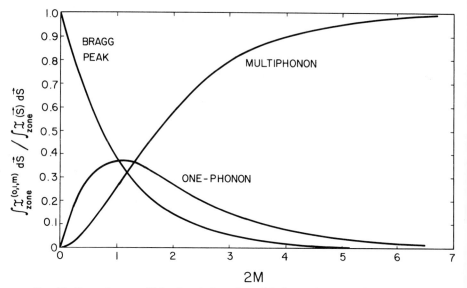

FIG. 19. Dependence on $2M$ or the relative values of the integrals over a Brillouin zone of the Bragg scattering, one-phonon thermal diffuse scattering, and multiphonon scattering. (From M. G. Lagally and M. B. Webb,[65] *In* "The Structure and Chemistry of Solid Surfaces" (G. A. Somorjai, ed.), Effects of plasma scattering and the atomic scattering factor in low-energy electron diffraction. Copyright 1969 John Wiley and Sons. Reprinted by permission of John Wiley and Sons, Ltd.)

make it practical to deflect the beam across a fixed detector.[68] Intensity contour measurements along a streak[57,68] indicate that visual observation is not adequate to determine streak length. The FWHM of the streak along its long direction is much less than the length observed visually. Nevertheless, a streak of considerable length is measured in many circumstances. We have already considered two possible contributions to the RHEED streak length, the finite instrument response and the cut of the Ewald sphere across the rod, but have discounted them as the principal cause of streak length for typical instruments and high-quality surfaces. Inelastic scattering will also contribute to streak length. RHEED measurements are usually made without any retarding bias; because no potential is applied to the fluorescent screen and, because it takes of the order of 1000 eV or more to excite the phosphor, the screen itself acts as a high-pass filter. Thus true secondary electrons do not contribute to the RHEED intensity. However, Kikuchi lines can cause significant intensity variations along or across a RHEED streak.[69] More importantly, it has been suggested that the length of RHEED streaks is due to the

---

[68] J. M. Van Hove, C. S. Lent, P. R. Pukite, and P. I. Cohen, *J. Vac. Sci. Technol.*, B **1**, 741 (1983).

[69] P. I. Cohen, personal communication.

5.4. SURFACE CRYSTALLOGRAPHY MEASUREMENTS

phonon scattering.[70] In this hypothesis, the extra momentum needed to get to a reciprocal-lattice rod when the S vector no longer contacts the rods is provided by a phonon. This clearly should contribute to the length of the streak. Measurements of the streak length at various temperatures indicate, however, that this effect is not significant.[68] Possibly the most significant contributions to streak length come from defects such as steps or long-range curvature of the crystal surface.[68,71-73] Steps can cause considerable broadening of rods (see Fig. 9). A cut with the Ewald sphere at the appropriate conditions will lead to long streaks, whereas at slightly different angles of incidence a short streak or spot will be observed.[68] Long-range curvature can contribute to streak length in the following way. Because the beam has a projected area on the sample that is quite large at grazing angles, areas with many slightly differing surface orientations (due to crystal mosaic) may participate in the diffraction. This will provide (Fig. 6b) a rod that gets increasingly broad with increasing $G_\perp$, allowing a long streak as the Ewald sphere passes through the rod. It was noted some years ago[71,72] that "good" surfaces gave short streaks or spots, a result that is being increasingly verified.[68] The various factors discussed above make it generally more difficult than in LEED to interpret RHEED angular profiles quantitatively and thus to exploit the greater resolving power that is available in RHEED.

5.4.4. Thermodynamics and Kinetics

As has been indicated already, the easiest information obtainable through LEED or RHEED, simply by observing the diffraction pattern, is the size and shape of the surface or overlayer unit mesh. Because many overlayers (as well as some clean surfaces) form superlattices, investigation of the change in intensity and the appearance or disappearance of superlattice reflections in LEED or RHEED patterns as a function of coverage, temperature, or time gives an easy way to study the thermodynamics of ordered phases and the kinetics of ordering or disordering. The most important measurement for these studies is simply the dependence of the diffracted-beam intensity on temperature. With the additional information on defects derived from the angular profile measurements discussed in the last section, a complete study of overlayer phase transformations and ordering kinetics can be made, and the results of such studies can in principle be interpreted in terms of adatom–adatom interactions.[41]

---

[70] S. Holloway and J. L. Beeby, *J. Phys. C* **11**, L247 (1978); J. L. Beeby, *Surf. Sci.* **80**, 56 (1979); S. Holloway, *Surf. Sci.* **80**, 62 (1979).

[71] J. F. Menadue, *Acta Crystallogr., Sect.* **A28**, 1 (1972).

[72] D. B. Dove, R. Ludeke, and L. L. Chang, *J. Appl. Phys.* **44**, 1897 (1973).

[73] J. M. Cowley, *Ultramicroscopy* **9**, 291 (1982); also personal communication (1983).

Figure 20 shows a generic overlayer phase diagram that illustrates the main features to be expected for overlayers that can be treated within a lattice-gas model, i.e., a model in which the ordered state consists of overlayer atoms adsorbed commensurately into regular substrate lattice sites and the disordered state consists of the random placement of the adsorbed atoms into these sites. Incommensurate layers and commensurate–incommensurate transitions require more complex treatment. As can be seen, the generic phase diagram for a commensurate overlayer is the same as for a simple binary alloy that undergoes phase separation, the atoms in the overlayer forming one component and the vacancies the other. Thus cover-

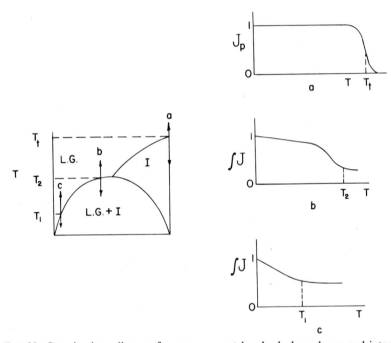

FIG. 20. Generic phase diagram for commensurately adsorbed overlayers and intensity-versus-temperature measurements. The phase diagram plots the allowed phases at any temperature as a function of coverage. I is an ordered phase with a particular structure. L. G. (lattice gas) is the disordered phase. Coexistence regions always occur if the adatoms have an attractive interaction. The dependence of the intensity on temperature at points a, b, and c is indicated schematically. The Debye–Waller factor has been removed. The integral is over the diffracted beam, which at b and c becomes increasingly broad as T increases. At a, the reflection remains sharp as long as there is any order, and the integral is equal to the peak intensity when measured with a typical instrument.

age is analogous to composition. The major features are two-phase regions, consisting of coexisting ordered and "disordered" phases (the disordered phase being an ordered phase of vacancies), and one-phase regions, either ordered or disordered in the sense just given. One-phase regions can have a width because a phase rich in one component can support a certain concentration of the other component before phase separation takes place. For overlayers, this means that at a given temperature the ordered overlayer can support a certain concentration of vacancies and the empty lattice (an ordered phase of vacancies) can support a certain concentration of overlayer atoms. Transitions from a two-phase region to a one-phase region or viceversa are, by necessity, first-order, whereas transitions between one-phase regions may be of first or second order. It is necessary to distinguish between actual order–disorder transitions, such as point a, where one phase goes continuously or discontinuously into another, and disappearing-phase transformations, such as point c, where, as the temperature is increased, an ever smaller amount of a given phase exists, but where the order in that phase does not significantly change with temperature because the phase boundary is nearly vertical. Intermediate between these cases is point b, where both the amount of the phase and the order in the phase change, with most of the change in order coming when only a little of the phase is left.

The most important measurement in phase transition experiments is the dependence on temperature of the Bragg intensity (the area under an angular profile) of a particular reflection, i.e., $\int_{\text{beam}} J_{hk}(\vartheta,E)\, d\Omega$ versus T, at various coverages. Three possible Bragg-intensity-versus-temperature plots for three different regions in the phase diagram are shown schematically in Fig. 20. It should be noted that the intensity can decay quite differently with temperature for different regions of the phase diagram. Angular distribution measurements are useful in interpreting the temperature behavior and in identifying transition temperatures or equilibrium distributions of ordered phase. In order–disorder transitions such as that shown at point a, no broadening of reflections should be observed until the transition temperature is reached. For disappearing-phase measurements, the reflections broaden continuously, but the broadening cannot be observed until the physical width of the line exceeds the resolving power of the instrument. Usually this occurs at temperatures considerably lower than the temperature at which the phase first precipitates (or, conversely, disappears) for a given coverage. Instrument response affects the measured dependence of the Bragg intensity on temperature because the angular profile changes as the order changes. The measured intensity at any diffraction geometry is the convolution of the true intensity $I(\vartheta,E)$ and the instrument response function $T(\vartheta,E)$ [see Eq. (5.3.2)]. Accurate determination of intensity–decay profiles requires the deconvolution of the instrument response function from the measurement.[74] Measurements of peak intensities are meaningless unless it can be demonstrated that the angular profile is narrow compared to the instrument response function at all measurement temperatures. This can be true only for points on the phase diagram such as point a, where long-range order is preserved up to the transition temperature, at which it vanishes.

Phase transition studies have general applicability to identification of the nature of the phase, including its geometric structure; the energetics that lead to the formation of a given phase, including adatom interactions and impurity stabilization; and the effect of defects or kinetic limitations on the formation or degree of order in a phase. Measurements of phase transitions between one-phase and two-phase regions are especially interesting from the point of view of adatom interactions, nucleation phenomena, and kinetics of ordering.

In the next section, the most important features of the instrumentation required to perform diffraction experiments are described, emphasizing the aspects of instrumental design that optimize a system for particular measurements.

5.5. Instrumentation and Sample Preparation

The major components of a diffractometer are a source, a sample goniometer, and a detector. In addition, for surface studies sample heating or cooling, a gas handling system or evaporation source, a mass spectrometer,

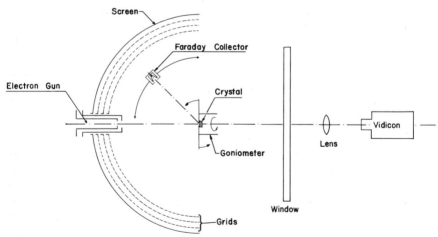

FIG. 21. Schematic diagram of a LEED diffractometer with Faraday cup and vidicon detectors.

[74] G.-C. Wang, T.-M. Lu, and M. G. Lagally, *J. Chem. Phys.* **69**, 479 (1978).

5.5. INSTRUMENTATION AND SAMPLE PREPARATION

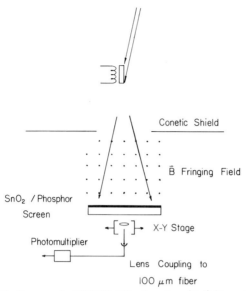

FIG. 22. Schematic diagram of a RHEED diffractometer with light-sensitive-diode detector and magnetic-field deflection of diffracted beams. (After Van Hove et al.[68])

and a separate gun for Auger electron spectroscopy are typically available. Schematic overall views of a LEED and a RHEED diffractometer are shown in Figs. 21 and 22. The focus of the discussion here will be on the gun, goniometer, and detector.* A brief discussion of sample preparation is also included.

5.5.1. Electron Guns

Electron guns used in diffractometers typically use electrostatic focusing and have a simple design consisting of a thermionic cathode, an extraction electrode, an array of focusing electrodes, and electrostatic deflection plates for guiding the beam. Filaments are usually made of W or thoriated W and may be hairpin wires or ribbons. Indirectly heated filaments (oxide, LaB_6, etc.) are also used. In order to avoid background light and contamination of the sample due to evaporation from the filament, some electron guns have off-axis filaments. The thermionic cathode is situated in a Wehnelt cylinder, a can that completely surrounds the cathode except for the beam extraction aperture. The potential on the Wehnelt cylinder can be adjusted positive or negative with respect to the cathode, and is typically at nearly the same

---

* For a detailed discussion of LEED instrumentation, see reference 45.

potential. Extraction of the beam is achieved with the first electrode of a unipotential lens that then focuses the initially divergent beam. This lens forms an image of the true electron source, which may be part of the hairpin or the aperture of the Wehnelt cylinder, depending on the filament type, the filament current, and the potentials on the Wehnelt cylinder and the extraction electrode. Wehnelt cylinder and first-lens element apertures in low-energy electron guns are typically about 1 mm in diameter or larger. The beam diameter can be decreased by reducing these aperture sizes. Small apertures give much reduced beam currents (e.g., 1 nA at 100 eV for a 0.4-mm-diameter first-lens aperture). For the 0.1-μA beam currents required with commonly used detector schemes (grids and phosphor screen, or a simple Faraday cup), a minimum beam size of 200 μm and a minimum divergence of 0.2° appear to be achievable at LEED energies with most common guns. Changing the bias on the Wehnelt cylinder with respect to the cathode drastically affects the beam current and its stability with energy, but does not significantly affect the smallest achievable beam diameter at the focus conditions for typical current densities (less than 0.2 mA cm^{-2}). At RHEED energies, spot sizes can be made much smaller.

Low-energy electron guns have not been optimized with respect to the parameters that give a high resolving power. Historically, beam currents of 0.1 μA in a spot of about $\frac{1}{4}$- to 1-mm diameter with a beam divergence of 0.25°–1° have been considered adequate for LEED. Wulfert and Henzler[75] have constructed a magnetically focused gun to improve this situation. They use the concept of focusing with a long solenoid to produce small-diameter low-energy beams with beam sizes of the order of 40 μm[75] in the imaging plane. The magnetic field is along the beam direction. The magnetic field acts as a 1 : 1 lens that images the crossover of the beam emerging from the Wehnelt cylinder of the gun onto the detector after reflection by the crystal. The corresponding minimum angle of resolution for normal incidence appears to be of the order of $\vartheta_{min} = 0.05°$, compared to values of the order of $\vartheta_{min} = 0.5°$ obtainable in systems equipped with standard guns. Because of the small beam currents achievable with this gun, a detector with gain is used, which in this system[75] is a Faraday cup with a channeltron electron multiplier.

Another development in low-energy electron guns is the use of field emission sources. The major advantages of a field emitter are its high luminosity and the fact that it is nearly a point source. The latter in principle makes a parallel beam a possibility. This can easily be visualized by considering a point source situated at the focus of a convergent lens. The high luminosity makes small beam sizes practical. Although field emitters have been used in

---

[75] F.-W. Wulfert, Ph.D. Thesis, Univ. of Hannover, 1982; see also ref. 43.

5.5. INSTRUMENTATION AND SAMPLE PREPARATION

high-energy-beam applications for some years, it has been difficult to make a low-energy beam without serious energy spread and angular divergence, because a high extraction potential is required with typical anode configurations. Because lenses are imperfect, the subsequent deceleration to low energies causes energy and angular spreading. An extraction anode design[12] that is conceptually different from standard anode configurations and takes advantage of the fact that high fields can be achieved at low potentials by reducing electrode separation allows field emission at potentials as low as 150 V, and perhaps lower. Currents as large a 1 nA at 150 eV with beam sizes of less than 5 μm have been achieved.[12]

Electron guns for RHEED applications operate at voltages anywhere from 5 kV to 100 kV. No particular effort has been made to optimize guns in this energy range for high resolving power, but even standard guns are likely to be better than low-energy guns because it is generally easier to produce a finely focused beam with small energy and angular spreads at higher energies. Various types of guns are in use; examples include guns from electron microscopes, guns in cylindrical-mirror analyzers used for scanning Auger microscopy, and high-energy CRT guns.

5.5.2. Detectors

Signal detection in diffractometers requires the measurement of an energy- and angle-resolved current. For LEED, a detector must be capable of (1) electron current measurement at energies 10 eV $< E_p <$ 1000 eV, where E_p is the incident-beam energy; (2) energy resolution, i.e., the separation of those electrons at or very nearly E_p from the inelastically scattered electrons; and (3) angular resolution, i.e., the ability to separate the current in one diffracted beam from all the others and to measure the angular distribution of current in one beam. For RHEED the energy range at which the detector operates is higher, but in principle its capabilities must be the same as for LEED detectors. Two types of detectors are in common use, a Faraday cup that is mechanically driven and a fluorescent screen with or without a set of hemispherical grids. The most common LEED detector is the fluorescent screen with a set of nested grids. The inner grid is operated at the same potential as the sample to provide a field-free region around the sample. The next grid (or two grids) is set at a negative bias to filter inelastically scattered electrons. The outermost grid is again operated at ground potential but is not needed for dc LEED operation. (If no intensities are to be measured, actually only two grids are required). The fluorescent screen is operated at several kilovolts positive potential to give the electrons sufficient energy to excite the phosphor. The major advantage of this detector is that it provides a visual display of essentially all the back-diffracted beams, which makes a rapid determination of the size and shape of the surface or overlayer unit mesh possible. On the other hand, the ability to give a visual display makes the fluorescent screen an inelegant detector from the point of view of signal processing. In many applications beam currents must be measured. Because the fluorescent screen changes an electron signal to an optical signal, reconversion to an electron signal is required.

Intensities in typical integrated-intensity-versus-energy profiles vary over three orders of magnitude, which makes necessary a similar dynamic range for the fluorescent-screen/detector combination. The response of a phosphor screen can be assumed to be linear over the ranges of beam currents used in LEED.[76] However, the dynamic range of the phosphor generally does not match that of detectors used to measure the optical intensity in a diffraction spot. Maxima in intensity-versus-energy profiles for typical incident-beam currents may saturate the detector. If the incident-beam current is reduced, the minima in intensity-versus-energy profiles become buried in fluorescent-screen noise, e.g., that due to stray light or inelastic scattering background. A fluorescent screen is therefore not ideal for measuring beam intensities quantitatively. In some applications, absolute intensities are not required. For example, angular distributions are independent of beam current as long as the phosphor and the detector are not saturated. For such measurements, the fluorescent screen represents a detector with a very good response. Because the average phosphor particle size is typically of the order of micrometers, the phosphor acts like a detector with a continuously movable, several-micrometer-wide aperture, which is so small that it contributes essentially nothing to the total instrument response. Of course, this optical signal must still be converted into electrons, and thus the aperture width of the light-sensitive detector must be included in the instrument response. The dynamic range must be high for this type of measurement.

An additional negative aspect of most standard fluorescent screens is that they are viewed in reflection, i.e., past the sample and through the grids. Aside from the fact that the sample blocks part of the field of view and that the grids cause a loss of more than half the light intensity from the screen because of their limited transmission (each grid has typically 0.8 to 0.9 transmission), viewing in reflection generally requires the detector to be 20–30 cm from the screen. This causes a significant loss of intensity. Transparent fluorescent screens (glass coated with SnO_2 and phosphor) have been used to avoid these problems. A light-sensitive detector can then be placed directly behind the screen.[57,77,78]

Spot telephotometers have commonly been used for recording the diffracted-beam intensity from fluorescent screens. It is difficult to follow the

---

[76] A. Bril and F. A. Kröger, *Philips Tech. Rev.* **12**, 120 (1950).

5.5. INSTRUMENTATION AND SAMPLE PREPARATION 285

motion of diffraction spots on the screen with a photometer as the energy is varied, and used in the dc mode and photometer lacks sensitivity. As a result, other methods of measuring the brightness of the fluorescent screen have been developed. To improve the sensitivity, Schrott et al.[79] have used a photodiode and synchronous detection, with modulation of the suppressor grid at 100 Hz. Background light is thus effectively removed. Stair et al.[80] recorded intensities by photographing the fluorescent screen at various diffraction conditions with high-speed 35mm film. In this way the intensities of all reflections are obtained at the same time under identical conditions, a method far preferable in terms of data reliability to measuring the intensity-versus-energy profiles sequentially. The film is subsequently mechanically scanned and digitized using a computerized microdensitometer. A computer program locates the diffracted beams and provides an integrated intensity for each beam at each incident-beam energy. The time to develop and digitize the film is long, which results in considerable delay between a measurement and the availability of the results of this measurement. A modification of the scanning procedure uses a vidicon camera interfaced to a minicomputer.[81,82] This reduces the delay time between measurement and availability of the results to about a day. An assessment of the sensitivity of this method has been made by Tommet et al.[82]

Photographing the screen leads to a reduction in total measurement time by introducing parallel detection rather than the serial detection used in a telephotometer. However photography introduces no detector gain. A reduction in total exposure of one or two orders of magnitude (from 10^{16} electrons mm^{-2} to 10^{14} electrons mm^{-2} for a set of intensity-versus-energy curves for all observable beams) is achieved because of the parallel detection. However, measurements of the intensity of each reflection require the same incident beam current and measurement time to achieve the same S/N ratio whether these measurements are obtained simultaneously or sequentially. Hence, in order to achieve a net gain in time, incident beam currents must be of the same order of magnitude as for photometers. The use of photography can be eliminated by using a vidicon camera to view the fluorescent screen directly.[7,9] In this detection scheme, the image on the screen is focused onto

[77] L. de Bersuder, *Rev. Sci. Instrum.* **45**, 1569 (1972).

[78] P. A. Bennett and M. B. Webb, personal communication (1978).

[79] A. G. Schrott, M. D. Chinn, C. G. Shaw, and S. C. Fain, Jr., *J. Vac. Sci. Technol.* **21**, 101 (1982).

[80] P. C. Stair, T. J. Kaminska, L. L. Kesmodel, and G. A. Somorjai, *Phys. Rev. B* **15**, 623 (1975).

[81] D. C. Frost, K. A. R. Mitchell, F. R. Shepherd, and P. R. Watson, *J. Vac. Sci. Technol.* **13**, 1196 (1976).

[82] T. N. Tommet, G. B. Olszewski, P. A. Chadwick, and S. L. Bernasek, *Rev. Sci. Instrum.* **50**, 147 (1979).

the sensitive element of the vidicon tube, which consists of tracks that are divided into a large number of channels. The distribution of intensity in the channels of one track is accumulated into a memory array, with simultaneous subtraction of background light stored previously in a second array. This background-light measurement is made, for example, by biasing the electron gun so that the beam cannot emerge or by turning off the screen voltage. The height of the track as well as the magnification of the optical system can be adjusted, effectively allowing changes in the detector dimensions relative to the intensity distribution on the screen. The contributions of the vidicon detector and lens to the instrument response have been measured and shown to be negligible compared to contributions from the electron beam.[9] Thus this type of detector is excellent from the point of view of resolving power, although, as has been noted, the fact that grids are used affects the sensitivity and achievable S/N ratios because the background due to inelastically scattered electrons is large. In a direct application, a vidicon can only measure diffracted-beam profiles sequentially, but these measurements can be made in real time. Despite the serial detection, the greater sensitivity of the vidicon relative to film gives about the same overall measurement time for a set of reflections. Delays in the availability of the data are eliminated because the data can be analyzed and displayed as they are taken.

The addition of a channel electron multiplier array[8] to the detector improves the sensitivity by introducing gain into the detector but also decreases the resolving power. The mean gain of a chevron (dual) channel electron multiplier array is of the order of 10^6. Thus a reduction in primary beam current of 10^6 gives, in principle, the same S/N at the same measurement times. Because of the finite channel width, however, the channel plates produce a spatial broadening of the input signal. For chevron arrays this broadening is accentuated because the signal coming into one channel in the first plate gets spread into several channels in the second. For a negligibly small beam incident on a typical channel size of 25 μm, the FWHM of the beam on the screen, after passing through the chevron plates, is about 75–100 μm, at least double that of the vidicon/optics/fluorescent-screen combination. If the resolving power is of no concern (e.g., in intensity-versus-energy profiles, for which an integral over the diffraction spot is in any case taken) this combination represents an excellent detector. In angular-profile measurements the limited resolving power that results with this detector becomes important. Although the detector response can, of course, be deconvoluted from the measured intensity distribution, it is clear that the increased sensitivity of channel plates brings a reduction in ultimate spatial resolution.

Because channel-plate arrays are usually flat, the distortion introduced in beams entering the plates at angles away from the normal to the plates must be taken into account.

5.5. INSTRUMENTATION AND SAMPLE PREPARATION

The sensitivity of the detector can be further increased by replacing the fluorescent screen with a position-sensitive pulse detector. With a fluorescent screen biased at typical energies of 5 keV, the minimum measurable current (using a vidicon) is estimated to be 1000 pulses s^{-1}. A position-sensitive detector can measure individual pulses. The first such detector that was constructed for LEED[10] consists of a resistive-anode encoder (RAE) preceded by a chevron channel electron multiplier array. An RAE is a continuous resistive film that acts as a current divider for an incoming electron pulse in order to determine the spatial position of the pulse. Although this type of detector determines the centroid of the arriving pulse, the spatial resolution is nevertheless not good. The spatial resolution is determined by the extent to which the thermal noise perturbs the pulse currents. Thus a trade-off between detector area and maximum allowable thermal noise occurs. For a 75-mm square RAE the lateral resolution is estimated to be between 300 and 400 μm.[83,84] The RAE can accommodate 50-kHz pulse rates. Data rates are therefore limited by the individual channel dead time in the channel-plate array and not by the RAE.

The ultimate attainable sensitivity and angular resolution are provided by a Faraday cup detector with a channel electron multiplier. With this arrangement individual pulses can be counted, and by making the aperture of the detector arbitrarily small, any degree of angular resolution can in principle be obtained. Such detectors are in common use in a variety of spectroscopic techniques, including LEED. Gronwald and Henzler[11] have described a Faraday cup detector that includes deflection plates in front of the aperture so that the beam profile can be measured without mechanical motion of the detector. Some Faraday cup detector designs contain no retarding grids but nevertheless provide much better energy resolution than is obtained with detectors incorporating retarding grids. Such detectors consist of a deep cup in close proximity to, but electrically isolated from, an aperture plate. The diameter of the cup is several times the diameter of the aperture, and the depth of the cup is several times its diameter. The cup is biased to within 1–2 eV of the energy of the elastically scattered electrons. This detector provides excellent energy resolution (of the order of the thermal spread in the incident beam) without significant loss of secondary electrons from the cup. Because the fields outside the aperture are negligible, the angular distribution of electrons in the diffracted beam being measured is not disturbed. The advantage of good energy resolution is that inelastically scattered electrons can be eliminated to a much greater degree, which makes the background in angular profile measurements less of a problem. Finally, a Faraday cup detector is the only means of measuring analog signals quantitatively. Hence

---

[83] M. Lampton and F. Paresce, *Rev. Sci. Instrum.* **45**, 1098 (1974).
[84] H. O. Anger, *Instrum. Soc. Am. Trans.* **5**, 311 (1966).

it is preferable for every diffractometer to have two detectors, the Faraday cup for quantitative current and high-angular-resolution measurements, and some form of position-sensitive parallel-output detector for the rapid data acquisition required for accurate structural determinations within the time or electron dose constraints of a typical surface crystallography experiment.

Detectors for RHEED have consisted in most cases simply of a fluorescent screen, although energy filtering and Faraday cups have been used.[71,72,85] RHEED measurements are always made in transmission through the screen, an advantage in terms of sensitivity. A light- and position-sensitive probe has been used to measure RHEED beam profiles.[57] To eliminate variations in screen response, magnetic deflection of the diffracted beam has been used to scan the beam across a fixed detector.[68,72,85] RHEED measurements in scanning transmission electron microscopes are made with somewhat more sophisticated detection schemes, such as a vidicon camera pickup from the fluorescent screen. In principle, all of the detection schemes discussed for LEED could as well be applied in RHEED.

5.5.3. Goniometers

The function of a goniometer is to position the crystal accurately with respect to the incident beam of electrons and the detector. The importance of such accuracy depends on the experiment being performed. For angular-profile measurements, absolute angles are not important. For intensity-versus-energy profiles, it is critical that absolute angles be well known and that the repeatability of setting these angles be excellent. Although they are not very precise, standard UHV manipulators are commonly used for all types of diffraction measurements.

Goniometers for special purposes have been built. An exceedingly precise one[86] was constructed to perform automatically constant-momentum-transfer averaging[61,62] of intensities. The goniometer is constructed to couple the motions of the Faraday cup and the crystal in colatitude, so that the momentum transfer vector automatically remains constant as the diffraction conditions are varied. Uncoupling of the motions is also possible to permit arbitrary angles of incidence and diffraction.

For fine-beam or scanning LEED applications it is also necessary that the goniometer be stable against vibrations. For such applications, modified versions of manipulators used for scanning Auger spectroscopy or similar techiques can be used.

---

[85] M. F. Tompsett and C. W. B. Grigson, *J. Instrum. Sci.* **43**, 430 (1966); M. F. Tompsett, *J. Mater. Sci.* **7**, 1069 (1972).

[86] P. A. Bennett, Ph.D. Thesis, Univ. of Wisconsin, Madison, 1980; P. A. Bennett and M. B. Webb, *Surf. Sci.* **104**, 74 (1981).

5.5.4. Sample Preparation

Sample preparation is extremely important in surface crystallography experiments. The magnitude of the problem can be appreciated when it is realized that the outer few atomic layers provide all the structural information. It may seem surprising that any surface can be prepared well enough to observe diffraction, until one remembers that the resolving power of typical instruments is only on the order of several hundred angstroms. Thus, surfaces that have ordered regions that are this large on the average appear perfect to the diffractometer. More importantly, ordered regions that are on the average much smaller than 100 Å still give good diffraction pictures. Ordered regions as small as a few atoms across give a measurable diffraction pattern if there are enough of them. Thus the surface order does not need to be very good if there is underlying crystallinity to provide orientational and translational coherence between the ordered regions. Much of the surface can be covered with scratches, etch pits, and so forth, and these will not (superficially at least) affect the diffraction pattern, especially when the instrument has a low resolving power.

Nevertheless, it is quite difficult to prepare surfaces for diffraction experiments. The best surfaces, as regards low defect density and cleanliness, are cleavage faces of crystals that can be cleaved in vacuum. These include a number of semiconductor crystals and the layer compounds. Such surfaces can be used to measure the instrument response. Most crystals do not cleave readily. Thus the materials and surface orientations that can be studied by cleaving in vacuum are quite limited. Most surfaces require extensive preparation. This includes orienting, polishing, and etching before further treatment in vacuum. Orientation to the desired axis to within 0.5°, using Laue back-reflection, is common practice; with the use of a stable multiaxis goniometer[87] and repeated attempts at polishing, accuracy to <0.1° can be achieved. Frequently, polishing results in surface curvature near the edges of the crystal, which results in an apparent misorientation effect there. Thus, care must be taken to use only the center of the crystal surface or to prepare a large-area surface for diffraction studies. The surface produced by polishing is so damaged that no diffraction pattern is observable, generally even with x rays. Subsequent etching removes the polishing damage to a sufficient degree to observe diffraction and also exposes the crystal mosaic, which may range in metals from a fraction of a degree to several degrees for poorly grown crystals. A wide variety of etching and polishing procedures exists for different materials.[88]

[87] W. L. Bond, *J. Sci. Instrum.* **38**, 63 (1961); J. F. Wendelken, S. P. Withrow, and C. A. Foster, *Rev. Sci. Instrum.* **48**, 1215 (1977).

[88] See, e.g., G. Petzow, "Metallographic Etching." Am. Soc. Met., Metals Park, Ohio, 1978; F. Rosebury, "Handbook of Electron Tube and Vacuum Techniques." Addison-Wesley, Reading, Massachusetts, 1965.

At the stage of introduction into the vacuum chamber there is still remanent surface damage and also a contaminant layer that frequently consists of an oxide or a carbonaceous deposit. These can be removed by thermal, chemical, or physical treatments, or a combination of any of them.[89] Thermal annealing is usually insufficient by itself to clean the surface, because of the tenacity of the contaminant layers. A combination of chemical and heat treatment is the gentlest and most satisfactory method of preparing the surface from the point of view of surface defects. It is also exceedingly slow. Generally, an oxidation for large periods of time is required to remove carbon from the surface and to deplete the near-surface regions of carbon, as it continues to diffuse to the sink that is provided by the surface. Oxidation leaves the surface with an oxide or at least a layer of chemisorbed oxygen. Oxygen may also have diffused into the lattice. A reduction in hydrogen can remove the oxygen. The hydrogen itself may be desorbed simply by heating. The resulting surface generally has good crystallinity, although it may still contain a large number of macroscopic defects, such as etch pits, scratches, etc. Thus the area that participates in the diffraction is less than the total surface area, as can be observed in noble-gas dosing experiments.[90] Repeated heat treatments, which are required, for example, to renew a surface after a chemisorption experiment, frequently accentuate gross defect structure, so that the surface may look exceedingly rough and nonspecular after a period of continued use.

The second major *in situ* surface preparation technique is physical removal of the surface contamination or damaged layers by sputter etching with subsequent annealing. Surfaces prepared in this manner always have remanent damage, with a defect structure that is of rather fine scale and frequently observable in the diffraction pattern as a broadening of the angular profiles or an increase in background. Generally, such damage cannot be annealed out in reasonable laboratory times. Noble gases that are used for sputtering are trapped in the lattice to depths of many tens of angstroms or more,[91] which causes displacement of atoms from regular lattice sites, strain, and dislocations. Surfaces are frequently left with a step structure and a nonequilibrium concentration of point defects that are difficult to eliminate. These defects can markedly affect a number of diffraction measurements, especially measurements of the thermodynamics and kinetics of overlayer ordering. Thus sputter etching should not be used on surfaces intended as

---

[89] A summary of vacuum cleaning treatments for wide range of materials is given by R. G. Musket, W. McLean, C. A. Colmenares, D. M. Makowiecki, and W. J. Siekhaus, *Appl. Surf. Sci.* **10**, 143 (1982).

[90] M. B. Webb, personal communication (1981).

[91] Trapping of Ar from the sputter etching process has been observed with Rutherford ion backscattering measurements. L. Feldman, personal communication (1980).

substrates for such experiments, unless the purpose is to study the influence of defects on ordering.

Surfaces can, of course, also be grown *in situ* by molecular beam epitaxy. These can be macroscopically much smoother than polished surfaces. Insufficient work has been done with diffraction on such surfaces to ascertain whether the density of atomic-scale defects can also be made lower than on polished or cleaved surfaces. The fact that many of these films are grown at conditions far from equilibrium suggests that they may have a relatively large concentration of structural defects.

5.6. Representative Experimental Results

In this section we very briefly illustrate the discussion of the different types of measurements with examples that are representative of the data that can be expected with typical LEED and RHEED instrumentation. Rather than presenting "finished" results, we emphasize data as they appear directly in the measurement.

Measurements of the integrated intensity versus energy in LEED have been made mainly with either a fluorescent-screen/grid detector or a Faraday cup. Figure 23 shows a comparison of intensity-versus-energy profiles of W(100) taken in both ways.[9] The agreement between the curves is typical for data taken at different times, with different detectors, or in different laboratories. The differences in peak positions or peak intensities are probably due to the sensitivity of integrated-intensity-versus-energy profiles to the angle of incidence, with even a fraction of a degree causing significant changes in a profile. The rising intensity in the peaks and background of the curve taken from the fluorescent screen is probably caused by the large inelastic-scattering background accepted by this detector relative to that accepted by the Faraday cup. Such curves are taken at a series of different incident angles and for several diffracted beams and then compared to dynamical calculations.[16,17] The order of magnitude of experimental uncertainties involved in equilibrium position determination is illustrated in Fig. 23. Added to this are the uncertainties in parameters entering into the calculations and the difficulty, mentioned earlier, of defining satisfactory reliability factors. The present status of equilibrium position determinations appears to be that, although the precision of calculations may be as good as 0.01 Å in the best cases, the absolute accuracy of structure determinations is probably no better than 0.03 Å.

Angular-profile measurements by LEED are illustrated in Fig. 24, which shows an angular scan over four orders of reflection from cleaved GaAs(110) and GaAs(110) that has been sputter-etched and only partially reordered by

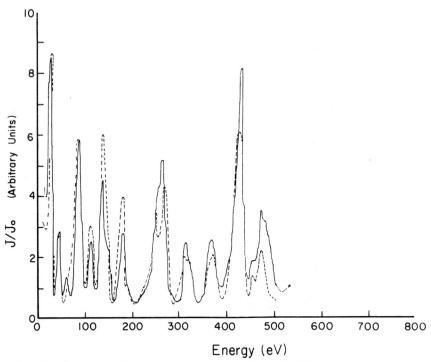

FIG. 23. Comparison of intensity-versus-energy profiles for W(100) taken with a Faraday cup (dashed curve) and with a vidicon detector (solid curve). The two curves are normalized at 80 eV. Measurements were made for a (00) beam at $\vartheta_0 = 22°$. (From Welkie and Lagally.[9] Copyright North-Holland Physics Publishing, Amsterdam, 1979.)

annealing and thus contains many steps.[92] The broadening of the reflections from the sputter-etched surface is evident. The spectra were taken at a diffraction condition at which the amplitudes scattered from the different terraces were out of phase (i.e., near a value of $G_\perp$ at which the rods in Fig. 9 are broadest). The curves represent an average of 30 scans requiring ~ 1 min.[92] Figure 25 shows expanded scans of two profiles taken over identical measurement times from the sputter-etched GaAs(110) surface annealed at two different temperatures.[92] The change in the profile width and shape is evident. The signal-to-noise ratio can differ markedly for the two curves, as expected. A complete two-dimensional angular profile of a diffraction spot[11] is shown in Fig. 26. It was determined by repeated scans, each slightly displaced from the previous one, of a very-small-aperture detector across the

---

[92] H. M. Clearfield, Ph.D. Thesis, Univ. of Wisconsin, Madison, 1984. See also H. M. Clearfield and M. G. Lagally, *J. Vac. Sci. Technol., A* **2,** 844 (1984).

5.6. REPRESENTATIVE EXPERIMENTAL RESULTS

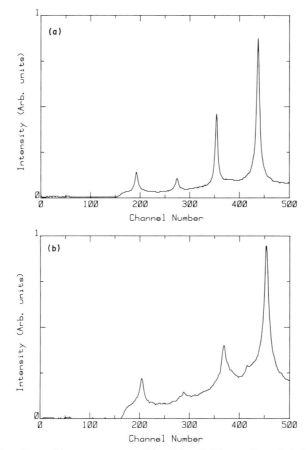

FIG. 24. Angular-profile measurements of the (01) to (04) beams from GaAs(110) surfaces. (a) Cleaved in vacuum, (b) sputter-etched and annealed. The energy, 180 eV, is near an out-of-phase condition for the stepped surface. The (01) beam is at the right. (From Clearfield.[92])

diffracted beam. Low-intensity wings are observed around the main peak. The full width at half-maximum of such angular profiles taken at a variety of $G_\perp$ values (i.e., at different energies) and for several beams is generally used to determine step heights and extract surface step densities,[25,26,43] average overlayer island sizes,[21,40,93,94] or other surface defects.[49] The complete angu-

---

[93] P. O. Hahn, J. Clabes, and M. Henzler, *J. Appl. Phys.* **51**, 2079 (1980).
[94] G. C. Wang, Ph.D. Thesis, Univ. of Wisconsin, Madison, 1978; M. G. Lagally, T.-M. Lu, and G.-C. Wang, *in* "Ordering in Two Dimensions" (S. Sinha, ed.), p. 113. Elsevier/North-Holland, New York, 1980.

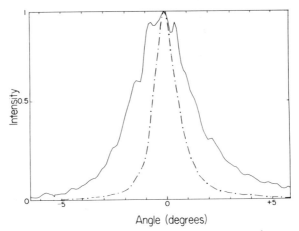

FIG. 25. Comparison of two angular profiles of the (01) beam from a sputter-etched GaAs(110) surface that has been annealed at two different temperatures. Solid curve: 350°C for 10 min. Dashed curve: 560°C for 10 min. Each curve is the average of 30 scans, and requires approximately one minute. The curves are normalized at their maxima. The actual peak intensities differ by a factor of 7. (From Clearfield.[92])

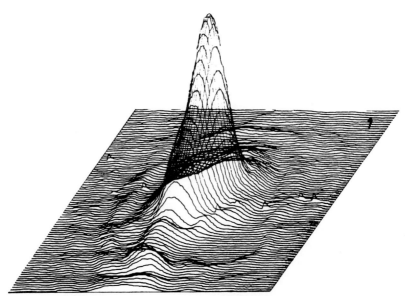

FIG. 26. Two-dimensional angular profile of the (00) reflection from Si(111) at 180eV. (From Gronwald and Henzler.[11] Copyright North-Holland Physics Publishing, Amsterdam, 1982.)

5.6. REPRESENTATIVE EXPERIMENTAL RESULTS

lar profile can be used to determine the size distribution of ordered islands or terraces rather than just the average size.[26,53] The uncertainty in angular profile measurements is typically not better than 5% of the signal, with the resolving power depending on the instrument response. In the best cases, the resolving power may be 5000–10,000 Å. The fitting of complete angular profiles with model calculations is still at a stage at which there can be only little confidence even in the uniqueness of overlayer and surface defect structure determinations. As better models are developed, a rapid improvement in the quantitative nature of structural defect determinations can be expected.

RHEED measurements of angular profiles are illustrated in Fig. 27, which shows contour maps of RHEED streaks for the ordering of a GaAs layer deposited onto a GaAs(001) surface.[57] The contour maps show that for longer ordering times the streaks become shorter and turn into spots, which indicates increased order in the overlayer. Similar measurements are being made on stepped surfaces.[68] Models being developed for defect structure analysis are, of course, equally applicable to RHEED data.

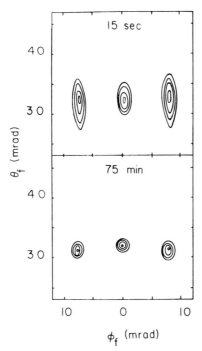

FIG. 27. Contour maps of RHEED profiles taken at two different times after deposition of Ga onto a GaAs(001) surface in the presence of an As background. With increased time the streaks sharpen to spots, demonstrating increasing order in the surface. (From Van Hove et al.[57])

Intensity–temperature measurements used to determine both mean square vibrational amplitudes in surfaces and the positions of phase boundaries are illustrated in Fig. 28. The temperature decay of the "peak" intensity of the $(\frac{1}{2},\frac{1}{2})$ superlattice reflection for saturation coverage of a W(110)p(2 × 1)–O layer[74] is plotted along with the change in width of the angular profile. Both the measured intensity and the intensity corrected for instrument response are shown. The Debye–Waller factor has not been accounted for. It is given by a straight-line fit to the data at low temperatures. The deviation from this straight line indicates a phase transition. Such measurements are repeated at various coverages (see Fig. 20) to establish the

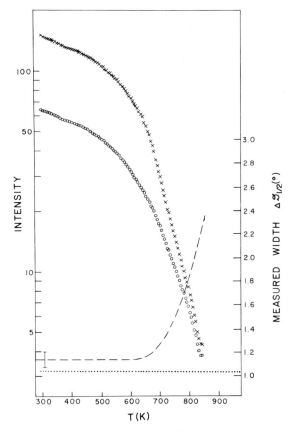

FIG. 28. Intensity–temperature measurement for the disordering of a saturation-coverage W(110)p(2×1)-O layer using $(\frac{1}{2},\frac{1}{2})$ beam at $E = 79$ eV; (○) Measured peak intensity; (x) intensity corrected for instrument response. The change in width of the angular profile (– –) and the instrument response function width (· · ·) are also shown. (From Wang et al.[74])

boundaries in the overlayer phase diagram. Measurements at various energies and for various beams give the $|S|$ dependence of the Debye–Waller factor but should all give the same position of the phase boundary at a given coverge. A careful analysis of the shape of the intensity decay curve can furthermore provide information on finite-size effects, correlation lengths, and the nature of the phase transition, including the determination of the order of the transition[86] and the values of critical exponents.[95] Peak intensities and angular profiles of diffracted beams can also be observed dynamically as a function of time at fixed coverage and temperature to follow the growth of an overlayer or surface phase. From the behavior of the peak intensity and width with time at different temperatures it is then in principle possible to determine activation energies and pre-exponential factors, as well as the growth law operative for the ordered structure.

Intensity–temperature measurements have been made for a number of years, initially to determine the Debye–Waller factor and thermal-diffuse-scattering intensity for clean surfaces and later to determine phase diagrams for overlayers. Debye–Waller factor measurements are difficult to interpret in terms of surface vibrational amplitudes because of the finite and exponentially decaying penetration of the electron beam. Few phase diagram measurements have been made. Difficulties lie in the interpretation of the temperature decay in terms of the position of the phase boundary at different coverages, as illustrated in Fig. 20, and the relationship between the diffracted intensity and the existing order on the surface.

5.7. Conclusions

In this chapter, we have attempted, in a tutorial fashion, to provide the basic elements necessary for an understanding of surface-sensitive diffraction techniques. The emphasis has been on those aspects that will help the reader evaluate the power, as well as the limitations, of diffraction techniques to study surface crystallography. Much of what has been said in the early sections of this chapter is also applicable to grazing-incidence x-ray diffraction and to atomic-beam diffraction, although they have not been explicitly mentioned. Of necessity, a number of other techniques have not been discussed. One of these, the measurement of surface extended-x-ray-absorption fine structure (EXAFS), a potentially quite powerful diffraction technique, can be used to determine the local structural environment (i.e., nearest-neighbor distances) around individual types of surfaces atoms. It is based on measuring the diffraction of a spherical wave that represents a photoelectron

[95] R. L. Park, T. L. Einstein, A. R. Kortan, and L. D. Roelofs, *in* "Ordering in Two Dimensions" (S. Sinha, ed.), p. 17, Elsevier/North-Holland, New York, 1980.

(or a core-level electron excited by any other technique) emitted from a particular type of atom. The resulting intensity as a function of momentum transfer can be evaluated as in the early parts of this chapter. The major difficulties with the technique appear to be the extremely small signal levels and the limited range in momentum transfer that is accessible, factors which lead to relatively large uncertainties in the determination of the positions of the nearest-neighbor atoms.

Limited surface crystallographic information is also provided by other techniques that are not related to diffraction. They are too numerous to mention here. The most important of them are ion-beam techniques, which are discussed elsewhere in this book.

Finally, a review of the experimental aspects of a technique or class of techniques should provide an evaluation of the state of the art as well as a prognosis for the future. The development of technology for surface-sensitive diffraction had lagged considerably behind that of other surface analysis techniques. This has begun to change as the importance of high-quality quantitative structural information in the interpretation and evaluation of the output of surface spectroscopies has become increasingly apparent. One can expect continuing advances in detector design and efficiency, and in the resolving power of diffractometers. One can also expect application of diffraction techniques to a widening range of problems, as well as increased recognition of the necessity of understanding instrumental parameters to obtain truly quantitative information. A further impetus to this development is the marriage of surface science and ultrahigh-vacuum technology with high-energy electron microscopy and diffraction, a field at the forefront of technology in which the ideas of instrumental response, data quality, and quantitative interpretation of results are firmly established. It is likely that this union will bring significant advances both in our understanding of surface crystallography and in the development of experimental techniques.

Acknowledgments

I would like to acknowledge discussions that have aided my understanding of various aspects of diffraction with D. Savage, D. Saloner, J. Martin, T.-M. Lu, P. I. Cohen, M. Henzler, and M. B. Webb. In addition, D. Savage performed a number of important calculations on finite-size effects and developed the visual approach to demonstrating the importance of various contributions to the instrument response that is illustrated in Fig. 14. Work described in this chapter that was performed here was supported in part by the National Science Foundation under Grant No. DMR 78-25754 and in part by the Office of Naval Research. I thank the several graduate students who have been responsible for the performance of this work. Finally, it is a pleasure to thank Ms. Lynn Kendall for typing this manuscript and its several progenitors, H. M. Clearfield, D. Savage, and D. Saloner for proofreading it, and J. Bohlman for making the drawings.

6. ION SCATTERING AND SECONDARY-ION MASS SPECTROMETRY

By W. Heiland

Universität Osnabrück
Osnabrück, Federal Republic of Germany

and

E. Taglauer

Max-Planck-Institut für Plasmaphysik
Munich, Federal Republic of Germany

6.1. Ion Scattering Spectrometry

6.1.1. Fundamental Aspects

The basic phenomenon of low-energy ion scattering for surface analysis is binary scattering. Although it is indeed a very simple concept, its validity was only realized a decade ago,[1-3] a few earlier publications having given some prior indication of this model for surface scattering.[4,5] Its rather late discovery was due to the need for ultrahigh vacuums and the fact that most previous experiments used alkali ions rather than rare-gas ions for surface scattering. Alkali ions are easier to produce and detect, and the backscattered-ion yield is high compared with the low ion yield for rare-gas ions. This has long been understood in terms of ionization energies relative to the work function of solids. But with alkalis it is more difficult to maintain clean surface conditions; only a few experiments with scattering of alkalis from clean surfaces have so far been reported.[6-9]

---

[1] D. P. Smith, *J. Appl. Phys.* **18,** 340 (1967).
[2] D. P. Smith, *Surf. Sci.* **25,** 171 (1971).
[3] W. Heiland and E. Taglauer, *J. Vac. Sci. Technol.* **9,** 620 (1972).
[4] B. V. Panin, *Sov. Phys.—JETP (Engl. Transl.)* **15,** 215 (1962).
[5] V. Walther and H. Hintenberger, *Z. Naturforsch., A* **18A,** 843 (1963).
[6] E. Hulpke, *Surf. Sci.* **52,** 615 (1975).
[7] I. Terzić, B. Goncil, D. Cirić, and J. Vukanić, *J. Phys. E* **10,** 420 (1977).
[8] I. Terzić, D. Cirić, and B. Perović, *Surf. Sci.* **85,** 149 (1979).
[9] E. Taglauer, W. Englert, W. Heiland, and D. P. Jackson, *Phys. Rev. Lett.* **45,** 740 (1980).

6. ION SCATTERING AND SECONDARY-ION MASS SPECTROMETRY

Another important point in this context is the energy range. Many experiments have been concerned with scattering of atoms or ions of thermal energies for the purpose of particle diffraction (Stern–Estermann type experiments). But these particle velocities are comparable with characteristic vibrational velocities of surface atoms, such that the particle energy can be coupled to the phonon spectrum. Binary scattering is not a good approximation for this case. It is thus only for particle velocities corresponding to energies above 10 eV that binary scattering can be expected to occur. For higher energies (above a few kilo-electron-volts) it is well known that particles penetrate matter and therefore the major part of the reflected flux is not backscattered from surface atoms, so that, without additional means, information about surface atoms is not readily obtained.

This introduction shows that there are three physical problems to be discussed: (i) the binary scattering model, (ii) neutralization effects, and (iii) the depth resolution. The first two are essential to establish the qualitative and quantitative aspects of ion scattering as an analytical method, the third point is important in comparison with other methods. The formal presentation of this part will not exactly follow this line, but rather will be experimentally oriented. The information in an ion scattering experiment is obtained from energy spectra of the reflected ions; i.e., two quantities, energy and yield (particles s^{-1} sr^{-1} eV^{-1}) are measured, these being discussed in the first two subsections. The following two subsections deal with the experimental hardware and applications. The idea is to carry through the physical problems defined above.

6.1.1.1. Binary Scattering. In Fig. 1 the results from two scattering experiments are shown, i.e., scattering of He^+ from Pb vapor and from a solid Pb target into a laboratory scattering angle of $\vartheta = 90°$.[10] For solid targets the impact angle ψ (relative to the surface) is also important. For single crystals an azimuthal angle φ has to be defined too (Fig. 2). The spectra shown are the result of a double differential measurement, i.e., angular and energy spectrometry. For the scattering from the vapor phase it can be concluded, since the vapor pressure is sufficiently low, that the scattering is due to binary collisions only. From conservation of energy and momentum it follows for the final energy E_1 of the particles with primary energy E_0 and mass M_1 scattered from atoms with mass M_2 that

$$E_1 = E_0 \left\{ \frac{M_1}{M_1 + M_2} \left[\cos \vartheta \pm \left(\frac{M_2^2}{M_1^2} - \sin^2 \vartheta \right)^{1/2} \right] \right\}^2, \quad (6.1.1)$$

with the positive sign for $M_2/M_1 > 1$ and both signs for $M_2/M_1 \leq 1$. The secondary energy as a function of the scattering angle ϑ_2 of the recoil atoms

---

[10] A. Zartner, Thesis, Tech. Univ. Munich, 1979.

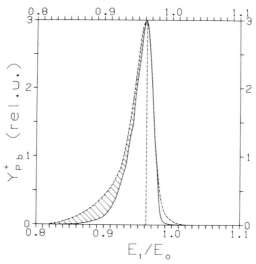

FIG. 1. Energy spectra of He+ ($E_0 = 700$ eV) scattered from solid Pb (solid curve) and atomic Pb in the vapor phase (dashed curve) into a laboratory scattering angle of $\vartheta = 90°$. The relative energy for a single binary collision of 0.962 is marked by the dashed line. (From Zartner.[10])

relative to the projectile direction is given by

$$E_2 = E_0 \frac{4M_1 M_2}{(M_1 + M_2)^2} \cos^2 \vartheta_2. \qquad (6.1.2)$$

For Eqs. (6.1.1) and (6.1.2) it is assumed that the target atom is at rest in the laboratory system. If the atoms are in motion, e.g., owing to thermal vibrations, addition of the velocities results in a broadening of the energy in proportion to $\sqrt{E_0 E_{\text{th}}}$. A test of these functional dependences is essential in any ion scattering experiment. Figures 3[1] and 4[11] give examples that show

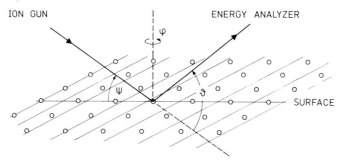

FIG. 2. Geometry of a surface scattering experiment with the laboratory scattering angle ϑ, impact angle ψ, and azimuthal angle φ.

[11] W. Heiland and E. Taglauer, *Nucl. Instrum. Methods* **132**, 535 (1976).

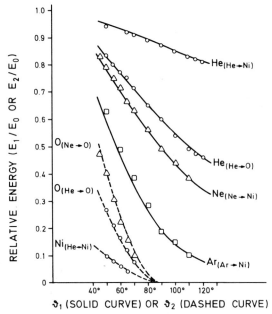

FIG. 3. Angular dependence of the secondary energy of backscattered ions. The curves are calculated assuming single binary collisions. (From Smith.[1])

good agreement between Eqs. (6.1.1) and (6.1.2) and the experimental energies of scattered and sputtered particles. These observations provide the experimental basis for the assumption that ion-surface scattering can be described by binary scattering in this range of energy, scattering angle and mass ratio ($A \geq M_2/M_1$). Multiple scattering and inelastic effects are neglected here (see Subsection 6.1.1.2.3.)

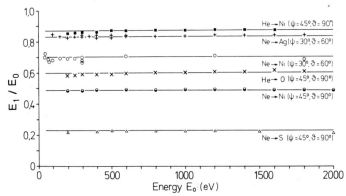

FIG. 4. Energy dependence of the secondary energy of backscattered ions. The straight lines are the theoretical values for single binary collisions. (From Heiland and Taglauer.[11] Copyright North-Holland Physics Publishing, Amsterdam, 1970.)

6.1. ION SCATTERING SPECTROMETRY

If we accept the binary scattering model, it has immediate consequences for the experiment, i.e., the mass and energy resolution are coupled as follows by differentiating Eq. (6.1.1):

$$\frac{M_2}{\Delta M_2} = \frac{E_1}{\Delta E_1} \frac{2A}{A+1} \frac{A + \sin^2 \vartheta - \cos \vartheta \, (A^2 - \sin^2 \vartheta)^{1/2}}{A^2 - \sin^2 \vartheta + \cos \vartheta \, (A^2 - \sin^2 \vartheta)^{1/2}} \quad (6.1.3)$$

for $A > 1$. Figure 5 shows this relation as a function of the scattering angle for different mass ratios and a fixed energy resolution of 1% or $E_1/\Delta E_1 = 100$. For a laboratory scattering angle $\vartheta = 90°$, Eq. (6.1.3) reduces to

$$\frac{M_2}{\Delta M_2} = \frac{E_1}{\Delta E_1} \frac{2A}{A^2 - 1}, \quad (6.1.4)$$

which is plotted in Fig. 6 as a function of M_2 for an energy resolution of $E_1/\Delta E_1 = 100$ and for the rare gases He, Ar, and Xe as primary ions. From Figs. 5 and 6 we conclude that mass resolution increases with increasing scattering angle and decreasing mass ratio A (for $A > 1$).

6.1.1.2. Scattering Cross-Sections and Neutralization.

6.1.1.2.1. ION YIELD. In the case of binary scattering, the intensity of ions backscattered into a solid angle $\Delta \Omega$ (sr) will be a function of the differ-

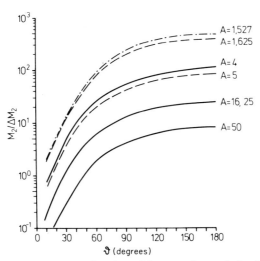

FIG. 5. Mass resolution of ISS as function of the scattering angle for $E_1/\Delta E_1 = 100$. The parameter A is the mass ratio M_2/M_1 (target mass/projectile mass). (From Taglauer et al.[12] Copyright 1975 The Institute of Physics.)

[12] E. Taglauer, W. Melchior, F. Schuster, and W. Heiland, *J. Phys. E* **8**, 768 (1975).

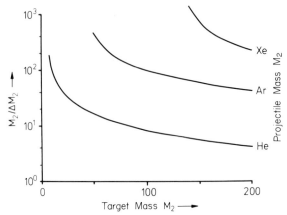

Fig. 6. Mass resolution of ISS as a function of the target mass M_2 for different projectiles. (From Taglauer et al.[12] Copyright 1975 The Institute of Physics.)

ential scattering cross section $d\sigma_s/d\Omega$ (cm² sr⁻¹), the survival probability of the ion P_s, and the density of atoms at the surface N_s (cm⁻²):

$$I_s = C \frac{d\sigma_s}{d\Omega} P_s N_s \, \Delta\Omega. \tag{6.1.5}$$

The factor C contains the primary current, the analyzer transmission, and the detection probability. If the surface is partly covered with adsorbed atoms N_A, the intensity can be written as[13,14]

$$I_s = C \frac{d\sigma_s}{d\Omega} P_s (N_s - \alpha N_A) \, \Delta\Omega, \tag{6.1.6}$$

where α is a shadowing coefficient. This relationship has been observed in many cases. The main assumption made is that the neutralization effect is affected little or not at all by the presence of other species on the same surface. An example is shown in Fig. 7,[15] demonstrating Eq. (6.1.5) for adsorbed atoms. This behavior cannot be generalized; in some cases it has been observed that, for example, the elastic scattering occurs from the substrate, but that the neutralization is affected by the adsorbate.[10,16]

6.1.1.2.2. NEUTRALIZATION. In general three neutralization mechanisms have been established: Auger neutralization, resonance neutralization from conduction electrons into excited levels, and quasi-resonant neutral-

[13] R. E. Honig and W. L. Harrington, *Thin Solid Films* **19**, 43 (1973).
[14] E. Taglauer and W. Heiland, *Surf. Sci.* **47**, 234 (1975).
[15] E. Taglauer and W. Heiland, *Appl. Phys. Lett.* **24**, 437 (1974).
[16] D. J. Godfrey and D. P. Woodruff, *Surf. Sci.* **89**, 76 (1979).

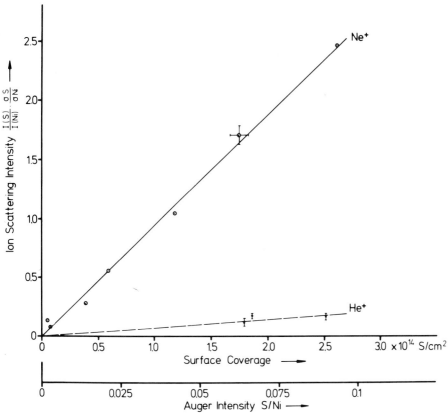

FIG. 7. Measurement of S adsorbed on Ni by ISS in comparison with AES. Primary ions are Ne[+] and He[+]. For ISS the parameters for the reaction Ne[+], He[+] → Ni(111) + S, are E_0 = 1000 eV, ϑ = 60°, and ψ = 30°. For AES relevant parameters are E_e = 2000 eV, i_e = 50 μA, E_{Ni} = 62 eV, and E_S = 150 eV. (From Taglauer and Heiland.[15])

ization from core levels into the ground state of the scattered ion (Fig. 8). The first two mechanisms cause an approximately exponential dependence of the ion-survival probability on the reciprocal-ion velocity,[17a-e] whereas the third mechanism leads to oscillations in the ion yield as a function of the ion velocity.[18-20] Theoretical results are in qualitative agreement with the experiments.[20,21]

---

[17] N. H. Tolk, J. C. Tully, W. Heiland, and C. W. White, eds., "Inelastic Ion–Surface Collisions." Academic Press, New York, 1977; contributions by (a) H. D. Hagstrum, p. 1; (b) W. Heiland and E. Taglauer, p. 27; (c) T. H. Buck, p. 43; (d) T. W. Rusch and R. L. Erickson, p. 73; (e) J. C. Tully and N. H. Tolk, p. 105.

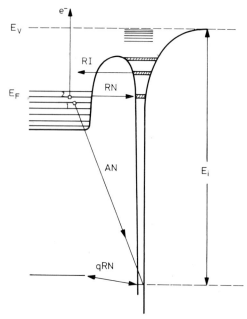

FIG. 8. Model for the electronic structure of an ion (or atom) close to a (metallic) surface. The solid is characterized by the Fermi energy E_F, the ion by the ionization energy E_i. Possible electron exchange processes are RI, resonance ionization (from an excited level), RN, resonance neutralization (into an excited level), AN, Auger neutralization (into the ground state), and qRN, quasi-resonant neutralization (between a core level and the ground state).

Absolute measurements of the ion yield or the charge-state fraction of rare-gas ions are not available below about 1 keV owing to experimental problems. Experiments at somewhat higher energies[17,23,24] allow an estimate; i.e., only about 10^{-2}–10^{-3} rare-gas ions survive in a surface-scattering event. For He^+ these experiments also show different yields for particles scattered from the surface compared with those scattered from the bulk (Fig. 9). A similar phenomenon may also come into play in the case of oxidized metal

---

[18] R. L. Erickson and D. P. Smith, *Phys. Rev. Lett.* **34**, 297 (1975).
[19] N. H. Tolk, J. C. Tully, J. S. Kraus, C. W. White, and S. N. Neff, *Phys. Rev. Lett.* **36**, 747 (1976).
[20] A. Zartner, E. Taglauer, and W. Heiland, *Phys. Rev. Lett.* **40**, 1259 (1978).
[21] J. C. Tully, *Phys. Rev. B* **10**, 4324 (1977).
[22] W. Bloss and D. Hone, *Surf. Sci.* **72**, 277 (1978).
[23] W. Eckstein, V. A. Molchanov, and H. Verbeek, *Nucl. Instrum. Methods* **149**, 599 (1978).
[24] L. K. Verhey, B. Poelsema and A. L. Boers, *Nucl. Instrum. Methods* **132**, 565 (1976).

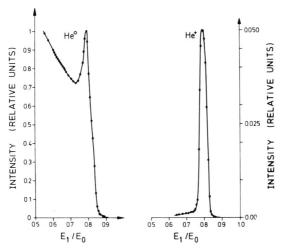

FIG. 9. Comparison of the spectra of neutral He and He⁺ backscattered from Ni. Parameters for the He⁺ experiment are $E_0 = 2$ keV, $\vartheta = 135°$, and $\psi = 90°$. The neutral spectrum shows particles scattered from the bulk. Neutralization of these particles is more effective than of those in the surface peak. (From Eckstein et al.[23] Copyright North-Holland Physics Publishing, Amsterdam, 1978.)

surfaces, where large low-energy tails in the energy spectra are observed[25] in contrast to the Gaussian peak shapes of Fig. 1.[10,17] In principle, these tails are caused by multiple scattering; the extent to which they contribute to the energy spectra depends on neutralization.

From the experimental point of view the neutralization effect is ambivalent. On the one hand, it provides the unique sensitivity of low-energy ion scattering to the top-most atomic layer. On the other, quantitative measurements are only possible with proper calibration, this sometimes being hard to achieve owing to the submonolayer sensitivity.

6.1.1.2.3. MULTIPLE SCATTERING. Multiple scattering is also observed from surface scattering only. In this case neutralization is less effective. The effect has been treated theoretically[26] and demonstrated experimentally first at rather high energies (50 keV).[27] In the low-energy range it mainly causes high-energy peaks in the scattering from single-crystal surfaces (Fig. 10)[28-31]

[25] W. L. Baun, *Appl. Surf. Sci.* **1**, 81 (1977).
[26] E. S. Parilis, N. Y. Turaev, and V. M. Kivilis, *Int. Conf. Phenom. Ioniz. Gases, Belgrade*, p. 47 (1967).
[27] S. Datz and C. Snoek, *Phys. Rev.* **134**, A347 (1964).
[28] E. Taglauer and W. Heiland, *Surf. Sci.* **33**, 27 (1972).
[29] E. S. Mashkova and V. A. Molchanov, *Radiat. Eff.* **13**, 183 (1972).
[30] S. H. A. Begeman and A. L. Boers, *Surf. Sci.* **30**, 134 (1972).
[31] V. E. Yurasova, V. I. Shulga, and D. S. Karpuzov, *Can. J. Phys.* **46**, 759 (1968).

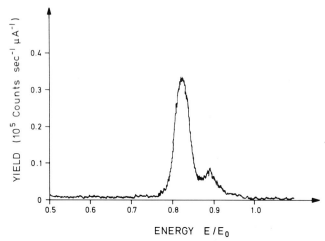

FIG. 10. Energy spectrum of Ne⁺ scattered from Ag(110). The plane of scattering is parallel to the (110) surface. Experimental parameters are $\psi = 30°$, $\vartheta = 60°$, $\varphi = 0°$, and $E_0 = 600$ eV. The high-energy peak is evidence for multiple scattering effects.

and shoulders for scattering from polycrystalline targets.[32] The effect has been exploited by means of computer simulations. Comparison of simulations and experiments have shown that the multiple scattering regime can be used to deduce information about the surface structure, the interaction potential, and the thermal vibrations of surface atoms.[33]

The simplest multiple scattering event is double binary scattering (Fig. 11). It demonstrates the essential aspects; i.e., that the scattering event is dependent on the relative position of the surface atoms and the impact angle, which play no role in a single binary collision. As can be deduced from Fig. 11, the impact parameter p_2 of the second collision depends on the impact

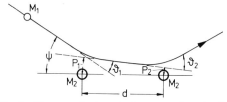

FIG. 11. Scheme of a double scattering event, demonstrating the dependence of the total scattering angle on the impact angle ψ and the surface lattice constant d given by Eq. (6.17). (From Heiland and Taglauer.[11] Copyright North-Holland Physics Publishing, Amsterdam, 1976.)

[32] S. P. Sharma and T. M. Buck, *J. Vac. Sci. Technol.* **7**, 72 (1970).
[33] W. Heiland and E. Taglauer, *Surf. Sci.* **68**, 96 (1971), and references therein.

parameter of the first collision p_1, the impact angle ψ, and the lattice constant d

$$p_2 = p_1 - d \sin(\psi - \vartheta_1), \quad (6.1.7)$$

where ϑ_1 is the laboratory scattering angle.[26] This relation also implies a dependence on thermal vibrations, which determine the actual d, since the vibrations are slow compared with the particle velocity.

Multiple-scattering models are based on an asymptotic approximation. Each collision is considered as an individual binary collision, the trajectories in between are straight lines, and binding forces between surface atoms are (with ample justification) neglected. Figure 12 shows trajectories calculated in a "string model".[34] The calculated energies (including inelastic losses) are

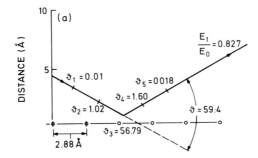

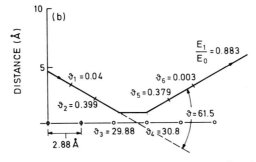

FIG. 12. Trajectories of Ne$^+$ ions scattered from a Ag(110) atomic string for two impact parameters leading to the same total scattering angle (within typical experimental resolution) but two different energies (see Fig. 10 and Table I). $\psi = 30°$, $\vartheta = 60°$, and $E_0 = 600$ eV. (a) $R_{min} = 0.87$ Å. (b) $R_{min} = 1.03$ Å. (From Heiland and Taglauer.[17b])

[34] W. Heiland, H. G. Schäffler, and E. Taglauer, "Atomic Collisions in Solids" (S. Datz, B. R. Appleton, and C. D. Moak, eds.), p. 599. Plenum, New York, 1975.

TABLE I. Energy of Backscattered Ne for Scattering of Ne from Ag(110)

Scattering angle	E_1/E_0 Single scattering	E_1/E_0 String model	E_1/E_0 Experiment $(60 \pm 1°)$
59.4	0.833	0.827 'single'	0.824
60.0	0.830	—	—
61.5	0.823	0.883 'double'	0.892

compared with single-binary-collision values and the experimental result (Fig. 10) in Table I.

The energy of the main peak at $E/E_0 = 0.824$ is in good agreement with the single-binary-collision values (neglecting inelastic losses) as well as the string model values. The inelastic losses are very small in the energy range considered here.[17] As far as the energy is concerned, either the single-binary-collision model or the multiple-binary-collision model, even in its simplest form (two-dimensional), is sufficient. The situation changes, however, if the intensity of the backscattered particles is to be evaluated. It has been shown[35,35a] that out-of-plane scattering, i.e., the three-dimensional structure of the solid, also has to be taken into account in order to get agreement between calculated and experimental energy spectra. On the other hand, the neutralization effect may also be different for the different trajectories, thus complicating analysis of the phenomenon.[9,17c] Nevertheless, under conditions where multiple scattering effects can be neglected, e.g., with He^+ ions under rather large impact and scattering angles, fair estimates of relative intensities can be obtained using Eq. (6.1.5) (see Subsection 6.1.2.2.)

As a method to determine positions of ordered adsorbate structures [O on Cu(110)] a particular aspect of multiple scattering has been used.[36] Downstream from a surface atom (Cu) there is a region of low primary-ion flux (shadow cone). This is compensated for by an increased flux at the edge of this cone. Under carefully selected geometrical conditions, this increased flux can lead to increased scattering from another surface atom (O). The variation of azimuthal and polar impact angles of the incoming beam then provides information about relative positions of surface atoms. The beauty of this approach is that it is only symmetry that is exploited, no knowledge of cross sections being necessary as long as the neutralization properties remain

---

[35] D. P. Jackson, W. Heiland, and E. Taglauer, *Phys. Rev. B* **24**, 4189 (1981).
[35a] W. Heiland, E. Taglauer, and M. T. Robinson, *Nucl. Instrum. Methods* **132**, 655 (1976).
[36] A. G. J. de Wit, R. P. N. Bronckers, T. M. Hupkens and J. M. Fluit, *Surf. Sci.* **82**, 177 (1979).

6.1. ION SCATTERING SPECTROMETRY

constant. So far this method has only been applied to one system, and its real value still has to be established.

6.1.1.2.4. CROSS SECTIONS: ATOMIC POTENTIALS. The basis for cross sections [Eq. (6.15)] and for calculations of trajectories and yields are the interaction potentials $V(r)$ between two atoms at a distance R. In the energy range under consideration the charge state of the projectile can usually be neglected,[37] and so atom–atom interaction potentials are used. The relation between the trajectory and the potential in the center of mass system is given by

$$\Theta_{\rm CM} = \pi - 2 \int_{R_{\min}}^{\infty} \frac{p\, dr}{r^2[1 - (p/r)^2 - V(r)/E_{\rm r}]^{1/2}}, \quad (6.1.8)$$

where p is the impact parameter, R the distance between ion and atom, $E_{\rm r}$ the relative kinetic energy $[E_{\rm r} = AE/(1 + A)]$, and $R_{\min}$ the distance of closest approach. In our energy range, screened Coulomb potentials are most widely used, e.g., the Thomas–Fermi (TF) potential

$$V(r) = \frac{Z_1 Z_2 e^2}{r} \Phi(r/a) \quad (6.1.9)$$

with the Molière approximation to the TF screening function:[37,38]

$$\Phi(r/a) = 0.35 \exp(-0.3\, r/a)$$
$$+ 0.55 \exp(-1.2\, r/a) + 0.10 \exp(-6\, r/a). \quad (6.1.10)$$

Here a is the screening length, which is very often chosen according to Firsov[39] as

$$a_{\rm F} = 0.885 a_0 / (Z_1^{1/2} + Z_2^{1/2})^{2/3}, \quad (6.1.11)$$

and $a_0 = 0.529 \times 10^{-8}$ cm is the Bohr radius. Experimental evidence[35,40] and also theoretical results[41] show that this potential is somewhat too strong. It may be adjusted by using, for instance, $a = 0.8 a_{\rm F}$. Differential scattering cross sections calculated on the basis of the TF potential are shown in Fig. 13.

These cross sections together with the approximate value of the ion-escape probability $P \sim 10^{-3}$ allow a first estimate of the yield to be expected in an experiment with double differentiation of scattered ions. For the estimate we use Eq. (6.1.5) inserting a monolayer density of 10^{15} atoms cm^{-2}, a primary current of 10^{13} ions s^{-1} ($\sim 10^{-6}$ A) and a solid angle of 10^{-3} sr (see Subsection

---

[37] I. M. Torrens, "Interatomic Potentials." Academic Press, New York, 1972.
[38] G. Molière, *Z. Naturforsch., A* **2A**, 133 (1947).
[39] O. B. Firsov, *Sov. Phys. —JETP (Engl. Transl.)* **36**, 1076 (1959).
[40] B. Poelsema, L. K. Verhey, and A. L. Boers, *Surf. Sci.* **64**, 554 (1977).
[41] K. Wilson, L. Haggmark, and J. Biersack, *Phys. Rev. B* **15**, 2458 (1977).

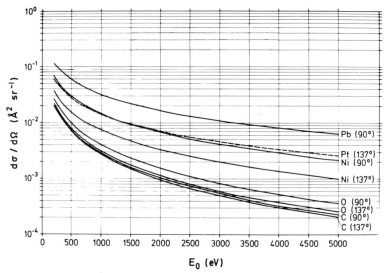

FIG. 13. Differential scattering cross section as function of the projectile energy calculated using a Thomas–Fermi–Molière interaction potential for different atom–atom pairs and laboratory scattering angles of 90° and 137°. The projectiles are He atoms.

6.1.2.1.2.). With $d\sigma/d\Omega$ of 10^{-17}–10^{-18} cm^2 sr^{-1}, this results in ion yields of about 10^4–10^5 ions s^{-1}. This means the experimental setup has to be carefully worked out with respect to analyzer transmission and particle detection.

6.1.2. Experiment

6.1.2.1. Apparatus: To Buy or Not to Buy, That is the Question. The general requirements for a system using ion beams for surface physics have to be in accordance with some basic needs. A given state of a surface has to be achieved, maintained, and monitored, where the term "state" may include structure on an atomic scale, composition, vibrational properties, etc. These qualities cannot, in general, be quantified by a single analytical technique, so that many surface physics experiments are indeed little (highly priced) factories which contain several analytical systems and also separate vacuum systems. The latter complication is caused by problems of sample preparation, which very often calls for a major expenditure of experimental time. On the other hand, the bare vacuum system is already a major engineering and capital investment, and therefore some foresight is needed for the planning of an experiment, i.e., which types of surface properties or interactions are of interest. For example, a study of the surface states of a semiconductor will require a different approach from that in research on catalytic reactions in a pressure range from UHV to 1 atm. A system built for the study of ion–surface interactions will probably also be different from an experiment where ion–surface scattering is added as a means of surface composition analysis, in the same sense that a system capable of detection of Auger electrons is not *a priori* suitable for the study of the physics of the Auger effect and, for example, research on catalysis. The following subsections therefore can only review some more or less successful approaches to ion scattering experiments. In practice, however, everybody has to make his own decision about what might be useful or feasible.

6.1.2.1.1. ION SOURCES. From the previous subsections it is clear that an ion beam for surface scattering experiments has to be defined with respect to energy (kinetic energy and energy width), mass, and angular width [Eqs. (6.1.1), (6.1.3), (6.1.5)]. For practical purposes, time stability with respect to current and position in space of the beam is also advantageous. For some applications, certainly depth profiling, beam rastering and signal gating are a necessity.

So far only rare-gas ions have been used for surface analysis. Other ions (alkali, hydrogen) are used in studies of the scattering mechanism[6-9] and charge-state formation.[42,43] Gas-discharge sources[1,42-44] or electron-impact sources[12,45-47] are equally suitable for the production of low-energy ion beams. Electron-impact sources are most readily built using commercial ionization gauges (Fig. 14). The changes to be made are simple, like cutting off the collector and the front part of the grid. A cylinder is used as electron reflector as well to improve the ionization efficiency. A lens system is used for ion extraction. This is followed by an imaging system. Figure 15 demonstrates the performance of an ion source like the one shown in Fig. 14.[48] The current density increases approximately as $U^{3/2}$. An example of a plasma ion source is shown in Fig. 16.[42,49,50] The advantages of electron-impact sources compared with other sources are in the stability of the current with time and the possibility to change the current without any changes in beam profile or energy. They have the same properties in these respects as an ionization

---

[42] W. Eckstein and H. Verbeek, *Vacuum* **23**, 159 (1973).
[43] S. A. Agamy and J. E. Robinson, *Nucl. Instrum. Methods* **149**, 595 (1978).
[44] D. J. Ball, T. M. Buck, D. McNair, and G. H. Wheatley, *Surf. Sci.* **30**, 69 (1972).
[45] 3M, Brand Model 515, Series CMA.
[46] H. Niehus and E. Bauer, *Rev. Sci. Instrum.* **46**, 1275 (1975).
[47] W. P. Ellis and T. N. Taylor, *J. Vac. Sci. Technol.* **15**, 679 (1978).
[48] J. Kirschner, personal communication, 1975.
[49] M. von Ardenne, *Z. Angew. Phys.* **1**, 619 (1962).
[50] H. Liebl and W. W. Harrison, *Int. J. Mass Spectrom. Ion Phys.* **22**, 237 (1976).

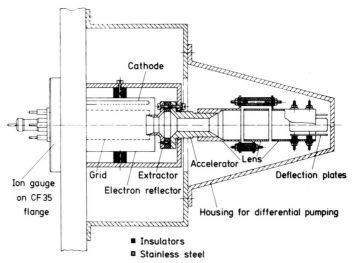

FIG. 14. Schematic of ion source for ISS using a commercial ionization gauge for electron impact ionization (see also Fig. 15).

gauge. It should be noted that with all ion sources the ion energy is not exactly equal to the voltage applied to the ion cage (impact source) or the anode (plasma source).

The sources discussed so far normally have a sufficiently low energy spread (Fig. 17).[51] This is caused mainly by the potential distribution within the source. Plasma sources may have an even narrower energy distribution if the ions are only extracted from the edge of the plasma.

Mass separation of the ions is not an absolute necessity, but it has the advantage of providing isotopically clean beams, which improves the quality of the spectra [Eq. (6.1.1.)]. A Wien filter, quadrupole, or magnetic sector field may be used. With the latter, direct line of sight between source and target is avoided, so that impurities that may evaporate from any part of the ion source will not be deposited on the target. Mass resolution does not have to be very high; e.g., to separate ^{20}Ne from ^{21}Ne, $M/\Delta M = 20$ is obviously sufficient.

6.1.2.1.2. ENERGY ANALYSIS. The energy and angular analysis are the essential parts of an ion scattering experiment [Eq. (6.1.1)]. The scattering angle ϑ has to be known to an accuracy which matches the accuracy of the energy analysis. The solid angle of acceptance also has to be defined, which enters into both the accuracy of the scattering angle ϑ as well as into the energy resolution.

---

[51] M. Grundner, W. Heiland, and E. Taglauer, unpublished work, 1973.

6.1. ION SCATTERING SPECTROMETRY

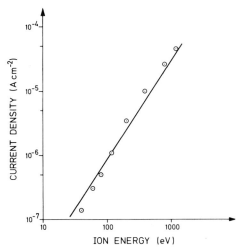

FIG. 15. Current density versus ion energy (acceleration voltage) of an ion source as shown in Fig. 14. The line gives the relation $i \propto U^{3/2}$, where i is current density and U is ion energy. (From Kirschner.[48])

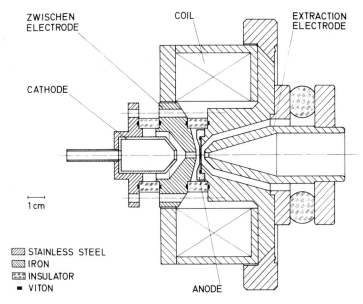

FIG. 16. Schematic of duoplasmatron ion source and extraction electrode for negative and positive reactive gases. Ions are generated in a hollow cathode discharge. (From Liebl and Harrison.[50] Reprinted by permission of the publisher from H. Liebl and W. W. Harrison, In *International Journal of Mass Spectrometry and Ion Physics*, Vol. 22, p. 237. Copyright 1976 by Elsevier Science Publishing Co.)

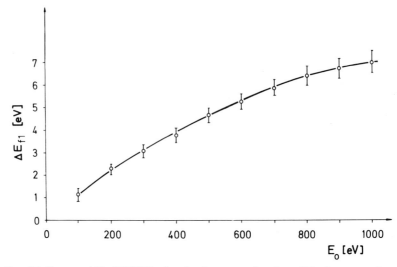

FIG. 17. Energy width (FWHM) of an ion beam as a function of the ion energy from an electron-impact ion source (e.g., like Fig. 14). (From Grundner et al.[51])

The laboratory scattering angle between the incident beam and the analyzer (Fig. 2) can be mechanically adjusted and checked by optical means. In systems with variable ϑ, i.e., with a rotatable energy analyzer, it is feasible to measure the incident beam directly at $\vartheta = 0$. The additional step needed is the measurement of the energy distribution of thermionic ions from a wire at a given potential located at the target position. Then the analyzer energy resolution and the ion-beam width can be evaluated. This is how Fig. 17 was obtained. In cases where direct adjustment of ϑ is not possible, the scattering angle can be evaluated with sufficient accuracy by measuring the binary scattering from a clean metal surface [using Eq. (6.1.1)].

The energy analysis can be done with electrostatic analyzers, i.e., cylindrical[1,12,44] or spherical[41] electrostatic prisms, cylindrical-mirror analyzers (CMAs), time-of-flight analyzers,[43,52] and magnetic spectrometers.[53]

There are some points that need special consideration in the context of surface-physics experiments. For basic research especially more than one analytical technique will be used, and thus compatibility is very important. Magnetic sector fields in one system with low-energy electron devices (LEED, ELS) may be problematic. Cylindrical-mirror analyzers usually allow little access for other techniques, so that the target manipulator must afford transport to different analyzing systems. Target positioning, however,

[52] T. M. Buck, Y. S. Chen, G. H. Wheatley, and W. F. van der Weg, Surf. Sci. 47, 244 (1975).
[53] L. L. Tongson and C. B. Cooper, Surf. Sci. 52, 263 (1975).

6.1. ION SCATTERING SPECTROMETRY

is rather critical for ion scattering for some types of measurements (e.g., multiple scattering studies). Furthermore, a variation of the laboratory scattering angle is practically impossible with a CMA, at least not without trading off its main advantage, i.e., the larger angle of acceptance compared with other analyzers.

The CMA is well documented in the literature[54,55] and also commercially available for ion or electron spectroscopy. In fact, the change from electron to ion spectroscopy is mainly a change of polarities.[46] We shall give here as an example some details of the design of a spherical electrostatic analyzer, which is a special case of the toroidal analyzer (Fig. 18). Inside the sector, cylindrical coordinates (ρ, ζ, φ) are used; outside (x, y, z) are used. Stray fields are neglected. The equation of the trajectory for a particle starting at ρ_1 with an angle α to the median is, for the spherical case,[56,57]

$$\rho = (1 + \rho_1) \alpha \sin \varphi + (\rho_1 - \delta) \cos \rho + \delta, \qquad (6.1.12)$$

where $\delta = \Delta E/E$ is the energy resolution. The decisions to be made are the choice of the median radius r_0 and the deflection angle ϕ. They are limited by the usual restrictions in size, but, obviously, larger dimensions allow good resolution with high intensity. The energy resolution δ has to be chosen according to the physical problems to be investigated; for standard analysis $\delta = 0.01$ is sufficient. The width s of the aperture then has to be calculated to reach this resolution according to $\Delta E/E = s/r_0$. The next step is to match the

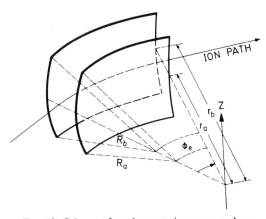

FIG. 18. Scheme of an electrostatic energy analyzer.

---

[54] E. W. Blauth, *Z. Phys.* **147**, 228 (1957).
[55] P. W. Palmberg, G. K. Bohn, and I. C. Tracy, *Appl. Phys. Lett.* **15**, 254 (1969).
[56] H. Ewald and H. Liebl, *Z. Naturforsch., A* **10A**, 872 (1955).
[57] H. Ewald and H. Liebl, *Z. Naturforsch.,* **12A**, 28 (1957).

acceptance angle α to the energy resolution, i.e., beams with a deviation α that pass through the aperture will be allowed. For the beams outside the analyzer we find the equation

$$y = r_0 K_{11} \alpha^2, \tag{6.1.13}$$

where

$$K_{11} = \frac{2}{3}\frac{l'}{r_0}\sin\phi + \left(\frac{2}{3} - \frac{1}{6}\frac{l'^2}{r_0^2}\right)\cos\phi + \frac{1}{3}\left(2 - \frac{1}{2}\frac{l'^2}{r_0^2}\right)\sin^2\phi$$
$$+ \frac{1}{3}\left(2\frac{l'^2}{r_0^2}\right)\cos^2\phi + \frac{3}{6}\frac{l'}{r_0}\sin^2\phi - \frac{3}{6}\frac{l'^2}{r_0^2} - \frac{3}{6}. \tag{6.1.14}$$

With α obtained from Eq. (6.1.14) and (6.1.13) we then calculate the beam width in the middle of the analyzer using Eq. (6.1.12). This number is needed for the definition of the radii r_a and r_b of the inner and outer plates.

If we choose, for example, $\phi = 90°$, a symmetric object–image configuration ($l' = r_0$), and (as above) $E/\Delta E = 100$ hence $s/r_0 = 0.01$, then we obtain $\alpha = 0.1 \approx 6°$. However, with respect to Eq. (6.1.1) adequate mass resolution requires smaller acceptance angles, of the order of 1°. From the first term of Eq. (6.1.12) one has $\rho_m \sim 0.1 \sin 45°$ or $\Delta r_m = 0.07 r_0$. In practice, a slightly larger value may be used as a safety measure. In Fig. 19 the experimental arrangement to test an analyzer is shown. With a median radius of $r_0 = 35$ mm, $r_a = 39$ mm, and $r_b = 31$ mm and the diaphragms placed at a distance r_0 with radii of 0.35 mm, we expect a $\Delta E/E = 0.01$ in first approximation. To terminate the stray fields at the entrance and the exit of the

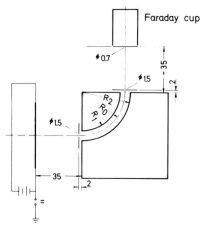

FIG. 19. Experimental setup for the measurement of the energy resolution of an energy analyzer. (From Grundner et al.[51])

6.1. ION SCATTERING SPECTROMETRY

analyzer, additional diaphragms, according to Herzog[58] are used. With a thermionic ion source the energy resolution is obtained from the measured FWHM, if we assume Gaussian peak shapes for the (thermal) ion energy distribution and the spectrometer function. Figure 20 shows the result. The slope being proportional to $(\Delta E/E)^2$, we find a value of $\Delta E/E = 1.16 \pm 0.07\%$, in good agreement with expectation. The deviations are mainly due to the aberrations, which are of the order of α^2.

The energy resolution of a time-of-flight (TOF) system was discussed by Agamy and Robinson[42] and Buck et al.[59] for scattering ions or neutrals. Since the energy is $E \propto v^2$ and the flight time is $t \propto 1/v$, the energy resolution is $\Delta E/E = 2\, \Delta t/t$. The energy resolution is therefore mainly defined by the length of the TOF tube. Since the physical size of particle detectors (channeltron, multiplier) is rather small (1 cm), the solid angle of detection for handy tube lengths (1 m) becomes of the order of 10^{-4}, compared with the 10^{-3} of electrostatic analyzers. For particle velocities between 10^6 and 10^8 cm s^{-1}, flight times are 10^{-4}–10^{-6} s for a tube length of 1 m. For a $\Delta E/E = 0.01$ the detection system has to have nanosecond capabilities. This is standard elec-

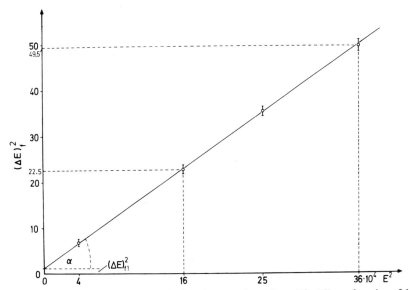

FIG. 20. Energy resolution (FWHM) of an electrostatic analyzer (Fig. 19) as a function of the ion energy. ΔE is the (constant) energy width of the thermal ion source used for the analysis. (From Grundner et al.[51])

[58] R. Herzog, Z. Phys. **41**, 18 (1940).
[59] T. M. Buck, G. H. Wheatley, G. L. Miller, C. A. H. Robinson, and Y.-S. Chen, Nucl. Instrum. Methods **149**, 591 (1978).

tronic technique. The pulsing of the primary beam, which is a standard mode of operation, entails experimental problems, which need careful attention. The time resolution is determined by the beam and aperture diameters, pulse rise time, beam energy spread, etc.[59] In general, the experimental requirements for TOF measurements are more complicated and expensive than for electrostatic analyzers. One advantage of the TOF is the possibility of measuring charged as well as neutral reflected particles if appropriate detectors are available. Moreover, it has the potential of drastically (i.e., by orders of magnitude) reducing the bombardment dose on the target owing to the fact that all scattered energies from one single pulse can be recorded,[59] whereas with electrostatic analyzers stepwise energy scans are necessary. The situation can be improved by combining an electrostatic analyzer with a position sensitive detector.

6.1.2.1.3. PARTICLE DETECTION. Experimental evidence (Fig. 1) and estimates [Subsection (6.1.1.2.4)] show particle fluxes beyond the energy analyzer of $10^4 - 10^5$ particles s^{-1}. Mainly open secondary-electron multipliers or channeltrons are therefore used in combination with particle counting techniques. Daly-type converters are also applicable.[60] Sufficient post acceleration for ions below 3 or 2 keV is necessary to ensure reproducible and sufficient detection probability. The point is to operate the detectors in the "kinetic" electron-emission range, in which the secondary-electron emission coefficient is larger than 1 electron per particle. For energies below 2 keV the electron emission steeply decreases to the range of the "potential" emission (i.e., the order of 1 electron per 10 incident particles).

Accurate neutral-particle detection in this energy range is only possible after very tedious calibration of the detector, or after reionization of the neutrals. This has only been successfully done by the stripping cell method.[42] The stripping cross sections are not very well known, so that the gas cell has to be calibrated at least for absolute measurements. Another problem is the unavoidable leaking through the orifices of the gas cell, giving rise to an additional gas load to the system. With gas-cell systems and electrostatic energy analysis, neutral particle detection has been extended down to about 500 eV for He. Neutral particle detection is, of course, essential for the study of charge states. It is also very useful for multiple-scattering studies but not necessarily for surface-composition or structure analysis. As mentioned above, the neutralization effect at the surface is beneficial in simplifying the spectra, at the expense of incurring problems in quantitative analysis.

6.1.2.1.4. BEAM-CURRENT PROBLEMS. In this subsection we discuss some problems that are rather specially related to the fact that ion beams are

---

[60] E. W. Blauth, W. M. Dräger, J. Kirschner, H. Liebl, N. Müller, and E. Taglauer, *J. Vac. Sci. Technol.* **8**, 384 (1971).

used. This fact has a bearing on the vacuum system requirements, target preparation and handling, and experimental procedure; i.e., the beam causes a neutral gas load, the targets ought to be "plane" and "clean," and they will be damaged depending on several parameters. These factors have to be taken into account when starting a surface study with ion beams.

The vacuum problem is straightforward: ion sources are operated at pressures of up to 10^{-4} mbar, whereas the target chamber should be in the 10^{-10}- or 10^{-11}-mbar range at least for the reactive gases. The problem is solved by differential pumping, which can be done relatively easily when a mass separator is used for the beam. If not, differential pumping is difficult since the limited space does not afford enough pumping speed. In this case differential pumping can be achieved by flooding the chamber, e.g., with He and pumping with, for example, a Ti sublimator and a liquid-nitrogen-cooled cryoshield. This allows the partial pressures of reactive gases to be kept sufficiently low.

The ion beam causes damage to clean surfaces by sputtering and desorption of adsorbed layers. In Fig. 21 the time needed to sputter $\frac{1}{10}$ of a monolayer of solid material (10^{14} atoms cm^{-2}) is estimated at a current density of $i = 10^{-6}$ A cm^{-2} or 6.24×10^{14} ions cm^{-2} s^{-1}. The example chosen is polycrystalline Ni with perpendicular incidence of the beam. Sputter yields are taken from Roth et al.[61] To stay with the examples discussed so far, this time

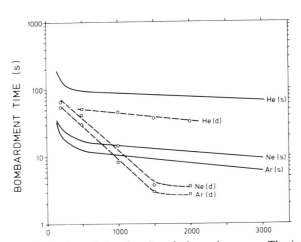

FIG. 21. Time available for analysis as function of primary-ion energy. The time is estimated for the desorption of $\frac{1}{10}$ of a monolayer (d) or sputtering of $\frac{1}{10}$ of a monolayer (s) assuming typical desorption cross sections, sputtering yields, and primary current densities (see text).

[61] J. Roth, J. Bohdansky, and W. Ottenberger, unpublished work, 1979.

allows easy estimates of the number of spectra that can be recorded within the damage limit estimated. For an energy resolution of 1% a full spectrum takes a minimum of 100 channels, which at a rate of 10^4 counts s^{-1} in the peak (see estimate above) could be run in 10 s with sufficient statistics. In many cases it is possible to take the interesting part of the spectrum (peak) to lengthen the time available.

In the case of desorption the situation is comparable at low energies, but at higher energies it is more difficult than in the case of sputtering (Fig. 21). The time for the desorption case corresponds to the time needed to desorb about 1/10 of an adsorbed layer. An exponential desorption behavior $N(t) = M(0) \exp(-\sigma_D i t)$ is assumed [N = particles cm^{-2}, σ_D = desorption cross section (cm^2), i = current density (A cm^{-2}), and t = time (s)]. The example chosen is sulfur on nickel (single crystal) with a current density of 10^{-6} A cm^{-2} (as above), but incident at $\psi = 30°$ to the surface. (In Fig. 21 the comparison is made for the same total current, i.e., for $\psi = 30°$ the current density is reduced to 5×10^{-7} A cm^{-2}.) The desorption cross sections are taken from Taglauer et al.[62] For adsorption studies He is certainly the ion to use, or Ne at very low energies. The philosophy concerning data acquisition is, of course,

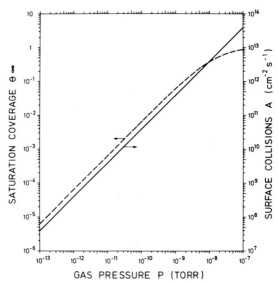

FIG. 22. Saturation coverage Θ_∞ (dashed) and surface collisions (solid) as function of the gas pressure. Θ_∞ is estimated for He$^+$ at 1 keV assuming the same parameters as in Fig. 21 and a sticking probability of 1 at zero coverage.

[62] E. Taglauer, W. Heiland, and J. Onsgaard, *Nucl. Instrum. Methods* **168**, 571 (1980).

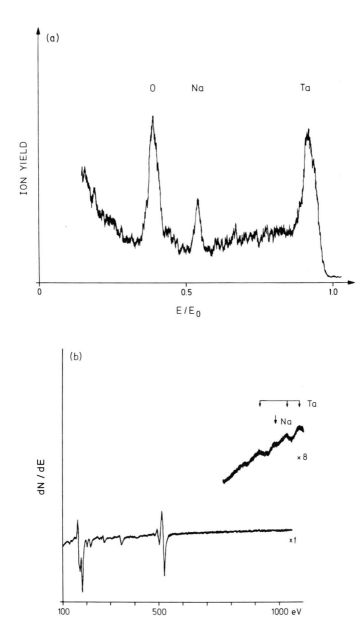

FIG. 23. ISS from a polycrystalline Ta_2O_5 surface contaminated with sodium (a) in comparison with AES from the same surface area (b). Energy analysis in both cases with (commercial) single pass CMA (cylindrical-mirror analyzer).

comparable with that discussed above for the case of damage to the clean surface.

The desorption process can also be used for surface cleaning. Especially with Ne, relatively high efficiencies with little damage or implantation can be obtained. Another aspect is the relation between pressure or molecular rate of arrival A at the surface and desorption. Figure 22 shows A (cm^{-2} s^{-1}) as a function of pressure. The other curve is the saturation coverage Θ_∞ calculated from $\Theta_\infty = A/N_0 \, (i\sigma_D + A/N_0)^{-1}$, where $N_0 = 10^{15}$ cm^{-2} (monolayer density), $i = 10^{-6}$ A cm^{-2} (current density), and $\sigma_D = 10^{-15}$ cm^{-2} (desorption cross section, typical value; see, e.g., Taglauer et al.[62]). A further assumption is that the sticking probability decreases linearly with coverage and is unity at zero coverage. At reasonable partial pressures of active gases, recontamination of the surface is not likely within the times for data acquisition estimated above. It should be noted that no evidence for CO dissociation on surfaces under low-energy ion bombardment has been observed,[62] i.e., experience shows that with ion beams carbon buildup is no problem in comparison with electron-beam systems.

6.1.2.2. Applications. The first example is a surface composition analysis in comparison with AES. The striking difference is the better identification of Na with ion scattering (ISS), whereas C is obviously easier to find with AES (Fig. 23). (It should be mentioned that Na is most easily detected with SIMS; see Chapter 6.2.) This result just demonstrates the different cross sections that come into play with ISS and AES. The next example demonstrates the different depth resolution of ISS and AES (Fig. 24).[63] The β-phase of CO on W(100) is dissociated with both C and O at the surface; in the β_3-phase the C diffuses into the surface, which is shown by the difference between the ISS and AES spectra. Note also that the W AES signal does not change from β to β_3, whereas in the ISS spectra an increase of the W signal appears, which can be understood in terms of Eq. (6.1.6).

This extreme surface sensitivity has been successfully exploited in surface segregation[64,65] and preferential sputtering studies.[66] Figure 25 shows the good agreement between theory and experiment when ISS is used.[64]

The same philosophy applies for the measurements on the different faces of polar crystals[67-69] or the determination of the orientation of adsorbed CO.[1,63] It is carried a bit further in the combination of ISS with LEED, in

[63] W. Heiland, W. Englert, and E. Taglauer, *J. Vac. Sci. Technol.* **15**, 419 (1978).
[64] H. H. Brongersma, M. J. Sparnaay, and T. M. Buck, *Surf. Sci.* **71**, 657 (1978).
[65] G. C. Nelson, *Surf. Sci.* **59**, 310 (1976).
[66] E. Taglauer and W. Heiland, *Appl. Phys. Lett.* **33**, 950 (1978).
[67] W. H. Strehlow and D. P. Smith, *Appl. Phys. Lett.* **13**, 34 (1968).
[68] H. H. Brongersmaa and P. Mul, *Chem. Phys. Lett.* **19**, 217 (1973).
[69] H. H. Brongersmaa and P. Mul, *Surf. Sci.* **35**, 393 (1973).

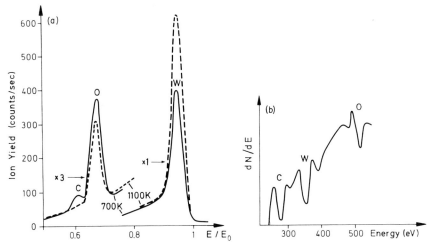

FIG. 24. ISS (a) of $^3He^+$ from W(100) covered with CO in comparison with AES (b) after annealing to 1100 K. Measurements by ISS were taken after annealing to 700 K and 1100 K with $\psi = 45°$, $\vartheta = 0°$, and $E = 1$ keV. Note the disappearance of the C-signal in the ISS, which shows the extreme surface sensitivity of ISS compared to AES. C is believed to diffuse into the W at 1100 K.

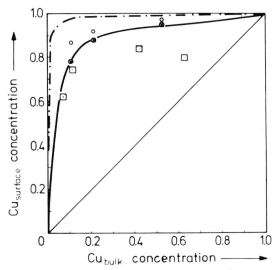

FIG. 25. Surface segregation on Cu–Ni alloys measured by ISS in comparison with H_2-adsorption measurements and theory (for details, see Brongersma et al.[64] Straight solid line valid if surface concentration equals bulk concentration. ⊗, 500°C LEIS; ⊙, 400°C LEIS; □, H-adsorption Sinfelt; —·—, theory A, 500°C; ———, theory B, 500°C.

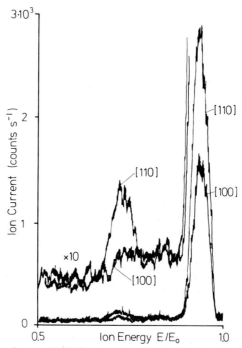

FIG. 26. ISS from O on Ag(110) for two azimuthal angles, i.e., the plane of scattering is parallel to (110) ($\varphi = 0$) and (100) ($\varphi = 90°$) surface directions respectively. The oxygen causes under the experimental conditions used a (2 × 1) LEED pattern (see Fig. 27). Experimental parameters are $E_0 = 600$ eV, $\psi = 30°$, $\vartheta = 60°$, and $i_T = 8 \times 10^{-9}$ A.

which generally ISS is mainly sensitive to next-nearest-neighbor position, and LEED to the surface periodicity. From the numerous examples,[33] we show O on Ag(100)[70] (Fig. 26). The highly anisotropic result of ISS with respect to the O-peak is most readily understood by the assumption that O sits in (110) troughs of the surface (Fig. 27). Anisotropy of the neutralization effect cannot be completely excluded (see above, Subsection 6.1.2.1.), so that the ISS information alone cannot be considered to be definitive.

In-depth analysis by sputtering and analyzing the remaining surface composition, ISS is, in principle, comparable with AES. Only in the case of analysis of the outermost layers may ISS be more successful, in some cases, as has been demonstrated for catalyst surfaces.[71] A similar case was studied by Knözinger et al.[72] (Fig. 28). The sequence of impregnation of a spinel matrix

[70] W. Heiland, F. Iberl, E. Taglauer, and D. Menzel, Surf. Sci. 53, 383 (1975).
[71] M. Shelef, M. A. Z. Weheeler, and H. C. Yao, Surf. Sci. 47, 697 (1975).
[72] H. Knözinger, H. Jeziorowski, and E. Taglauer, Proc. 7th Int. Congr. Catal., p. 604, Tokyo, (1980).

6.1. ION SCATTERING SPECTROMETRY

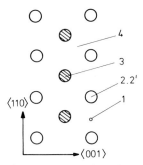

FIG. 27. Structure of a (110) fcc surface with possible adsorption sites. Open circles top layer, dashed circles second layer. Site 4 most probable for the (2 × 1) O adsorption on Ag(110).

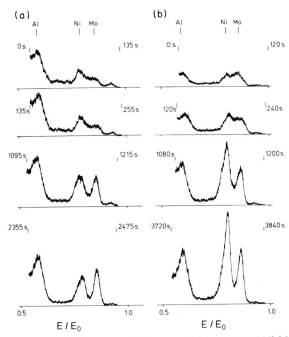

FIG. 28. ISS "depth analysis" from two Al_2O_3 supported catalysts: (a) Ni 3 Mo 12 Al 6 and (b) Mo 12 Ni 3 Al 6. Scattering is of He^+ at 500 eV. (Sputtering time (s) is not converted to depth.) Note the different surface concentrations of Ni relative to Mo reflecting the difference in the impregnation sequence in a monolayer depth scale. This also causes different catalytic activities. (From Knözinger et al.[72])

with Ni and Mo solutions is well represented in the ISS spectra. They also support the suggestion that nickel ions are located in the molybdate monolayers and the supporting matrix as well. The depth resolution necessary here seems to be of the order of one or two atomic layers.

Multiple scattering has been widely studied as a phenomenon per se. The few applications are a surface-structure determination of GaP,[69] evaluation of the parameters of the interaction potential by comparison with computer calculations,[35,40] and rather qualitative estimates of surface Debye temperatures.[35] An interesting experiment was concerned with the adsorption of O on Cu(110),[36] where the "two atom blocking model" or focusing effect was used, to pin down the adsorption site of the oxygen. In accordance with the estimates of Fig. 21, the experiment was, however, a dynamic one, i.e., the oxygen desorbed by the Ne beam had to be replenished by adsorption from the gas phase to reach an equilibrium coverage.

To summarize, we conclude that ion scattering is a technique especially suitable for studies of the outermost layer of a solid. Quantitative results are difficult to obtain owing to a lack of knowledge about the basic physics of ion neutralization on surfaces. The technique is compatible with other surface-physics experiments and instrumentation. By a combination appropriate to the problems to be studied it is possible to overcome the weaknesses of ISS, i.e., quantitative analysis and positive mass identification. One of the techniques with excellent mass identification — secondary-ion mass spectrometry (SIMS) — will be discussed in the next subsections.

6.2. Secondary-Ion Mass Spectrometry

6.2.1. Fundamental Aspects

Secondary ion mass spectrometry (SIMS) is based on sputtering, a long-known process[73] in which ion impact causes secondary-particle emission from solid surfaces. Within the last twenty years this process has been exploited for surface analytical techniques. Different approaches have been developed, which may be classified by their instrumentation, i.e., the detection of the secondary particles. This is done in the ion probes by standard double focusing mass spectrometry[74,75] or the use of quadrupole mass filters.[76,77] Other mass spectrometric techniques such as time of flight are also used. Some part in research and application is also played by instruments

---

[73] For reviews about sputtering, see appendix.
[74] R. F. K. Herzog and F. P. Viehböck, *Phys. Rev.* **76**, 855 (1949).
[75] H. Liebl, *J. Appl. Phys.* **38**, 5277 (1967).
[76] A. Benninghoven and E. Loebach, *Rev. Sci. Instrum.* **42**, 49 (1971).
[77] K. Wittmaack, *Rev. Sci. Instrum.* **47**, 157 (1976).

using the radiation emitted from excited sputtered particles.[78-80] Still another approach is the use of plasmas to ionize or excite sputtered particles, in which the plasma can be used as ion source and for excitation,[81,82] i.e., as ion source and ionizer [glow-discharge mass spectrometry (GDMS)][83,84] or as an ionizer only [secondary neutral mass spectrometry (SNMS)].[85]

The latter techniques are a clear indication of some analogy to ion-scattering spectrometry. Besides the elastic part—scattering and sputtering—the inelastic events leading to the formation of ions, excited particles, or particles in their ground states are a major scientific problem in both fields. The final-state formation has other dimensions in SIMS owing to the greater complexity of the processes. The sputtering process is not one phenomenon; at least two very different events, "collision cascades" and "thermal spikes", can cause material release from surfaces. This material may consist not only of atomic particles (positive, negative and excited ions, and excited and ground-state neutrals) but also of molecules, again in all possible states of excitation. We do not review the whole field, (see, e.g., Wittmaack[86]) but rather try to work out the more general results of sputtering and final-state formation under the aspects of its impact on instrumentation in this field. A few examples to demonstrate the possibilities of SIMS close the chapter.

6.2.1.1. Sputtering. Physical sputtering, i.e., the removal of surface atoms by ion impact, is caused by the energy and momentum deposited by the primary ions in the solid.[87] Starting the discussion at very low energies (as in the treatment of ISS), the primary ions are mostly reflected at the surface, depending on the angle of incidence. The energy and momentum are primarily exchanged with surface atoms. If there is enough energy, these surface atoms may bounce back after a single collision (or a sequence) and overcome the surface binding energy (Fig. 28). This class of sputtered particles may have rather high kinetic energies [up to the order of $E_2 = M_1 M_2 (M_1 + M_2)^{-2} \cos^2 \vartheta_2$, Eq. (6.1.2)]. They will also rather closely reflect structural properties of the surface as well as other geometrical features, such as impact angle, etc. A comparable class of particles originates from reflected primary

---

[78] R. Goutte, C. Guilland, R. Javelas, and J.-P. Meriaux, *Optik* **26**, 575 (1977).
[79] N. H. Tolk, D. L. Simms, E. B. Foley, and C. W. White, *Radiat. Eff.* **18**, 221 (1973).
[80] G. E. Thomas and E. E. de Khuizenaar, *Int. J. Mass Spectrom. Ion Phys.* **15**, 165 (1974).
[81] G. K. Wehner, *Phys. Rev.* **114**, 1203 (1959).
[82] J. E. Greene and J. M. Whelan, *J. Appl. Phys.* **44**, 2509 (1973).
[83] J. W. Coburn and E. Kay, *Appl. Phys. Lett.* **18**, 435 (1971).
[84] J. W. Coburn, E. Taglauer, and E. Kay, *J. Appl. Phys.* **45**, 1779 (1974).
[85] H. Oechsner and W. Gerhard, *Phys. Lett. A* **40**, 211 (1972).
[86] K. Wittmaack, *in* "Inelastic Ion–Surface Collisions" (N. H. Tolk, J. C. Tully, W. Heiland, and C. W. White, eds.), p. 153. Academic Press, New York, 1977, 1977.
[87] P. Sigmund, *in* "Inelastic Ion–Surface Collisions" (N. H. Tolk, J. C. Tully, W. Heiland, and C. W. White, eds.), p. 121. Academic Press, New York, 1977.

ions that have penetrated the solid (Fig. 29). These processes are relevant at somewhat higher energies and mostly for light ions.[61,88,89] With increasing energy the so-called "primary knock-on" may receive enough energy (E_2 again) to set a rather large number of target atoms in motion, thus forming a "collision cascade" (Fig. 29). If the collision cascade intersects the surface, atoms will be released, but on the average with lower energies than from the previous processes.

Qualitatively, and to a good measure also quantitatively, these sputtering processes can be understood in terms of the collision cross sections. That is, with increasing energy, it becomes less and less probable that a primary knock-on with high energy will be created near the surface owing to the decreasing cross section for ion–atom scattering. The size of a collision cascade formed will increase much more slowly since it is governed by low-energy (large cross section) collisions. Sputter yields, for example, therefore show a maximum at rather moderate energies, which shifts with decreasing mass ratio M_2/M_1 to higher energies (Fig. 30), reflecting the fact that collision cascades formed too deep inside the material will not contribute to sputtering. On the other hand, the collision cascade regime provides a sputtered-particle energy distribution that is relatively independent of many parameters, such as primary-ion mass, ion energy, impact angle, etc.; i.e., the collision cascades are decoupled from the primary event.

The latter philosophy is even more realistic in the case of so-called thermal spikes (Fig. 29). These occur when the collision cascade develops in such a way that the energy is dissipated in a small volume where many atoms are in

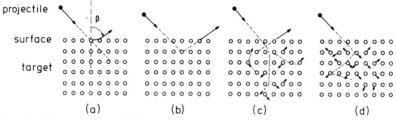

FIG. 29. Schematic of ion solid interactions corresponding to e.g., (a) low-energy projectile of medium mass ($\leq$ 500 eV Ne$^+$), (b) medium- to high-energy projectile of low mass (1-keV to 1-MeV for H$^+$ and He$^+$; penetration depth will increase with energy), (c) medium-energy projectile of medium mass (10-keV Ar$^+$), (d) medium- to high-energy projectile of heavy mass onto target with comparable mass/atom (100-keV Au$^+$ → Au, 8-keV Xe$^+$ → AuAg) or schematics for (a) surface scattering, (b) scattering from the bulk, (c) collision cascade, and (d) thermal spike.

[88] R. Weissmann and R. Behrisch, *Radiat. Eff.* **18**, 55 (1973).
[89] R. Behrisch, G. Maderlechner, B. M. U. Scherzer, and M. T. Robinson, *Appl. Phys.* **18**, 391 (1979).

6.2. SECONDARY-ION MASS SPECTROMETRY

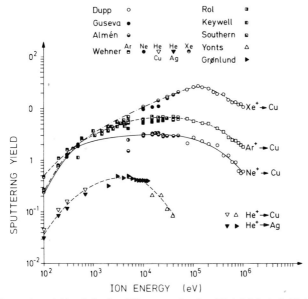

FIG. 30. Sputtering yields of Cu for different projectiles (He$^+$, Ne$^+$, Ar$^+$, Xe$^+$) versus ion energy. For He$^+$, interpolation with Ag data is made. (Data from Keywell; Lagreid and Wehner; Grønlund and Moore; Almen and Bruce; Rosenberg and Wehner; Rol *et al.*; Guseva; Yonts *et al.*; Southern *et al.*; Dupp and Scharmann.[90])

motion at the same time. As a consequence, there is no longer linear superposition of particle interactions. The effect is most clearly demonstrated in the 100-keV region by using heavy atomic and molecular primary ions (Te$^+$, Te$_2^+$) on heavy substrates (Au), where nonlinear enhancement of the sputtering yield is observed for the molecular projectiles.[91] It should be noted that the excitation in the cascade or the spike seems to be within the kinetic energy of the atoms only. The electronic excitation as measured by the secondary-electron emission does not change.[92,93] This is relevant in the

[90] F. Keywell, *Phys. Rev.* **97**, 1611 (1955); N. Lagreid and G. K. Wehner, *J. Appl. Phys.* **32**, 365 (1961); F. Grønlund and W. J. Moore, *J. Chem. Phys.* **32**, 1540 (1960); O. Almén and G. Bruce, *Nucl. Instrum. Methods* **11**, 257, 279 (1961); D. Rosenberg and G. K. Wehner, *J. Appl. Phys.* **33**, 1842 (1961); P. K. Rol, J. M. Fluit, and J. Kistemaker, *Physica (Amsterdam)* **26**, 1000 (1960); M. I. Guseva, *Sov. Phys. — Solid State (Engl. Transl.)* **1**, 1410 (1959); O. C. Yonts, C. E. Norman, and D. E. Harrison, Jr., *J. Appl. Phys.* **31**, 447 (1960); A. L. Southern, W. R. Willis, and M. T. Robinson, *J. Appl. Phys.* **34**, 153 (1963); G. Dupp and A. Scharmann, *Z. Phys.* **192**, 284 (1966).
[91] H. H. Andersen and H. L. Bay, *J. Appl. Phys.* **45**, 953 (1974); **46**, 1569 (1975).
[92] G. Staudenmaier, W. O. Hofer, and H. Liebl, *Int. J. Mass Spectrom. Ion Phys.* **11**, 103 (1976).
[93] F. Thum and W. O. Hofer, *Surf. Sci.* **90**, 331 (1979).

context of charge-state formation. Even if the quantitative importance of thermal spikes to the sputtering is still under dispute,[87,94] there is no question that in the discussion of actual experimental results they have to be considered, as well as in the planning of experiments. One of the main consequences of the thermal spike contribution is the influence on the energy distribution of the secondary particles. It has been shown[95] that the low-energy part of the secondary energy distribution of sputtered neutrals measured with a time-of-flight technique is indeed influenced by thermal spike effects (Fig. 31).

The energy distribution has two main consequences for SIMS. Any instrument that is to be operated over some range of primary energies and a larger range of masses has to be designed in a way that no discrimination of particles due to their energy occurs; otherwise even qualitative surface composition analysis may be erroneous. Energy distributions are very different

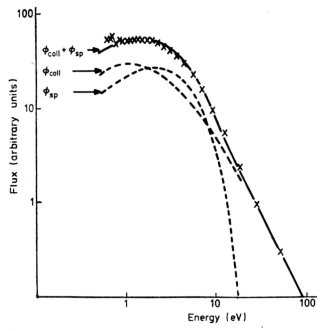

FIG. 31. Secondary-energy distribution of Ag atoms sputtered from an Ag–Au alloy under 6-keV Xe$^+$ impact. The experimental result (crosses) is analyzed as being due to collision cascades and thermal spikes (dashed lines), which add up to the theoretical distribution (solid line). (From Szymonski et al.[95a])

[94] R. Kelly, Radiat. Eff. 32, 91 (1977).
[95] H. Overeijnder, A. Haring, and A. E. de Vries, Radiat. Eff. 37, 205 (1978).
[95a] M. Szymonski, R. S. Bhattacharya, H. Overeijnder, and A. deVries, J. Phys. D 11, 751 (1978).

6.2. SECONDARY-ION MASS SPECTROMETRY

for monatomic and polyatomic species,[86] the molecules usually having a lower contribution at higher energies. Secondly, the final-state formation is different for the particles originating from different processes. Analogous considerations apply to the angular dependence. The thermal-spike-produced particles very probably have a cosinelike distribution in space. This is also true of the collision cascade particles at normal incidence. Forward peaking is observed in the case of glancing incidence.[61] Furthermore, the angular distribution is different for different species sputtered from the same surface. Again, if, for example, only charged particles are measured, care has to be taken not to discriminate against some classes of particles.

Another aspect of sputtering has been discussed i.e., radiation damage (see Subsection 6.1.2.1.4.). This discussion has to be carried further here since it is mostly heavy ions (Cs^+, Ar^+, O^+) at energies between 3 and 20 keV that are used in SIMS. This involves not only damage in the form of surface defects but also rather serious changes of surface structure and surface composition. The structural changes can be estimated as in Fig. 21. Surface-composition changes are due to "preferential sputtering" and "atomic mixing." Both processes are qualitatively understood within the collision cascade model.[87] Figure 32 shows the change of the surface composition of Ta_2O_5 under ion

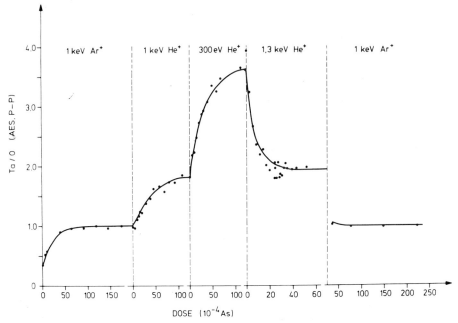

FIG. 32. Surface composition of Ta_2O_5 during bombardment with Ar^+ and He^+ analyzed by ISS. Note the reversibility of the process ("preferential sputtering"). (From Taglauer and Heiland.[66])

bombardment. In a simple model, the steady-state yields of sputtered particles will reflect the bulk concentrations C_1^β and C_2^β, whereas the surface concentrations C_1^s and C_2^s will be given by the sputter yields S_1 and S_2 of the two components; i.e.,

$$\frac{C_1^s}{C_2^s} = \frac{S_2}{S_1} \frac{C_1^\beta}{C_2^\beta}. \tag{6.2.1}$$

There will be a transition regime in the composition of the sputtered flux. This may be further complicated if, for example, charged particles are measured, since the final-state formation depends on the surface composition (see Subsection 6.2.2.2.). The atomic mixing process is based on similar effects. The energy and momentum transfer is mass-ratio dependent, so that different components within the range of the primary ions will react differently.

A final remark concerns the possible macroscopic changes of the surface structure (Fig. 33). Rather bizarre structures, viz., ridges, cones, holes, pyramids, have been observed,[96] which in most cases reflect the crystallinity of the bombardment material. The effect may be initiated by surface impurities or other surface defects (e.g., dislocations). The implanted gas may also cause macroscopic surface changes due to blistering.[97] In general, these effects can be avoided by a proper choice of parameters, especially the type and energy of the primary ion. Low-energy heavy ions (e.g., Cs$^+$) incident at grazing angles cause a minimum of surface damage and implantation and give best results in depth profiling.[98] Bombardment with reactive ions apparently develops less surface structure than Ar$^+$ bombardment.[99]

6.2.1.2. Ion Yields. For a fixed projectile energy the yield of sputtered particles of a state i will be given by

$$dI^i(E, \beta) = Ay(E, \beta)\alpha^i(E, \beta)\gamma C \, \Delta\Omega \, dE, \tag{6.2.2}$$

where E is the secondary energy, β the outgoing polar angle, A a constant containing the primary-ion flux, analyzer transmission, and detector efficiency, $y(E, \beta)$ the differential sputtering yield, $\alpha^i(E, \beta)$ the spectral probability for the formation of the state i, γ the isotope abundance, C the concentration in the escape layer of the species analyzed, and $\Delta\Omega = \sin \beta \, d\beta \, d\varphi$ the solid angle of the detector. The total sputtering yield is given by

$$Y = \int_0^{E_{max}} \int_0^{2\pi} \int_0^{\pi} \sum_i \alpha^i(E, \beta)y(E, \beta) \sin \beta \, dE \, d\varphi \, d\beta, \tag{6.2.3}$$

[96] J. L. Whitton, in "The Physics of Ionized Gases" (R. K. Janev, ed.), p. 335. The Institute of Physics, Belgrade, 1978.
[97] J. Roth, Conf. Ser.—Inst. Phys. No. 28, p. 280 (1976).
[98] C. W. Magee, W. L. Harrington, and R. E. Honig, Rev. Sci. Instrum. **49**, 477 (1978).
[99] W. O. Hofer and H. Liebl, Appl. Phys. **8**, 359 (1975).

6.2. SECONDARY-ION MASS SPECTROMETRY

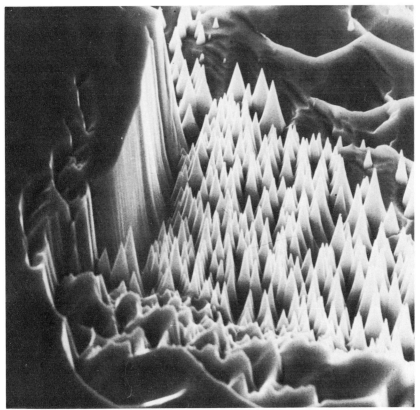

FIG. 33. Structure of a polycrystalline Cu surface after 40-keV Ar$^+$ bombardment (total dose 10^{19} ions cm^{-2}) Magnification 20,000×. (From J. L. Whitton, private communication.)

where $E_{\max}$ is the maximum energy of the sputtered particles. This quantity is found in the literature and it is most easily measured by weight-loss techniques. In comparison with ISS [Eq. (6.1.5)] the differences in the physics are described, respectively, by $d\sigma/d\Omega$, y, P, and α. For the approximation of binary scattering $d\sigma/d\Omega$ is a function of the primary energy, the masses and the scattering angle only. In the SIMS case y depends not only on the energy, masses, and outgoing angle, but also on the impact angle, the composition and the surface structure. The probability P is a function of the outgoing velocity (and angle) of the species involved but in most cases not of the surface composition, whereas α^i depends very strongly on the surface composition besides the other parameters. The quantity α^i describes the transition of a surface atom from a bound state in the solid into a state i of the free atom or ion, whereas P describes to a good approximation a charge

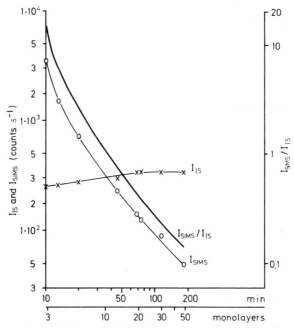

FIG. 34. Comparison of ISS and SIMS (Cu$^+$ signal from polycrystalline Cu as a function of bombardment time. SIMS bombardment with Ar ions at $E_0 = 800$ eV and $i_T = 5 \times 10^{-7}$ A cm^{-2}. The increase of the ISS signal is attributed to the decrease of the blocking of the Cu atoms by adsorbed species, whereas the decrease of the SIMS signal is due to the change of the ion yield with decreasing surface contamination (matrix effect connected with oxygen). (From Grundner et al.[100])

exchange process between a rare-gas ion and a surface atom or the free electrons of the solid. Figure 34 shows as an example a direct comparison of ISS and SIMS for positive, monatomic Cu ions sputtered from a surface cleaned of oxygen during bombardment.[100] It can be shown that for the fast sputtered particles, especially excited neutrals, $\alpha^* \propto \alpha'^*$. The function f is suitable for describing some experimental observations[79,101] where one has $f \propto \exp(-A/av)$ as in the case of ion scattering. It can be shown, however, by simple estimates using experimental A/a values such as 10^7 cm s^{-1} [101] that practically no low-energy positive ions could survive if α^i were to be factorized for sputtered ions in this way. In general, α^i is thus not a physical but only a formal analog to P.

[100] M. Grundner, W. Heiland, and E. Taglauer, *Appl. Phys.* **14**, 243 (1974).
[101] C. W. White, E. W. Thomas, W. F. van der Weg, and N. H. Tolk, in "Inelastic Ion–Surface Collisions" (N. H. Tolk, J. C. Tully, W. Heiland, and C. W. White, eds.), p. 201. Academic Press, New York, 1977.

6.2. SECONDARY-ION MASS SPECTROMETRY

So far there exists no quantitative theory for the final state formation during sputtering. Some experimental findings can be generalized as empirical rules that can be very useful for a limited class of experiments.[86] They are based on observations such as the increase of the positive-ion yield in the presence of electronegative species (oxygen) or the increase of the negative-ion yield in the presence of electropositive species (Cs). In general, there also seems to be some relation between electron affinity and ion formation: species with high electron affinity form mainly negative ions (C, O), whereas for some species forming positive ions the ionization potential seems to be a comparable parameter.[102] Owing to the large changes of α^i with, for example, oxygen present at the surface the control of vacuum conditions is very important in SIMS work. "Sputter cleaning" is not always the solution to the problem since oxygen or other species are implanted in the solid under ion bombardment and show up again at the surface during further sputtering, e.g., in the case of depth analysis. The situation is less severe if only relative measurements are intended and these are for a limited case involving substances of equal chemical character, typical "finger print" analyses. In all other cases extremely careful instrumentation and operation is necessary to avoid the pitfalls, all the more so since ion mass spectra are the more easily obtained the dirtier the surface, in contrast to ion-scattering spectra.

6.2.2. SIMS Experiments

In this subsection discussion is restricted to secondary-ion mass analysis. Experiments using neutral detection are, in principle, different in that not only are the ions generated in the sputtering process detected, but post-ionization by electrons or photons is used. After ionization of the neutrals the apparatus is similar.[83-85] Detection of excited particles by optical spectroscopy is a similar approach. In this case it is comparably straightforward, by proper imaging, to sample over all emitted particles (space and velocity); the transmission of the spectrometer can be measured, but the interpretation of the results may be more complicated owing to the wealth of lines observed. For quantitative results, adequate handling of the data is rather elaborate. (For details, see Tolk et al.[79] and White et al.[101]). The conditions concerning the primary beam are the same for these experiments.

6.2.2.1. Apparatus. By analogy with ISS experiments we have three parts: (i) ion-beam formation, (ii) target handling, and (iii) mass analysis.

6.2.2.1.1. Ion Source. Primary-beam formation is not too different from the requirements for ISS experiments. In general, most experiments are done at somewhat higher energies (1–20 keV), and higher current densities are useful for depth analysis. Plasma-discharge ion sources (e.g., duoplasma-

---

[102] V. R. Deline, C. A. Evans, Jr., and P. Williams, *Appl. Phys. Lett.* **33**, 578 (1978).

tron, Fig. 16) are therefore mostly used. Mass separation is necessary for trace analysis to avoid contamination (isotope separation is not necessary). To separate out the neutral part of the beam some beam deflection is needed. The neutral beam is a nuisance in the case of beam rastering, which is a necessity in most SIMS experiments to avoid edge or crater effects caused by concentration gradients. Accordingly, the signal has to be gated.[103]

With the higher beam energy, focusing of the beam also becomes possible into a range that allows a lateral resolution below 1 μm.[104] The ion-beam optics is shown in Fig. 35. Several other ion microprobes are described in the literature.[75,105–107] These devices are very powerful for three-dimensional composition analysis. For multicomponent systems some consideration has to be given to sample consumption. With the apparatus shown in Fig. 35 with 8-keV Ar$^+$, a beam current density of 10^{-2} A cm^{-2} is used, the total current being 10^{-12} A in a beam with a diameter of $\frac{1}{10}$ μm. Further improve-

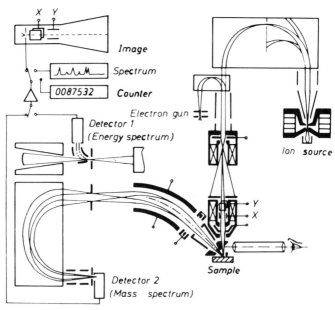

FIG. 35. Scheme of an ion microprobe with primary-ion and electron-beam system and energy and mass analyzer (COALA). (From Liebl.[104])

[103] W. O. Hofer, H. Liebl, G. Roos, and G. Staudenmaier, *Int. J. Mass Spectrom. Ion Phys.* **19**, 327 (1976).
[104] H. Liebl, *Adv. Mass Spectrom.* **7A**, 751 (1978).
[105] T. Kondo, H. Tamura, and H. Hirose, *Shitsuryo Bunseki* **22**, 229 (1974).
[106] J. Vastel, *Proc. Int. Conf. X-Ray Opt. Microanal., 7th, Moscow*, p. 118 (1974).
[107] F. G. Rüdenauer, and W. Steiger, *Proc. 6th Int. Vac. Congr., Kyoto*, p. 383 (1974).

ments of the lateral resolution are possible if field-ion sources that provide enough brightness are used. Furthermore, the ion optics has to be carefully designed to exploit fully the possibilities offered by the new sources.[108] Optimal matching of the secondary-beam emittance is then a logical consequence (see Subsection 6.2.2.1.3 and Fig. 36).[104]

6.2.2.1.2. TARGET HANDLING. As already stated, target preparation depends on the aim of the study. The considerations based on Figs. 21 and 22 are valid here as well. Most SIMS work, however, is done at energies between 1 and 20 keV with heavier ions such as Ar^+, O^+, N^+, etc., and with current densities in the region of 10^{-6} A cm^{-2}, with a few exceptions such as "static" SIMS.[76] Target damage is therefore rather high. On the other hand, high current densities are needed to reach reasonable etch rates in the case of depth analysis.

A well-known problem with quantitative SIMS analyses is the so-called "matrix effect," i.e., the extreme sensitivity of the ionization probability to the chemical surface composition as mentioned above. One way of obtaining fairly constant ion yields during depth profiling is to offer a sufficient amount of oxygen to the target surface, e.g., through a gas jet.[98]

Some thought has to be given to the sputtered material that will be deposited near the beam spot on the target itself and in the neighborhood in general. The first effect is relatively easy to understand even though not much can be done about it; if different spots on the same target are analyzed, the larger the distance between these spots the better. The second effect

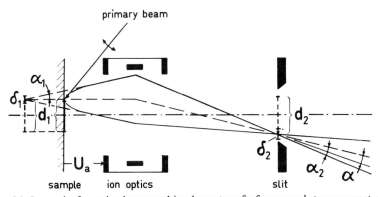

FIG. 36. Ion optics for optimal sputtered-ion-beam transfer from sample to mass spectrometer entrance slit. d_1 = surface area dimension to be analyzed, $\alpha_1 = (U_i/U_a)^{1/2}$ = half angle of sputtered beam of ions with initial energy eU_i after acceleration eU_a, d_2 = width of image of d_1, δ_2 = width of least confusion (cross over), δ_1 = virtual spot size, α_2 = half-angle at the spectrometer slit, α = half-angle accepted by the spectrometer. (From Liebl.[104])

[108] H. Liebl, "Adv. Mass Spectrom. 7A, 807 (1978).

causes contamination of any target with material from previous samples via resputtering. This is caused by high-energy reflected neutrals that hit the neighborhood of the target. The effect can be minimized by working at energies, where the sputter yields are low. For the same reason material with high sputter yields such as Cu should be avoided as a construction material in the neighborhood of the target.

6.2.2.1.3. MASS SPECTROMETRY. In comparison with standard mass spectrometry, SIMS from solids has two major problems to cope with experimentally: the energy distribution and the angular distribution of the sputtered ions (neutrals). To obtain reliable and quantitative measurements, energy analysis in conjunction with mass analysis is a necessity. This facility is incorporated in the double focusing mass spectrometers of ion microprobes.[75] Some quadrupole instruments are also equipped with energy analyzers.[77,98,100,109] There are two philosophies: (1) The energy analyzer is used with a rather low energy resolution as a low-energy band pass filter. This is achieved by, for example, postacceleration of the secondary ions (an analyzer with $\Delta E/E = 0.1$ will pass a band of 100 eV at an acceleration of 1 keV). (2) For each species (mass line) a complete energy scan is done, corrected for analyzer transmission, and then integrated. This second way is the most correct one, but, as can be seen from previous estimates of experimental time scales, not practicable in many cases.

For the first case we give some estimates for the design of a SIMS experiment.[104] As a rule of thumb, the path length L of a double focusing spectrometer (energy analyzer plus magnetic-sector field), the slit width s of the entrance aperture, and the mass resolution $M/\Delta M$ are related by

$$L = 20sM/\Delta M \tag{6.2.4}$$

If the radii of the electric and magnetic fields are comparable (e.g., Fig. 34). For an α^2 corrected instrument (α is the angle of acceptance, Fig. 37) the acceptance angle can then be estimated from

$$\alpha = (s/2L)^{1/3} = (4M/\Delta M)^{-1/3},$$

which allows simple estimates for instrumental designs.

The application of SIMS has rapidly increased in many laboratories since the introduction of the quadrupole mass filter[110] to secondary-ion mass spectrometry. This was first used by Krohn,[111] and commercial systems have been available since 1971.

[109] A. R. Krauss and D. M. Gruen, *Appl. Phys.* **14**, 89 (1977).
[110] W. Paul and H. Steinwedel, *Z. Naturforsch., A* **8A**, 448 (1953).
[111] V. E. Krohn, *J. Appl. Phys.* **33**, 3523 (1962).

6.2. SECONDARY-ION MASS SPECTROMETRY

Quadrupoles are very convenient SIMS components for a variety of reasons:

(i) Only voltages are applied, no magnet being necessary, thus avoiding heavy construction and stray fields.
(ii) The mass scan yields a linear mass scale.
(iii) Quadrupoles have a virtually energy-independent transmission for energies below about 100 eV, which almost exactly matches secondary-ion energy distributions.
(iv) It is easily mounted and UHV compatible.
(v) Within certain limits there are no crucial restrictions in the angular distributions at the entrance (the transverse ion-velocity components must fit the geometry of the instrument).
(vi) Mass resolution can be adjusted externally within certain limits.

There are a few points that have to be taken into account for proper application of quadrupoles to SIMS work. It has become common to adopt the very favorable practice of having a (low-resolution) energy analyzer in front of the target. This eliminates backscattered primary particles, fast or metastable neutrals, and photons. It also allows the pass energy and width to be set. In this way very low background and a high dynamic range (up to 6 orders of magnitude difference in the count rate for adjacent masses) have been achieved. Usually the multiplier detector is also placed off-axis to reduce the background intensities. To switch from positive- to negative-ion detection, only the detector voltages have to be changed accordingly. A

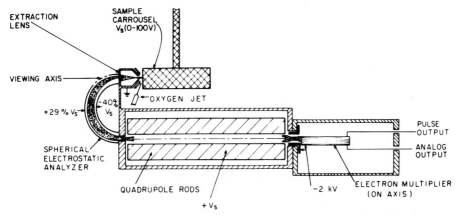

FIG. 37. Scheme of a SIMS experiment. A Colutron ion source is combined with a raster ion-beam system. The ion-analysis system includes a 180° spherical energy analyzer and a quadrupole mass filter. Note the oxygen jet providing for ion-yield enhancement by using a "matrix effect." (From Magee et al.[98])

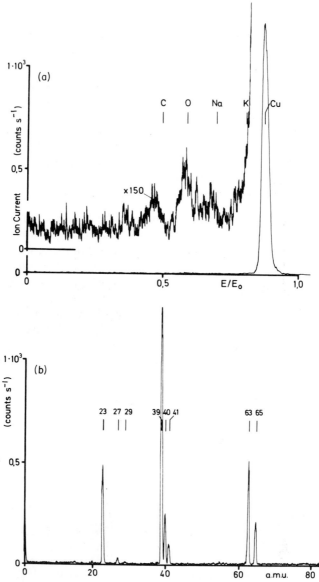

FIG. 38. (a) "Positive SIMS" spectrum of a polycrystalline Cu surface (Fig. 34) in comparison *(in situ)* with (b) ISS. (a) He → Cu with $E_0 = 1$ keV, $i_T = 3.5 \times 10^{-8}$ A cm^{-2}, $\vartheta = 90°$. (b) Ar → Cu with $E_0 = 1.5$ keV, $i_T = 5 \times 10^{-7}$ A cm^{-2}. Note the differences in mass resolution (^{63}Cu, ^{65}Cu) and sensitivity especially with respect to Na. (From Grundner et al.[100]) (For a qualitative comparison with AES see Fig. 23.)

6.2. SECONDARY-ION MASS SPECTROMETRY

typical set-up is shown in Fig. 37.[98] Results obtained with this apparatus are discussed in Subsection (6.2.2.2).

6.2.2.2. Applications. The first example (Fig. 38) shows a mass spectrum and an ISS spectrum taken simultaneously from a Cu surface.[100] They provide the raw data for Fig. 34. The differences in mass resolution are obvious (cf. Fig. 6) as are the differences in ion yield. In the SIMS spectrum the alkali (Na, K), alkaline earth (Ca) and Al demonstrate the high ion yield α^+ of these species, whereas O is clearly suppressed, having a very low α^+. Both techniques have a superior yield for the alkali compared with AES (Fig. 23).

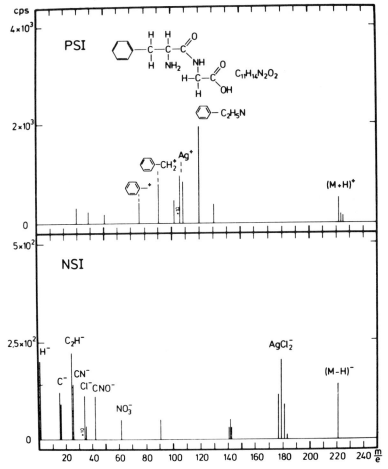

FIG. 39. "Positive" (PSI) and "negative" (NSI) SIMS from an organic compound, phenylalanylglycine ($M = 222$) (adsorbed on Ag). Bombardment with 3-keV Ar$^+$, analysis with a quadrupole mass filter. (Reprinted with permission from Benninghoven and Sichtermann.[112] Copyright 1978 American Chemical Society.)

TABLE II. Intensities (Arbitrary Units) of SIMS (Positive) for Ni Ions and Clusters of Ni(100)–CO[a]

Surface condition	CO coverage	Ion yields				
		Ni^+	$NiCO^+$	Ni_2^+	Ni_2CO^+	$NiCO^+/Ni_2CO^+$
Clean	0	0.3	0	<0.1	0	—
Low coverage, 293 K	~0.25	9.3	3.1	4.1	0.5	6
Saturation, 293 K	~0.40	50	30	20	3	10
10^{-8} Torr, 293 K	>0.40	90	60	30	6	10
c(2 × 2), 77 K	0.50	40	20	15	2	10
Hexagonal, 77 K	0.60	230	110	50	10	11

[a] From Hopster and Brundle.[113]

The extremely high mass resolution provided by mass spectrometry is shown in Fig. 39[112] with the positive and negative spectra of phenylalanylglycine (mass 222) adsorbed on silver. These studies allow interesting conclusions about the chemistry of biologically important substances.

On the other hand, the relation between ion yields and surface chemistry is not straightforward, as demonstrated by, for example, Hopster and Brundle[113] (Table II). When CO is adsorbed on a Ni(100) single-crystalline surface, different surface structures are observed by LEED (low electron energy diffraction). By electron-loss spectroscopy,[114] it has been shown that in the

TABLE III. Secondary-Ion Yields Y (Secondary Ions/Primary Ion) for 3-keV Ar^+ Bombardment of Sputter-Cleaned and Oxidized Metal Surfaces ($\psi = 20°$)[a]

Metal	Y (cleaned surface)	Y (oxidized surface)
Mg^+	0.0085	0.65
Al^+	0.02	2.0
V^+	0.0013	1.2
Cr^+	0.005	1.2
Fe^+	0.001	0.38
Ni^+	0.0003	0.02
Cu^+	0.00013	0.0045
Sr^+	0.0002	0.13

[112] A. Benninghoven and W. K. Sichtermann, *Anal. Chem.* **50**, 1180 (1978).
[113] H. Hopster and C. R. Brundle, *J. Vac. Sci. Technol.* **16**, 548 (1979).
[114] S. Anderson, *Proc. 7th Int. Vac. Congr., Vienna*, p. 1019 (1977).

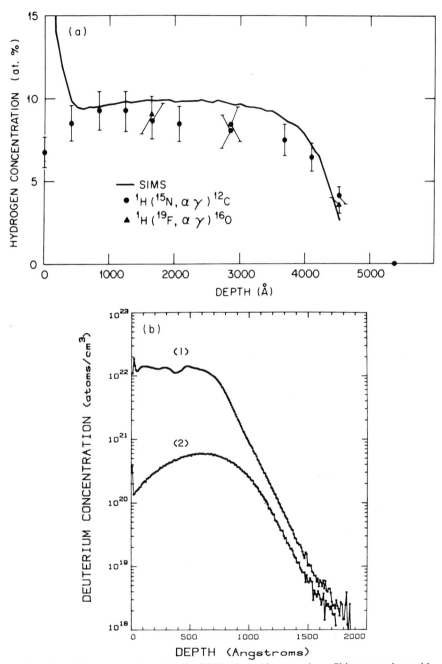

FIG. 40. (a) Depth analysis by SIMS of H implanted in amorphous Si in comparison with analysis by nuclear reactions using a high-energy mega-election-volt ion accelerator. ——, SIMS; ●, $^1H(^{15}N, \alpha\gamma)^{12}C$; ▲, $^1H(^{19}F, \alpha\gamma)^{16}O$. (From Clark et al.[116]) (b) SIMS analysis of D implanted in Si with D^+ energy 2 keV at total doses of (1) 10^{18} atoms cm^{-2} and (2) 10^{16} atoms cm^{-2}. (From Magee et al.[117])

low-coverage phase the CO forms bridge bonds and linear bonds, whereas for the c(2 × 2) at $T = 77$ K and with a coverage of $\Theta = 0.5$ only linear bonds are found. In the region of the hexagonal phase both bond types are again present. However, the positive-ion yields do not show a significant change in $NiCO^+$ to Ni_2CO^+ yield, as might be expected from the bonding of the CO to either one or two surface Ni atoms.

As mentioned before, the ion yield and therefore the sensitivity for a certain atomic (or molecular) species depends on its electronic properties as well as the chemical bonds to the surrounding atoms, "matrix effects." Therefore, some atoms are more abundant in the positive-, others in the negative-ion spectra ("positive" or "negative" SIMS). Nevertheless, there have been many attempts to standardize ion yield measurements in order to get a general idea of the sensitivities that are to be expected. In Table III[115] a few examples are given, but they can only be taken as a guideline with order of magnitude accuracy.

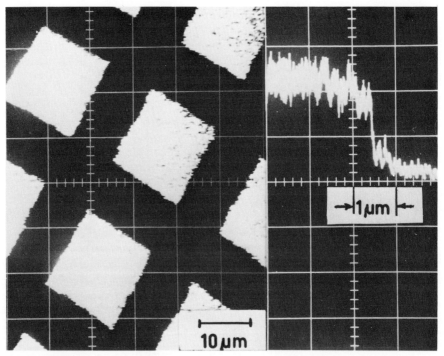

FIG. 41. Scanning image (left) and line scan (right) of a test mesh (nickel grid) using the ion microprobe of Fig. 35 with an 8-keV Ar^+ beam of 10^{-12} A. (From Liebl.[104])

[115] A. Benninghoven, *Surf. Sci.* **35**, 427, (1973).

The next example is the result of a quantitative depth analysis of H in Si in comparison with nuclear techniques (Fig. 40).[116,117] The SIMS data were obtained using an Ar^+ primary beam and analyzing the H^- yield in the apparatus shown in Fig. 37. The nuclear data are measured by using the nuclear reactions $^1H(^{15}N, \alpha\gamma)^{12}C$ and $^1H(^{19}F, \alpha\gamma)^{16}O$, which are excited by 16.4 MeV and 6.41 MeV $^{19}F^{+3}$ and $^{15}N^{+2}$ ion beams. The resulting α particles are counted by standard nuclear techniques, their numbers giving the concentration and their energy loss the depth of the H. Besides the surface H peak in the SIMS experiment, the agreement between the two techniques is excellent. Figure 40b demonstrates the extreme sensitivity achieved by SIMS, here for H in Si, which has a range of concentration covering nearly four orders of magnitude.[117] The main advantage of SIMS here is, of course, the comparatively small effort in manpower and capital.

The last example shows results from an ion microprobe[104] (Fig. 41). The apparatus is schematically shown in Fig. 35. The oscilloscope trace indicates a beam diameter of 0.1 μm. This is estimated to be the limit using a duoplasmatron ion source. Further progress may be possible with field-ionization sources to the final limit of about 10 nm given by the sputtering process.

6.3. Synopsis

Both techniques that have been presented in this chapter, ISS and SIMS, are definitely very useful for surface analysis, both having their specific advantages and drawbacks. Ion-scattering spectroscopy has probably the highest and most exclusive sensitivity to the outermost atomic layer. The spectra are simple and basically easy to interpret, ion yields not being too different for various masses. For increasing mass number, sensitivity generally increases, but mass resolution decreases. Therefore, positive mass identification is sometimes a problem. Quantitative analysis is also difficult, but can be achieved in cases with proper calibration. Secondary-ion mass spectroscopy techniques have found applications to a much higher degree and the instrumentations are much more elaborate. The positive identification of atomic and molecular masses, the extreme sensitivity and the potential for hydrogen detection are the main assets for its many applications. Problems arise for quantitative analysis due to the matrix effect and also due to the abundance of polyatomic species. Both methods are extremely powerful in combination with other techniques.

[116] G. J. Clark, C. W. White, D. D. Allred, B. R. Appleton, F. B. Kock, and C. W. Magee, *Nucl. Instrum. Methods* **149**, 9 (1978).

[117] C. W. Magee, S. A. Cohen, D. E. Voss, and D. K. Brice, Lab. Rep. PPPL-1575. Princeton Univ., Princeton, New Jersey (1979).

The increasing number of applications to a wide variety of problems in surface physics and surface chemistry will certainly also advance the understanding and widen the applicability of both methods.

Additional References

Some review articles or monographs on ion–solid interaction:

Aono, M. (1984). *Nucl. Instrum. Methods* **B2**, 374, (1984). Quantitative surface structure analysis by low-energy ion scattering.

Behrisch, R. (1964). Festkörperzerstäubung durch Ionenbeschuss. *Ergebn. Exakten Naturwiss.* **35**.

Behrisch, R., ed. (1981). Sputtering by particle bombardment I. *Top. Appl. Phys.* **47**.

Carter, G., and Colligon, J. S. (1968). "Ion Bombardment of Surfaces." Heinemann, London.

Kaminsky, M. (1965). "Atomic and Ionic Impact Phenomena on Metal Surfaces." Springer-Verlag, Berlin and New York.

McCracken, G. M. (1975). The behaviour of surfaces under ion bombardment. *Rep. Prog. Phys.* **38**, 241–327.

MacDonald, R. J. (1970). The ejection of atomic particles from ion bombarded solids. *Adv. Phys.* **19**, 457–524.

Sigmund, P. (1972). Collision theory of displacement damage, ion ranges and sputtering. *Rev. Roum. Phys.* **17**, 823–870, 969–1000, 1079–1106.

Taglauer, E., and Heiland, W. (1981). Inelastic particle–surface collisions. *Springer Ser. Chem. Phys.* **17**.

7. HIGH-FIELD TECHNIQUES

By J. A. Panitz

Surface Science Division
Sandia National Laboratories
Albuquerque, New Mexico

"It seems as if the evasive atoms still hide from the curious eye of the casual sightseer and reveal themselves accordingly only to the serious researcher. . . ."[1]

7.1. Field-Electron-Emission Microscopy

It has been noted that "in the absence of lenses, illuminating devices, and automatic controls, a field-emission microscope is less of an apparatus and more of a direct aid to the eye and brain."[2] Perhaps for this reason field-emission microscopy made possible a direct observation of adsorption, desorption, and diffusion phenomena on atomically clean, single crystal metal surfaces by 1940, more than two decades before modern techniques of surface analysis became routinely available.

The foundation for field-electron-emission microscopy was established in the early 1900s with the discovery that an electron current could be measured between two room-temperature electrodes in high vacuum whenever a critical electric field strength was established between them.[3-6] In 1928 Eyring et al.[7] performed a classic experiment, which used sharply pointed cathodes of tungsten, nickel, and platinum placed opposite a planar tungsten anode. They found that the logarithm of the electron current striking the anode depended linearly on the reciprocal of the cathode potential. A sharply pointed cathode (or "tip") produced the same current as a dull tip, but at a lower voltage. This suggested that the electron emission process depended upon the electric field strength generated at the tip surface. In

---

[1] E. W. Müller, *Science (Washington, D.C.)* **149**, 591 (1965).
[2] T. G. Rochow and E. G. Rochow, "An Introduction to Microscopy by Means of Light, Electrons, X-Rays, or Ultrasound," p. 302. Plenum, New York, 1978.
[3] G. M. Hobbs, *Philos. Mag.* **10**, 17 (1905).
[4] R. A. Millikan and Shackelford, *Phys. Rev.* **12**, 167 (1918).
[5] R. A. Millikan and Shackelford, *Phys. Rev.* **15**, 239 (1920).
[6] R. A. Millikan and C. F. Eyring, *Phys. Rev.* **27**, 51 (1926).
[7] C. F. Eyring, S. Mackeown, and R. A. Millikan, *Phys. Rev.* **31**, 900 (1928).

order to estimate the field strength at the tip, Laplace's equation was solved using a geometry chosen to approximate the actual configuration of the electrodes. A family of equipotential contours was generated. One was matched to the contour of the tip measured from an optical micrograph and one to the planar anode. From the calculation, the electric field strength at the tip surface was shown to be several volts per nanometer. The method is still in use today, having been refined by many investigators.[8-10]

Eyring et al. thermally annealed each emitter tip in order to clean its surface. They baked their vacuum chamber in order to achieve very low pressures and had the option of selecting one of many tips by using a magnetically coupled manipulator. If they had deposited a fluorescent screen onto their tungsten anode, the modern field-electron-emission microscope would have been developed almost a decade earlier.

Although experimental evidence for field emission of electrons was well documented by 1928, theoretical predictions of the measured current–voltage characteristics were not particularly successful.[11,12] Until 1928, field emission was seen as a purely classical process in which electrons were thermally activated over a field-reduced energy barrier.[13] This model did not correctly predict the empirical relation[14] between the field emission current I and the voltage V applied to the tip

$$I = C \exp(-b/V), \qquad (7.1.1)$$

where C and b are constants. A quantum mechanical treatment of the emission process was required in order to explain the empirical result.[15-17]

7.1.1. The Fowler–Nordheim Equation

In order for an electron to escape from a metal it must acquire sufficient energy to overcome the potential barrier at the surface presented by the work function Φ. Classically, the electron must be activated over the work function barrier. However, in the presence of a sufficiently large electric field the barrier will be distorted. In this case the probability for electron tunneling directly through the barrier will become significant as depicted in Fig. 1.

---

[8] J. A. Becker, *Bell Syst. Tech. J.* **30**, 907 (1951).
[9] W. P. Dyke, J. K. Trolin, W. W. Dolan, and G. J. Barnes, *J. Appl. Phys.* **24**, 570 (1953).
[10] M. Drechsler and E. Henkel, *Z. Angew. Phys.* **6**, 341 (1954).
[11] O. W. Richardson, *Proc. R. Soc. London, Ser. A* **117**, 719 (1928).
[12] J. M. Houston, *Z. Phys.* **47**, 33 (1928).
[13] W. Schottky, *Z. Phys.* **14**, 63 (1923).
[14] C. C. Lauritsen and R. A. Millikan, *Proc. Natl. Acad. Sci. U.S.A.* **14**, 45 (1928).
[15] R. H. Fowler and L. W. Nordheim, *Proc. R. Soc. London, Ser. A* **119**, 173 (1928).
[16] L. W. Nordheim, *Proc. R. Soc. London, Ser. A* **121**, 626 (1928).
[17] T. E. Stern, B. S. Gossling, and R. H. Fowler, *Proc. R. Soc. London, Ser. A* **124**, 699 (1929).

7.1. FIELD-ELECTRON-EMISSION MICROSCOPY

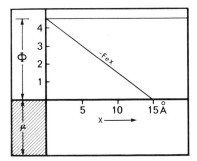

FIG. 1. A schematic energy diagram showing the work function barrier (here, equal to 4.5 eV) that an electron must overcome in order to escape from a metal surface. An applied electric field F distorts the barrier. This increases the probability of electron tunneling.

Field emission of electrons at low temperatures (in which direct thermal activation over the barrier can be neglected) is a direct manifestation of quantum mechanical tunneling, and as such has no classical analog. In fact, the success of the Fowler–Nordheim theory in predicting the current–voltage characteristic of field-emitted electrons was an early confirmation of the newly developed quantum theory of matter.

The field strength required for tunneling can be estimated from the Heisenberg uncertainty principle, as suggested by Gomer.[18] Appreciable tunneling will occur only when the uncertainty in the position of the electron becomes comparable to the width of the barrier Φ/Fe (see Fig. 1). If the uncertainty in the kinetic energy of the electron is taken to be of the order of the barrier height Φ, then

$$\Delta x \, \Delta p = (\Phi/Fe)(2m\Phi)^{1/2} \approx \hbar/2 \qquad (7.1.2)$$

or

$$F \approx (8m\Phi^3/\hbar^2 e^2)^{1/2} \approx 10^8 \quad \text{V cm}^{-1}. \qquad (7.1.3)$$

The field strength predicted by Eq. (7.1.3) is about an order of magnitude larger than required experimentally to detect a field emission current. The problem is the simplified triangular barrier of Fig. 1. It does not include the contribution of the coulombic attraction of the electron to its image charge in the metal $-e^2/4x$. Including the image potential will change the shape of the barrier as depicted in Fig. 2. The image potential is a classical concept. Strictly speaking, it must be inaccurate, at least for distances close to the surface where electron exchange and correlation effects are significant. However, it must be a reasonable approximation to the actual potential because

---

[18] R. Gomer, "Field Emission and Field Ionization." Harvard Univ. Press, Cambridge, Massachusetts, 1961.

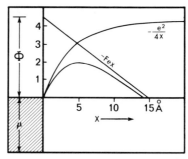

Fig. 2. A schematic energy diagram showing the shape of the potential barrier that results from applying an external field F to a metal if the classical coulombic image potential of the electron in the metal $-e^2/4x$ is included.

its use in the Fowler–Nordheim theory results in an accurate prediction of the current–voltage characteristic of field-emitted electrons.

In addition to the assumption of a classical image potential, the Fowler–Nordheim theory makes several other simplifying assumptions. These include

(1) the assumption that the tip behaves as though it were a simple, one-dimensional, free-electron-like metal, subject to Fermi–Dirac statistics at $T = 0$;

(2) the assumption that the tip surface can be approximated as a smooth, infinite plane (surface morphology is allowed provided that all features are much smaller than the width of the potential barrier); and

(3) the assumption that the work function Φ is uniform and isotropic over the tip surface.

Under these assumptions, if the barrier-penetration probability is calculated from the WKB approximation and multiplied by the arrival rate of electrons at the barrier, the result (in amperes per square centimeter) is the Fowler–Nordheim equation:

$$J = I/A = (1.54 \times 10^{-6}) \exp[-(6.83 \times 10^7)e^{3/2}\Phi^{3/2}f(y)/F][F^2/e\Phi t^2(y)], \quad (7.1.4)$$

where A is the emitting area, $t(y)$ and $f(y)$ are tabulated functions[19,20] of $y = (eF)^{1/2}/\Phi$, and the electric field strength is given by

$$F = V/KR. \quad (7.1.5)$$

[19] R. H. Good and E. W. Müller, in "Handbuch der Physik" (S. Flügge, ed.), Vol. 21, p. 176. Springer-Verlag, Berlin, and New York, 1956.
[20] A. G. J. Van Oostrom, *Philips Res. Rep., Suppl.* No. 1, p. 102 (1966).

For an isolated sphere in space $K = 1$. For actual tip geometries $3 \leq K \leq 6$.[21] Equation (7.1.4) is of the form

$$I/V^2 = a \exp(-b\Phi^{3/2}/V).$$

If the logarithm of I/V^2 is plotted against $1/V$, a straight line is obtained. From the Fowler–Nordheim equation, the slope of this line is found to be

$$S = -b\Phi^{3/2} = 6.8 \times 10^7 e^{3/2}\Phi^{3/2}f(y)KR. \qquad (7.1.6)$$

Assuming that the tip radius is known from a micrograph of the tip profile, and $K \approx 5$, the average work function of the emitter Φ can be obtained. On the other hand, if the average work function of the tip is known, KR can be directly determined.

7.1.2. The Field-Electron-Emission Microscope

In 1936 Johnson and Shockley[22] introduced the first microscope that used field-emitted electrons to form an image. It employed a cylindrical geometry in which a long thin wire served as the field emission cathode (see Fig. 3). The wire was stretched along the axis of a cylindrical glass envelope whose inside surface was covered with a transparent anode at ground potential. The anode was coated with a material that fluoresced under electron bombardment. By applying several thousand volts to the cathode, field emission was initiated. The resulting electrons were accelerated to the fluorescent screen where they

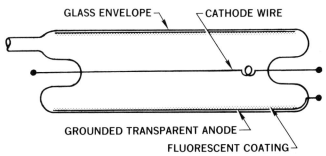

FIG. 3. A cylindrical field emission microscope developed by Johnson and Shockley in 1936. A fine-wire cathode is biased with respect to a fluorescent screen deposited on a transparent conductive anode that covers the inside diameter of an evacuated glass envelope. Axial magnification is unity. Radial magnification is determined by the ratio of wire diameter to screen diameter and is limited in practice to ~2000. The resolution of the microscope is usually worse than 10 nm.

[21] W. P. Dyke and W. W. Dolan, *Adv. Electron. Electron Phys.* **8,** 89 (1956).
[22] R. P. Johnson and W. Shockley, *Phys. Rev.* **49,** 436 (1936).

produced a pattern characteristic of the condition of the wire surface. By electrically heating the wire to a high temperature, the emission pattern became well defined and stable. The effect was correctly interpreted as an example of surface cleaning by thermal desorption of contaminant species.[22]

Crystal growth could be observed by uniformly coating the cathode wire with cesium. Cesium produces an almost contrastless emission pattern. As the temperature of the wire was increased to several thousand degrees Kelvin, the pattern changed in a way that suggested that individual crystallites were developing. Their size and shape could be studied as a function of wire temperature by observing the room-temperature emission pattern after heat treatment. Thermal activation of thoriated tungsten[23] and molybdenum[24,25] was studied in a similar manner. The results helped to clarify the nature of the activation process, which was commercially important to the developing vacuum tube industry.

The magnification and resolution of the cylindrical microscope are different in the axial and radial directions because of its nonspherical symmetry. As a result, the field emission image is distorted, making image interpretation difficult. In order to correct the problem, Erwin Müller[26,27] designed a spherically symmetric field-electron-emission microscope (or FEEM) that is still used today. Müller's microscope, introduced in 1936, was virtually identical to the design of Erying and co-workers. Field-emitted electrons were projected from a tip to an anode covered with a fluorescent material. The apparatus is shown schematically in Fig. 4. A fine wire is formed into a triangular loop with the protruding end etched to a needlelike point. One side of the loop bridges two supporting electrodes. The electrodes can be outgassed by passing an electric current through them. If the current is suddenly increased, the bridge will "burn out," and the tip assembly can be independently outgassed. By controlling the current, the tip can be thermally annealed at an elevated temperature. The conductive anode is formed by vapor depositing tin oxide onto the glass envelope of the microscope. The anode is coated with zinc oxide or Willemite ($ZnSiO_4$: Mn) in order to make the emission pattern visible.

Side arms contain sublimation sources aimed at the tip, and a barium getter is used to achieve ultrahigh-vacuum conditions. The pressure in the microscope body can be deduced from the change in the appearance of the field emission pattern with time. As residual gas contaminants are adsorbed onto the tip apex, they will become visible in the emission pattern. If a

---

[23] A. J. Ahearn and J. A. Becker, *Phys. Rev.* **49,** 879 (1936).
[24] E. Bruche and H. Mahl, *Z. Tech. Phys.* **16,** 623 (1935).
[25] E. Bruche and H. Mahl, *Z. Tech. Phys.* **17,** 81 (1936).
[26] E. W. Müller, *Z. Phys.* **37,** 838 (1936); **102,** 734 (1936).
[27] E. W. Müller, *Z. Tech. Phys.* **17,** 412 (1936).

7.1. FIELD-ELECTRON-EMISSION MICROSCOPY

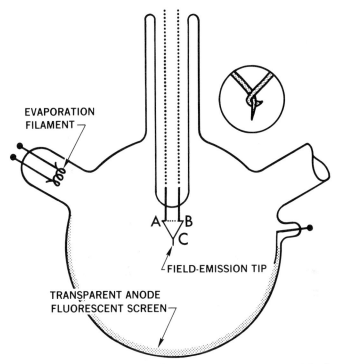

FIG. 4. A point-projection field emission microscope developed by E. W. Müller in 1936 and in use today. Magnifications of $\sim 10^6$ at a resolution of ~ 20 nm are routinely achieved. A flat fluorescent screen is usually used in order to achieve better quality photographs of the field emission pattern. The insert shows the tip assembly (C) in more detail.

sticking probability of unity is assumed, the number of adsorption events per unit time measured from the emission pattern is directly proportional to the pressure in the microscope. Since all adsorbates do not cause visible changes in an emission pattern, the method is not entirely reliable. However, it can be more accurate than a hot-filament pressure gauge, particularly at very low pressures, for which outgassing, selective adsorption, or x-ray production may present a problem.

The size of a tip apex is usually smaller than the size of the individual crystallites in most polycrystalline materials. This means that a tip etched from polycrystalline material will usually be formed from a single perfect crystal. (Occasionally, a tip will be etched from two or more single crystallites of the polycrystalline material separated by well-defined grain boundaries.) The etching process exposes at the surface of the tip apex many different crystal planes of low and high Miller indices, smoothly joined into an approximately hemispherical contour. A typical tip profile is shown in the transmission electron micrograph of Fig. 5. Individual crystal planes may have different packing densities of atoms resulting in inherently different charge distributions. As a result, the work function of a clean tip will vary over its surface depending on local crystallography. The field emission image will reflect the local crystallography of the surface because the probability of field-emitting an electron will depend on the local work function. Since surface adsorbates will tend to change the local work function, they can often be seen in the field emission pattern as a localized contrast variation. The simplicity of the FEEM technique allowed Müller[28] to study adsorption and desorption phenomena on clean metal surfaces in ultrahigh vacuum, as early as 1937.

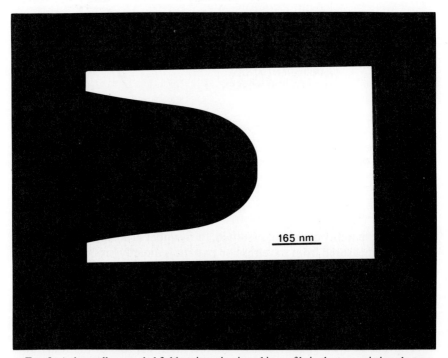

FIG. 5. A thermally annealed field-emitter tip viewed in profile in the transmission electron microscope. The tip was prepared by electrochemical polishing of a fine wire (typically 0.1 mm in diameter). Following the electropolish, the tip was heated in vacuum close to its melting point.

[27a] J. A. Panitz, *J. Phys. E* **15,** 1281 (1982).
[28] E. W. Müller, *Z. Phys.* **106,** 132, 541 (1937).

7.1.3. The Magnification of a Point-Projection Microscope

A microscope that magnifies by projecting a charged species from a highly curved surface to a detector is called a "point-projection" microscope. The FEEM is such a device. Because lenses are not used, the magnification M of the image is determined purely by geometric considerations (see Fig. 6). If a tip of radius R is placed a distance D from a fluorescent screen, the magnification will be

$$M = S/s_0 = (D + \beta R)/\beta R \simeq D/\beta R, \tag{7.1.7}$$

where $\beta = 1.5$ is an "image compression factor." It describes the deviation from pure radial projection ($\beta = 1$) caused by the presence of the tip wire and its supporting assembly. Since the tip radius is typically less than a few hundred nanometers, a very high magnification is easy to achieve.

It is important to emphasize that the magnification of a point-projection microscope relies only on geometric considerations. Unlike other microscopes a probing beam of charged particles is not required. Lenses are unnecessary and the substrate cannot move relative to a probing species. As a result, image aberrations are minimal, and external vibrations will not be magnified. This means that a useful magnification of several million times can be achieved in practice.

The radius of a tip can be accurately determined by viewing the tip in

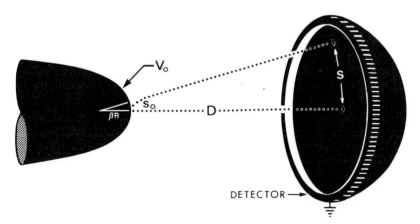

FIG. 6. A schematic representation of the geometric process used to obtain magnification in a point-projection microscope. A specimen top is biased to a positive potential V_0. Charged particles formed in the vicinity of the tip apex are accelerated almost radially into space. If the trajectories from the tip are projected backward, they will tend to intersect within the tip at a distance βR from its surface. βR can be treated as an effective tip radius. R is the measured radius of the tip and β is an "image compression" factor. For an isolated sphere in space $\beta = 1.0$. (From Panitz.[27a] Copyright 1982 The Institute of Physics.)

profile in the transmission electron microscope (Fig. 5). Since the tip-to-detector distance can also be accurately measured, the uncertainty in the magnification of a point-projection microscope will be determined by the uncertainty in the image compression factor. Unfortunately, the image compression factor is tip dependent. Although β can be determined for each tip from a trajectory calculation employing an idealized tip of comparable shape and dimensions,[29] the procedure is only an approximation. Furthermore, the actual magnification of an image will vary locally across the image because of local variations in the radius of curvature of the tip. Although this is generally a small effect, the uncertainty in β means that the magnification of an FEEM image is only known to within thirty or forty percent.

7.1.4. The Resolution of a Point-Projection Microscope

The resolution of a point-projection microscope is limited by the velocity component of the charged species parallel to the tip surface as it accelerates into space. Consider two adjacent species at the surface of the tip apex. If T is the time that it takes for each species to travel from the tip to the fluorescent screen, and v_T is their transverse velocity, then the species will have separated by a distance $= 2v_T T$ when they reach the screen. The resolution of the microscope cannot be better than this distance δ projected back onto the tip surface, where

$$\delta = 2v_T T\beta R/D \quad (7.1.8)$$

and $D/\beta R$ is the magnification. Since a charged particle acquires its full kinetic energy very close to the tip, its travel time is approximately equal to the time required for it to drift over the tip-to-screen distance with maximum velocity v_R. In other words

$$T \approx D/v_R = D(m/2neV)^{1/2}, \quad (7.1.9)$$

where m/ne is the mass-to-charge ratio of the species, and V is the potential applied to the tip. If Eqs. (7.1.3) and (7.1.4) are combined, the resolution becomes

$$\delta = 2\beta R v_T/v_R = v_T(2m\beta^2 R^2/neV)^{1/2}. \quad (7.1.10)$$

During field emission, there are two contributions to v_T. The first is just the average transverse velocity that the electrons possess as they leave the metal surface, essentially equal to their Fermi velocity. The second is caused by their finite de Broglie wavelength. Consider an electron localized to a region of the tip surface Δx. From the Heisenberg uncertainty principle, its corre-

---

[29] L. W. Swanson, *Proc. Int. Field-Emiss. Symp., 29th,* Goteborg, Sweden, p. 101. (1982). Edited by H. Nordén and H. O. Andren (Almquist and Wiksell, International, Stockholm.).

sponding momentum parallel to the tip surface will be uncertain by $\hbar/2\,\Delta x$. Consequently, the uncertainty in the average transverse velocity of field-emitted electrons due to diffraction will be

$$v_D = \hbar/2m\,\Delta x. \tag{7.1.11}$$

Müller[30] combined both contributions quadratically to obtain a resolution of about twenty angstroms, in agreement with experiment. Rose,[31] and later Brodie,[32] argued that, under very special conditions, atomic resolution ($\delta \approx$ 0.4 nm) might also be possible.

7.1.5. Adsorption Studies Using the FEEM

If the average work function of the clean tip is known, the change in work function corresponding to adsorption on the surface can be determined. Experimentally, one measures the ratio of the Fowler–Nordheim slope before and after adsorption and assumes that KR remains constant.[18] From Eq. (7.1.11):

$$\Phi_a = (S_a/S_c)^{3/2}(f_c/f_a)^{2/3}\Phi_c, \tag{7.1.12}$$

where the subscript a means *after adsorption,* and the subscript c denotes the clean tip surface *prior to adsorption.* Since $(f_c/f_a)^{2/3} \approx 1$, the relative change in the average work function due to adsorption can be determined. The accuracy of Eq. (7.1.12) depends upon the assumption that the regions of the emitter that contribute most strongly to emission prior to adsorption will contribute most strongly after adsorption. Since this is not always true, it is best to measure the work function change of small regions of the emitter such as single crystal planes. Although an apertured photomultiplier can be used to examine selected areas of the field emission pattern, it is difficult to eliminate contributions due to light scattered from high-emission areas of the tip. Therefore, a "probe hole" technique is usually used.[32-36]

In probe-hole field emission microscopy, the tip is manipulated so that the field emission image can be shifted with respect to a small aperture (or probe hole) in the fluorescent screen. A Faraday collector or electron multiplier is placed behind the probe hole in order to measure the field emission current. In principle, if the aperture is made much smaller than the area of the region being examined, a reasonably accurate measurement of local work function

---

[30] E. W. Müller, *Ergeb. Exakten Naturwiss.* **27**, 290 (1953).
[31] D. J. Rose, *J. Appl. Phys.* **27**, 215 (1956).
[32] I. Brodie, *Surf. Sci.* **70**, 186 (1978).
[33] E. W. Müller, *Z. Phys.* **120**, 261, 270 (1943).
[34] W. M. H. Sachtler, *Angew. Chem.* **80**, 673 (1968).
[35] B. E. Nieuwenhuys and W. M. H. Sachtler, *Surf. Sci.* **34**, 317 (1973).
[36] M. Domke, G. Jahnig, and M. Drechsler, *Surf. Sci.* **42**, 389 (1974).

can be made. Of course, one is never certain that all electrons from adjacent regions of the emitter will be excluded from the measurement.

The probe-hole technique has made a major contribution to our understanding of the work function concept. If nitrogen is adsorbed onto macroscopic samples of polycrystalline tungsten, a large variation in the work function (0–0.12 eV) is observed with microscopic techniques.[37,38] Probe-hole field emission experiments[39,40] demonstrate that both the magnitude and the direction of the polarization (the sign of Φ) depend on the crystal plane under examination. Since macroscopic measurements average the work function over a large area of the specimen, the wide variation in work function measured for polycrystalline samples can be explained.

Although individual adsorbates cannot always be observed in an FEEM image, the local work function of the tip surface is very sensitive to their presence. As a result, the fractional coverage of an adsorbate on the tip is usually obtained indirectly from a change in work function on adsorption. As a first approximation, a dipole moment P can be associated with each adsorbed species.[18] In this case, the work function change due to adsorption can be written as

$$\Delta\Phi = CN_T P, \tag{7.1.13}$$

where N_T is the total number of adsorbed species, and C is a constant of proportionality. If the coverage θ is defined as the number of species adsorbed on the surface N_T divided by the number of sites available on the surface for adsorption N, then

$$\Delta\Phi = CNP\theta. \tag{7.1.14}$$

Although $C \approx 4\pi^3$, its actual value is not known, so that an absolute coverage cannot be obtained from Eq. (7.1.14). If the dipole moment associated with each adsorbate is independent of coverage, a relative coverage in terms of some maximum coverage θ_{max} can be obtained from Eq. (7.1.14), where

$$\theta_{max} = \Delta\Phi/\Delta\Phi_{max}. \tag{7.1.15}$$

Strictly speaking, Eq. (7.1.15) is valid only in small, localized regions of the surface where crystallographic variations in the dipole moment for each adsorbate can be neglected.

It should be apparent from the previous discussion that qualitative information concerning an adsorbate–substrate interaction is very easy to obtain

---

[37] M. P. Hill and B. A. Pethica, *J. Chem. Phys.* **38**, 567 (1963).
[38] C. M. Quinn and M. W. Roberts, *J. Chem. Phys.* **40**, 237 (1964).
[39] A. A. Holscher, *J. Chem. Phys.* **41**, 579 (1964).
[40] A. G. J. Van Oostrom, *J. Chem. Phys.* **47**, 761 (1967).

from the appearance of the field emission pattern and its current–voltage characteristic. Although relative values of work function and coverage can be easily obtained, absolute values are difficult or impossible to measure. Fortunately, there is another type of field emission measurement that can provide detailed information on the nature of the adsorbate–substrate interaction. The measurement involves a determination of the kinetic energy distribution of the electrons field-emitted from the tip. Intuitively, one would expect that the energy distribution would be sensitive to the electronic properties of a clean metal surface and reflect any modification of the tunneling process resulting from a change in surface charge distribution on adsorption.

7.1.6. Field Emission Energy Distributions

The energy distribution of field-emitted electrons was measured before 1940 by several investigators.[26,41,42] These measurements were constrained by poor energy resolution. In 1959 the first measurement of the true narrow half-width of the total energy distribution was made by Young.[43,44] Prior to Young's observations, it was assumed that the field emission energy distribution (FEED) measured only those electrons that had a component of velocity normal to the tip surface.[19,45] Actually, electrons having both perpendicular *and* transverse energy components to the emitter surface contribute to the measured current. It is experimentally impossible to distinguish between the two. Transverse momentum is quickly dominated by the forward momentum that an electron acquires in the radially diverging field above the emitter surface.[46]

Many devices have been constructed in order to measure the total energy distribution of field-emitted electrons. Most operate as retarding-potential analyzers.[20,44,47–50] A typical analyzer consists of three electrodes. The field-emitter tip held at potential V_T, is the first. The second consists of an apertured counter–electrode that acts as the anode. It is held at potential V_A. The third electrode is a collector located immediately behind the aperture in the anode. It is held at potential V_C. The retarding analyzers that appear in the

---

[41] J. E. Henderson and R. E. Badgley, *Phys. Rev.* **38**, 540 (1931).
[42] R. K. Dahlstrom, K. V. Mackenzie, and J. E. Henderson, *Phys. Rev.* **48**, 484 (1935).
[43] R. D. Young, *Phys. Rev.* **113**, 110 (1959).
[44] R. D. Young and E. W. Müller, *Phys. Rev.* **113**, 115 (1959).
[45] E. L. Murphy and R. H. Good, Jr., *Phys. Rev.* **102**, 1464 (1956).
[46] J. W. Gadzuk and E. W. Plummer, *Rev. Mod. Phys.* **45**, 487 (1973).
[47] L. W. Swanson and A. E. Bell, *Adv. Electron. Electron Phys.* **32**, 193 (1973).
[48] R. D. B. Whitcutt and B. H. Blott, *Phys. Rev. Lett.* **23**, 639 (1969).
[49] E. W. Plummer and R. D. Young, *Phys. Rev. B* **1**, 2088 (1970).
[50] C. Lea and R. Gomer, *J. Chem. Phys.* **54**, 3349 (1971).

literature differ primarily in their geometry, the method that they use to manipulate a selected region of the field emission image over the anode aperture, and the method used to focus the electrons onto the collector. In all cases, the field strength required for electron emission is established by the potential difference that is applied between the emitter tip cathode and the anode ($V_A - V_T$). Only electrons that have sufficient kinetic energy to overcome the retarding potential difference established between the collector and the tip ($V_C - V_T$) will be measured. The variation in the collector current as a function of retarding potential is the integral of the total energy distribution of the field-emitted electrons. Energy distributions are obtained by graphical or electronic differentiation.

A typical retarding-potential analyzer has an energy resolution of $\Delta E/E \approx 30$ mV, and a signal-to-noise ratio of $\Delta I/I \approx 0.01$. To improve performance, a spherical electrostatic deflection analyzer was developed by Kuyatt and Plummer.[51] A computer-optimized lens system maximized the collected current and provided an energy resolution of ~20 mV at a signal-to-noise ratio of about 1000:1.

A clean tip surface characteristically produces a single peak in the energy distribution of field-emitted electrons. Swanson and Crouser were the first to observe two peaks in the energy distribution of electrons emitted from the (100) plane of tungsten.[47] The height of the peak (at 0.37 eV) was shown to be a sensitive function of the cleanliness of the surface. This feature (which has become known as the *Swanson hump*) is easily attenuated by the adsorption of a partial monolayer of CO, H_2, N_2, O_2, or even Kr or Xe.[49,50,52,53] Conflicting theories of the origin of the Swanson hump appear in the literature. Gadzuk and Plummer[46] appear to have demonstrated that the structure is due to a surface state in the $\langle 100 \rangle$ spin–orbit-split gap of tungsten. Figure 7 shows a comparison of Gadzuk's calculation with the measured energy distribution from the (100) plane of tungsten.[54] The energy scale in Fig. 7 has been shifted so that the zero of energy lies at the Fermi energy E_F.

Field-electron energy distribution instrumentation and the interpretation of FEED data in terms of band structure, surface states, and adsorption have been discussed in several excellent reviews.[45,55,56] It is clear that FEED data is sensitive to such phenomena. However, it is not always a simple matter to

---

[51] C. E. Kuyatt and E. W. Plummer, *Rev. Sci, Instrum.* **43**, 108 (1972).
[52] E. W. Plummer and J. W. Gadzuk, *Phys. Rev. Lett.* **25**, 1493 (1970).
[53] E. W. Plummer and A. E. Bell, *J. Vac. Sci. Technol.* **9**, 583 (1972).
[54] J. W. Gadzuk, *J. Vac. Sci. Technol.* **9**, 591 (1972).
[55] J. W. Gadzuk and E. W. Plummer, *Solid State Surf. Sci.* **3**, 165 (1973).
[56] L. W. Swanson and A. E. Bell, *Adv. Electron. Electron Phys.* **32**, 193 (1973).

7.2. APPLICATIONS OF FIELD-ELECTRON-EMISSION MICROSCOPY

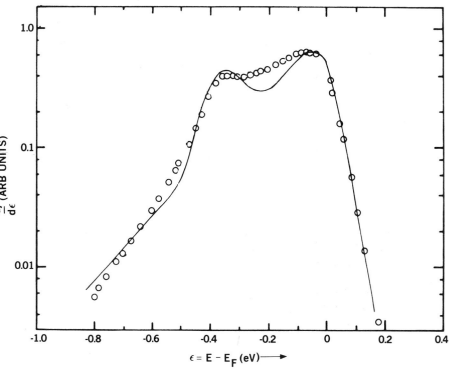

FIG. 7. A comparison of the experimental (O), and theoretical (−) total energy distribution of field-emitted electrons from the (100) plane of tungsten. Experimental parameters are $\Phi = 4.65$ eV, $F = 0.368$ V Å^{-1}, $T = 300$ K. [Figure courtesy of J. W. Gadzuk (The National Bureau of Standards, Washington, D.C.).]

interpret structure in the experimental energy distributions in terms of such phenomena.[46] In addition, it is difficult to probe very far from the Fermi energy, which limits the utility of this powerful technique.

7.2. Applications of Field-Electron-Emission Microscopy

7.2.1. Surface Diffusion

One of the most direct applications of field emission microscopy is the study of adsorbate diffusion on clean metal surfaces as a function of surface coverage. The usual method relies on observing changes in the intensity of a field emission pattern as an adsorbate diffuses over the surface at some well-defined temperature T. To minimize the possibility of the electric field affecting the measurement, diffusion is usually initiated at elevated temperatures in the absence of a field. The field emission pattern after diffusion is observed in the presence of the field but at a lower temperature for which surface diffusion is negligible. Since the resolution of the FEEM does not usually allow individual adsorbate motion to be observed, only the motion of an ensemble of adsorbates can be easily seen. Such diffusion is usually characterized by the movement of a sharp boundary, which delineates the adsorbate layer from the clean metal surface. From the observed movement of the boundary as a function of diffusion temperature and time, the diffusion coefficient and the activation energy for surface diffusion can be obtained.

One interesting system involves the diffusion of a mobile physisorbed layer on top of an immobile chemisorbed layer. Although such phenomena can be observed by cooling the emitter tip to below the multilayer condensation temperature of the adsorbate, it is preferable to cool the entire microscope in a cryostat. This technique, pioneered by Gomer,[18,57] ensures that all surfaces are held below the condensation temperature of the adsorbate. Since each molecule that arrives at a surface will be effectively removed from the gas phase, only those regions of the tip in a direct line of sight with the effusion source of the adsorbate will be covered.

Gomer et al.[58-60] demonstrated that three distinct types of diffusion could occur for hydrogen and oxygen on tungsten. The particular type of diffusion depended on the morphology of the surface. If more than a monolayer of adsorbate was deposited, uniform diffusion with a sharp boundary was observed below 20 K for hydrogen, and below 30 K for oxygen. Gomer[18] explained the observations by assuming that a mobile physisorbed layer was diffusing on top of an immobile chemisorbed layer. When the molecules in the physisorbed layer reached the edge of the chemisorbed layer, they "precipitated" onto the clean surface, where they chemisorbed, thereby extending the sharp boundary. Gomer characterized this type of diffusion by an average diffusion constant

$$D = \bar{x}^2/t, \tag{7.2.1}$$

where $\bar{x}$ is the average distance traversed by the boundary at time t. He calculated an activation energy for diffusion by assuming a diffusion coefficient of the form

$$D = A^2 v \exp(-E_d/kT), \tag{7.2.2}$$

[57] R. Gomer and J. K. Hulm, *J. Am. Chem. Soc.* **75,** 4114 (1953).
[58] B. Halpern and R. Gomer, *J. Chem. Phys.* **51,** 3043 (1969).
[59] R. Gomer, R. Wortman, and R. Lundy, *J. Chem. Phys.* **26,** 1147 (1957); R. Wortman, R. Gomer, and R. Lundy, *J. Chem. Phys.* **27,** 1099 (1957).
[60] R. Gomer and J. K. Hulm, *J. Chem. Phys.* **27,** 1363 (1957).

where $v \approx 10^{12}$ s^{-1}, and $A \approx 0.3$ nm. Typical activation energies for diffusion ranged from 24.8 ± 1 kcal for O_2 on W(110) to 5.9 ± 1 kcal for H_2 on W(110).

For monolayer coverages a second type of diffusion was observed at temperatures above 180 K for hydrogen, and above 500 K for oxygen. This type of diffusion was characterized by a diffusion boundary that spread outward from the smooth close-packed (110) plane, advancing most rapidly along crystallographic zones characterized by an atomically smooth surface. Gomer reasoned that on the smooth regions of the tip, the activation energy for surface migration is high. However, upon reaching a rougher surface topography for which the activation energy for diffusion is lower, the adsorbate would become interstitially trapped.

At coverages much less than a monolayer, a third type of diffusion was observed. At very low coverages, the rate of diffusion is not limited by migration over the smooth (110)-like regions of surface but by the number and size of "trap" sites on the rougher adjacent regions of the surface. In this case, an activation energy of 30 ± 1.5 kcal was obtained for oxygen, whereas for hydrogen 9.6 to 16 ± 3 kcal was obtained.

Many other adsorption systems have been studied by using the field emission technique. These include O_2 on tungsten,[61] CO on tungsten,[62-64] CO_2 on tungsten,[65] and Kr, Xe, Ne, and Ar on tungsten and titanium.[66-68] At submonolayer coverages, inert gas diffusion is characterized by an immobile boundary with enhanced emission appearing on initially clean regions of the tip. Since chemisorption cannot occur, the first layer of inert gas must be physically adsorbed to the substrate. Enhanced emission suggests that the inert gas layer decreases the local work function of the substrate. Under conditions of multilayer adsorption, diffusion is observed to proceed in "waves" or steps. At high temperatures, thermal desorption of the multilayer appears in the field emission pattern as a series of almost concentric boundaries that shrink toward the center of the tip at different rates.[18] These observations suggest that the inert gas boundary can be characterized by a liquid-like flow of the adsorbate where line tension forces (the two-dimension analog of surface tension forces) predominate. The outermost layers appear to behave as two-dimensional anisotropic liquids in which evaporation of

---

[61] E. W. Müller, *Z. Elektrochem.* **59**, 372 (1955).
[62] G. Ehrlich, T. W. Hickmott, and F. G. Hudda, *J. Chem. Phys.* **28**, 506 (1958).
[63] R. Gomer, *J. Chem. Phys.* **28**, 168 (1958).
[64] R. Klein, *J. Chem. Phys.* **31**, 1306 (1959).
[65] D. O. Hayward and R. Gomer, *J. Chem. Phys.* **30**, 1617 (1959).
[66] R. Gomer, *J. Chem. Phys.* **29**, 441 (1958).
[67] R. Gomer, *J. Phys. Chem.* **63**, 468 (1959).
[68] G. Ehrlich and F. G. Hudda, *J. Chem. Phys.* **30**, 493 (1959).

adsorbate atoms is most probable at the layer edge, where binding is the weakest.[18]

Isosteric heats of adsorption can be determined from field emission measurements. By using a probe-hole technique, these can be measured as a function of local surface crystallography.[36] With some sophistication, the absolute coverage of an adsorbate can also be obtained. Gomer and co-workers[69,70] have measured the absolute coverages of CO and O_2 on tungsten (110), and the sticking coefficient of CO, O_2, and Xe on tungsten (110) and (100) by using a novel field emission detector and a calibrated effusion source.

It is clear that field emission microscopy can provide a unique view of the diffusion process on an atomic scale. However, the technique suffers from two major difficulties. Since visual observations can be made only if the adsorption process causes a change in image contrast, it may be difficult to obtain information on a particular diffusion process of interest. In addition, diffusion observed in one region of the emitter surface will generally occur only after an adsorbate has migrated over adjacent regions of the tip. Since the adsorbate may encounter a variety of different surface conditions before it diffuses into the region of interest, it is difficult to correlate a diffusion constant with a unique regional morphology. In an attempt to overcome these difficulties, Gomer[71] has suggested another measurement technique that relies on local fluctuations in the field emission current of a small region of the adsorbate-covered surface. Fluctuations in the field emission current are assumed to reflect diffusion-limited concentration fluctuations in the region of the surface under observation. Gomer has been able to relate the time correlation function of these fluctuations with the local surface diffusion coefficient through a characteristic decay time that can be measured experimentally.[71] This method has been used to provide direct evidence for tunneling in the surface diffusion of hydrogen and deuterium on the (110) plane of tungsten.[72,73]

7.2.2. Sputtering, Nucleation, and Electron Sources

The field emission technique can observe and characterize surface processes on a nanometer scale as a function of local crystallography. This unique capability has been used to investigate a number of difficult problems. For example, field emission microscopy has been used to determine the absolute abundance of chemisorbed species removed by the electron-

---

[69] C. Kohrt and R. Gomer, *Surf. Sci.* **24**, 77 (1971); **40**, 71 (1973).
[70] C. Wang and R. Gomer, *Surf. Sci.* **74**, 389 (1978).
[71] R. Gomer, *Surf. Sci.* **38**, 373 (1973).
[72] R. DiFoggio and R. Gomer, *Phys. Rev. Lett.* **44**, 1258 (1980).
[73] R. DiFoggio and R. Gomer, *Phys. Rev. B* **25**, 3490 (1982).

stimulated-desorption (ESD) process.[56,74-76] A conventional ESD measurement uses a mass spectrometer to detect ions desorbed from the surface. The measurement technique discriminates against neutral species. An FEEM measurement determines the amount of adsorbate remaining on the surface after ESD by measuring a change in work function. As a result, it does not discriminate against neutral species. Field emission is very sensitive to a change in work function. This means that ESD cross sections as low as 10^{-21} cm^2 can be determined. If a probe-hole FEEM is employed, ESD cross sections for a variety of individual crystal planes can be determined during a single experiment.

Sputtering yields, sputtering rates, and threshold energies for sputtering can also be investigated with field emission techniques.[56,77-79] Threshold energies are determined by measuring the roughness of a field-emitter surface as a function of sputtering energy. The onset of surface roughness appears as a change in the field voltage proportionality factor KR [Eq. (7.1.5)], which can be determined from the slope of the Fowler–Nordheim curve. Sputtering yields and sputtering rates are difficult to assess quantitatively but may be inferred from work function changes associated with the removal of a previously adsorbed metallic layer during the sputtering process.

The process of surface self-diffusion can be examined by measuring the blunting rate of a field-emitter tip as a function of temperature.[80-83] Transmission electron microscopy is used to obtain the emitter profile and the rate of change of the emitter-tip radius. The change in the field emission pattern due to the removal of atomic planes by thermal migration is used to monitor the rate at which an emitter tip changes length.[81]

Nucleation phenomena can be examined by field emission microscopy. The Cu–W, Ag–W, Fe–W, and Zr–W systems have been investigated[84-86] and have yielded new information about this important process.

[74] C. J. Bennette and L. W. Swanson, *J. Appl. Phys.* **39,** 2749 (1968).
[75] D. Menzel and R. Gomer, *J. Chem. Phys.* **41,** 3311, (1974).
[76] W. Ermrich, *Philips Res. Rep.* **20,** 94 (1965).
[77] R. W. Strayer, E. C. Cooper, and L. W. Swanson, *Proc. 25th Annu. Conf. Phys. Electron., Cambridge Mass.,* p. 150 (1965).
[78] C. J. Bennette *NASA Res. Rep.* **NA53-8900** (1967).
[79] H. Vernickel, *Z. Naturforsch., A* **21A,** 1308 (1966).
[80] J. L. Boling and W. W. Dolan, *J. Appl. Phys.* **29,** 556 (1958).
[81] J. P. Barbour, F. M. Charbonnier, W. W. Dolan, W. P. Dyke, E. E. Martin, and J. K. Trolan, *Phys. Rev.* **117,** 1452 (1960).
[82] A. J. Melmed, *J. Appl. Phys.* **38,** 1885 (1967).
[83] P. C. Bettler and F. M. Charbonnier, *Phys. Rev.* **119,** 85 (1960).
[84] A. J. Melmed, *J. Appl. Phys.* **36,** 3585 (1965).
[85] E. Sugata and K. Takeda, *Phys. Status Solidi* **38,** 549 (1970).
[86] A. J. Melmed, *Surf. Sci.* **7,** 478 (1967).

Since a field-electron emitter is a high-brightness, point source of near-monochromatic electrons, it is an ideal source for electron optical instruments. The success of field emission sources in this regard is well documented by the commercial availability of high-resolution scanning electron microscopes using field-emitter tips.[87,88] It is interesting to note that the clean, high-vacuum conditions necessary for long emitter lifetimes and maximum current stability have led to a significant improvement in the vacuum environment within commercial instruments.

Further application of field-emitter technology to submicron photolithography, Auger and x-ray microprobe analyzers, laser-irradiated microwave diodes, microwave amplifiers, and flash x-ray devices seems quite promising.[89,90] Macroscopic field emission cathodes consisting of an ordered array of many thousand field emission tips have been used as a source of relativistic electrons in prototype inertial confinement fusion reactors[91] and for pulsed e-beam annealing of semiconductor surfaces.[92]

7.2.3. Electrical Breakdown in High Vacuum

Two metal electrodes placed a fixed distance apart in high vacuum will withstand a certain potential difference V_B applied between them. The quantity V_B is called the "breakdown voltage" of the vacuum gap, and depends upon the geometry of the electrodes and their surface characteristics. If V_B is exceeded, the gap will "break down," a phenomenon characterized by a rapid decrease in the gap impedance as a plasma rapidly expands from the cathode to the anode. Although the breakdown process is not entirely understood, it appears to be initiated by field-electron emission from microscopic protrusions on the cathode surface.[93] Such asperities are probably formed during fabrication, and may actually exist as extrusions of electrode material in the plane of the surface.[94] As V_B is approached, the electrostatic field stress at the cathode surface can erect the extrusions, so that field emission from their apexes may be initiated.[94] Statistically, the asperities should have a range of apex radii, so that each will begin to emit at a different

---

[87] A. V. Crewe, J. Wall, and L. M. Welter, *J. Appl. Phys.* **39**, 5861 (1968).
[88] A. V. Crewe, M. Isaacson, and D. Johnson, *Rev. Sci. Instrum.* **40**, 241 (1969).
[89] L. O. Hocker, D. R. Sokoloff, V. Danev, A. Szoke, and A. Javan, *Appl. Phys. Lett.* **12**, 401 (1968).
[90] W. P. Dyke, *IRE Trans. Mil. Electron.* **MIL-4**, 38 (1960).
[91] G. Yonas, *Sci. Am.* **239**, 50 (1978).
[92] A. C. Greenwald, A. R. Kirkpatrick, R. G. Little, and J. A. Minnucci, *J. Appl. Phys.* **50**, 783 (1979).
[93] D. W. Williams and W. T. Williams, *J. Phys. D* **5**, 280 (1972).
[94] R. P. Little and S. T. Smith, *J. Appl. Phys.* **36**, 1502 (1965).

7.2. APPLICATIONS OF FIELD-ELECTRON-EMISSION MICROSCOPY

applied potential. The prebreakdown current[95] measured between the electrodes as V_B is approached reflects the integrated current from a collection of asperities with the smallest radii. It should be noted that several asperities of different radii, emitting electrons simultaneously, will produce a current–voltage characteristic that will obey the Fowler–Nordheim equation.

Field-emitted electrons having acquired an energy of 80–100 eV can efficiently ionize residual gas atoms in the electrode gap. The field emission current that strikes the anode can also desorb weakly bound surface species that can subsequently be ionized, producing a current of positive ions in the gap. If a dc voltage is applied to the gap, these ions will have time to accelerate to the cathode where they can enhance the field emission process by reducing space charge or by sputter desorbing cathode material. If a short-duration voltage pulse is applied to the gap, then the ions formed in the gap may not have sufficient time to reach the cathode before the voltage terminates. The result may be an apparent increase in the breakdown voltage of the gap. If the voltage pulse vaporizes cathode asperities by intense joule heating, a rapid increase in field emission current during the pulse can be observed.[96]

A breakdown event may be beneficial by "conditioning" the electrodes to withstand subsequently higher voltages. Presumably, the effect is caused by destroying cathode asperities of small radii, or by cleaning the anode surface by electron bombardment. Unfortunately, a breakdown event may also be harmful by sputter-damaging the cathode, thereby producing sharper asperities and a lower breakdown voltage. The breakdown mechanism in vacuum is obviously quite complicated and is critically dependent on the surface condition of the electrodes.

Since field emission appears to play a dominant role in the breakdown mechanism, it is not surprising that field emission microscopy has been used to characterize prebreakdown phenomena.[20] The rationale for such studies is based on the premise that cathode protrusions are responsible for initiating breakdown, and a field-emitter tip is an excellent approximation to an isolated cathode protrusion. It should be noted that electrical breakdown in a field emission microscope is simulated by biasing the tip to a positive potential. The electrostatic stress generated at the tip surface by the applied field can cause the emitter to fracture along crystallographic imperfections. The energy stored in the capacitance of the tip leads then vaporizes the tip surface leading to damage that can be observed in the FEEM.

It is interesting to speculate if and how breakdown would occur if all cathode asperities could be eliminated. A clue may appear in a study of anode phenomena prior to breakdown.[97] Under prebreakdown conditions,

---

[95] T. Kelsey, *J. Phys. D* **5**, 569 (1972).
[96] N. V. Belkin and E. A. Avilov, *Sov. Phys.—Tech. Phys. (Engl. Transl.)* **15**, 1339 (1971).
[97] J. A. Panitz, *J. Appl. Phys.* **44**, 372 (1973).

anode species are often observed in the gap. These species are presumably created at very sharp anode protrusions by the process of field ionization or field desorption, which will be discussed shortly. In the absence of cathode asperities the breakdown process will be anode dominated and will occur at field strengths of about an order of magnitude higher than conventionally observed.

7.2.4. Molecular Imaging

The simplicity of the field-electron-emission microscope, its intrinsically high magnification, and its nanometer resolution are ideal qualifications for imaging individual molecules. Successful imaging of a single molecule presupposes that the molecules change the local work function of the surface in such a way as to reflect the true contour of the adsorbed species in an FEEM image. Since any adsorbed species can affect the local work function, it is important to eliminate contaminant species during imaging. This precludes the possibility of depositing the molecule of interest by any procedure that requires removing the tip from its ultrahigh-vacuum environment. Nevertheless, such procedures have been suggested[98] and even attempted.[99,100]

The first experiments with molecular imaging in the FEEM were reported by Müller[101,102] in 1950. He used the organic dye copper-phthalocyanine ($C_{32}H_{16}N_8Cu$) because it could be sublimed onto a clean field-emitter tip in high vacuum. Striking high-contrast images were obtained (Fig. 8a) in which fourfold symmetry, characteristic of the planar phthalocyanine molecule (Fig. 8b), could be seen. Müller's conclusion that the fourfold symmetric images represented the shape of the phthalocyanine molecule received widespread attention.[103,104] Surprisingly, the images were 10–20 times larger than predicted from the most optimistic magnification calculations (Eq. 7.1.7). Furthermore, the apparent resolution was almost an order of magnitude better than expected. Müller[102] explained both effects in terms of a locally enhanced magnification, caused by a divergence of electron trajectories in the vicinity of the adsorbed molecule.[105] Later, Rose[31] showed that a perfectly conducting hemispherical protrusion of radius r_0 on a tip of apex

---

[98] H. Montague-Pollack, *Abstr. 11th Int. Field-Emiss. Symp.*, Cambridge, Engl., p. 23 (1964).

[99] R. C. Abbott and W. Livingston, Jr., *Abstr. University Park, Pa.*, p. 59 (1965). Unpublished.

[100] R. C. Abbott, *Rev. Sci. Instr.* **36**, 1233 (1965).

[101] E. W. Müller, *Naturwissenschaften* **14**, 333 (1950).

[102] E. W. Müller, *Z. Naturforsch., A* **5A**, 473 (1950).

[103] E. W. Müller, *Life* **28**, (June 19), 67 (1950).

[104] E. W. Müller, *Sci. Am.* **186**, 58 (May 1952).

[105] Since most organic molecules are insulators or, at best, semiconductors, this explanation is only qualitatively correct.

7.2. APPLICATIONS OF FIELD-ELECTRON-EMISSION MICROSCOPY

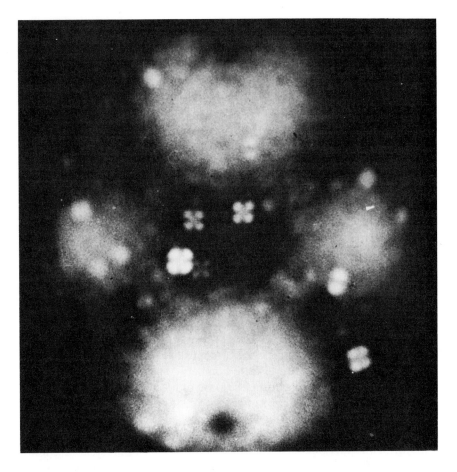

FIG. 8. Top: Field-electron-emission image obtained by vapor depositing copper phthalocyanine ($C_{32}H_{16}N_{18}Cu$) on a thermally annealed, tungsten field-emitter tip at 77 K. [Courtesy of A. J. Melmed, National Bureau of Standards, Washington, D.C.] Bottom: The structure of the copper phthalocyanine molecule ($C_{32}H_{16}N_{18}Cu$). (From Melmed.[108])

radius R would cause an increase in the normal magnification of the field emission microscope by an amount M_p, where

$$M_p \approx 1.1(R/r_0)^{1/2} M. \qquad (7.2.3)$$

For a semiconducting phthalocyanine molecule $r_0 \approx 0.5$ nm, on a tip 160 nm in radius, $M_p = 20M$, in apparent agreement with the FEEM image size. Figure 8a shows that a range of discrete image sizes and intensities is actually seen. Other observations indicated that these effects were due to the number of phthalocyanine molecules adsorbed at a single location on the surface.[106,107] The larger, more intense "quadruplet" images were interpreted as a planar stacking of many individual molecules of common orientation to form a phthalocyanine "crystallite." The smaller, less intense, images were thought to reflect only a few stacked molecules. Twofold symmetric images, or "doublets," were associated with a single molecule standing on edge.[106-108]

Gomer and Speer[109] measured the current–voltage characteristic of individual quadruplets caused by the adsorption of zinc-phthalocyanine. When compared with theory, their measurements gave a reasonable ionization potential for the molecule (7 ± 0.5 eV). This further supported Müller's contention that individual molecules were being imaged. Additional support came from imaging a twofold symmetric molecule, flavanthrene ($C_{28}H_{12}N_2O_2$), which could be vapor deposited on the tip surface. More than 99% of the image features were doublets,[106,108] (Fig. 9a), which apparently reflected the known symmetry of the molecule (Fig. 9b).

Unfortunately, when a large number of organic molecules of other shapes were examined, only quadruplet and doublet images were seen.[108,110-113] It appeared that field-electron-emission images were not displaying the actual contour of an individual adsorbate but rather some characteristic common to all of them. Müller[108] suggested that the images might result from a complicated diffraction phenomenon caused by an interaction between the adsorbed molecule, the substrate, and the tunneling electrons. Still later, he associated the appearance of the image with the presence of π-electrons in the molecule. He suggested that the magnetic moment of the π-electrons should

---

[106] A. J. Melmed and E. W. Müller, *J. Chem. Phys.* **29**, 1037 (1958).

[107] A. J. Melmed, USAF Tech. Rep. AFOSR TN 58-646 (1958) (ASTIA AD No. 162 178).

[108] J. Melmed, in "Field-Ion Microscopy" (J. J. Hren and S. Ranganathan, eds.) p. 211 (Plenum Press, New York, 1968).

[109] R. Gomer and D. A. Speer, *J. Chem Phys.* **21**, 73 (1953).

[110] P. Wolf, *Z. Angew. Phys.* **6**, 529 (1954).

[111] E. Hörl and F. Strangler, *Acta Phys. Austriaca* **10**, 1 (1956).

[112] J. A. Becker and R. G. Brandes, *J. Appl. Phys.* **27**, 221 (1956).

[113] R. Haefer, *Acta Phys. Austriaca* **8**, 105 (1953).

7.2. APPLICATIONS OF FIELD-ELECTRON-EMISSION MICROSCOPY

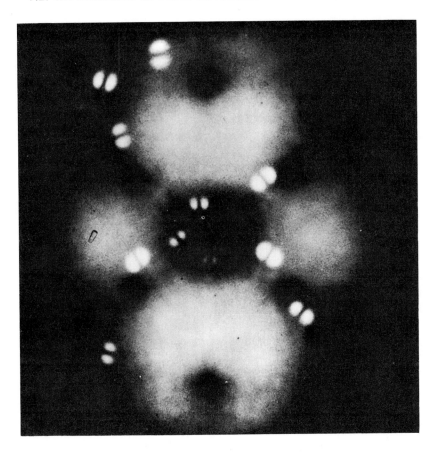

FIG. 9. Top: Field-electron-emission image obtained by vapor depositing flavanthrene ($C_{28}H_{12}N_2O_2$) on a thermally annealed tungsten field-emitter tip at 77 K. [Figure courtesy of A. J. Melmed (The National Bureau of Standards, Washington, D.C.).] Bottom: The structure of the flavanthrene molecule. (From Melmed.[108])

interact strongly with the spin of the slow-moving, field-emitted electrons.[114] The resulting interaction could cause a splitting of the field-emitted-electron beam. However, Müller's model could not explain the formation of a fourfold symmetric beam required for the production of quadruplet images.

Hörl and Strangler[111] proposed that the images had nothing to do with an adsorbed molecule but were caused by clusters of tungsten atoms displaced from the substrate during molecular deposition in the presence of a high field. Since identical image features were also obtained after deposition in the absence of a field, Hörl and Strangler's explanation was largely ignored. It is conceivable that the images could result from a decomposition product common to all of the imaged species, but this is unlikely in view of the large number of different molecular structures and substrates that were examined.

Melmed[115] proposed that an adsorbed molecule or a molecular crystallite could act as an "aperture" in which the probability of electron emission would be larger than that of the surrounding clean surface. He qualitatively argued that field-emitted electrons might interact with an excess negative charge distribution associated with the polarized molecule, splitting the electron trajectories into two or four equally intense bundles. The exact mechanism responsible for splitting the beam was not explained.

An alternate possibility, consistent with the hypothesis that the adsorbed molecules stack to form an ordered crystallite, was advanced by Giaever.[116] He suggested that doublet images are caused by electrons emitted from single planar molecules standing on edge, whereas quadruplet, and more infrequently, esoteric patterns are caused by emission from the ends of small molecular crystallites (Fig. 10). Giaever believed that under certain conditions a field emission pattern could reflect the presence of an individual molecule but not its actual morphology. He interpreted a common symmetry in the emission pattern of different molecules in terms of a common symmetry in the electric field distribution surrounding any planar adsorbed species.

7.2.5. Single-Atom Imaging

The possibility that the FEEM may have sufficient resolution to image single atoms was raised during the time that molecular imaging was in vogue. Müller[102] proposed that "granulation" observed in an FEEM image of a tungsten following barium deposition was due to enhanced emission at individual barium atoms. Becker[117,118] questioned this interpretation, pre-

---

[114] E. W. Müller, Electron Microsc. Soc. Am., 1953 (unpublished lecture). See ref. 21, p. 139.
[115] A. J. Melmed, Ph.D. Thesis, Pennsylvania State Univ., University Park, 1958.
[116] I. Giaever, *Surf. Sci.* **29**, 1 (1972).
[117] J. A. Becker, *Bell Syst. Techn. J.* **30**, 907 (1951).
[118] J. A. Becker, *Prgm. 1st Int. Field-Emiss. Symp.*, McMinville, Or. (1952). Unpublished.

7.2. APPLICATIONS OF FIELD-ELECTRON-EMISSION MICROSCOPY

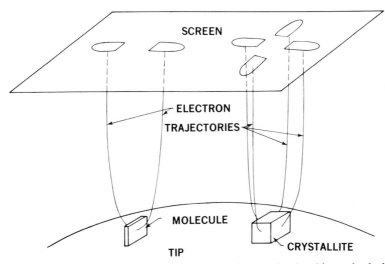

FIG. 10. "A schematic drawing of electron emission from molecular objects adsorbed on a field-emitter tip. For clarity, the sizes of the molecular objects and the tip itself have been drawn grossly magnified. The separation of the electron trajectories is due to the local field distortion around the molecules." (Figure and caption from Giaever.[116] Copyright North-Holland Physics Publishing, Amsterdam.)

ferring to regard the granulation as enhanced emission from clusters of several barium atoms. Ashworth[119] found that if hydrogen was introduced into a well-baked FEEM, single image spots could be observed on flat smooth crystal faces that had a low work function. Eventually, each spot would appear to split into two halves. These remained close together, and would usually rotate about each other before disappearing. Alternatively, the individual image spots would not disappear but would move randomly away from each other over the surface. If argon was adsorbed, only single image spots were observed. Ashworth believed that the FEEM images of molecular hydrogen reflected dissociation of the adsorbate. After dissociation, two image spots were observed, suggesting that two hydrogen atoms (visible as separate entities) moved randomly across the surface. Since argon does not adsorb as a molecular species, only single image spots would be seen. A smooth flat surface of low work function was required to see an image, because on such a surface an adsorbate would produce the maximum perturbation in the surface potential. This would lead to enhanced emission and image contrast and an increase in local magnification and resolution.[30–32]

The goal of imaging individual atoms in the FEEM was never completely

---

[119] F. Ashworth, Ph.D. Thesis, Univ. of Bristol, Bristol, England, 1948.

achieved. Nevertheless, it was this challenge that ultimately led Müller to develop the field-ion microscope in which single atoms can unambiguously be observed.

7.3. Field-Ion Microscopy

In theory, any ionization mechanism that can be localized to the immediate vicinity of a surface species can be used to image that species. By applying a dc bias to the tip surface, the resulting ions can be forced to follow almost radially directed field lines, which leads to a magnified image of their point of origin. As early as 1941, Müller observed [120] that adsorbed barium atoms could be removed from a tungsten field emission tip by reversing the polarity of the tip bias and increasing the field strength beyond 8 V nm^{-1}. He later reasoned [121] that if the desorption process involved the formation of positive barium ions, these ions might yield a magnified image of their adsorption sites on the surface. (If neutrals were formed by the desorption process they would not be accelerated in the applied field and no magnification would

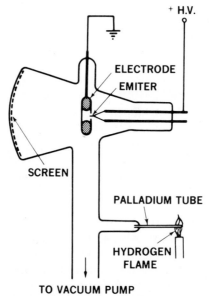

FIG. 11. A schematic drawing of the original room-temperature field-ion microscope introduced by E. W. Müller in 1951. (From E. W. Müller and T. T. Tsong,[126] "Field-Ion Microscopy: Principles and Applications." Copyright 1969 by American Elsevier Publishing Co.)

[120] E. W. Müller, *Naturwissenschaften* **29**, 533 (1941).
[121] E. W. Müller, *Z. Phys.* **131**, 136 (1951).

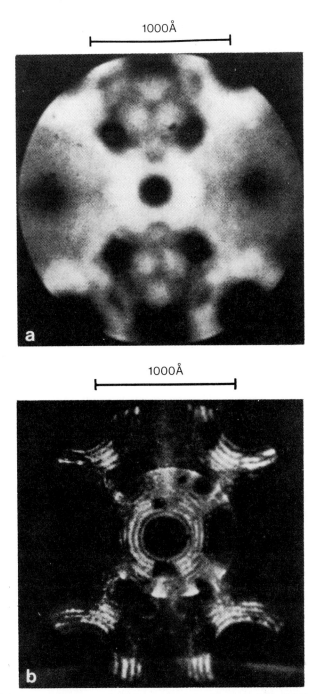

FIG. 12. (a) A room-temperature field-electron-emission image, and (b) a field ion image of a tungsten tip ($R = 940$ Å) taken by E. W. Müller in 1951. The almost concentric, diffuse circle of light at approximately one-half of the image diameter was attributed to secondary electrons released by hydrogen-ion bombardment of an annular cathode placed opposite the tip. (From Müller.[121])

result.) The desorption process produces very small ion currents. Even a monolayer of surface species desorbed as positive ions from the entire surface of the tip would yield only 10^{-14} C of charge.

Müller[121] observed that for a barely discernable image on a fluorescent screen an image intensification of at least 10^6 would be required. Since modern methods of image intensification were not available, this could be achieved only by a rapid succession of adsorption–desorption sequences, with the resulting images photographically integrated. It would be convenient to supply the imaging species from the gas phase so that a continuous supply would be available for adsorption onto the tip surface. Hydrogen was selected for this purpose and introduced into an existing FEEM by diffusion through a heated palladium tube (Fig. 11). Müller believed that if the emitter tip was kept at a sufficiently high positive bias in the presence of ambient hydrogen, an adsorption–desorption cycle could be indefinitely maintained. We now know that field ionization of gas-phase hydrogen in free space was actually responsible for the images that Müller observed (Fig. 12). But, fortuitously, the right conditions were established, and a truly revolutionary microscopy was born.[121]

7.3.1. Field Ionization

Field ionization of an isolated atom in free space by electron tunneling was first proposed by Oppenheimer[122] in 1928. By formulating the problem quantum mechanically, he was able to demonstrate that the lifetime of an electron in an isolated hydrogen atom depended upon the magnitude of an external electric field. At fields of the order of 1 V cm^{-1}, the calculated lifetime was greater than any estimate of the age of the universe, but at fields approaching 15 V nm^{-1}, ionization would occur within a time frame of less than one second. Although Erying et al.[7] applied positive fields to their emitter tips in 1928, they did not observe a positive ion current because their experiments were performed in high vacuum.

In 1931, Lanczos[123] explained field-induced quenching of spectral lines in the Stark effect by quantum mechanical tunneling from an excited state of the hydrogen atom. Guerney[124] considered the tunneling process to be an explanation of the neutralization of hydrogen ions at the cathode of an

---

[122] J. R. Oppenheimer, *Phys. Rev.* **31,** 67 (1928).
[123] C. Z. Lanczos, *Z. Phys.* **68,** 204 (1931).
[124] R. W. Gurney, *Proc. R. Soc. London, Ser. A* **134,** 137 (1932).

electrolytic cell. He noted that tunneling should occur at a critical distance from the electrode surface where the ground state energy of the atom was equal to the Fermi level of the metal. However, the field-ion imaging process could not be explained until Inghram and Gomer [125] developed a theoretical description of field ionization in the vicinity of a metal surface. They clearly showed that hydrogen was not ionized on the emitter surface but in free space. Field-ion images were produced by a gas-phase ionization process and not by an adsorption–desorption cycle at the surface of the tip.

In free space, the valence electron of a neutral atom can be considered as bound in a potential well. An energy I (equal to the ionization potential) must be supplied in order to ionize the atom. In an electric field, the potential well will be distorted, with the side of the well toward the anode reduced in width. As the electric field strength is increased, this width will become comparable to the de Broglie wavelength of the valence electron in the atom, and the probability for tunneling through the potential barrier will greatly increase. As the distance between the atom and the anode is decreased, the barrier width will decrease, due to short-range contributions to the potential from atom–surface interactions.

Since tunneling can occur only into an unoccupied state of the metal, the ground state of the valence electron in the atom must lie above the Fermi level of the metal. This condition (shown schematically in Fig. 13) is satisfied at some critical distance X_c from the metal surface, where

$$eFX_c = I - \Phi. \tag{7.3.1}$$

The image potential due to the tunneling electron (and the change in the polarization energy of the atom upon ionization) have been omitted in Eq. (7.3.1) because their combined contribution is usually negligible. For hydrogen $I = 13.5$ eV. In a field of 2 V nm^{-1} above a tungsten surface (for which $\Phi = 4.5$ eV), the critical distance for ionization $X_c = 0.45$ nm.

A reasonably accurate estimate of the ionization probability of an atom at the critical distance can be obtained by multiplying the probability of an electron tunneling from the atom by the frequency with which it strikes the tunneling barrier. The former can be estimated from the WKB approximation using a simplified triangular barrier. The latter can be estimated from the Bohr model using an "effective" nuclear charge appropriate to the atom in question.[126] The field-ion current that is measured should then be the

---

[125] M. G. Inghram and R. J. Gomer, *J. Chem. Phys.* **22**, 1279 (1954).

[126] E. W. Müller and T. T. Tsong, "Field-Ion Microscopy: Principles and Applications." Am. Elsevier, New York, 1969.

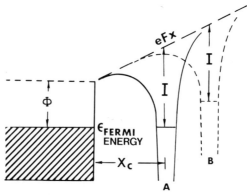

FIG. 13. A schematic drawing of the field-ionization process. At position A, the atom is located at a distance X_c from the metal surface. X_c is the critical distance for field ionization, and corresponds to the position at which the energy level of the valence electron in the atom is equal to the Fermi energy of the metal. At closer distances to the surface, the energy level of the electron will lie below the Fermi energy so that tunneling will not occur. For distances far from the surface (e.g., position B) the barrier width will be large, so that tunneling will be improbable. (From Panitz.[27a] Copyright 1982 The Institute of Physics.)

product of the ionization probability of an atom (assumed to be isotropic) and the number of atoms that arrive in the ionization zone each second. It was noticed quite early that the measured field-ion current was at least an order of magnitude larger than expected if a supply of imaging gas molecules governed by the kinetic theory of gases and an ionization probability of 100% was assumed. The paradox was resolved when it was realized that the number of gas-phase molecules in the vicinity of the tip is artificially enhanced by field-induced dipole attraction to the tip apex. When calculated values for the enhancement factor due to the field-induced dipole energy αF^2 (α = the polarizability) were performed,[127-129] field-ion currents in good agreement with the measured values were obtained.

7.3.2. Field-Ion Energy Distributions

The energy distribution of field-ionized gas atoms can provide valuable insight into the nature of the ionization process. If an ion is created at the emitter-tip surface, it will acquire a kinetic energy corresponding to its acceleration through the full potential difference that is applied between the tip and a suitable collector. If the ion is created in space at some distance from

[127] M. J. Southon, Ph.D. Thesis, University of Cambridge (1963).
[128] E. W. Müller and K. Bahadur, *Phys. Rev.* **102**, 624 (1956).
[129] E. W. Müller, *Adv. Electron. Electron Phys.* **13**, 83 (1960).

the surface, it will acquire less kinetic energy in traveling to the collector. Therefore, the threshold energy in a field-ion energy distribution will indicate the position at which the ions are formed, and the width of the energy distribution will reflect the spatial extent of the ionization region.

The first field-ion energy distribution was measured in 1954 by Inghram and Gomer[125] using a mass spectrometer that incorporated a field-ionization source. From the width of their mass peaks, they concluded that the width of the energy distribution for field-ionized hydrogen was less than 20 eV, the resolution of their instrument. At very high fields they noticed that the distribution broadened, indicating spatial ionization far from the tip surface. By using an improved analyzer with an energy resolution of ~2 eV, Müller and Bahadur[128,130] found that the half-width of the energy distribution for field-ionized argon was ~2 eV, with a sharp threshold indicating spatial ionization ~0.5 nm above their tungsten surface. These measurements were extended by Tsong and Müller[131] who investigated a number of gases including helium. Figure 14 shows several typical energy distributions for field-ionized helium as a function of the potential applied to the tip. The width of the distribution at 20 kV (corresponding to $F = 45$ V nm^{-1} and the best visual image) indicates that the region of ionization is less than 0.02 nm

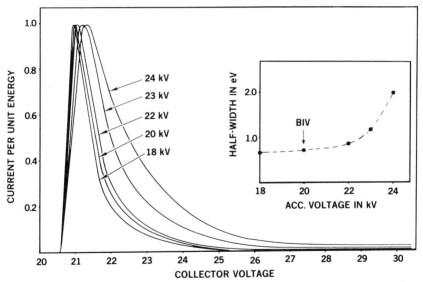

FIG. 14. An energy distribution for helium ions, field-ionized above a tungsten field-emitter tip. (From Tsong and Müller.[131])

[130] E. W. Müller and K. Bahadur, *Phys. Rev.* **99**, 1651 (1955).
[131] T. T. Tsong and E. W. Müller, *J. Chem. Phys.* **41**, 3279 (1964).

thick or only a few tenths of the diameter of the ionizing gas atom. Similar results were found for neon, argon, and hydrogen. As the field strength was increased, the distributions broadened toward lower energies, in agreement with Inghram and Gomer's earlier observations. Jason et al.[132] subsequently discovered a weak periodic structure in the low-energy tail of hydrogen and neon distributions. Jason[133] and Aiferieff and Duke[134] interpreted the multiply peaked structure as a resonance associated with the tunneling of an electron between the ionizing atom and the nearby metal surface. The resonance model predicted an increase in peak spacing with applied field, as well as a progressive decrease in peak spacing toward lower energies. Both effects were observed experimentally but at a significantly higher field strength than predicted. Lucas (unsuccessfully) attempted to explain the experimental observations in terms of an excitation of multiple surface plasmons.[135,136]

In order to examine what had become known as the "Jason effect" in more detail, Müller and Krishnaswamy[137] measured the field-ion energy distributions caused by a single atom on smooth and rough crystal planes. Jason had averaged the ion current over a large region of the tip surface so that the effect of surface morphology could not be determined. By using a probe hole to limit the field of view to a single atomic site and a high-resolution energy analyzer, Müller and Krishnaswamy found as many as seven Jason peaks above the close-packed planes of tungsten and iridium but only two or three above the more open "structured" planes [such as the (111)]. For highly disordered surfaces of carbon and silicon, it was difficult to detect more than one peak in the energy distribution.

These measurements indicated that the atomic roughness of a surface could reduce or even eliminate the Jason effect. Intuitively, the phenomenon is easy to understand. Imagine that the electron that tunnels from an atom ionizing in space has a finite probability of being reflected from the nearby metal surface and essentially zero probability of tunneling back into the resulting ion. Under such conditions, standing waves will exist in the potential well defined by the ionizing atom and the surface. The strength of the resulting resonances (which are reflected by the Jason peaks) will depend on the uniformity of the surface potential that governs the electron reflectivity at the metal surface. The smoother the surface, the more uniform the potential, and the greater will be the probability for electron reflection necessary to produce the Jason effect.

[132] A. J. Jason, R. P. Burns, and M. G. Inghram, *J. Chem. Phys.* **43**, 3762 (1965).
[133] A. J. Jason, *Phys. Rev.* **156**, 156 (1967).
[134] M. E. Aiferieff and C. B. Duke, *J. Chem. Phys.* **46**, 938 (1967).
[135] A. A. Lucas, *Phys. Rev. Lett.* **26**, 813(1971); *Phys. Rev. B* **4**, 2939 (1971).
[136] A. A. Lucas and M. Sunjic, *J. Vac. Sci. Technol.* **9**, 725 (1972).
[137] E. W. Müller and S. V. Krishnaswamy, *Surf. Sci.* **36**, 29 (1973).

At very high field strengths (above 50 V nm^{-1}) Müller and Krishnaswamy[137] found that a few percent of the field-ionized noble gas atoms had energies higher than the threshold energy of the distribution. It appeared as though some imaging gas molecules were being ionized within the critical distance from the surface — within the so-called "forbidden zone." In order to ionize an atom within the forbidden zone, energy must be supplied to raise the level of the valence electron in the atom above the Fermi level of the metal. Müller and Krishnaswamy believed that this energy was supplied by a continuous current of low-energy electrons striking the tip surface. The electron current is presumably produced by spatial ionization of imaging gas atoms far from the tip.[137] Imaging gas atoms are found on the tip because they are bonded to its surface by a short-range field-induced dipole–dipole bond.[138] This unique type of bonding occurs only at fields greater than 10 V nm^{-1}. The process is called "field adsorption." It was discovered experimentally by Müller et al.[139] in 1969 and explained theoretically by Tsong.[140]

7.3.3. The Hopping Gas Model

Field-ion energy distribution measurements localized the region of greatest ionization probability to a volume of space located above the field-emitter surface. At a certain electric field strength (called the best-image field), the ionization region separates into "disks" of high ionization probability associated with each of the more protruding atoms of the surface. At these positions the local electric field strength is greatest. Field-ion energy distributions indicate that the ionization disks are of the order of 0.02 nm thick and are approximately equal in diameter to an imaging gas atom. At the best-image field, the field-ion image will display maximum surface detail and definition over the widest field of view. The best-image field is established by applying a potential difference between the emitter tip and the screen, which is known as the "best-image voltage" or BIV. Although BIV is subjectively chosen, different observers always select the same BIV to within a few percent.

Since the imaging gas molecules are polarized in the inhomogeneous field above the emitter surface, they will eventually strike the emitter with a dipole attraction velocity $(\alpha/m)^{1/2}F$, which is much larger than kT. If they do not condense on the tip surface, the imaging gas molecules will lose some fraction of this energy (determined by their thermal accommodation coefficient) and rebound, only to be attracted once again by polarization forces. At BIV, a reasonable fraction of the emitter-tip surface will be covered with field-ad-

[138] E. W. Müller, *J. Less-Common. Met.* **28**, 37 (1972).
[139] E. W. Müller, S. B. McLane, and J. A. Panitz, *Surf. Sci.* **17**, 430 (1969).
[140] T. T. Tsong, *Phys. Rev. Lett.* **25**, 911 (1970).

sorbed imaging gas molecules, which greatly enhances the thermal accommodation process.

One can picture the polarized gas-phase molecules as eventually being trapped in a region of space close to the emitter apex while slowly diffusing over the surface of the tip in a random "hopping" motion.[126] At some point, each molecule will pass through the ionization disk above a protruding atom of the surface. If its velocity is sufficiently reduced by the thermal accommodation process, it will dwell in the ionization region and have a finite probability of losing an electron to the metal by the tunneling process. The tunneling probability is enhanced by the presence of field-adsorbed gas atoms at protruding sites directly below the ionization region. If tunneling occurs, the resulting ion will accelerate rapidly away from the tip surface in an almost radial direction. At an imaging gas pressure of several millitorr, some 10^5 ions s^{-1} will be created above each protruding surface atom. This is equivalent to an ion current of the order of 10^{-14} A. The intensity of the resulting field-ion image is roughly equivalent to that of the Milky Way viewed with dark-adapted eyes on a moonless night.

7.3.4. The Low-Temperature Field-Ion Microscope

In thermally accommodating to the tip temperature, an imaging gas species will lose its field-induced dipole energy $\frac{1}{2}\alpha F^2$, so that after ionization its kinetic energy parallel to the tip surface will be

$$\tfrac{1}{2}mv_T^2 \approx kT. \qquad (7.3.2)$$

The resolution of the FIM is obtained by combining Eqs. (7.1.5), (7.1.10), and (7.3.2) and noting that the resolution will be ultimately limited by the spatial extent of the ionization zone δ_0, that is,

$$\delta = \delta_0 + (4kT\beta^2 R/KeF)^{1/2} = \delta_0 + A(RT/F)^{1/2}, \qquad (7.3.3)$$

where A is essentially constant and δ_0 is of the order of the diameter of the imaging gas ion. All other things being equal, cooling the tip from room temperature to liquid-nitrogen temperature will improve the image resolution by roughly $(80/300)^{1/2} \approx 50\%$.

As early as 1951, Müller[121] noted that "experiments were in progress with helium desorption from strongly cooled tips." However, in 1952 Gomer[141] reported that immersing an entire FIM in liquid nitrogen did not improve the resolution of the image. Apparently, observing the FIM pattern through the cryogenic liquid prevented an improvement in resolution from being noticed. Perhaps as a result, low-temperature experiments were not actually

---

[141] R. Gomer, *J. Chem. Phys.* **20**, 1772 (1952).

pursued. Fortunately, in 1956 Müller[142,143] noted the great improvement in resolution that could be obtained by cooling the tip. Atomic resolution had been achieved simply by using helium as an imaging gas and cooling only the tip with liquid nitrogen.

For optimum resolution at any tip radius, Eq. (7.3.3) predicts that the tip temperature should be kept as low as possible, the diameter of the imaging gas ion should be chosen to be as small as possible, and the electric field at BIV should be made as large as possible. With a condensation temperature below 5 K, an effective diameter of 0.19 nm and a best-image field of 45 V nm^{-1}, helium is by far the best choice for an imaging gas. However, the mechanical stress exerted on the emitter by the field at helium BIV ($F^2/8\pi = 10^{11}$ dyn cm^{-1}) will elastically deform most materials. Only those that have a relatively high Young's modulus and are free of major lattice defects will survive. With some sacrifice in resolution, other imaging gases or gas mixtures can be selected in order to reduce the field stress at BIV. For example, hydrogen produces a field-ion image at about one-half of helium BIV, reducing the field stress by 50%. By a careful choice of imaging conditions a number of materials have been successfully imaged.[126]

If the tip voltage is raised beyond BIV, the ionization region in front of the tip is extended further into space and becomes laterally delocalized so that the FIM image becomes progressively blurred. As the electric field strength increases it will reach a value sufficiently high to ionize weakly adsorbed surface species and displaced lattice atoms. These will be desorbed from the surface as positive ions. Eventually, even the lattice will begin to dissolve as lattice species become ionized. At a given tip potential and temperature, the surface will eventually assume a stable topography or "end form" of minimum free energy. This surface will be atomically smooth, and can be as crystallographically perfect as the bulk. The removal of lattice atoms in a high electric field is called "field evaporation," in analogy with thermal evaporation, which occurs at high temperatures. Field evaporation was discovered by Müller[142,144] in 1956. It is the only method that can prepare an emitter surface with the atomic regularity required for high-resolution lattice imaging. Field evaporation is a specific case of "field desorption," a term applied to the removal of *any* species in a high electric field. Since a material of interest may field-evaporate at a field strength that is equivalent to (or below) BIV, field evaporation can also prevent stable FIM images from being obtained.

Figure 15 shows a ball model of the apex of a (011)-oriented, bcc field-

---

[142] E. W. Müller, *Z. Naturforsch.*, **11A**, 87 (1956).
[143] E. W. Müller, *J. Appl. Phys.* **27**, 474 (1956).
[144] E. W. Müller, *Phys. Rev.* **102**, 618 (1956).

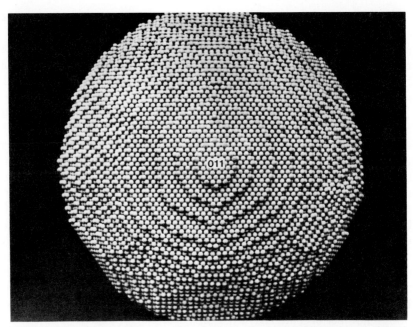

Fig. 15. A ball model of the apex region of a (011)-oriented, bcc field-emitter tip following field evaporation. (From Panitz.[27a] Copyright 1982 The Institute of Physics.)

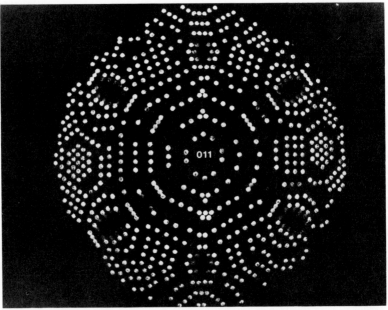

Fig. 16. The ball model of Fig. 15 with the more protruding edge atoms at lattice steps painted with fluorescent dye and photographed in the dark. (From Panitz.[27a] Copyright 1982 The Institute of Physics.)

emitter tip as it would appear after field evaporation if each surface atom could be seen.[129] Since the more protruding atoms exposed at the corners of the lattice steps will generate the highest local fields, these atoms will appear brightest in the field-ion image. In order to simulate a field-ion micrograph, atoms with four next-nearest neighbors can be painted with a bright fluorescent dye, and atoms on slightly less protruding sites can be painted with a weak fluorescent dye.[129] When photographed in the dark, the resulting image (shown in Fig. 16) accurately represents an image of tungsten observed in an FIM (Fig. 17).

Once a stable field-evaporated endform is reached, the tip voltage can be lowered to BIV for imaging, or can be increased gradually to controllably dissect the near-surface region of the lattice, atomic layer by atomic layer. Since the lattice dissolves during field evaporation, bulk artifacts (naturally occurring in the near-surface region or purposely introduced) can be exposed for imaging or removed from the lattice. Field evaporation is the only known method that will produce atomically clean surfaces with complete confidence.

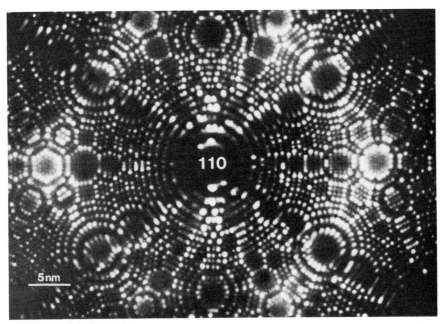

FIG. 17. A helium field-ion micrograph of a (110)-oriented tungsten tip taken at 18 K using a chevron channel-plate image intensifier located within the FIM. (From Panitz.[27a] Copyright 1982 The Institute of Physics.)

7.3.5. The Magnification of a Field-Ion Image

The magnification of a field-ion image will vary slightly across the imaged area because of local variations in the radius of curvature of the tip. Fortunately, a very precise local magnification can be obtained if the FIM image exhibits crystallographic order.[126] If γ is the known apex angle between two imaged, crystallographic directions and n is the number of net plane rings of known step height s that are resolved in the image, then, from Fig. 18, the local radius of curvature is

$$R_L = ns/(1 - \cos \gamma). \tag{7.3.4}$$

For cubic crystals, the step heights of the $\{hkl\}$ poles are given by

$$s = a/\delta(h^2 + k^2 + l^2)^{1/2}, \tag{7.3.5}$$

where a is the lattice constant and $\delta = 1$ or 2 as shown in Table I. For tungsten, an average tip radius (in nanometers) can be obtained by multiplying the number of net plane rings seen in the image between the centers of the (001) and (121) planes by 0.16. Unlike an FEEM image or an electron

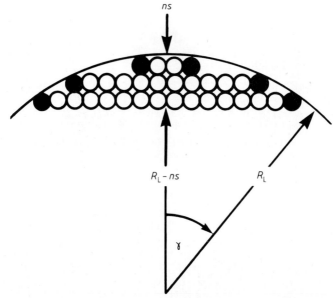

FIG. 18. A schematic drawing of a tip in profile showing two crystallographic directions separated by a known angle γ. R_L is the local radius of curvature of the tip, n is the number of net plane rings seen in the image, and s the known step height per ring (see text). The dark atoms at the edge are those that are seen in an FIM image. (From Panitz.[27a] Copyright 1982 The Institute of Physics.)

TABLE I. The Scaling Factor δ Used to Determine the Step Heights of the hkl Poles in a Field-Ion Image[a,b]

δ	Simple Cubic Lattice	bcc Lattice	fcc Lattice
1	All values of h, k, l	$h + k + l$ even	h, k, l odd
2		$h + k + l$ odd	h, k, l mixed parity

[a] From Panitz.[27a] Copyright 1982 The Institute of Physics.
[b] See Eq. (7.3.5).

micrograph, each resolved field-ion image contains its own magnification standard, which is the crystal lattice itself.

7.3.6. The Modern Field-Ion Microscope

A low-temperature field-ion microscope is a conceptually simple device. A generic form of the instrument is shown schematically in Fig. 19. A cryogenically cooled specimen tip that is biased to a high positive potential is located in an ultrahigh-vacuum environment that can be selectively backfilled to about 10^{-5} Torr with a very pure imaging gas. If helium or neon are used for imaging, a liquid-nitrogen-cooled titanium sublimator provides effective pumping of all background contaminants during the imaging process. Helium or hydrogen can be conveniently introduced into the vacuum chamber by diffusion through a thin-walled electrically heated quartz or palladium thimble. It is not necessary to provide a very pure supply of these species, since larger molecules and other impurities cannot diffuse into the vacuum system.

A fiber-optic faceplate coated with a fluorescent material (such as P1 phosphor) provides a convenient way to record the field-ion image. The faceplate will transmit the image from the vacuum environment to laboratory ambient, where it can be easily photographed.[145] If a contact print is made by pressing a photographic emulsion directly onto the faceplate, optical lenses with their inherent slow speed and distortion can be avoided. Using Polaroid® Type 55 film, a print and a negative can be immediately obtained without the need for darkroom facilities.[146] In many instances images can be quickly recorded without the use of additional image intensification.

One disadvantage of a simple FIM is the relatively long time required to change specimen tips, particularly if ultrahigh-vacuum conditions must be reestablished prior to imaging. In order to overcome this difficulty, the tip can be placed in a prepumped specimen chamber and transferred under vacuum through an isolation valve, to the base of the cryogenic reservoir.[147]

---

[145] J. J. Hren and R. W. Newman, *Rev. Sci. Instrum.* **38**, 869 (1967).
[146] J. A. Panitz, *Rev. Sci. Instrum.* **44**, 1034 (1973).
[147] J. A. Panitz, *J. Vac. Sci. Technol.* **14**, 502 (1977).

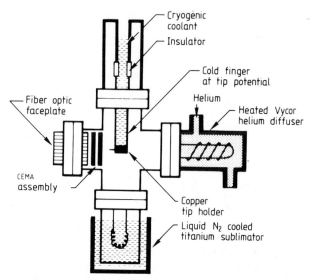

FIG. 19. A schematic drawing of a simple generic form of a modern field-ion microscope. The device is typically constructed from interchangeable stainless steel vacuum components. The chamber is attached to a conventional ultrahigh-vacuum system that is not shown. With this simple microscope, metallic surfaces can be imaged with atomic resolution, and surface processes directly examined as a function of surface crystallography and morphology. (From Panitz.[27a] Copyright 1982 The Institute of Physics.)

Alternatively, a multiple tip holder can be employed so that ten or fifteen tips can be inserted into the vacuum chamber at one time, with each tip individually selectable for imaging.[148]

Methods for introducing a number of very pure gases, for heating the tip, for changing its orientation in space, and for cooling the tip below 60 K (in order to improve image resolution) introduce additional complexity. Currently, commercial field-ion microscopes are not produced, although at least one large manufacturer of vacuum equipment (Vacuum Generators, England) will assemble custom instruments on an individual basis.

7.3.7. Microchannel-Plate Image Intensification

A convenient way of providing internal image intensification in an FIM relies on the use of microchannel plates.[149] These devices were first employed in field-ion microscopes in 1969.[150] Today, microchannel plate image intensification is routinely used.

[148] M. Khoshnevisan and C. H. Stephan, *J. Phys. E* **6**, 10 (1973).
[149] M. Lampton, *Sci. Am.* **245**, 62 (November 1981).
[150] P. J. Turner, P. Cartwright, M. J. Southon, A. Van Oostrom, and B. W. Manley, *J. Phys. E.* **2**, 731 (1969).

7.3. FIELD-ION MICROSCOPY

A microchannel plate (or MCP) is composed of a myriad of glass capillaries arranged in an ordered array that is produced by selectively etching a fiber-optic bundle. The plate, which can have a diameter of many centimeters, is usually about 1.5 mm thick and can be planar or spherically curved. A planar MCP viewed normal to its surface in an optical microscope is shown in Fig. 20. The capillary tubes are typically less than 40 μm in diameter, and usually spaced on 50-μm centers. Each capillary acts as an individual electron multiplier when a potential difference of several hundred volts is applied across the conductive coatings that are deposited on each side of the MCP. Incoming ions (or electrons) that strike a capillary wall will produce, on the average, more than one secondary electron. These electrons are propagated through the channel by the electric field produced by the applied potential. Since the secondary electrons collide with the capillary wall several times during transit, an electron avalanche will result. One ion striking the entrance to a capillary can easily produce a thousand electrons at its end. By properly biasing the front surface of the MCP, particles that strike the intercapillary area between adjacent channels can be collected.[151] Although the efficiency of the MCP can approach 100% in this mode of operation, its spatial resolution will be correspondingly degraded.

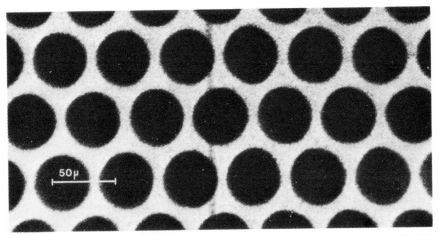

FIG. 20. An optical micrograph of the central region of a microchannel plate, or "MCP," which is also known as a CEMA (channel electron multiplier array). Each dark circular area is the entrance to a microcapillary tube that is typically 40 μm in diameter. A CEMA provides a convenient way to intensify a weak field-ion image. By combining two MCPs in tandem a gain of 10^6 can be achieved, which is sufficient to image single ions or electrons. (From Panitz.[27a] Copyright 1982 The Institute of Physics.)

---

[151] J. A. Panitz and J. A. Foesch, *Rev. Sci. Instrum.* **47**, 44 (1976).

The secondary electrons that emerge from each channel of an MCP are usually electrostatically focused onto a transparent phosphor-coated plate, where they produce a visible spot of light. If a field-ion image has sufficient magnification that the ions produced above adjacent surface species are separated at the MCP by several channel diameters, the entire weak field-ion pattern will be visible as a high-resolution image in ambient room light. By cascading two microchannel plates in tandem, a gain of 10^6 can be achieved. With MCP detectors, even the images resulting from individual field-desorbed ions can be recorded.[152]

The widespread use of a microchannel plate image intensification has greatly extended the utility of the field-ion microscope. Previously, many materials of interest could not be imaged because their lattice field evaporated before BIV was reached. With internal MCP detectors, materials that slowly field-evaporate during imaging can be photographed because exposure times less than a tenth of a second are usually adequate.

7.3.8. Surface Diffusion Studies Using the FIM

The ability of the field-ion microscope to observe individual metallic atoms adsorbed on atomically clean surfaces was recognized by Müller[153] as an important asset in performing surface diffusion experiments. The first quantitative measurements of diffusion parameters were made in 1966 when Ehrlich and Hudda[154] investigated the migration of tungsten atoms adsorbed on individual crystal planes of a tungsten tip. Later, Ehrlich and Kirk[155] field-desorbed tungsten adatoms from tungsten in order to obtain a binding energy for the metallic adsorbate. Other experiments followed that investigated the motion of individual adatoms,[156-160] adatom clusters and their dissociation,[161-163] pairs of adatoms cooperatively diffusing through a

---

[152] R. J. Walko and E. W. Müller, *Phys. Status Solidi, A* **9**, K9 (1972).
[153] E. W. Müller, *Z. Electrochem.* **61**, 43 (1957).
[154] G. Ehrlich and F. G. Hudda, *J. Chem. Phys.* **44**, 1039 (1966).
[155] G. Ehrlich and C. F. Kirk, *J. Chem. Phys.* **48**, 1465 (1968).
[156] E. W. Plummer and T. N. Rhodin, *J. Chem. Phys.* **49**, 3479 (1968).
[157] D. W. Bassett and M. J. Parsely, *J. Phys. D* **2**, 13 (1969).
[158] D. W. Bassett and M. J. Parsely, *J. Phys. D* **3**, 707 (1970).
[159] T. T. Tsong, *J. Chem. Phys.* **54**, 4205 (1971).
[160] P. G. Plative and W. R. Graham, *Thin Solid Films* **51**, 175 (1978).
[161] D. W. Bassett and M. J. Parsely, *Nature (London)* **221**, 1046 (1969).
[162] D. W. Bassett, *Surf. Sci.* **23**, 240 (1970).
[163] D. W. Bassett and D. R. Tice, *Surf. Sci.* **40**, 499 (1973).

long-range interaction,[164-173] and surface rearrangement with and without an electric field.[165,174,175]

The procedure for performing a diffusion experiment in the FIM is straightforward but tedious. Field evaporation is used to obtain a clean and atomically perfect surface that is then characterized by low-temperature field-ion microscopy. For adatom studies, a metallic adsorbate is deposited onto the tip in the absence of the field (usually by thermal evaporation from a heated filament). If field-ion imaging shows that the adsorbate coverage is too great, the field can be raised in order to remove some of the adsorbed species. Eventually, the desired number of adsorbates in the required lattice location will be obtained.

Surface diffusion is initiated by heating the tip to a known temperature (usually 200–500 K) for a fixed interval of time. Generally, the tip is not imaged during the diffusion interval in order to allow the diffusion process to occur in the absence of a field. Some measurements of diffusion in the presence of a field have been made. They yield information on the polarizability and the surface-induced dipole moment of an adatom.[176-179] Following the diffusion interval, the tip is cooled and a low-temperature field-ion image is taken of the new adatom position on the surface. The process is repeated, often hundreds of times. From a series of sequential field-ion images, the mean square displacement of an adatom over the surface can be measured as a function of diffusion temperature. A measurement accuracy of 0.01 nm is possible because adatom displacements can be directly compared to known interatomic spacings on well-resolved crystal planes in the FIM image.

A diffusion experiment will typically involve many temperature cycles and may last for many hours. As a result, ultrahigh-vacuum conditions and pressures below 10^{-10} Torr are mandatory if a contaminant-free surface is to

[164] T. T. Tsong, *J. Chem. Phys.* **55**, 4658 (1971).
[165] T. T. Tsong, *Phys. Rev. B* **6**, 417 (1972).
[166] W. R. Graham and G. Ehrlich, *Phys. Rev. Lett.* **31**, 1407 (1973).
[167] G. Ayrault and G. Ehrlich, *J. Chem. Phys.* **60**, 281 (1974).
[168] W. R. Graham and G. Ehrlich, *J. Phys. F.* **4**, L212 (1974).
[169] T. T. Tsong, P. Cowan, and G. Kellogg, *Thin Solid Films* **25**, 97 (1975).
[170] W. R. Graham and G. Ehrlich, *Thin Solid Films* **25**, 85 (1975).
[171] T. Sakata and S. Nakamura, *Surf. Sci.* **51**, 313 (1975).
[172] P. A. Reed and G. Ehrlich, *Philos. Mag.* **32**, 1095 (1975).
[173] K. Stolt, W. R. Graham, and G. Ehrlich, *J. Chem. Phys.* **65**, 3206 (1976).
[174] D. W. Bassett, *Surf. Sci.* **53**, 74 (1975).
[175] S. Nishigaki and S. Nakamura, *Jpn. J. Appl. Phys.* **15**, 19 (1976).
[176] T. T. Tsong and R. J. Walko, *Phys. Status Solidi, A* **12**, 111 (1972).
[177] W. R. Graham and G. Ehrlich, *Surf. Sci.* **45**, 530 (1974).
[178] T. T. Tsong and G. L. Kellogg, *Phys. Rev. B* **12**, 1343 (1975).
[179] G. L. Kellogg and T. T. Tsong, *Surf. Sci.* **62**, 343 (1977).

be preserved. The main difficulty that is encountered is the elimination of ambient hydrogen. The concentration of hydrogen will be increased in the vicinity of the tip during imaging by polarization forces and increased on the surface by field adsorption when the field is present.

The mean square displacement of an adatom $\langle x^2 \rangle$, measured in a diffusion experiment in a time interval γ, is related to the diffusion coefficient by the Einstein[180] equation

$$D = \langle x^2 \rangle / 2\alpha\gamma, \qquad (7.3.6)$$

where D is the diffusion coefficient, and $\alpha = 1$ for the one-dimensional diffusion [viz., diffusion between two rows of atoms, e.g., diffusion on the (211) plane of tungsten]. For two-dimensional diffusion, $\alpha = 2$ and the one-dimensional displacement of Eq. (7.3.6) must be replaced with $\langle x^2 \rangle + \langle y^2 \rangle$. The mean square displacement can be related to the number of jumps N that the adatom experiences. For the one-dimensional case, random walk theory[181] predicts

$$\langle x^2 \rangle = NL^2, \qquad (7.3.7)$$

where L is the mean jump distance. If a mean jump frequency Γ is defined as the number of adatom jumps per jump interval, the diffusion coefficient becomes

$$D = L^2\Gamma/2. \qquad (7.3.8)$$

If we assume that diffusion occurs by thermal activation (with the adatom hopping from site to site across an atomically smooth surface), the mean jump frequency will be related to the activation energy for surface diffusion E_d by an expression of the form

$$\Gamma = v_0 \exp(-E_d/kT), \qquad (7.3.9)$$

where k is Boltzmann's constant, T is the temperature, and v is a pre-exponential frequency factor

$$v = v_0 \exp(\Delta S/K). \qquad (7.3.10)$$

In Eq. (7.3.10), ΔS is the activation entropy and v_0 is the atomic vibrational frequency. Although v_0 is not known, its magnitude can be estimated from the uncertainty principle for energy and time. That is

$$v_0 = \Delta E/h \approx kT/h. \qquad (7.3.11)$$

[180] A. Einstein, *Ann. Phys. (Leipzig)* **17**, 549 (1905); **19**, 371 (1906).
[181] See, e.g., W. Teller, "An Introduction to Probability Theory and Its Applications," 3rd Ed. Wiley, New York, 1967.

7.3. FIELD-ION MICROSCOPY

Experimental results are usually expressed in terms of the activation energy for surface diffusion E_d and the diffusivity D_0. The diffusivity is defined in terms of the diffusion coefficient. Combining Eqs. (7.3.8)–(7.3.11) yields

$$D = D_0 \exp(-E_d/kT), \qquad (7.3.12)$$

where

$$D_0 \equiv \nu L^2/2 = (kTL^2/2h) \exp(\Delta S/k) \approx kTL^2/2h. \qquad (7.3.13)$$

Since the activation entropy is usually negligible (at least for single-atom diffusion), the diffusivity at room temperature can be estimated by assuming a typical jump distance, say $L = 0.3$ nm. In this case, $D_0 = 3 \times 10^{-3}$ cm^2 s^{-1}.

To obtain a value for the diffusivity and the activation energy in terms of experimentally measurable parameters, Eqs. (7.3.6) and (7.3.12) are combined to give

$$\langle x^2 \rangle/2\gamma = D_0 \exp(-E_d/kT). \qquad (7.3.14)$$

If the mean square displacement of the adatom $\langle x^2 \rangle$ is measured in a time interval γ as a function of temperature, then an Arrhenius plot of $\langle x^2 \rangle/2$ versus $1/T$ will yield a straight line of slope $-E_d/k$ and intercept $\ln D_0$. Figure 21 shows a typical example in which individual tungsten adatoms and single diatomic tungsten clusters were diffused on the (110) plane of tungsten. Typical activation energies range from 0.16 eV for a rhodium adatom on Rh(111),[167] where $D_0 = 2 \times 10^{-4}$ eV, to 1.04 eV for a rhenium adatom on W(110),[157,158] where $D_0 = 1.5 \times 10^{-2}$. Diffusivities average about 0.3 cm^2 s^{-1}, which is typical for a rhodium adatom on Rh(110),[167] where $E_d = 0.6$ eV. More complete tabulations of the results of FIM diffusion experiments appear in excellent reviews by Kellogg et al.[182] and others.[183-187] A bibliography of FIM surface diffusion studies between 1951 and 1978 is included in a bibliography of field-ion microscopy and related techniques.[188]

---

[182] G. L. Kellogg, T. T. Tsong, and P. Cowan, Surf. Sci. **70**, 485 (1978).

[183] D. W. Bassett, in "Surface and Defect Properties of Solids" (M. W. Roberts and J. M. Thomas, eds.), Vol. 2, p. 34. Chem. Soc., London, 1973.

[184] G. Ehrlich, Surf. Sci. **63**, 422 (1977).

[185] M. W. Roberts and C. S. McKee, "Chemistry of the Metal–Gas Interface" Oxford Univ. Press (Clarendon), London and New York, 1978.

[186] T. T. Tsong and P. Cowan, in "Chemistry and Physics of Solid Surfaces" (R. Vanselow, ed.), Vol. 2, p. 209. CRC Press, Boca Raton, Florida, 1977.

[187] D. A. King, J. Vac. Sci. Technol. **17**, 241 (1980).

[188] R. E. Thurstans and J. M. Walls, "Field-Ion Microscopy and Related Techniques: A Bibliography 1951–1978," p. 123. Warwick, Birmingham, England, 1980.

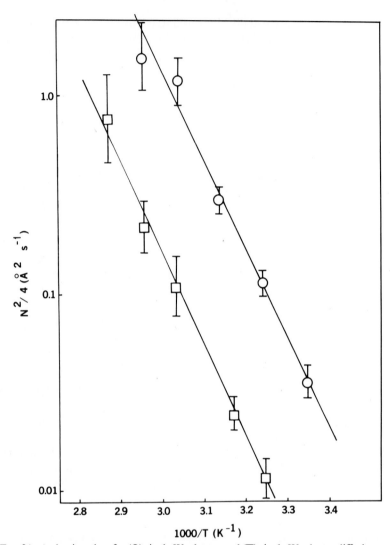

Fig. 21. Arrhenius plots for (○) single W adatom and (□) single W_2 cluster diffusion on the W(110) plane. From the plots the activation energy of surface diffusion is derived to be 0.90 ± 0.07 eV and 0.92 ± 0.14 eV for W and W_2, respectively. The diffusivity D_0 derived is 6.2×10^{-3} and 1.4×10^{-3} cm^2 s^{-1} for W and W_2, respectively. (Figure courtesy of Sandia National Laboratories, Albuquerque, New Mexico.)

7.4. Field Evaporation

At the present time there is no theory that accurately predicts all facets of the field evaporation phenomenon despite extensive discussions in the literature[129,189-197] and many excellent reviews.[28,126,198,199] Müller's original "image-force" theory of field evaporation described the process in terms of thermal activation over a field-reduced energy barrier. In this model, a neutral species is desorbed from the surface and then ionized at the critical distance given by Eq. (7.3.2). Despite its simplicity, the image-force theory predicts many of the experimental observations quite accurately.

In the image-force theory, a neutral species is removed from its surface site by supplying a sublimation energy Λ, which is assumed to be site independent. At the critical distance ionization occurs, which requires an additional energy $\Sigma_n I_n$ equal to the sum of ionization potential up to the nth value. Since the ion finds itself in an electric field, it can be polarized, which gives us an additional energy $(\alpha_a - \alpha_i)F^2/2$ where α_a and α_i are the polarizabilities of the neutral atom and the ion, respectively. But the tunneling electrons travel back to the metal, resulting in an energy gain of $n\Phi$, where Φ is a crystallographically dependent work function and n is the charge state of the ion. The total energy corresponding to all of these contributions can be considered as a binding energy Q, where

$$Q = \Lambda + \sum_n I_n - n\Phi - F^2(\alpha_a - \alpha_i)/2. \quad (7.4.1)$$

Although the last term of Eq. (7.5.1) can be a significant contribution to the total binding energy, it is usually ignored in the basic theory because the polarizability of the surface atom α_a is not reliably known.

In the presence of an applied field, the energy barrier represented by Eq. (7.4.1) is lowered by the amplitude of the so-called "Schottky hump." The magnitude of the effect can be found by considering the potential experienced by an ion of charge ne near the surface

$$V = -F(ne)x - (ne)^2/4x, \quad (7.4.2)$$

[189] R. J. Gomer, *J. Chem. Phys.* **31**, 341 (1959).
[190] E. W. Müller, *IV Int. Kongr. Elektronenmikrosk., Berlin, 1958,* **1**, 820 (1960).
[191] R. Gomer and L. W. Swanson, *J. Chem. Phys.* **38**, 1613 (1963).
[192] L. W. Swanson and R. Gomer, *J. Chem. Phys.* **29**, 2813 (1963).
[193] D. G. Brandon, *Br. J. Appl. Phys.* **14**, 474 (1963).
[194] D. G. Brandon, *Philos. Mag.* **14**, 803 (1966).
[195] T. T. Tsong, *Surf. Sci.* **10**, 102 (1968).
[196] T. T. Tsong and E. W. Müller, *Phys. Status Solidi A* **1**, 513 (1970).
[197] D. McKinstry, *Surf. Sci.* **29**, 37 (1972).
[198] D. G. Brandon, *Surf. Sci.* **3**, 1 (1965).
[199] E. W. Müller and T. T. Tsong, *Prog. Surf. Sci.* **4**, 11 (1973).

where the last term is the classical image potential mentioned previously in connection with the Fowler–Nordheim theory of field-electron emission. Once again, it should be stressed that a classically derived image potential is strictly incorrect, at least close to the surface where quantum effects will become dominant.

The amplitude of the Schottky hump is found by setting the derivative of Eq. (7.4.2) equal to zero. The result, $(n^3e^3F)^{1/2}$, reduces the energy barrier to desorption presented by the binding energy of Eq. (7.4.1). As a result, the activation energy for desorption is just

$$Q = \Lambda + \sum_n I_n - n\Phi - (n^3e^3F)^{1/2}. \tag{7.4.3}$$

Equation (7.4.3) specifically ignores the binding energy due to field-induced polarization. The characteristic time required for thermal activation over the barrier is of the form

$$\gamma = \gamma_0 \exp(-Q/kT), \tag{7.4.4}$$

where γ_0 is the reciprocal of the characteristic vibrational frequency of the atom bound to the surface at temperature T. The image-force theory predicts from Eqs. (7.4.3) and (7.4.4) a desorption field strength F_d given by

$$F_d = \left(\Lambda + \sum_n I_n - n\Phi - kT \ln \gamma/\gamma_0\right)^2 / n^3e^3. \tag{7.4.5}$$

At low temperatures, the last term in Eq. (7.4.3) is usually negligible, so that

$$F_d = \left(\Lambda + \sum_n I_n - n\Phi\right)^2 / n^3e^3. \tag{7.4.6}$$

For $n = 1$, the evaporation fields predicted by Eq. (7.4.6) differ significantly from those observed experimentally for most metals. However, when doubly charged ions are assumed, the agreement with experiment is reasonably good. Brandon[198] found that the calculated evaporation field for most metals was a minimum for $n = 2$, suggesting that most metals should evaporate as a doubly charged ion. An identical result is obtained from Gomer and Swanson's more detailed treatment of the field-desorption process.[192] They predict an evaporation field

$$F_d = (\Lambda + I_n - n\Phi - n^2e^2/X_c)/ne_c, \tag{7.4.7}$$

where X_c is the critical distance for ionization. In practice, X_c is assumed to be equal to the single-bond atomic radius of the metal atom.

Mass spectrometric measurements of iron and nickel made by Barofsky and Müller[200,201] supported the general conclusion of an $n = 2$ species, but for other materials the charge state of the field-evaporated ion depended critically on the emitter temperature. For example, beryllium was shown to evaporate exclusively as Be^{2+} at 21 K. However, at the same ion flux, Be^+ appeared as the emitter temperature was raised, and increased in abundance to become the dominant species at room temperature. As the temperature of the emitter is increased, the electric field strength required to maintain a constant ion flux (or field-evaporation rate), decreases. Therefore, by changing the emitter temperature, one actually measures the relative charge-state abundances as a function of both the electric field strength and the temperature. Ernst[202] has suggested that a decrease in the abundance of Rh^{2+} ions with decreasing field strength and increasing temperature is due to post-ionization Rh^+ desorbed from the surface. In the post-ionization model, a species is assumed to leave the tip surface in a low charge state. Higher charge states are acquired by subsequent ionization in space. One or more electrons are lost to the metal by tunneling as the ion moves rapidly away from the tip. The process is strongly dependent on the electric field strength and may also depend on the local crystallography of the surface. By using a one-dimensional tunneling probability, Ernst was able to predict the relative abundance of singly and doubly charged rhodium ions as a function of field strength to within a factor of two.[202] A more refined, three-dimensional calculation by Haydock and Kingman[203] strongly supports the post-ionization model. Haydock and Kingman suggest that multiple post-ionizations may be responsible for the higher charge states of molybdenum and tungsten that are observed experimentally. This conclusion is supported by a study in which Kellogg[204,205] observed a charge-state abundance for a number of materials, over a wide field range, that agreed with Haydock and Kingman's predictions. Post-ionization appears to be the dominant mechanism for producing the higher charge states of field-evaporated metal ions.

7.5. Field-Ion Mass Spectroscopy

In addition to providing valuable insight into the physical mechanisms responsible for the ionization event, ions produced by a high electric field can be used in an analytical capacity to determine the composition of un-

---

[200] P. F. Barofsky and E. W. Müller, *Surf. Sci.* **10**, 177 (1968).
[201] D. F. Barofsky and E. W. Müller, *Int. J. Mass Spectrom. Ion Phys.* **2**, 125 (1969).
[202] N. Ernst, *Surf. Sci.* **87**, 469 (1979).
[203] R. Haydock and D. R. Kingman, *Phys. Rev. Lett.* **44**, 1520 (1980).
[204] G. L. Kellogg, *Phys. Rev. B* **24**, 1848 (1981).
[205] G. L. Kellogg, *Surf. Sci.* **120**, 319 (1982).

known samples. Field-ionization mass spectroscopy began modestly in 1954 when Inghram and Gomer[125] published the first quantitative study of field-ionized hydrogen, nitrogen, oxygen, methyl alcohol, and hydrocarbon species. Later, they were able to demonstrate that some species (such as CH_3O^+ in the spectrum of methyl alcohol) did not originate in the gas phase but were produced at the tip surface and supplied by continuous migration of a condensed phase on the tip shank.[206]

One advantage of using high electric fields to induce ionization lies in the ionization process itself. Ionization in an electric field produces a fragmentation pattern different from (and often simpler than) a conventional electron-bombardment ionization source.[207,208] As a result, a gas-phase spectrum taken with a combined electron-bombardment and field-ionization source[209-212] can provide more information than either source alone and is often easier to interpret. Since vibrational or electronic excitations are not generally induced under high-field conditions, metastable complexes can often be identified in the field-ionization mass spectra. Unfortunately, these advantages can be offset by a complex field-induced chemistry,[207] whose interpretation may require a detailed knowledge of the ion formation mechanism.[207,213] Systematic application of field desorption mass spectrometry to a wide range of practical problems, including medical diagnosis,[214-217] coal research,[218] drug[219] and environmental pollutant detection,[216,220,221] and problems of catalysis,[222,223] have established the technique as a useful molec-

---

[206] M. G. Inghram and R. Gomer, *Z. Naturforsch. A* **10A**, 863 (1955).
[207] H. D. Beckey, "Field Ionization Mass Spectrometry." Pergamon, Oxford, 1971.
[208] H. R. Schulten and H. D. Beckey, *Tetrahedron* **29**, 3861 (1973).
[209] H. H. Gierlich, A. Heindric, and H. D. Beckey, *Rev. Sci. Instrum.* **45**, 1208 (1974).
[210] M. Anbar and G. A. St. John, *Anal. Chem.* **48**, 198 (1976).
[211] H. R. Schulten and N. M. M. Nibbering, *Biomed. Mass Spectrom.* **4**, 55 (1977).
[212] A. M. Hogg and J. D. Payzant, *Int. J. Mass Spectrom. Ion Phys.* **27**, 291 (1978).
[213] F. W. Rollgen, V. Giessmann, H. J. Heinen, and H. D. Beckey, *Ultramicroscopy* **4**, 375 (1979).
[214] H. R. Schulten, W. D. Lehmann, and Z. Ziskoven, *Z. Naturforsch., C* **33C**, 484 (1978).
[215] H. Miyazaki, E. Shirai, M. Ishibashi, K. Hosoi, S. Shibata, and M. Iwanaga, *Biomed. Mass Spectrom.* **5**, 559 (1978).
[216] H. R. Schulten, *Int. J. Mass Spectrom. Ion Phys.* **32**, 97 (1979).
[217] H. R. Schulten and D. Kummler, *Anal. Chim. Acta* **113**, 253 (1980).
[218] T. Yoshida, R. Yoshida, Y. Maekawa, Y. Yoshida, and Y. Itagaki, *Fuel* **58**, 153 (1979).
[219] D. A. Brent, *J. Assoc. Off. Anal. Chem.* **59**, 1009 (1976).
[220] H. R. Schulten, *J. Agric. Food Chem.* **24**, 743 (1976).
[221] Y. Yamato, M. Suzuki, and T. Watanabe, *Biomed. Mass Spectrom.* **6**, 205 (1979).
[222] W. A. Schmidt, O. Frank, and A. W. Czanderna, *Phys. Status Solidi A* **16**, 127 (1973).
[223] W. A. Schmidt, O. Frank, and J. H. Bloch, *Surf. Sci.* **44**, 185 (1974).

7.5. FIELD-ION MASS SPECTROSCOPY

ular probe. Several excellent reviews[224-229] summarize advances in this rapidly expanding area.

7.5.1. Field-Ionization Sources

A gas-phase field-ionization source is unique in that it can supply a reasonably monoenergetic beam of positive ions that appear to originate from a point in space when a single emitter tip is used. If an extended ionization region can be used, multiple tip structures, produced by microfabrication techniques,[230] grown on filaments of platinum or carbon, or occurring naturally on extended metal edges (such as razor blades), can be employed.[207] Their advantage lies in their large surface area and the correspondingly high ion currents that they are capable of generating.

The nearly monoenergetic nature of the ions produced by the field-ionization process is illustrated by the 0.8 eV FWHM helium-ion energy distribution at BIV above a single emitter tip, as shown in Fig. 14. Monoenergetic ions, like monochromatic light, can be focused to a small diameter spot because chromatic aberration will be absent from the image. The ability to focus the ions from a field-ion source is important, not only in mass spectroscopy but also in other applications. For example, if a field-ion beam can be focused to submicron dimensions, it can be used as a microanalytical probe[231] or as a tool to fabricate submicron structures by direct ion sputtering.

In order to estimate the ion current produced by a single emitter tip, the apex of the tip can be treated as a sphere of radius R. In this case, the surface area of the tip apex that actually contributes to the ionization process will be $2R^2\theta$, where θ defines the angle in radians over which field ionization can occur. Since the electric field strength at the surface of a tip decreases with increasing angle from the apex,[21] there will be a maximum angle beyond which field ionization will not occur. Experimentally, $\theta_{max} \approx \pi/4$, so that the total surface area of a field emitter that contributes to the ionization current will be of the order of πR^2. If we assume that 10^5 ions s^{-1} are produced over

---

[224] H. D. Beckey, *Res./Dev.* **20**, 26 (1969).
[225] J. A. B. Robertson, *J. Phys. E* **7**, 321 (1974).
[226] P. J. Derrick, *Mass Spectrom.* **4**, 132 (1977).
[227] H. D. Beckey, K. Levsen, F. W. Rollgen, and H. R. Schulten, *Surf. Sci.* **70**, 325 (1978).
[228] D. L. Cocke and J. H. Bloch, *Surf. Sci.* **70**, 363 (1978).
[229] W. D. Reynolds, *Anal. Chem.* **51**, 283A (1979).
[230] C. A. Spindt, I. Brodie, L. Humphrey, and E. R. Westerberg, *J. Appl. Phys.* **47**, 5248 (1976).
[231] G. R. Hansen and B. M. Siegel, *J. Vac. Sci. Technol.* **16**, 1875 (1979).

each surface atom in this region, a typical field-ion source (with $R = 100$ nm) will produce a total field-ion current of 10^{11} ions s^{-1}, or about 10 nA.

One difficulty in developing such a source lies in focusing the entire ion beam, which diverges into a solid angle of approximately π sr. A typical lens system will accept ions within a solid angle of perhaps $\pi/10$ sr, so that the usable current from a field-ion source may be only of the order of a few nanoamperes. Another difficulty results from the variation of the field-ion current with time. Variations in the source current will appear as a low-frequency "flicker noise" in the focused spot. Orloff and Swanson[232] have developed low-noise field ionization sources with angular current densities approaching 10 μA sr^{-1}. Hansen and Siegel[231] have shown that 60 μA sr^{-1} can be achieved by field-desorbing hydrogen physisorbed on tungsten at liquid-helium temperatures. Their source uses ions that originate at one of the stable, localized, and anomalously bright regions that were previously observed in hydrogen field-ion images of tungsten at 4.2 K.[233] By using a variety of gases, Hansen and Siegel hope to increase the versatility of such a source. They propose to create and focus a beam of light ions (such as H_2^+ or He^+) in order to expose patterns on a semiconductor surface coated with a suitable resist. Heavier ions, such as Ar^+ and Xe^+, would be used to sputter such patterns directly into the surface.

7.5.2. Liquid-Metal Field Desorption Sources

The production of a focused beam of metallic species could be used to selectively dope small regions of a semiconductor surface by ion implantation or as a probe in a scanning ion microscope. One way to produce a continuous beam of metallic ions is to place a supply of weakly bound metallic adatoms on the surface of a tip biased to high voltage. If the adatoms were mobile on the surface, polarization forces would draw them into the high-field region of the tip apex where field desorption would occur. Since most metals are not mobile at room temperature, such a scheme would require heating the tip to an elevated temperature. A continuously operating source would also require a reservoir of metal atoms to replenish those lost by the desorption process.

Krohn and Ringo[234] investigated such a source and measured ion currents of several microamperes. The ion current persisted as long as a thick layer of liquid metal was present on the tip surface. The half-width of a typical energy distribution was found to be ~12 eV FWHM, a value later measured with

---

[232] J. H. Orloff and L. W. Swanson, *J. Vac. Sci. Technol.* **12**, 1209 (1975); **15**, 845 (1978); *J. Appl. Phys.* **50**, 6026 (1969).

[233] A. Jason, B. Halpern, M. G. Inghram, and R. Gomer, *J. Chem. Phys.* **52**, 2227 (1970).

[234] V. E. Krohn and G. R. Ringo, *Appl. Phys. Lett.* **27**, 479 (1975).

greater precision by Culbertson et al.[235] for a gallium source. At very low currents (of the order of 10 nA) Culbertson and co-workers[235] found that the Ga^+ energy distribution sharpened to ~1.5 eV FWHM. This suggested that under conditions of low emission the ionization process was occurring within a thin layer of gallium atoms on the surface.

The metallic layer on the tip surface appears to form a stable, pointed protrusion at the tip apex in the presence of a field. This protrusion is responsible for the ion current that is observed. It is called a "Taylor cone," and is created by a balance of electrostatic and surface tension forces at the surface of the molten metal. The cone behaves as though it were a continuously field-evaporating tip. In 1964, Sir George Taylor[236] showed that a cone of half-angle 49.3° will be produced at the surface of a liquid under the influence of a critical electric field strength that depends on the geometry of the source. In principle, a Taylor cone, once created, is dynamically stable and infinitely sharp. In practice, instabilities can occur that may lead to the formation of a liquid jet or a stream of isolated droplets. A liquid-metal field-ion source can produce a single or multiple Taylor cone structure on the tip apex. Electron microscope images of an operating source have been obtained, but the resolution is not sufficient to show the shape of each structure in detail. Gomer[237] has presented a comprehensive theoretical treatment of liquid-metal sources. Experimental sources have been constructed using a variety of liquid metals. These include gallium,[234,238] the alkali metals,[235,239,240] gold,[241] alloys such as Wood's metal,[242] and mercury.[241] A focused liquid-gallium source capable of producing submicron gallium structures has been described by Seliger.[243] Levi-Setti[244] has shown that a liquid-gallium source can be used to produce scanning ion micrographs with excellent resolution. Waugh[245] has shown that elemental mapping of surface composition can be achieved by using a liquid-gallium source for secondary-ion mass spectroscopy (SIMS).

[235] R. J. Culbertson, T. Sakurai, and G. H. Robertson, *J. Vac. Sci. Technol.* **16**, 574 (1979).
[236] G. I. Taylor, *Proc. R. Soc. London, Ser. A* **280**, 383 (1964).
[237] R. Gomer, *J. Appl. Phys.* **19**, 365 (1979).
[238] T. Sakurai, R. J. Culbertson, and J. H. Robertson, *Appl. Phys. Lett.* **34**, 11 (1979).
[239] J. F. Mahoney, A. T. Yahiku, H. L. Daley. R. D. Moore, and J. Perel, *J. Appl. Phys.* **40**, 5101 (1969).
[240] R. Clampitt, K. L. Aitken, and D. K. Jefferies, *J. Vac. Sci. Technol.* **12**, 1208 (1975).
[241] R. Clampitt and D. K. Jefferies, *Nucl. Instrum. Methods* **149**, 739 (1978).
[242] B. W. Colby and C. A. Evans, Jr., *Anal. Chem.* **45**, 1887 (1973).
[243] R. D. Seliger, *Ultramicroscopy* **4**, 361 (1979).
[244] R. Levi-Setti, P. H. LaMarche, and K. Lam, *Proc. Int. Field-Emiss. Symp., 29th, Göteborg, Sweden*, p. 417, H. Nordén and H. O. Andren, eds. (Almquist and Wiksell, Stockholm, 1982).
[245] A. R. Waugh, *Proc. Int. Field-Emiss. Symp., 29th, Göteborg, Sweden*, p. 409. H. Nordén and H. O. Andren, eds. (Almquist and Wiksell, Stockholm, 1982).

7.6. Atom-Probe Mass Spectroscopy

Barofsky and Müller's early measurements[192,193] suggested that a unique microanalytic capability could be provided by a mass spectrometer capable of identifying a small number of field-evaporating lattice constituents. In 1967, Müller and Panitz[246] proposed to make the entrance aperture of a spectrometer equal in area to a single magnified image spot in a field-ion micrograph. In this way they hoped to provide an ultimate microanalytic capability: the unambiguous identification of a single field-evaporated surface species. A magnetic sector spectrometer can be used for analysis, but it is not ideal since only a limited mass range can be observed at one time. If the mass-to-charge ratio of a single field-evaporated species fell outside of the preselected mass interval, it would not be detected. In order to overcome this difficulty, Barofsky[247] suggested the use of a time-of-flight (TOF) mass spectrometer. Such a spectrometer is simple to construct, has a wide mass range, and easily discriminates between a field-evaporated species and a random event. As a result, the TOF atom-probe is the most common form of this instrument in use today. Although many atom-probes are in use throughout the world, a commercial instrument is not currently available.

7.6.1. The Voltage-Pulsed Atom-Probe

The atom-probe[248] was named by analogy with the electron microprobe developed previously by Castaing.[249] Castaing's instrument could identify the constituents of a small macroscopic region of a sample containing perhaps 10^{11} atoms. With the atom-probe one could, in principle, "determine the nature of one single atom seen on a metal surface and selected from neighboring atoms at the discretion of the observer."[248] In order to achieve this goal, the field-ion image of a surface species is positioned over the entrance aperture of the time-of-flight mass spectrometer. This is accomplished by mechanically moving the tip assembly. Field evaporation is initiated by applying a short-duration high-voltage pulse to the tip. If the pulse amplitude is carefully adjusted, a few atoms of the surface layer or many layers of the near surface region, can be controllably removed.

On application of the high-voltage pulse, surface species are ionized and acquire a kinetic energy

$$mv^2/2 = neV_0, \qquad (7.6.1)$$

---

[246] E. W. Müller and J. A. Panitz, *Abstr. 14th Int. Field-Emiss. Symp., Washington, D.C.* (1967). Unpublished.

[247] J. A. Panitz, *J. Phys. E.* **15**, 1281 (1982).

[248] E. W. Müller, J. A. Panitz, and S. B. McClane, *Rev. Sci. Instrum.* **39**, 83 (1968).

[249] R. Castaing, *Adv. Electron. Electron Phys.* **13**, 317 (1960).

7.6. ATOM-PROBE MASS SPECTROSCOPY

where ne is the ion's charge and V_0 is the tip potential produced by the pulse. It is usually convenient to apply a dc bias to the tip so that $V_0 = V_{dc} + V_{pulse}$. To ensure adequate mass resolution, the high-voltage pulse must be sufficiently short so that the field-evaporation interval is much shorter than the arrival time at the detector of two neighboring species that are to be separated. A fixed point on the leading edge of the pulse can be used as a marker to identify the instant of time corresponding to the field desorption event. If we use this as the "zero" of time, the ion's travel time to a detector can be accurately measured.

The field-evaporated ions will acquire their full kinetic energy within a few tip radii from the surface. In order to define the acceleration region very precisely, a grounded plate containing an aperture is placed very close to the tip. The pulse duration must be sufficiently long to ensure that the ions have left this region before the pulse terminates. After the ions traverse the aperture, they enter a field-free "drift" region of length D. At some point within the drift region the ions encounter a fluorescent screen or MCP detector of the field-ion microscope. The screen or MCP contains a probe hole. Only the surface species whose image was positioned over the probe hole will continue to travel at constant velocity to a detector positioned at the end of the drift region. If T is the travel time of the ion measured from the instant of desorption, its mass-to-charge ratio will be given by Eq. (7.1.9). If m/n is expressed in atomic mass units, $V_{dc} + V_{pulse}$ is measured in kilovolts, the travel time is measured in microseconds, and the length of the drift region is measured in meters, Eq. (7.1.9) becomes

$$m/n = 0.193(V_{dc} + V_{pulse})T^2/D^2. \qquad (7.6.2)$$

A typical mass resolution of $\Delta m/n \approx 1\%$ can be routinely achieved with the TOF atom-probe.[250] The mass resolution is limited primarily by a spread in the kinetic energy of the field-evaporating ions. This quantity can easily approach several hundred electron volts. The energy spread results from field evaporation during the time that the high-voltage pulse is changing in amplitude.[251,252] In order to improve the mass resolution, Müller and Krishnaswamy[253] used isochronal focusing to electrostatically compensate for the initial energy spread of the field-evaporated ions. An energy resolution of ~0.1% was achieved but at the expense of detection efficiency and instrumental simplicity. Another way to achieve high mass resolution is to use a magnetic-sector spectrometer. In this case, field desorption is continuous and produced by a dc bias applied to the tip. The resulting ions have a

---

[250] J. A. Panitz, *Prog. Surf. Sci.* **8**, 219 (1978).
[251] E. W. Müller, *Lab. Pract.* **22**, 408 (1973).
[252] S. V. Krishnaswamy and E. W. Müller, *Rev. Sci. Instrum.* **45**, 1049 (1974).
[253] E. W. Müller and S. V. Krishnaswamy, *Rev. Sci. Instrum.* **45**, 1053 (1974).

constant energy. A magnetic sector atom probe was constructed by Müller and Sakurai[254] in 1974. Another type of atom-probe, the *pulsed-laser atom-probe,* was suggested by Tsong[255] in 1978.

7.6.2. The Pulsed-Laser Atom-Probe

Field desorption is a thermally activated process. Tsong reasoned that a short-duration thermal pulse from a laser could be used instead of a voltage pulse to initiate field evaporation in a TOF spectrometer. Since the tip could be kept at a constant dc potential, the energy spread of the resulting ions should be of the order of kT, or less than 0.1 eV. An additional benefit of laser-induced desorption was foreseen. It is difficult to use the voltage pulsed atom-probe to examine insulators or high-purity semiconductors because of their high intrinsic resistance. The resistance of the tip and the capacitance between tip and ground form a low-pass filter. To propagate a pulse to the tip apex the pulse must have duration much greater than the time constant of the system. Any material with an intrinsic resistance of 10^8 Ω will produce a time constant of the order of milliseconds. This means that very long ion travel times (or very low tip voltages) would have to be used to assure good mass resolution in a TOF instrument. Since a high voltage must be applied to establish the desorption field, the ions would have to be decelerated. Focusing then becomes a problem. On the other hand, a tip can be easily pulsed to a high temperature with a laser. The pulse can be made several nanoseconds or less in duration, and a constant tip potential can be used. Furthermore, since field and temperature can be varied, the electric field strength applied to the tip can be reduced by using higher laser powers to establish a given evaporation rate. This would reduce the possibility of producing field-induced artifacts, or allow the production of such artifacts to be studied as a function of field strength. Since the surface temperature can be arbitrarily elevated during the desorption event, the effect of surface mobility can also be studied.

The feasibility of using a laser to thermally induce field evaporation in a TOF atom probe has been demonstrated by Kellogg and Tsong[256] and Drachsel *et al.*[257] The latter group believes they have also seen photon-induced (nonthermal) desorption during an examination of ethylene on silver.[258] Kellogg[259] has shown that the H^+ species normally observed in the

---

[254] E. W. Müller and T. Sakurai, *J. Vac. Sci. Technol.* **11**, 878 (1974).

[255] T. T. Tsong, *Surf. Sci.* **70**, 211 (1978).

[256] G. L. Kellogg and T. T. Tsong, *J. Appl. Phys.* **51**, 1184 (1980).

[257] W. Drachsel, S. Nishigaki, and J. H. Block, *Int. J. Mass Spectrom. Ion Phys.* **32**, 333 (1980).

[258] S. Nishigaki, W. Drachsel, and J. H. Block, *Surf. Sci.* **87**, 389 (1979).

[259] G. L. Kellogg, *J. Chem. Phys.* **74**, 1479 (1981).

voltage-pulsed atom-probe is an artifact, caused by field dissociation of adsorbed molecular hydrogen in the high electric field above the tip surface. He also demonstrated that silicon could be examined by atom-probe mass spectroscopy. Measurements by Tsong[260] have shown that the pulsed-laser atom-probe can achieve excellent mass resolution. However, the thermal response of the tip may actually be much slower than the duration of the laser pulse. Since desorption will be governed by the entire thermal cycling of the tip, its thermal time constant will ultimately determine the highest mass resolution that can be achieved.

7.6.3. Atom-Probe Measurements

Essentially three different measurements can be made with an atom-probe. Although spectra are recorded, the three measurements differ in procedure and ultimate aim. The first type of measurement is concerned with determining the relative abundances of species that reside *on* the first atomic layer of the specimen lattice. Since this layer will primarily consist of weakly bound residual gas species, desorption of this layer must be achieved without disturbing the underlying substrate lattice.

The second type of measurement is concerned with determining the relative abundance of species contained within each individual atomic layer of the substrate lattice. By field-evaporating the lattice one layer at a time and recording the abundance of all species within each layer, a depth profile of the near-surface region can be obtained.

The third type of measurement is concerned with determining an average composition for the near-surface region of the lattice. In this case, the abundance of each species is averaged over many atomic layers.

7.6.4. First-Layer Composition

In the absence of an electric field, species characteristic of the ambient environment will be adsorbed on the tip surface with a probability that will depend on their partial pressure in the gas phase, their abundance on the surface, and the temperature of the tip. In addition, weakly bound displaced lattice atoms may be present. At equilibrium, each of these species will exist on the surface at a unique position fixed by chemical or stearic interactions with the lattice. In a very real sense, the first layer of the tip surface is a microcosm of the ambient environment and the reaction of the tip material to it.

In order to determine the composition of the first layer, one assumes that the most tightly bound species of the first layer will desorb at a field strength

---

[260] T. T. Tsong, *Rev. Sci. Instrum.* **53**, 1442 (1982).

that is less than that necessary to field-evaporate the underlying lattice. In the case of strong chemical interactions with lattice atoms (as in chemisorption, or on the formation of chemical compounds) this assumption may fail. Experimentally, one tests the validity of this assumption by observing changes in a field-ion image of the lattice before and after desorption. Since the species that characterize the first layer will generally be more weakly bound than those of the bulk, they will be desorbed at field strengths that are well below the evaporation field of the lattice. Therefore, the analysis of the "first layer" begins by applying a very low field (typically less than 10 V nm^{-1}) where the probability of removing any species from the surface is known (from experience) to be vanishingly small. A mass spectrum is recorded at this field, and the tip bias (V_{dc}) is incremented by 50–100 V. Using a small increment in specimen bias increases the probability that, during the next pulse, species will be desorbed very close to the maximum amplitude of the desorption pulse. This, in turn, assures a minimum energy spread in the ions being desorbed and a maximum mass resolution for each spectrum recorded.

For each increment in tip bias a mass spectrum is recorded until a tip bias V_m is reached that is 50–100 V less than the lattice. It is assumed that at this point all species that resided on the first layer of lattice atoms have been desorbed. The composition of the first layer is obtained by summing the abundances that were recorded in the individual mass spectra during each desorption event.

Alternatively, the abundance of a species can be obtained as a function of desorption field. Crudely speaking, this is a measure of the binding of a species to the surface if the evaporation field for the lattice F_E and the corresponding evaporation bias V_E are known; the desorption field F, at specimen potential V can be calculated from the expression

$$F = (V/V_E)F_E. \qquad (7.6.3)$$

By linearly increasing the tip potential with time and recording the total ion current, an analog to thermal-desorption mass spectrometry can be obtained.[261] Since the tip is kept at low temperature, thermal migration over the surface is eliminated, but field-induced migration remains a serious concern.

7.6.5. Composition of One Atomic Layer

The composition of a single atomic layer of the lattice as sampled by the probe hole can be obtained by summing the abundance obtained for each species recorded during the successive evaporation events required to remove the layer. It is assumed that all species that are contained within the

---

[261] J. A. Panitz, *J. Vac. Sci. Technol.* **16**, 868 (1979).

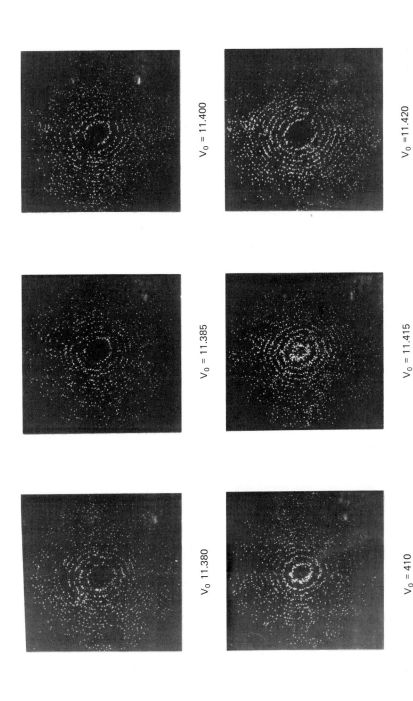

FIG. 22. A sequence of field-desorption images of (110)-oriented tungsten at 80 K showing the collapse of the (110) plane, which occurs at a tip bias $V_0 = 11.415$ V. $p \cong 5 \times 10^{-9}$ Torr. (From Panitz.[250])

layer are removed as positive ions and detected with the same efficiency. For a refractory metal substrate containing easily ionized low-Z impurities, the high field strength required to evaporate the lattice will probably guarantee the ionization of all species within the layer. For low-Z substrates containing high-Z impurities, the situation is less certain. Field evaporation will guarantee the removal of all species as the lattice dissolves, but the low field strength may not be sufficient to ensure complete ionization of all high-Z impurities. Since little is known about the ionization efficiency in high fields, the possibility of selective ionization cannot be ignored. Furthermore, since the single-ion detectors used in atom-probe devices are not particularly sensitive to low-energy neutrals, these species will go undetected. Only when all species are ionized with the same efficiency will the measured composition of each atomic layer accurately reflect its composition prior to desorption.

The removal of one atomic layer of the specimen is monitored by observing the collapse of a prominent net plane in the field-ion image of the lattice as shown in Fig. 22. If the layer spacing is known, the depth to which the lattice is probed will be uniquely determined. This "built-in" depth scale makes depth profiling with the atom probe as accurate as the known interplanar spacings of the lattice.

Unlike sputter-depth profiling, field evaporation will completely remove a monolayer without the possibility of driving some species of that layer further into the lattice or mixing species within the sampled region.[262] As a result, the atom probe is a unique tool that can characterize the depth distribution of low-energy species in metals,[147,263] and can be used as the basis of an extensive program of material characterization.[247-287] As a metallurgical tool, the unique spatial and depth resolution of the atom-probe provides a unique complement to existing microanalytic techniques. Numerous reviews[250,288-294] clearly demonstrate the growing importance of atom-probe analysis in the materials sciences.

---

[262] Z. L. Liau, B. Y. Tsaur, and J. W. Mayer, *J. Vac. Sci. Technol.* **16**, 121 (1979).
[263] A. Wagner and D. N. Seidman, *Phys. Rev. Lett.* **42**, 515 (1979).
[264] D. N. Seidman, *Proc. Microsc. Soc. Can.* **3**, 36 (1976).
[265] T. M. Hall, A. Wagner, A. S. Berger, and D. N. Seidman, *Metallography* **10**, 485 (1976).
[266] Y. S. Ng, S. B. McLane, and T. T. Tsong, *J. Appl. Phys.* **49**, 2517 (1978).
[267] M. K. Miller and G. D. Smith, *Met. Sci.* **11**, 249 (1977).
[268] D. N. Seidman, *Surf. Sci.* **70**, 532 (1978).
[269] S. Nakamura and T. Kuroda, *Surf. Sci.* **70**, 452 (1978).
[270] A. J. Watts and B. Ralph, *Surf. Sci.* **70**, 459 (1978).
[271] Y. Ohno, T. Kuroda, and S. Nakamura, *Surf. Sci.* **75**, 689 (1978).
[272] M. K. Miller, P. A. Beaven, R. J. Lewis, and G. D. W. Smith, *Surf. Sci.* **70**, 470 (1978).
[273] T. T. Tsong, Y. Ng, and S. V. Krishnawwamy, *Appl. Phys. Lett.* **32**, 778 (1978).
[274] A. Wagner, T. M. Hall, and D. N. Seidman, *J. Nucl. Mater.* **69**, 413 (1978).

7.6.6. Average Composition of the Near-Surface Region

By measuring the abundance of all species in each atomic layer, and summing over many layers, an average composition of a volume of the near-surface region is obtained. Since the tip potential must be increased as the lattice is evaporated (to maintain a constant evaporation field as the tip radius increases), the travel times of a given species will become progressively shorter.

Unlike Auger spectroscopy, electron spectroscopy for chemical analysis (ESCA), or photoemission, the atom-probe technique is truly an invasive probe of the surface. The surface must be completely destroyed in order to identify its constituents. This means that any phenomenon that occurs on the surface can be sampled only once during its evolution. In order to extend the number of observations for statistical reliability (or to examine a phenomenon at a different point in its evolution) an identical surface must be recreated many times. If field-ion microscopy can be used to image the tip, and if field evaporation can be employed to reestablish an identical surface morphology, then reestablishing the phenomenon will be a question only of patience.

---

[275] T. T. Tsong, Y. S. Ng, and A. J. Melmed, *Surf. Sci.* **77**, L187 (1978).

[276] Y. S. Ng and T. T. Tsong, *Surf. Sci.* **78**, 419 (1979).

[277] Y. S. Ng, T. T. Tsong, and S. B. McLane, *Surf. Sci.* **84**, 31 (1979).

[278] G. K. L. Cranstoun, D. R. Pyke, and G. D. W. Smith, *Appl. Surf. Sci.* **2**, 375 (1979).

[279] S. Nakamura, Y. S. Ng, T. T. Tsong, and S. B. McLane, *Surf. Sci.* **87**, 656 (1979).

[280] S. V. Krishnaswmay, R. Messier, Y. S. Ng, and T. T. Tsong, *Appl. Phys. Lett.* **35**, 870 (1979).

[281] D. M. Seidman, *J. Met.* **31**, 116 (1979).

[282] P. A. Beaven, K. M. Delargy, M. K. Miller, P. R. Williams, and G. D. W. Smith, *Ultramicroscopy* **4**, 385 (1979).

[283] M. Leisch, *Ultramicroscopy* **4**, 381 (1979).

[284] S. V. Krishnaswamy, R. Messier, Y. S. Ng, T. T. Tsong, and S. B. McLane, *J. Non-Cryst. Solids* **35**, 351 (1980).

[285] P. W. Bach, J. Beyer, and C. A. Verbraak, *Scr. Metall.* **14**, 205 (1980).

[286] J. V. Wood, P. F. Mills, J. K. Bingham, and J. V. Bee, *Metall. Trans., A.* **10A**, 575 (1979).

[287] E. W. Müller and T. T. Tsong, *Prog. Surf. Sci.* **4**, 50 (1973).

[288] E. W. Müller, *J. Microsc. (Oxford)* **100**, 121 (1974).

[289] A. Van Oostrom, *CRC Crit. Rev. Solid State Sci.* **4**, 353 (1974).

[290] J. A. Panitz, *CRC Crit. Rev. Solid State Sci.* **5**, 153 (1975).

[291] E. W. Müller and S. V. Krishnaswamy, in "Characterization of Metal and Polymer Surfaces" (L.-H. Lee, ed.), Vol. 1, p. 21. Academic Press, New York, 1977.

[292] T. T. Tsong, *Chem Scr.* **14**, 7 (1979).

[293] A. R. Waugh and M. J. Southon, *Surf. Sci.* **89**, 718 (1979).

[294] P. A. Beaven, M. K. Miller, and G. D. W. Smith, *Inst. Metall. (London), Phase Transform* **2**, I/12 (1979).

7.7. Field Desorption Microscopy

Both the field-electron and field-ion microscope provide nondestructive images of a tip surface by mapping local changes in its work function or the electric field distribution immediately above it. In principle, it is also possible to image the tip surface by desorbing a species from it. Müller[121] proposed to use barium for such a purpose as early as 1951, and later, with Bahadur,[128] he attempted to use nitrogen. The barium experiments were not reported in the literature, and the attempt to use nitrogen failed, probably because field-induced corrosion[126] was constantly changing the surface morphology of the tip.

Another imaging procedure was attempted by George[295] in 1965. He used a three-stage image intensifier in an attempt to record the field-evaporation process in vacuum. He hoped to record an image showing the position of atoms in the dissolving lattice. Inexplicably, no recognizable image was seen.

In 1972, Walko and Müller[152] repeated this experiment using a proximity-focused channel-plate image intensifier (with a gain of 10^6) located within the microscope chamber. They obtained both room-temperature and low-temperature images of field-evaporating tungsten. The images resembled their field-ion counterparts. Figure 22 shows a sequence of desorption images taken during field evaporation of tungsten at 80 K showing the gradual collapse of the (110) plane. At a tip bias of $V_0 = 11.415$ kV, the last atoms of the plane evaporate. In the next image, taken at 11.420 kV, underlying planes are imaged. At this point one atomic layer of the lattice [in the (110) region] has been removed.

If the individual desorption images of several net plane collapses are integrated, one would expect to obtain a contrastless image. Since the lattice atoms of each underlying layer do not necessarily occupy identical spatial positions, the image should "average out." Instead, the pattern is highly structured. These "multilayer images" were first observed by Waugh *et al.*[296] who found that they were characteristic of the material but not of its crystal structure. The cause of prominent features within a multilayer desorption image (such as shown in Fig. 23) is not completely understood. It is clear that the appearance of such features must be associated with the trajectories of the species being desorbed. Various explanations for image features involving ion-trajectory variations have appeared in the literature. These include directed surface migration immediately prior to desorption[297] and focusing by the polyhedral topography of the real tip surface.[298]

---

[295] T. H. George, *Bull. Am. Phys. Soc.* **10**, 493 (1965).
[296] A. R. Waugh, E. D. Boyes, and M. J. Southon, *Nature (London)* **253**, 342 (1975).
[297] A. R. Waugh, E. D. Boyes, and M. J. Southon, *Surf. Sci.* **61**, 109 (1976).
[298] S. V. Krishnaswamy, M. Martinka, and E. W. Müller, *Surf. Sci.* **64**, 23 (1977).

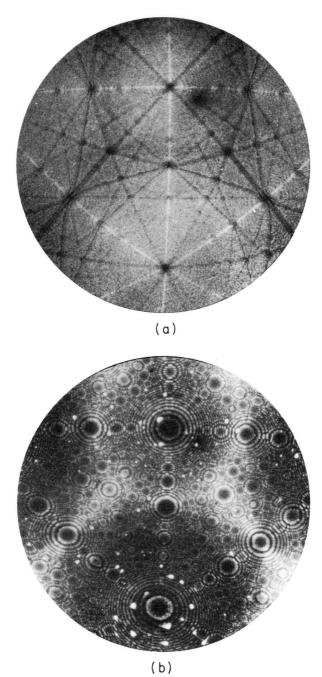

FIG. 23. (a) A multilayer desorption image of rhodium at 28 K in high vacuum. Approximately 30 atomic layers have been removed. (b) The corresponding helium-ion image is shown. (From Krishnaswamy et al.[298] Copyright North-Holland Physics Publishing, Amsterdam, 1977.)

7.7.1. The 10-cm Atom-Probe

Multilayer desorption images have raised a serious question about the reliability of atom-by-atom analysis in an atom-probe experiment. The atom-probe technique relies upon the assumption that a gas-phase species ionized above a surface atom during imaging and the desorbed surface atom itself will follow identical trajectories. Multilayer images show that the two trajectories can be quite different. This means that there can be a significant "aiming error" that can prevent an unambiguous identification of a species preselected in a field-ion image. The aiming problem was recognized as early as 1970. Tungsten atoms on the (110) plane could be detected only in the atom-probe if the probe hole was slightly shifted toward the center of the plane.[299]

In order to minimize the effect of the aiming error, the size of the probe hole can be increased. Because of the long draft distance in the TOF atom-probe, the probe hole can be eliminated and the entrance aperture of the detector used to define the region of the surface under examination.[300] By decreasing the length of the drift region, a large percentage of the surface can be examined. Panitz[146] showed that a drift region about ten centimeters in length would allow an appreciable portion of the tip to be examined with acceptable mass resolution. The resulting instrument was called a "10-cm" atom-probe.

7.7.2. The Imaging Atom-Probe

There are several advantages in examining a large area of the tip surface. A species having a small surface abundance produces a large easily identified mass peak, and species that do not survive field-ion imaging fields can be identified by recording a mass spectrum at their characteristic desorption field. These species can be imaged at the detector of the 10-cm atom-probe in order to deduce their distribution on the surface prior to desorption. The difficulty is that a desorption image cannot distinguish one desorbed species from another. Since several different species could desorb at one time, a unique crystallographic map of a preselected species is impossible to obtain.

An obvious solution to this problem is to activate the MCP detector of a desorption microscope at a fixed time after applying the high-voltage desorption pulse to the tip. By allowing the detector to have gain coincidentally with the arrival of a preselected species, only the preselected species will be imaged. Panitz,[301] developed a workable device based on this principle. It is

---

[299] S. S. Brenner and J. T. McKinney, *Surf. Sci.* **23**, 88 (1970).
[300] R. S. Chambers and G. Ehrlich, *J. Vac. Sci. Technol.* **13**, 273 (1976).
[301] J. A. Panitz, *J. Vac. Sci. Technol.* **11**, 206 (1974).

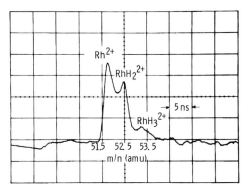

FIG. 24. Mass spectrum of pulse-field-desorbed rhodium in the presence of ~8 × 10⁻¹⁰ Torr of hydrogen. Because chemically stable hydrides of rhodium are unknown at zero field, it is assumed that rhodium hydride formation is favored in the high electric field generated at the tip surface. $V_{dc} = 7.1$ kV, $V_{pulse} \approx 1$ kV, $T \approx 80$ K. (From Panitz.[250])

known as the imaging atom-probe, or IAP. For the first time, the crystallographic distribution of a preselected surface species could be mapped over a large area of the tip surface. The mass resolution that can be achieved by the IAP is shown in Fig. 24. Since the mass resolution is ultimately limited by the energy spread of the desorbing ions, the imaging atom-probe will have a mass resolution as good as those of its counterparts that have much longer drift distances. All that is required is a correspondingly more accurate measurement of the shorter ion travel times. The imaging atom-probe has been used for metallurgical investigations,[302-309] surface studies,[305-307] as a tokamak diagnostic,[308-311] and for depth profiling of low-energy deuterium in tungsten.[310,311] A complete description of the experimental technique has been reviewed elsewhere.[250]

[302] A. R. Waugh, *J. Phys. E* **11**, 49 (1978).
[303] M. K. Miller, T. J. Godfrey, P. A. Beaven, P. R. Williams, K. M. Delargy, and G. D. W. Smith, *Ultramicroscopy* **4**, 383 (1979).
[304] A. R. Waugh, P. F. Mills, and M. J. Southon, *Philos. Trans. R. Soc. London Ser. A* **295**, 133 (1980).
[305] A. R. Waugh and M. J. Southon, *Surf. Sci.* **68**, 79 (1977).
[306] J. A. Panitz, *J. Vac. Sci. Technol.* **12**, 210 (1975).
[307] J. A. Panitz, *1979 DIE Workshop Hydrogen Met., Albuquerque* (available from NTIS, SAND79-0247C, 1979).
[308] G. L. Kellogg and J. A. Panitz, *J. Nucl. Mater.* **85**, 951 (1979).
[309] G. L. Kellogg and J. A. Panitz, *Appl. Surf. Sci.* **3**, 13 (1979).
[310] J. A. Panitz, *J. Vac. Sci. Technol.* **14**, 502 (1977).
[311] G. L. Kellogg, *Bull. Am. Phys. Soc.* **24**, 272 (1979).

7.8. Molecular Imaging with Ions

With the introduction of the low-temperature FIM in 1956, the resolution required to determine the tertiary structure of a biomolecule was available. Perhaps motivated by the success of the structure Watson and Crick[312] proposed for DNA, attempts to image biomolecules in the FIM began in the late 1960s. Two problems had to be considered. The first was to find a way to deposit a molecule of interest onto the apex of a tip. The second was to develop a scheme to ensure that the electric field strength required for field-ion imaging did not distort or destroy the adsorbed species. Of course, the act of removing a molecule from solution and drying it on the surface of a tip in high vacuum will probably distort the structure of the molecule. This problem is an inevitable consequence of any imaging technique that requires a high-vacuum environment.

In earlier field emission experiments, organic molecules were directly sublimed from the gas phase onto the surface of a tip cleaned by thermal annealing or field evaporation. Surface cleanliness was important because contrast in an FEEM image results from a local change in work function caused by an adsorbed molecule. It is not possible to determine from an FEEM image if a particular feature of interest is caused by the presence of the desired species or by an inadvertent contaminant. The organic molecules that were used in the FEEM experiments were very small, typically only five or ten lattice constants in extent. As a result, these species might be expected to be stearically constrained by substrate morphology. This means that surface morphology is important if it varies over distances comparable to the dimension of the molecule. On the other hand, most biological molecules are very large. They can easily span several hundred lattice constants. For these molecules, stearic effects should not be particularly important, and surface conditions could be much less critical. This is an important consideration because biological molecules cannot be sublimed in high vacuum onto a clean surface. They must be deposited from aqueous solution in the laboratory ambient.

Although many different deposition procedures were developed to place molecules on the surface of a tip,[98-100] their success could be evaluated only in terms of the FEEM and FIM images that they produced. An alternative approach to evaluate the success of deposition was introduced by Panitz and Giaever.[313] They used the transmission electron microscope (TEM) to observe thick protein layers formed on the tip surface by the immune reaction. It was found that a tip, when viewed in profile in the TEM, is an excellent substrate on which to observe thick protein layers. Because the tip scatters

---

[312] J. D. Watson and F. H. C. Crick, *Nature (London)* **171**, 737 (1953).
[313] J. A. Panitz and I. Giaever, *Surf. Sci.* **97**, 25 (1980).

electrons very effectively, it appears opaque in an electron micrograph. Although a single protein layer is barely detectable as a contrast variation along the tip contour, double and triple layers produced by the immune reaction are clearly visible. By correlating changes in the appearance of protein multilayers with corresponding changes in a dosing parameter, a simple aqueous deposition scheme was developed. By extending the TEM studies to a spherical protein molecule containing a core rich in iron, an aqueous dosing procedure for placing isolated molecules on a tip was subsequently developed.[314] Figure 25 shows the effectiveness of the technique by displaying a TEM micrograph of individual molecules of ferritin deposited on the contour of a large-radius tungsten tip. Ferritin is a nearly spherical molecule about 13 nm in diameter. It is a hollow shell assembled from twenty-four identical protein subunits. Its core contains about five thousand iron atoms incorporated in a complex ferric hydroxyphosphate polymer. Ferritin is found in the liver and spleen of higher organisms, where it acts as a reservoir for iron, an important but toxic element.

The electric field required for field-ion imaging is very large. Even if hydrogen is used as an imaging gas, the surface is subjected to a field of the order of 20 V nm^{-1}. Such fields produce a large outward-directed electrostatic stress that acts on all surface species, including the adsorbed molecules. It has the effect of tearing organic species apart and desorbing them from the tip surface. Abbott[315] used the FEEM to measure the average dc field strength required to desorb copper phthalocyanine, transfer RNA, and light meromyosin from a tip surface. He found values of 11.9, 10.3, and 7 V nm^{-1}, respectively. Melmed and Müller[106], using the FEEM, found that copper phthalocyanine was desorbed between 12.6 and 1.54 V nm^{-1}, depending on

FIG. 25. A TEM image of the apex of a large-radius tungsten field-emitter tip, thermally annealed in high vacuum and covered with ferritin. Ferritin is a nearly spherical protein (~ 13 nm in diameter) that contains about 5000 iron atoms at its center. Only the iron-rich core of each ferritin molecule is visible in a TEM image. (After Panitz and Giaever.[314] Copyright North-Holland Physics Publishing, Amsterdam, 1981.)

[314] J. A. Panitz and I. Giaever, *Ultramicroscopy* **6**, 3 (1981).
[315] R. C. Abbott, *J. Chem. Phys.* **34**, 4533 (1965).

whether a quadruplet or doublet image was observed. More recently, Panitz[316] reported desorption field strengths for bovine serum albumine (BSA), BSA-antibody complexes, and ferritin. These species appear to be desorbed well below the best-image field for hydrogen.

In order to prevent a molecule from desorbing prior to field-ion imaging, Gurney et al.[317] proposed an ingenious imaging scheme. They suggested that one should embed a molecule on the tip surface within a thin deposit of a refractory metal. By increasing the electric field strength, the molecule might desorb from the surface, but it should leave a cavity in the deposited metal film. In principle, the shape of the cavity would be visible in a field-ion image and reflect the shape of the molecule prior to desorption. Graham et al.[318] tried to image molecules in this way but first annealed the layer in order to improve its uniformity. To improve image contrast, a series of helium field-ion images were taken at 20 K and superimposed. Each image was taken after one atomic layer had been removed by field evaporation. Several features suggestive of a molecular contour were obtained (Fig. 26), but reproducibility was poor.

Müller and Rendulic[319] used a variation of the shadow technique in an attempt to image vitamin B_{12} and coenzyme I (diphosphopyridine nucleotide). Their procedure involved incorporating the molecules in a polycrystalline platinum film, electrolytically deposited onto a tungsten tip. The deposited layer was alternately field-evaporated and imaged in helium in an attempt to expose and then observe randomly dispersed molecules. Single bright image spots and doublets were seen, but convincing images were not obtained. In particular, the expected fourfold symmetry of the vitamin B_{12} molecule was not seen.

In 1971, Machlin et al.[320] tried a novel method to image biological molecules. He hoped to observe the molecule in the process of desorbing. In his approach, a molecule-coated tip was placed in a field-ion microscope equipped with an image intensifier. The tip was cooled to about 5 K and imaged with hydrogen. The imaging procedure consisted of raising the dc tip voltage while recording the ion image on movie film. Machlin hoped to capture transient images that might appear before hydrogen BIV was reached. These images might reflect molecules or molecular fragments on the tip surface in the process of desorbing. A double blind procedure was used in an attempt to objectively evaluate the images. Many intriguing

---

[316] J. A. Panitz, *Bull. Am. Phys. Soc.* **24**, 272 (1979).
[317] T. Gurney, Jr., F. Hutchinson, and R. D. Young, *J. Chem. Phys.* **42**, 3939 (1965).
[318] W. R. Graham, R. Hutchinson, and D. A. Reed, *J. Appl. Phys.* **44**, 5155 (1973).
[319] E. W. Müller and K. D. Rendulic, *Science (Washington, D.C.)* **156**, 961 (1967).
[320] E. S. Machlin, A. Freilich, D. C. Agrawal, J. J. Burton, and C. L. Briant, *J. Microsc. (Oxford)* **104**, 127 (1975).

7.8. MOLECULAR IMAGING WITH IONS 419

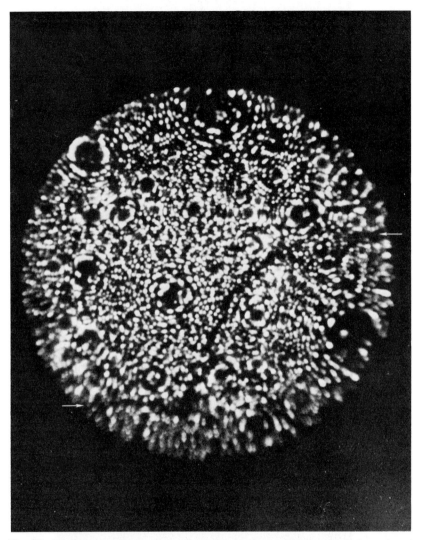

FIG. 26. A helium-ion image at 25 K of several (superimposed) layers of a platinum depositon an iridium tip. The tip was dipped into an aqueous solution of DNA (from micrococcus luteus) prior to platinum deposition in vacuum at 400 K. The dark feature indicated by arrows is thought to represent a DNA molecule. (From Graham et al.[318])

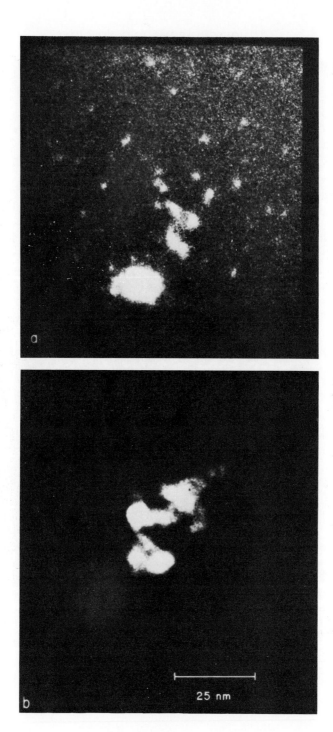

7.8. MOLECULAR IMAGING WITH IONS

features were seen (Fig. 27), but like all previous FIM imaging attempts, definitive images were never obtained.

In 1977 another imaging scheme was conceived.[321] Its virtue lies in its demonstrated ability to obtain reproducible nondestructive images of unstained biological molecules.[322-324] The technique is called field-ion tomography because it produces a series of contour slice (or "tomographic") images that show the distribution of molecules on a tip at various elevations above the tip surface. From these images, a three-dimensional reconstruction of the imaged species can be obtained.[325] The molecule of interest is deposited onto a field-emitter tip from aqueous solution.[314,326-327] After the tip is placed in a modified FIM, a thick layer of amorphous benzene ice is condensed from the gas phase onto the tip surface. Since the tip is cooled to a temperature below 30 K, the molecules of interest are effectively embedded within a frozen immobile matrix that completely surrounds and covers them. By increasing the tip voltage, the benzene layer can be controllably desorbed at a field strength below 4 V nm^{-1}. This means that the molecules of interest will not be removed during the imaging process.[323] Provided the benzene layer covers the deposited molecules, the benzene ion image at the detector will appear bright and relatively contrastless. However, as a molecule becomes exposed by the receding benzene layer, a dark region will appear in the image. Its contour will reflect the contour of the molecule defined by its intersection with the remaining benzene layer.

From a series of successive contour slice images taken during the removal of the benzene layer, a series of molecular contours are obtained. These can be digitally reconstructed in an objective way to produce a three-dimensional image of the molecular distribution on the surface that produced them.[325] An example of a field-ion tomographic image of ferritin is shown in Fig. 28. The image on the left is a field-ion tomographic image of unstained

---

FIG. 27. Transient images recorded during field-ion microscopy of an iridium tip on which a solution containing a single-stranded polynucleotide (poly-U) was freeze dried. The images were recorded using hydrogen as an imaging gas at a field strength of about (a) 5 V nm^{-1} and (b) 6.5 V nm^{-1}. The tip temperature was 5 K. (From Machlin et al.[320])

---

[321] J. A. Panitz and I. Giaever, *Abstr. 25th Int. Field-Emiss. Symp., Albuquerque, N.M., 1978*, published in *Ultramicroscopy* **4**, 366 (1979); *Abstr. 26th Field-Emiss. Symp., West Berlin, 1979*, published in *Ultramicroscopy* **5**, 248 (1980).
[322] J. A. Panitz, *J. Microsc. (Oxford)* **125**, 3 (1982).
[323] J. A. Panitz, *Ultramicroscopy* **7**, 241 (1982).
[324] J. A. Panitz and D. C. Ghiglia *J. Microsc. (Oxford)* **127**, 259 (1982).
[325] D. C. Ghiglia and M. Flickner, *Opt. Lett.* **7**, 116 (1982).
[326] J. A. Panitz, C. L. Andrews, and D. C. Beur, *JEMT* (1985). In press.
[327] J. A. Panitz, *Rev. Sci. Instrum.* (1985). In press.

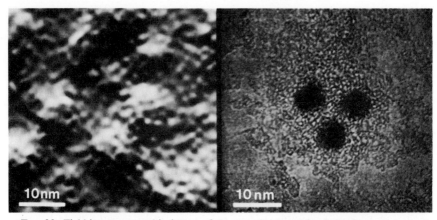

FIG. 28. Field-ion tomographic image of the unstained protein shell of several isolated ferritin molecules deposited onto the surface of a tungsten field-emitter tip (left). This image was digitally created from a series of contour slice images taken during controlled field desorption of a thick benzene layer condensed onto a molecule-coated tip at 20 K. The image at the right is a transmission electron micrograph of ferritin on a thin carbon substrate. Only the iron-rich cores of individual ferritin molecules can be seen. Their outer protein shell is invisible and is probably destroyed by electron bombardment during imaging. (From Panitz.[247] Copyright 1982 The Institute of Physics.)

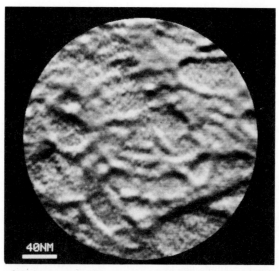

FIG. 29. A field-ion tomographic image of poly(GC) DNA on tungsten at 20 K.

ferritin molecules on a tungsten substrate.[324] The image on the right is a transmission electron micrograph of unstained ferritin on a thin carbon substrate. Unlike field-ion tomography, which provides a picture of the entire molecule, transmission electron microscopy can only provide an image of its iron-rich, electron-opaque core. An attempt to image DNA by field-ion tomography has recently been reported.[328] Figure 29 shows a tomographic image of DNA in which the characteristic stranded appearance of the molecule is evident. By combining field-ion tomography with the atom-probe techniques discussed previously, a real possibility exists for determining the molecular weight of a single, preselected macromolecule and visually mapping its constituent elements with a spatial resolution better than 1 nm.

Acknowledgments

The author gratefully acknowledges the United States Department of Energy, which has supported the author's research under Sandia Contract DE-AC04-76DP00789.

[328] J. A. Panitz, *Ultramicroscopy* **11**, 161 (1983).

8. THE THERMAL DESORPTION OF ADSORBED SPECIES

By John T. Yates, Jr.*

Surface Science Division
National Bureau of Standards
Washington, D.C.

8.1. Introduction

One of the simplest qualitative or quantitative experiments that may be carried out on a surface containing an adsorbed layer involves the thermal desorption of species from the surface. An adsorbed species, bound to the surface by van der Waals forces (physisorption) or by covalent bonding (chemisorption) resides on the surface in its adsorption potential well. At a characteristic temperature it will begin to desorb from the surface via an activated process. The study of thermally activated desorption phenomena represents a natural combination of chemical kinetics, statistical thermodynamics, and molecular dynamics. Knowledge of the nature of the desorption process has fundamental implications to understanding the nature of elementary chemical processes in the layer, to the energetics of bonding and the specification of the chemical nature of the bound species, and to the nature and magnitude of interactional effects between adsorbate species. For example, the kinetic order of a desorption process may suggest the nature of the elementary step(s) governing the process. Zero-order kinetics are often indicative of desorption from a multilayer where the rate of desorption is independent of surface coverage. Zero-order kinetics may also be evidence for equilibrium between different surface phases with one phase maintaining a constant surface concentration of the desorbing phase. First-order kinetics may be indicative of the presence of a single surface species. Second-order kinetics are an indication of adsorbate atom recombination processes leading to the production of a diatomic molecule that is then evolved.

The common observation of multiple desorption processes (multiple

---

* Present address: Surface Science Center, Department of Chemistry, University of Pittsburgh, Pittsburgh, Pennsylvania.

binding states) that occur at different characteristic temperatures may indicate that a mixture of different kinds of adsorbed species exist together, or that the act of desorption induces an interconversion from one species (or desorption process) to another as coverage changes.

The adsorption and subsequent desorption of mixtures of isotopic species (i.e., H_2 and D_2, $^{14}N_2$ and $^{15}N_2$, and $^{12}C\ ^{18}O$ and $^{13}C\ ^{16}O$) provides information about the stability of molecular adsorbate species and about dissociative chemisorption.

Unusual coverage-dependent variations in the kinetic parameters measured (pre-exponential factor and activation energy) often suggest desorption mechanisms involving transient intermediate species that would be difficult to observe directly. Unusual absolute values for these parameters often suggest entropic models for the adsorbate or for the desorption process.

The extension of thermal desorption methods to the study of the interaction of more complex molecules with surfaces has now occurred. Studies of desorption kinetics from these systems can suggest the nature of surface-stabilized molecular fragments of importance to a fundamental understanding of catalytic chemistry at surfaces.

We can anticipate future developments involving more detailed quantum mechanical specification of the thermal desorption process. Questions regarding the kinetic energy distribution and the spatial distribution of desorbing molecules from single crystal surfaces are just now being addressed theoretically and experimentally. The experimental determination of the internal energy distribution of desorbing molecules is just ahead and will involve the use of laser spectroscopy to investigate desorbing species in exquisite detail.

In surface science today a number of new measurement techniques involving various spectroscopies and diffraction methods are being focused on the determination of the detailed nature of adsorbed layers. A common element to many of these studies is the concomitant investigation of the thermal desorption characteristics of the adsorbate. Thus thermal desorption methods may be useful in identifying and adjusting the surface population of species of interest as well as determining both relative and absolute surface coverages of these species. Because of the ease of doing thermal desorption experiments and because of the fundamental nature of the information obtained, it is appropriate to suggest that the omission of thermal desorption information in a research report is a sign of incompleteness.

The immense practical importance of fundamental knowledge about thermal desorption is very evident. All heterogeneous catalytic processes must proceed via a thermal desorption step in which product molecules are evolved, leaving the catalytic sites ready for the next reactant molecules to be adsorbed. Desorption effects from the walls of vacuum systems determine the quality of the vacuum environment, which determines the surface purity

8.2. DESORPTION FROM POLYCRYSTALLINE SUBSTRATES

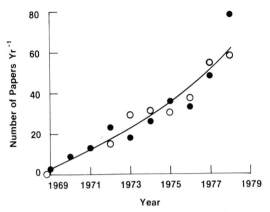

FIG. 1. Rate of scientific publication on thermal desorption and temperature programmed desorption. Data was obtained using computer search procedures in *Chemical Abstracts, Electrical and Electronics Abstracts, Physics Abstracts,* and *Computer and Control Abstracts.*

of materials being handled or studied. Some kinds of vacuum pumps and vacuum gauges depend on the minimization of thermal desorption phenomena for proper operation, whereas others depend on adsorption processes for their pumping action. The production of high-purity gases is dependent on minimizing desorption from container walls and vacuum system walls. Indeed, one of the major limitations to doing meaningful surface experiments in the vacuum environment of outer space is the desorption of impurities from the space hardware itself. The properties of surfaces and adsorbed layers on surfaces are thus of fundamental importance to many technologies, and thermal desorption phenomena are therefore inherently involved in many practical areas. A literature study of the growth of research papers concerned with thermal desorption is summarized in Fig. 1. Data from both Chemical Abstracts and Physics Abstracts through 1979 attest to the growing interest in thermal desorption phenomena and to the diversity of applications involving thermal desorption. This review covers selected aspects of thermal desorption through 1979.

8.2. Early Studies of Desorption from Polycrystalline Substrates

8.2.1. Introduction

Polycrystalline tungsten filaments were first employed as substrates for thermal desorption studies by Taylor and Langmuir[1] over 40 years ago. They

[1] J. B. Taylor and I. Langmuir, *Phys. Rev.* **44**, 4236 (1933).

used a "flash filament" method to determine the relative surface coverage of Cs chemisorbed on W. Twenty years later, Ehrlich pioneered the flash desorption method as a general means for studying the adsorption of gases such as N_2, CO, and Xe on polycrystalline surfaces (mainly tungsten).[2]

The basic principle employed by Ehrlich was to expose a clean polycrystalline filament to a gas at *low pressure* in a flowing gas environment and to rapidly flash the filament at a high rate of heating ($\sim 10^3$ K s^{-1}) following adsorption for a known time. By varying the initial time and pressure of adsorption, the initial coverage of the adsorbate could be varied. Mercury diffusion pumps were employed and rigorous sample and vacuum system processing procedures were followed. The use of a *flowing* low-pressure stream of gas was devised earlier by Becker and Hartman[3] to attempt to maintain gas purity and therefore adsorbate purity in the experiments. Ehrlich's original flash desorption studies resulted in the discovery of multiple surface binding states for N_2, CO, and H_2 adsorbates on polycrystalline tungsten, and in the characterization of desorption kinetics from these states.

Fast rates of temperature programming were employed by Ehrlich on polycrystalline surfaces to reduce the possibility of surface diffusion processes from one crystal plane to another interfering with the desorption process. Thus, if on a polycrystalline surface different binding states were attributed to different crystal planes, at low rates of temperature programming, crossover from one plane to another could, in principle, wash out the discrimination of separate desorption processes from the separate planes. The temperature programming was achieved by application of a constant voltage to the filament, resulting in a hyperbolic dependence of temperature on time.

Redhead[4] carried out temperature-programmed desorption on similar adsorption systems to that of Ehrlich but instead employed very much slower rates of heating having a linear dependence of temperature on time. This simple change in experimental procedure led to an increase in resolution of the method and to the discovery of additional binding states unresolved by rapid flash desorption. (For an illustration of this, see the case of nitrogen chemisorbed on polycrystalline tungsten.[5]) A much-quoted example of the Redhead method for temperature-programmed desorption concerns CO chemisorbed on polycrystalline tungsten[4] as shown in Fig. 2. Here

---

[2] An excellent review of the method as applied by Ehrlich may be found in G. Ehrlich, *Adv. Catal.* **14**, 255 (1963).

[3] J. A. Becker and C. D. Hartman, *J. Phys. Chem.* **57**, 153 (1953).

[4] P. A. Redhead, *Trans. Faraday Soc.* **57**, 641 (1961).

[5] (a) J. T. Yates, Jr. and T. E. Madey, *J. Chem. Phys.* **43**, 1055 (1965); (b) T. E. Madey and J. T. Yates, Jr., *J. Chem. Phys.* **44**, 1675 (1966).

8.2. DESORPTION FROM POLYCRYSTALLINE SUBSTRATES

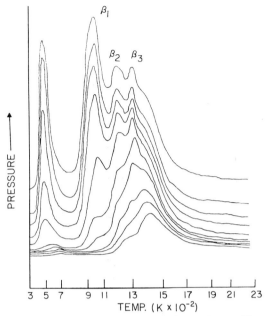

FIG. 2. Carbon monoxide thermal desorption spectra from polycrystalline tungsten wire. For increasing exposure time the intensity of the various desorption states α, β_1, β_2, and β_3 change. (From Redhead.[4])

at least four binding states of CO are clearly observed, and are designated α, β_1, β_2, and β_3. Thus it is seen that we are dealing with a form of kinetic spectroscopy. It was the observation of multiple binding states in desorption that led to three general postulates concerning their origin:

(1) The binding states represent different modes of bonding of the molecule to different crystal planes.

(2) The binding states represent different structural species present together in the chemisorbed layer on a single crystal plane.

(3) The binding states represent different modes of bonding which are produced as thermal desorption takes place, causing coverage dependent interactional effects to vary.

Many subsequent studies carried out in combination with other surface measurement methods have shown that all three postulates are valid in different cases. Indeed the explanation of the origin of the thermal desorption binding states has been a driving force for many of the research efforts in chemisorption within the past 15 years.

8.2.2. Material Balance Equation for Thermal Desorption in a Vacuum System

The basis for the measurement of thermal desorption spectra (using gas evolution) is given below.

The rate of desorption of a species from a surface R_d should be a function of the species' surface concentration N and the temperature T. The Arrhenius form is

$$-\frac{dN}{dt} = R_d = v_0^{(n)} N^n e^{-E_d/kT}, \quad (8.2.1)$$

where $v_0^{(n)}$ is the pre-exponential factor, n the desorption order (usually 0, 1, 2), and E_d the activation energy for desorption. It is often assumed as a first approximation that $v_0^{(n)}$ and E_d are constants, but in fact it has been shown that in many cases these kinetic parameters vary with coverage N. If one is using the pressure rise ΔP above the base pressure P_0 to measure desorption kinetics under conditions for which the pumping speed, $\mathcal{S}$, is constant and no adsorption/desorption processes occur on extreneous surfaces, then, following Redhead[6]

$$\frac{d(\Delta P)}{dt} = \frac{kT}{V}\left(-A\frac{dN}{dt}\right) - \frac{\mathcal{S}\Delta P}{V}$$

$$= \text{(rate of gas evolution)} - \begin{pmatrix}\text{incremental pumping} \\ \text{rate above steady} \\ \text{state background rate}\end{pmatrix}. \quad (8.2.2)$$

Therefore,

$$R_d = \frac{V}{AkT}\frac{d(\Delta P)}{dt} + \frac{\mathcal{S}\,\Delta P}{AkT}, \quad (8.2.3)$$

which may be rewritten as

$$\frac{AkT}{V}R_d = \frac{d(\Delta P)}{dt} + \frac{\Delta P}{\tau}, \quad (8.2.4)$$

where $\tau = V/\mathcal{S}$ = characteristic pumping time of the system. If we define a "characteristic time" δt for the desorption of a binding state, then two limits exist in Eq. (8.2.4), namely $\tau \gg \delta t$ or $\tau \ll \delta t$. In the former case, (as carried out by Ehrlich)

$$R_d = \frac{V}{AkT}\frac{d(\Delta P)}{dt}, \quad (8.2.5)$$

---

[6] P. A. Redhead, *Vacuum* **12**, 203 (1962).

whereas in the latter case

$$R_d = \frac{V}{AkT}\frac{\Delta P}{\tau}. \quad (8.2.6)$$

For most situations involving fast-pumping-speed vacuum systems ($\tau \leq 0.25$ s) and $\delta t \geq 2$ s, Eq. (8.2.6) is applied conveniently, and the pressure rise ΔP is a direct measure of the desorption rate R_d. In the intermediate case, both terms in Eq. (8.2.4) could be measured as was done by Madey and Yates.[5,6]

In a glass or metal ultrahigh-vacuum system, it is possible to measure τ by rapidly opening or closing a gas admission valve or by producing a pulse of gas by rapid desorption from a filament and subsequently following the decay of pressure with time. Frequently, it is found that τ is not constant. This may be due to wall effects or to re-emission of gas from ion pumps or other elements within the system. Redhead[4] has demonstrated methods for determining that wall effects are absent in adsorption/desorption experiments under flow conditions. It should be remembered that the pumping speed of a gas under molecular flow conditions (with no wall effects) is directly proportional to molecular mean velocity and for a Boltzmann distribution therefore proportional to $(mT)^{-1/2}$, where m is molecular mass.

8.3. Thermal Desorption from Single Crystals

8.3.1. Introduction

The use of polycrystalline filament substrates for studies of chemisorption is a convenient way of exploring the general features of a new adsorption system. Using mass spectrometric detection of desorbing products, one often gains insights into the basic chemical processes that occur during heating, and this may form the basis for more sophisticated experiments on single crystals. For example, the detailed study of tungsten oxide(s) desorption from polycrystalline tungsten surfaces has led to general insights concerning multilayer oxide formation.[7]

The extension of temperature-programmed desorption to single crystal substrates is now a standard procedure in many surface experiments. The first attempts to do this were by Delchar and Ehrlich,[8] who studied N_2 adsorption on a tungsten crystal exposing a mixture of surface planes. Soon after this, experiments on thin single crystal slices exposing 90% of a chosen single crystal plane became common with the study of hydrogen chemisorp-

---

[7] D. A. King, T. E. Madey, and J. T. Yates, Jr., *J. Chem. Phys.* **55**, 3236, 3247 (1971).
[8] T. A. Delchar and G. Ehrlich, *J. Chem. Phys.* **42**, 2686 (1965).

tion on W(100) single crystals being one of the first examples.[9,10] Studies of atomic oxygen desorption from oxygen covered W(100) were also carried out by Ageev and Ionov[11] at about the same time.

8.3.2. Preparation, Mounting, and Temperature Programming of Single Crystals

Oriented single crystals of metals may be cut into thin slices using either spark erosion or a string saw. This is best done on an adjustable crystal holder that can be fitted into an x-ray apparatus set up to carry out Laue back reflection orientation. A convenient aid in crystal orientation is the Laue Atlas.[12] The crystal slice must then be ground and metallographically polished using an adjustable crystal holder also suitable for final x-ray adjustment of the crystal orientation. A good design has been given by Wendelken et al.[13] The grinding of the crystal permits exact adjustment of crystal orientation as well as adjustment of the crystal thickness to a proper value (usually several tenths of a millimeter). Metallographic polishing is often carried out using diamond paste or alumina polishing compounds with particle sizes ranging down to the submicron level. Chemical and electrochemical polishing procedures have also been utilized for final crystal preparation.[14] In general for thermal desorption procedures in which both faces of the crystal could contribute to the measurements, it is appropriate to grind and polish both faces of the crystal. In order to avoid spurious contributions from the edge of the crystal slice, which will expose various crystal planes, Tamm and Schmidt[15] have shown that it is possible to cut the edges preferentially such that the edges have the same orientation as the faces, as shown in Fig. 3.

Single crystals of metals are generally mounted by welding to support wires of the same composition as the crystal to be studied. Two general methods for heating are employed: (1) The crystal may be heated by *radiation* and/or electron bombardment using a close-coupled tungsten or other refractory metal filament. This is particularly convenient if the heater filament and the crystal are of the same composition, since alloying of evaporated filament material with the crystal is avoided. Thorium oxide coatings to reduce emitter work functions should be avoided on electron bombard-

---

[9] P. W. Tamm and L. D. Schmidt, *J. Chem. Phys.* **51**, 5352 (1969).

[10] T. E. Madey and J. T. Yates, Jr., "Structure et Properties des Surfaces des Solides," No. 187, pp. 155–162. CNRS, Paris, 1970.

[11] V. N. Ageev and N. I. Ionov, *Zh. Tekh. Fiz.* **39**, 1523 (1969).

[12] E. Preuss, B. Krahl-Urban, and R. Butz, "Laue Atlas." Wiley, New York, 1974.

[13] J. F. Wendelken, S. P. Withrow, and C. A. Foster, *Rev. Sci. Instrum.* **48**, 1215 (1977).

[14] R. L. Anderson, "Revealing Microstructures in Metals," Sci. Pap. 425-C000-P2. Westinghouse Res. Lab., Pittsburgh, Pennsylvania, 1961.

[15] P. W. Tamm and L. D. Schmidt, *Surf. Sci.* **26**, 286 (1971).

8.3. THERMAL DESORPTION FROM SINGLE CRYSTALS

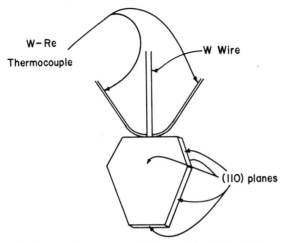

FIG. 3. Schematic diagram of a tungsten single crystal in which all faces expose the (110) plane. The crystal is heated by electron bombardment. (From Tamm and Schmidt.[15] Copyright North-Holland Physics Publishing, Amsterdam, 1971.)

ment sources used for heating crystals because of the emission of impurities that may be collected by the single crystal.[16] Radiative heating of a W crystal by a close-coupled W filament may be used to reach temperatures near ~1200 K.[17] The crystal may be heated by *conduction* from hot mounting leads welded to the edge or back side of the crystal. This method does not generally rely on direct ohmic heating of the crystal, since its resistance is usually low compared to the mounting leads. The crystal heating occurs by means of conduction from the support wires. Often support wires are of the same composition as the crystal, although W and Ta wires are frequently used for crystals of lower melting point. In this case, the refractory support wires are never heated to temperatures at which their metal vapor pressures are high. However, since W and Ta wires when handled in the atmosphere or when oxidized in a vacuum system often contain appreciable thicknesses of their (relatively high-vapor-pressure) oxides[7], one must avoid direct line of sight between the support and the crystal face being studied, or oxide contamination may occur.

A variation on the radiative heating of a crystal for thermal desorption was first employed by Tamm and Schmidt[9] and by Madey and Yates[10] and later by Yates and King.[17] Here a focused beam of light from a quartz–halogen tungsten filament source (300–500 W) was directed to the crystal through

---

[16] J. T. Yates, Jr. and D. A. King, *J. Vac. Sci. Technol.* **9**, 1256 (1972).
[17] J. T. Yates, Jr. and D. A. King, *Surf. Sci.* **32**, 479 (1972).

transparent vacuum system walls. Temperature programming up to ~900 K was easily achieved when thermal conduction between the ℓ-N_2 bath used for cooling the crystal was minimized using ~0.25-mm-diameter support wires.

The crystal temperature during thermal desorption is generally measured by means of a thermocouple junction made of small-diameter (<0.1-mm) wire, which is welded to the crystal face or edge. A number of thermocouple materials are commonly used; the research worker should employ a reference junction[18] at the ice point for the best work. One of the favorite couples for work below room temperature is the W–26%-Re versus W–3%-Re; the low-temperature calibration of this couple has been carried out by Sandstrom and Withrow[19] down to 77 K.

Temperature programming of a single crystal is generally carried out by the application of a constant ac or dc voltage to the crystal and its support wires. Ohmic heating of the support wires plus conduction leads to uniform heating of the crystal. The voltage applied is adjusted with a variable-transformer/filament-transformer combination or with a dc power supply to achieve approximately linear heating rates in the region of interest. Feedback circuitry has been used to linearize the temperature programming. Since the $\int P \, dt$ is proportional to the change in coverage during desorption, it is usual practice to plot the pressure of gas evolved versus time, and to then employ the function $T(t)$ to determine the temperature scale applicable to the experiment. This is a convenient operation for a minicomputer, and one may find examples of this method in the work of the Weinberg group.[20]

A comparison of thermal desorption from two single crystal planes of tungsten is shown in Fig. 4. Here a series of thermal desorption spectra are shown for increasing initial hydrogen coverage.[21,22] Two major desorption states (β_1 and β_2) are seen on W(100) whereas four β states are observed on W(111). The small γ-H_2 states seen on W(100) probably originate from the crystal edge or from the tungsten support assembly. The tungsten crystal structure for each system is also shown. For the case of hydrogen on W(100) it has been found that hydrogen chemisorption promotes reconstruction of the W(100) surface.[23,24] At present we have no models to explain the complex

[18] A. I. Dahl, ed., "Temperature," Vol. 3, Part 2. Reinhold, New York, 1962.
[19] D. R. Sandstrom and S. P. Withrow, *J. Vac. Sci. Technol.* **14**, 748 (1977).
[20] J. L. Taylor, D. E. Ibbotson, and W. H. Weinberg, *Surf. Sci.* **79**, 349 (1979); see also J. L. Taylor, Ph.D. Thesis, California Inst. Technol., Pasadena, 1978.
[21] T. E. Madey, *Surf. Sci.* **29**, 571 (1972); T. E. Madey and J. T. Yates, Jr., *Surf. Sci.* **63**, 203 (1977).
[22] P. W. Tamm and L. D. Schmidt, *J. Chem. Phys.* **54**, 4775 (1971).
[23] M. K. Debe and D. A. King, *Surf. Sci.* **81**, 193 (1979).
[24] T. E. Felter, R. A. Barker, and P. J. Estrup, *Phys. Rev. Lett.* **38**, 1138 (1977).

8.3. THERMAL DESORPTION FROM SINGLE CRYSTALS

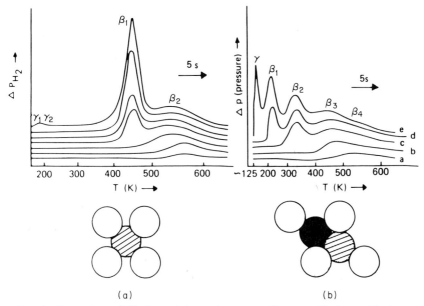

FIG. 4. Comparison of H_2 thermal desorption spectra from two single crystal planes of tungsten. (a) Development of H_2 desorption states on W(100) for increasing initial coverages of hydrogen. The small γ-H_2 states probably originate from the crystal edge or the tungsten support assembly. (b) Same, except for W(111) where the molecular γ-H_2 originates from the (111) plane. Crystal planes are shown below the corresponding spectra. Deeper shading indicates W atoms at successively deeper levels below the W surface atoms. (From Madey and Yates.[21] Copyright North-Holland Physics Publishing, Amsterdam, 1977.)

H_2 desorption behavior observed from W(111), nor which consider the influence on desorption kinetics of crystal structure changes on W(100). The more complex desorption of H_2 from W(111) may be related to the greater number of types of adsorption sites present on W(111) compared to W(100). It is also undoubtedly true that adsorbate–adsorbate interactional effects are of importance in determining the distribution of binding states.

It is in fact not necessary to continuously program the crystal temperature in order to carry out temperature programmed thermal desorption measurements. Kohrt and Gomer[25] have used step temperature programming to study both O_2 and CO desorption from W(110) using a field emission detector for measuring the incremental quantity of gas desorbed at each step. In

---

[25] C. Kohrt and R. Gomer, *J. Chem. Phys.* **52**, 3283 (1970); *Surf. Sci.* **24**, 77 (1971); **40**, 71 (1973).

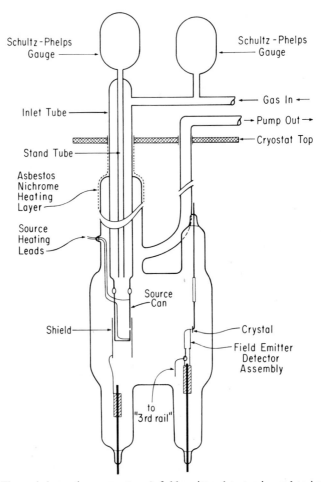

FIG. 5. Thermal desorption apparatus. A field emitter detector is used to intercept and measure pulses of desorbing gas from the crystal under study. The source can is used to deposit the gas initially on the single crystal. (From Kohrt and Gomer.[25])

this elegant work, the entire apparatus (Fig. 5) is immersed under ℓ-H_2 to avoid wall effects and to condense all desorbed gas that misses the field emission tip during desorption.

8.3.3. Isothermal Desorption Measurements

Thermal desorption may be carried out at selected constant temperatures for the purpose of making kinetic measurements. Isothermal desorption

8.3. THERMAL DESORPTION FROM SINGLE CRYSTALS

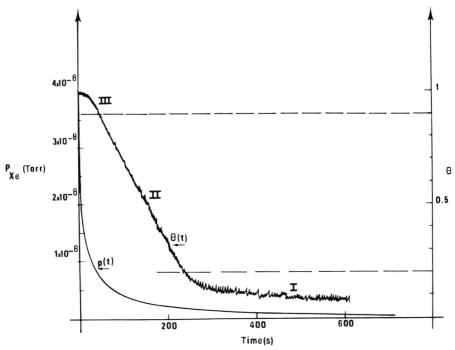

FIG. 6. Isothermal desorption kinetics for Xe physically adsorbed on (0001) graphite. Desorption took place at 76 ± 0.1 K with $p_1 = 3 \times 10^{-8}$ Torr. The surface coverage of Xe is followed by Auger spectroscopy. (From Bienfait and Venables.[26] Copyright North-Holland Physics Publishing, Amsterdam, 1977.)

kinetics on a single crystal were first employed by Kohrt and Gomer[25] for O_2 and CO desorption studies on W(110). A more recent application is best illustrated by studies of Bienfait and Venables who used Auger spectroscopy to follow the surface coverage of Xe physically adsorbed on (0001) graphite as the Xe desorbed isothermally.[26] Typical data are shown in Fig. 6, in which three regions of desorption kinetics are observed. Region II in which the rate of desorption is independent of surface coverage corresponds to *zero*-order kinetics, which results from an equilibrium between migrating Xe adsorbate atoms and two-dimensional Xe crystals on the (0001) surface. The isothermal desorption method has been extended to the study of Au desorption from Si(111). In Fig. 7, Auger signals from Au and Si are plotted versus time for a particular isothermal desorption experiment.[27] Three zero-order re-

[26] M. Bienfait and J. A. Venables, *Surf. Sci,* **64**, 425 (1977).
[27] G. Le Lay, M. Manneville, and R. Kem, *Surf. Sci.* **65**, 261 (1977).

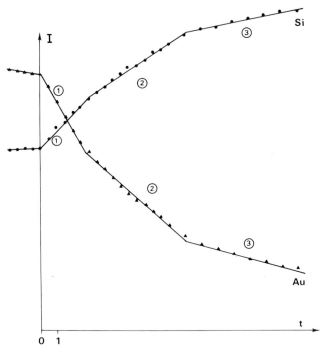

FIG. 7. Isothermal desorption kinetics for Au on Si(111) at 865°C as followed by Auger spectroscopy. The intensity of the 69-eV Au peak and the 92-eV Si peak change according to zero-order kinetics and show breaks corresponding to three desorption processes. (From Le Lay et al.[27] Copyright North-Holland Physics Publishing, Amsterdam, 1977.)

gions are noted, separated by sharp breaks seen simultaneously in both the Au and Si Auger signals. These three zero-order desorption regions correspond to three different condensed two-dimensional phases of Au on Si(111), which desorb in three successive kinetic steps. The three phases exhibit different cohesive energies leading to different activation energies of desorption. It should be pointed out here that although the isothermal desorption method involving Auger detection of surface coverage is very simple in concept, one must not consider it to be a general method for desorption studies, since electron-beam-stimulated desorption or decomposition of the adsorbate can also occur for many adsorbates, thereby confusing the kinetic measurements.

8.3.4. Isosteric Heat of Adsorption

One of the main objectives of measuring the kinetics of thermal desorption is to obtain the activation energy for the desorption process. For *revers-*

8.3. THERMAL DESORPTION FROM SINGLE CRYSTALS

ible adsorption/desorption processes, the energetics of the adsorption state may be measured by thermodynamic means in the same fashion as one would measure the enthalpy of vaporization of a condensed phase by means of the temperature dependence of the vapor pressure. In both cases, the Clausius–Clapeyron equation is employed. Since for the desorption case, the enthalpy of desorption would be expected to vary slightly with coverage due to interactional effects, the measurement of the temperature dependence of the equilibrium pressure must be carried out at constant surface coverage. The Clausius–Clapeyron equation for the desorption processes is then

$$\left[\frac{d \ln P}{d(1/T)}\right]_{N=\text{const}} = \frac{-\Delta H_{\text{des}}}{R} = \frac{-Q_{\text{iso}}}{R} \quad (8.3.1)$$

where Q_{iso} is the isosteric heat of desorption (a positive quantity) and R the gas constant = 1.986 cal K^{-1} mole^{-1} = 8.314 J K^{-1} mole^{-1}. Since by definition

$$H = E + PV, \quad (8.3.2)$$

then

$$Q_{\text{iso}} = \Delta E_{\text{des}} + RT, \quad (8.3.3)$$

where ΔE_{des} is the energy absorbed by the system during desorption, and RT is the reversible expansion work done during desorption.

In principle then, Q_{iso} and ΔE_{des} differ by RT. In practice, for most experimental measurements the accuracy of measurement of the energetics of desorption makes this difference academic.

In order to apply Eq. (8.3.1) it is necessary to be able to measure the equilibrium pressure of the adsorbate as a function of temperature at constant coverage N. Usually some measurable surface property that varies in a well-defined way with surface coverage is employed. Often the work function is used, although low-energy electra diffraction (LEED) intensity, Auger intensity, or other spectroscopic intensities can be used if the measurement process can be carried out in the gas pressure range of interest without disturbing the state of the adsorption system appreciably.

The use of work function as a monitor of surface coverage in the measurement of isosteric heats of adsorption was applied in early studies by Palmberg[28] in the study of Xe physisorption on Pd(100). Calibrations were made for the work function versus surface coverage (as deduced from Auger measurements and LEED observation of the final close-packed hexagonal Xe layer). More recently, adsorption isotherms for CO chemisorbed on Pd

---

[28] P. W. Palmberg, *Surf. Sci.* **25**, 598 (1971).

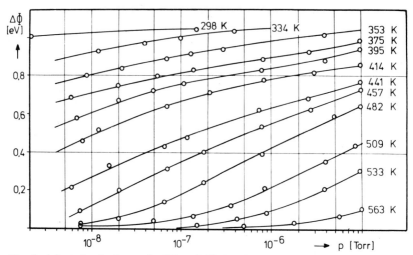

FIG. 8. Adsorption isotherms for CO on Pd(210) using work function as a monitor of CO surface coverage. (From Conrad et al.[29] Copyright North-Holland Physics Publishing, Amsterdam, 1974.)

single crystals have been measured by Conrad et al.[29] and a representative display of the data plotting work function change versus equilibrium CO pressure at various temperatures for Pd(210) is shown in Fig. 8. The derived energy of adsorption as a function of work function increase is shown in Fig. 9. In this range of coverage the thermal desorption spectra exhibit only a single peak whose position on the temperature scale continuously shifts to lower temperature as initial CO coverage increases, reflecting the continuous variation in adsorption energy. Studies of this system have shown that two CO binding states are populated.[30] Studies by ESDIAD[30] when coupled with IR studies on this surface have suggested an atomic site location for a bridging-CO species responsible for the single desorption state seen. Comparison of the *initial* isosteric heats of adsorption for various Pd single crystal surfaces indicates that about 15% variation exists, and suggests that no pronounced plane specificity for CO chemisorption occurs, i.e., that a major structural factor leading to large variation in CO binding energy is absent.[29]

---

[29] H. Conrad, G. Ertl, J. Koch, and E. E. Latta, *Surf. Sci.* **43**, 462 (1974).
[30] T. E. Madey, J. T. Yates, Jr., A. M. Bradshaw, and F. M. Hoffmann, *Surf. Sci.* **89**, 370 (1979).

8.3. THERMAL DESORPTION FROM SINGLE CRYSTALS

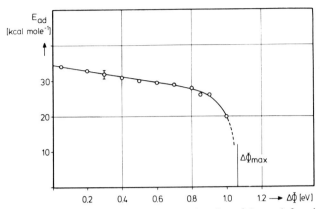

FIG. 9. Adsorption energy for CO on Pd(210) as a function of the work function increase. (From Conrad et al.[29] Copyright North-Holland Physics Publishing, Amsterdam, 1974.)

8.3.5. Absolute Coverage Measurements

The thermal desorption methods can in principle measure relative coverages of an adsorbate by evaluation of $\int P \, dt$. This measurement hinges on minimization of error factors due to wall effects, gain changes in detectors, changes in pumping speed, desorption from extraneous surfaces, etc. Although some attempts at the *estimation* of *absolute* surface coverages from thermal desorption data have been made,[2,4] the errors involved are so large that the results are almost useless at present.

Three methods for estimating absolute surface coverages of adsorbates on single crystals are readily available: (1) molecular beam methods involving precise flux measurements, (2) thermal desorption studies coupled with LEED measurements of overlayer periodicity, (3) radioactivity measurements from adsorbed layers.

A molecular-beam method was devised at NBS[31] and the apparatus is shown schematically in Fig. 10. It consists of an accurately made thin plate effusion orifice having a thickness/radius ratio of 1.0. The angular distribution of gas emitted from this orifice may be calculated[32] and was verified experimentally by intercepting and adsorbing gas on the front surface of a circular single crystal disk at various distances from the orifice. A known total flux of gas from the effusion source was achieved using a capillary whose conductance was accurately calibrated by standard volumetric methods. If the total flux and its angular distribution are both known, and

[31] T. E. Madey, *Surf. Sci.* **33**, 355 (1972).
[32] B. B. Dayton, *Trans. 3rd AVS Natl. Vac. Symp., 1956,* Chicago, p. 5 (1957).

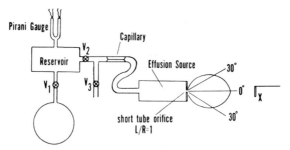

FIG. 10. Schematic gas dosing system for accurate absolute measurement of surface coverage and sticking coefficient. Gas stored in the reservoir at a known pressure transfers through the calibrated capillary at a known rate into the effusion source. The angular distribution of molecules leaving the short tube orifice may be calculated and is indicated schematically. X is the single-crystal adsorbent. (From Madey.[31] Copyright North-Holland Physics Publishing, Amsterdam, 1972.)

the crystal size and distance are known, then the absolute surface coverage and sticking coefficient may be continuously determined from measurements of *relative* pressure during adsorption. This method requires neither absolute knowledge of detector sensitivity, nor system volume, nor pumping speed although all such parameters must remain constant during the measurement. A small correction must be made for multiple collisions of reflected molecules from the crystal that strike the front surface of the effusion plate and then return to the crystal. The absolute saturation coverage of hydrogen and oxygen on W(100) was determined[31,33] and has formed the basis for other absolute measurements of coverage. In addition, Wang and Gomer[34] have employed an apparatus similar to Fig. 5 containing a calibrated effusion source. Their absolute measurements of the saturation oxygen coverage on W(100) at 300 K [1.0×10^{15} O atoms cm^{-2}] agree well with the NBS results[31] of 1.14×10^{15} O atoms cm^{-2}. In this work the crystal temperature and the gas temperature may be independently varied.

Low-energy electron diffraction provides a convenient method for determining adsorbate overlayer periodicity. By measuring diffracted beam intensities as a function of exposure it is often found that intermediate surface structures yield maxima in intensity at exposures below saturation. If one *assumes* that these maxima in intensity correspond to a crystal surface completely filled with the ordered overlayer, then the absolute coverage at this point is assumed to be known. Comparison of the sequence of overlayer

---

[33] T. E. Madey, *Surf. Sci.* **36**, 281 (1973).
[34] C. Wang and R. Gomer, *Surf. Sci.* **84**, 329 (1979).

8.3. THERMAL DESORPTION FROM SINGLE CRYSTALS

structures achieved at various exposures with thermal desorption data can then lead to an absolute coverage scale. This comparison of LEED and thermal desorption coverages has been made by Erley et al.[35] for CO on Ni(111) and by Thiel et al.[36] for CO on Rh(111). An intermediate ($\sqrt{3} \times \sqrt{3}$)R30° structure corresponding theoretically to 0.33 CO/Rh evolved 0.38 CO/Rh in thermal desorption. For Rh(111) the final LEED pattern (2 × 2) (0.75 CO/Rh) corresponds to overlayer structures in which two modes of CO bonding (linear and bridged) are present.[36] Confirmation of this dual mode of CO bonding to Rh(111) has recently been made using electron energy loss spectroscopy, in which two vibrational frequencies for the C=O stretching motion have been observed.[37] This procedure involving LEED assignment of absolute surface coverages should be used cautiously, preferably in conjunction with other methods, including thermal desorption spectroscopy.

Several attempts have been made to measure absolute surface coverages of adsorbed layers on metals using radioactive isotopes and counting techniques involving Geiger–Müller counters.[38-42] Klier et al.[43a] employed ^{14}CO absorbate on Ni(110) and Ni(100) surfaces. On both surfaces the saturation CO coverage measured was about 1.1×10^{15} molecules cm^{-2} whereas the Ni surface density varied from 1.14×10^{15} Ni atoms cm^{-2} (110) to 1.61×10^{15} Ni atoms cm^{-2}(100). Klier et al. concluded that the saturation coverage of CO was dictated by the lateral repulsive CO–CO interactions rather than the number of underlying Ni atoms. They also employed the technique to measure the kinetics of adsorption and the rate of exchange between CO(ads) and CO(g). The isotopic counting technique has been checked by LEED measurements on Ni(100),[43b] giving 1.1×10^{15} CO molecules cm^{-2} as the saturation CO coverage, in remarkable agreement with the radioactive counting measurement.

Davies and Norton[44] have reported on nuclear reaction methods for determining absolute surface coverages of N, C, O, and D on single crystal metals. The accuracy of these methods appears to be comparable to other methods now in use.

[35] W. Erley, K. Besocke, and H. Wagner, *J. Chem. Phys.* **66**, (1977) 5269.
[36] P. A. Thiel, E. D. Williams, J. T. Yates, Jr., and W. H. Weinberg, *Surf. Sci.* **84**, 54 (1979).
[37] L. H. Dubois and G. A. Somorjai, *Surf. Sci.* **91**, 514 (1980).
[38] J. A. Dillon and H. E. Farnsworth, *Rev. Sci. Instrum.* **25**, 96 (1954).
[39] J. A. Dillon and H. E. Farnsworth, *J. Chem. Phys.* **22**, 1601 (1954).
[40] A. D. Crowell, *J. Chem. Phys.* **32**, 1576 (1960).
[41] A. D. Crowell and L. D. Matthews, *Surf. Sci.* **7**, 79 (1967).
[42] K. Klier, *Rev. Sci. Instrum.* **40**, 372 (1969).
[43] (a) K. Klier, A. C. Zettlemoyer, and H. Leidheiser, Jr., *J. Chem. Phys.* **52**, 589 (1970); (b) J. C. Tracy, *J. Chem. Phys.* **56**, 2736 (1972).
[44] J. A. Davies and P. R. Norton, *Nucl. Instrum. Methods* **168**, 611 (1980).

8.3.6. Detectors for the Study of Thermal Desorption

Two general methods for the study of the kinetics of desorption are commonly used. The first method, involving the measurement of gas density (and possibly gas composition) during desorption, was used by early workers and continues to be popular because of the ease of measurement. The second method, involving the measurement of surface coverage during desorption, was also used in early investigations of thermal desorption and has become more widely used because of certain advantages.

8.3.6.1. Detectors of Gas Evolution. 8.3.6.1.1. BAYARD–ALPERT GAUGE. The Bayard–Alpert ionization gauge was first used by Ehrlich[2] as a detector for thermal desorption measurements. Because of the rapid rate of gas evolution in his measurements, coupled with small volume and low pumping speed, large pressure pulses were evolved in times of the order of 1 s. To avoid ac noise, dc heating of the gauge filament was employed. In addition to this modification, the aperture between the glass vacuum system and the gauge tube was enlarged to prevent the formation of density gradients during the desorption. Accurate gauge constants were measured for various gases by calibration against an (absolute) McLeod gauge of special design. The design and calibration of ionization gauges and other gauges is well-reviewed by Redhead *et al.*[45] and will not be covered here.

8.3.6.1.2. MASS SPECTROMETERS. Mass spectrometers are now commonly used as detectors of desorption products in thermal desorption. The obvious advantage of qualitative and semiquantitative analysis of the desorbing gas assumes high importance when a mixture of desorption products is involved. A variety of types of mass spectrometers have been employed for thermal desorption mass spectroscopy including the omegatron spectrometer,[5,46] which works on a cyclotron resonance principle, the magnetic sector mass spectrometer,[47] the time-of-flight mass spectrometer,[48–50] and the quadrupole mass spectrometer. All of these mass spectrometers and others are discussed by Redhead.[45] Measurement of the time rate of change of several mass peaks simultaneously (multiplexing) has been achieved by interfacing a quadrupole mass spectrometer with a minicomputer.[51a,b] This is a particu-

---

[45] P. A. Redhead, J. P. Hobson, and E. V. Kornelsen, "The Physical Basis of Ultrahigh Vacuum." Chapman & Hall, London, 1968.

[46] T. E. Madey, J. T. Yates, Jr., and R. C. Stern, *J. Chem. Phys.* **42**, 1372 (1965).

[47] W. D. Davis and T. A. Vanderslice, *Trans. AVS Nat. Vac. Symp, 7th,* Washington, D. C., 417 (1960).

[48] B. McCarroll, *J. Chem. Phys.* **46**, 863 (1967).

[49] B. McCarroll, *J. Chem. Phys.* **41**, 5077 (1967).

[50] V. N. Ageev, N. I. Ionov, and Y. K. Ustinov, *Zh. Tekh. Fiz.* **34**, 546 (1964).

[51] (a) M. Bowker and R. Madix, *Surf. Sci.* **95**, 190 (1980); (b) J. L. Taylor, D. E. Ibbotson, and W. H. Weinberg, *Surf. Sci.* **90**, 37 (1979).

8.3. THERMAL DESORPTION FROM SINGLE CRYSTALS

larly convenient procedure when studying complex adsorbate molecules that decompose to yield several desorbing products.

One of the problems in thermal desorption in large metal vacuum systems is the degradation in resolution caused by desorbing gas interaction with the walls and pumps in the system. This may lead to changes in desorption peak shapes, changes in apparent yield of desorbing species, and even changes in the chemical identity of desorbing species. A clever modification[52] to partially eliminate some of these effects is shown in Fig. 11. A glass envelope surrounds the ionization source of a quadrupole mass spectrometer, partially shielding it from the random flux of gas contained in the vacuum chamber as a result of desorption from the single crystal. A small orifice, about one-half the diameter of the single crystal is placed about 0.2 cm from the front surface of the crystal. Desorbing gas from the center portion of the crystal enters the mass spectrometer directly, whereas random flux reflected from the vacuum system walls is excluded to a high degree. Thus, this method permits sampling desorbing gas by line-of-sight collection from the uniform center portion of a single crystal. The method provides improved resolution of closely spaced desorption states. Because of the small pumping speed out of the glass envelope, heating rates must be kept low to avoid pressure buildup in the envelope in order to apply Eq. (8.2.6). Alternatively, the full mass balance Eq. (8.2.4) could be applied to determine desorption kinetics accurately.

8.3.6.2. Detectors of Surface Coverage Changes. The direct measurement of surface coverage changes during isothermal desorption was first

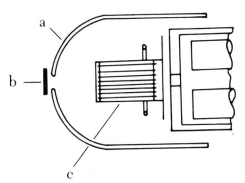

FIG. 11. Arrangement for counteracting the effects of gas with walls and pumps of system. The glass shield enhances direct transmission of desorbing gas species to the mass spectrometer. (a) Glass shield, (b) single crystal, (c) ionization source of quadrupole mass spectrometer. (From Feulner and Menzel.[52])

[52] P. Feulner and D. Menzel, *J. Vac. Sci. Technol.* **17**, 662 (1980).

employed on a single crystal by Tamm and Schmidt,[9] who studied H_2 desorption from W(100). They simply heated the fully covered crystal to a given steady temperature, allowed isothermal desorption to occur for various times, and then raised the temperature to desorb the remaining H_2. Plots of the coverage versus time gave kinetic parameters (pre-exponential factor, activation energy) in good agreement with those obtained by temperature programmed desorption measurements. Yates and King[53] used electron stimulated desorption (ESD) of CO^+ from adsorbed CO on W(100) to derive the desorption kinetics for a CO state designated α_1-CO. In Fig. 12 it may be seen that the behavior of the CO^+ signal may be used to infer the desorption kinetics for the α_1-CO state. Thermal conversion of α_1-CO to an

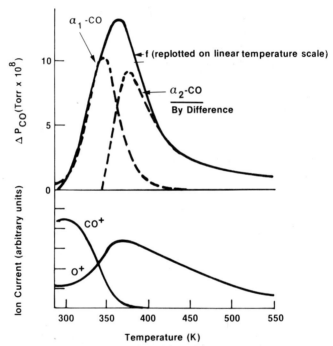

FIG. 12. Comparison of temperature-programmed desorption and electron-stimulated desorption behavior of CO on W(100). The desorption peak for α_1-CO has been calculated from the CO^+ time dependence. The parameters for this plot are $\nu = 3.0 \times 10^9$ s^{-1}, $E_d = 63,200$ J mole^{-1}, $\beta = dT/dt = 9.1$ K s^{-1}, and $\tau = V/\mathcal{S} = 1.76$ s. (The kinetic parameters calculated for the α_1-CO state may not be strictly valid, since it has not been shown that the α_1-CO state desorbs with constant E_d over its total coverage range.) (From Yates and King.[53] Copyright North-Holland Physics Publishing, Amsterdam, 1972.)

[53] J. T. Yates, Jr. and D. A. King, *Surf. Sci.* **32**, 479 (1972).

α_2-CO state may be followed by the production of an O^+-yielding state as shown.

Auger spectroscopy has been used to monitor the isothermal desorption of Pb from various Cu surfaces.[54] Zero-order kinetics are observed and interpreted as being due to the involvement of Pb islands down to low coverages. Pfnur et al.[55] have employed a rapid work function measurement technique to measure CO surface coverage changes in a temperature-programmed desorption experiment from Ru(001). A vibrating capacitor technique was employed[56] and care was taken to demonstrate that the work function was truly a measure of surface coverage under the conditions of the experiment. The method was also employed to measure isothermal desorption behavior as well as isosteric heats of adsorption under equilibrium conditions.[55] As will be discussed in a later section, results with the work function detector have yielded kinetic parameters that differ from those reported previously.

8.4. Treatment of Experimental Desorption Data

8.4.1. Desorption Kinetics Using Gas Evolution Measurements (Constant Rate of Heating)

This treatment follows closely the pioneering work of Redhead.[6] We consider a surface desorbing a gas into a pumped ultrahigh-vacuum system, in which the average residence time for a gas molecule is τ. Equation (8.2.4) may be written in dimensionless form as

$$\frac{d\mathcal{P}}{dt} + \frac{\mathcal{P}}{\tau} = -\frac{1}{N_0}\frac{dN}{dt} \equiv g(T) \qquad (8.4.1)$$

where $\mathcal{P}$ is a dimensionless quantity equal to $P/P_{\infty,\max}$ where $P_{\infty,\max}$ would be the maximum pressure observed for desorption of N_0 molecules cm^{-2} into the same vacuum system with zero pumping speed ($\tau = \infty$). The number N_0 is the initial coverage on the surface prior to desorption.

For $\beta = dT/dt$, the linear temperature sweep rate, one obtains

$$P(T) = \exp\left(\frac{-T}{\beta\tau}\right)\frac{v_0^{(1)}}{\beta}$$

$$\left[\int_0^T \exp\left[\frac{T}{\beta\tau} - \frac{E_d}{RT} - \frac{v_0^{(1)}R}{\beta E_d}T^2\exp\left(\frac{-E_d}{RT}\right)\right]dT, \quad (8.4.2)\right.$$

---

[54] M. G. Barthes and G. E. Rhead, *Surf. Sci.* **80**, 421 (1979).

[55] H. Pfnur, P. Feulner, H. A. Engelhardt, and D. Menzel, *Chem. Phys. Lett.* **59**, 481 (1978); see also *Ned. Tijdschr. Vacuumtech.* **16**, 232 (1978).

[56] H. A. Engelhardt, P. Feulner, H. Pfnur, and D. Menzel, *J. Phys. E* **10**, 1133 (1977).

where $v_0^{(1)}$ is the first-order pre-exponential term (s^{-1}), E_d the activation energy for desorption, and R the gas constant. Here $P(T)$ is the normalized profile of a desorption spectrum characterized by β, τ, $v_0^{(1)}$ and E_d.

In order to determine the kinetic parameters E_d and $v_0^{(1)}$, it is convenient to refer to a simpler relationship derived by Redhead.[6] The temperature T_p, at which the rate of desorption is a maximum is related to E_d and $v_0^{(1)}$ as

$$\frac{E_d}{RT_p} = \ln \frac{v_0^{(1)} T_p}{\beta} - \ln \frac{E_d}{RT_p} \qquad (8.4.3)$$

Since the experimental desorption peak maximum $T_m \approx T_p$, one may determine *pairs* of the parameters $v_0^{(1)}$ and E_d which satisfy Eq. (8.4.3). Then desorption peak shapes may be generated from Eq. (8.4.2) using each pair of values for E_d and $v_0^{(1)}$. By matching the theoretical peaks to the experimental peak shape, one may deduce the best values of E_d and $v_0^{(1)}$ fitting the data. An example of this fit is shown in Fig. 13 for Xe desorption for W(111) at various

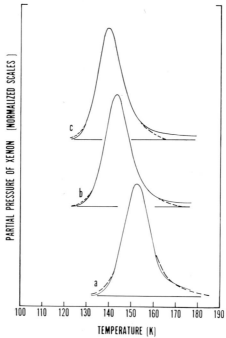

FIG. 13. Comparison of theoretical first-order desorption curves for Xe with experiment [Xe physisorbed on W(111)]. Initial fractional coverage of xenon θ_{Xe} = (a) 0.0025, (b) 0.021, (c) 0.134. E_d (kcal mole^{-1}) = (a) 10.23, (b) 9.61, (c) 9.34. $v = 10^{15}$ s^{-1}. Dotted line is for theory, solid for experiment. (From Dresser et al.[64] Copyright North-Holland Physics Publishing, Amsterdam, 1974.)

8.4. TREATMENT OF EXPERIMENTAL DESORPTION DATA

initial coverages. Chan et al.[57] have developed a method for determining both $v_0^{(n)}$ and E_d from measurement of T_p and the peak width, $\Delta T_{1/2}$, for both first- and second-order kinetics. It should be pointed out here that any fit of a thermal desorption peak to a kinetic model necessarily assumes that $v_0^{(n)}$ and E_d are *coverage independent* in the range of the coverage covered during the desorption experiment. This is often not the case and one should exercise care in interpreting thermal desorption data using the above methods if large adsorbate interactional effects are expected. This may be done by carefully measuring desorption kinetic parameters in a series of experiments done at different initial coverages.

8.4.2. Desorption Kinetics Using Gas Evolution Measurements (Variable Rate of Heating)

Taylor and Weinberg[58] have perfected a method for determining kinetic parameters in thermal desorption that involves changing the temperature program over a wide range of values of $\beta = dT/dt$. This method is illustrated for CO chemisorbed on Ir(110), as shown in Fig. 14. Here it can be seen that

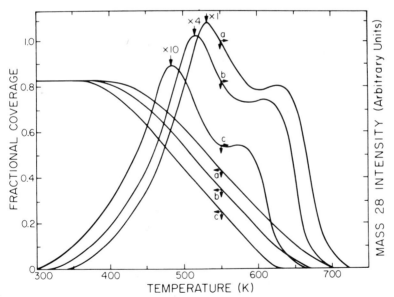

FIG. 14. Thermal desorption spectra for CO chemisorbed on Ir(110) for variable temperature sweep rates, starting at a fixed initial CO coverage. $T_0 = 300$ K. Sweep rates; (a) 212 K s^{-1}, (b) 78 K s^{-1}, (c) 13 K s^{-1}. (From Taylor and Weinberg.[58] Copyright North-Holland Physics Publishing, Amsterdam, 1978.)

[57] C. M. Chan, R. Aris, and W. H. Weinberg, *Appl. Surf. Sci.* **1**, 360 (1978).
[58] J. L. Taylor and W. H. Weinberg, *Surf. Sci.* **78**, 259 (1978).

450   8. THE THERMAL DESORPTION OF ADSORBED SPECIES

higher values of β lead to a shifting of the two desorption features to higher temperature. From integration of each of the thermal desorption curves, one may determine the fractional coverage on the surface as a function of temperature for each heating rate, as also is shown in Fig. 14. For a rapidly pumped vacuum system, the ordinate in the thermal desorption curves is proportional to the rate of desorption R_d. Thus, assuming a value for the

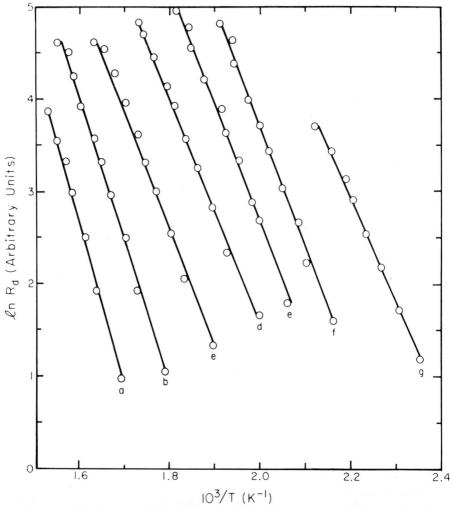

FIG. 15. Arrhenius plots for CO desorption from Ir(110) for various coverages: (a) 0.1, (b) 0.2, (c) 0.3, (d) 0.4, (e) 0.5, (f) 0.6, (g) 0.7. Data points were obtained from thermal desorption spectra like those in Fig. 14. (From Taylor and Weinberg.[58] Copyright North-Holland Publishing, Amsterdam, 1978.)

8.4. TREATMENT OF EXPERIMENTAL DESORPTION DATA

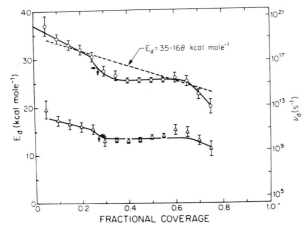

FIG. 16. Variation in the desorption energy and the first-order pre-exponential factor with CO surface coverage on Ir(110). (From Taylor and Weinberg.[58] Copyright North-Holland Physics Publishing, Amsterdam, 1978.)

desorption order n allows a direct determination of the pre-exponential factor $v_0^{(n)}$ and the activation energy E_d in Eq. (8.2.1), by plotting at constant coverage $\ln R_d$ versus $1/T$ in the usual Arrhenius fashion. This is illustrated in Fig. 15 for the CO/Ir(110) case, where each line represents a constant surface coverage. A plot of E_d and $v_0^{(1)}$ versus CO coverage is shown in Fig. 16. It may be seen that compensatory variations in E_d and $v_0^{(1)}$ occur over the coverage range investigated. It is believed in this case that the two desorption features are due to interactional effects between CO molecules rather than to more profound chemical or site selection effects.

8.4.3. Coverage Measurements Made during Programmed Desorption

Pfnur et al.[55] have used the vibrating capacitor method for measuring the instantaneous surface coverage of CO on Ru(001) during temperature programming. It was shown that the work function change $\Delta\phi$ was directly proportional to coverage of CO and is virtually independent of temperature. Therefore,

$$\dot\phi = \frac{d(\Delta\phi)}{dt} = k\frac{dN}{dt},$$

and one may employ Eq. (8.2.1) to determine E_d and $v_0^{(1)}$ as a function of coverage for this system by plotting $\ln \dot\phi/\phi$ versus $1/T$ for various constant coverages. Over a 1000-fold variation in the heating rate was used. Results of this study are described at the end of this article.

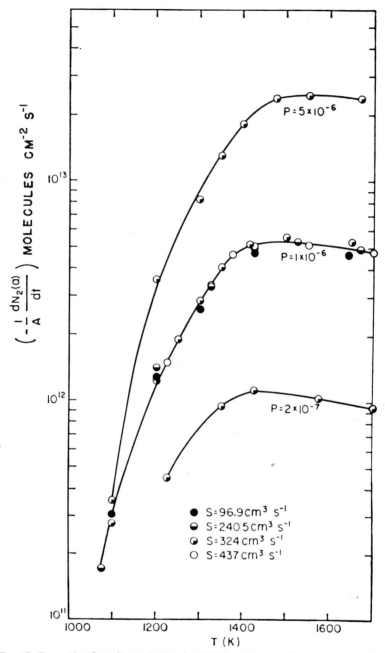

Fig. 17. Desorption flux of isotopically mixed N_2 from polycrystalline tungsten as a function of W temperature. The figure demonstrates that the measured flux was independent of flow rate S. In addition the flux of desorbing N_2 is directly proportional to the steady state pressure of N_2 about 1200 K. (From Madey and Yates.[5b])

8.4.4. Chemical Methods for Measuring Desorption Kinetics

In certain cases, chemical methods may be employed to measure desorption rates. This is illustrated by the work of Madey and Yates[5b] in which the rate of isotopic exchange between $^{14}N_2$ and $^{15}N_2$ was quantitatively studied on a polycrystalline tungsten surface. It was first demonstrated that N_2 desorbed from mixtures of $^{14}N_2$ and $^{15}N_2$ with complete isotopic mixing to yield statistical quantities of $^{14}N_2(g)$, $^{15}N\ ^{14}N(g)$, and $^{15}N_2(g)$. Under *flow* conditions in which $^{14}N_2(g)$ and $^{15}N_2(g)$ are admitted to the apparatus with the tungsten surface held at constant temperature, the steady-state mole

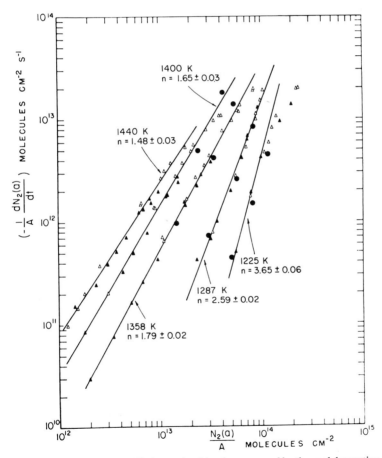

FIG. 18. Comparison between N_2 desorption kinetics measured by thermal desorption spectroscopy and kinetics measured by isotopic scrambling of $^{14}N_2$ and $^{15}N_2$. Triangles are TDS results; circles are isotopic mixing results. (From Madey and Yates.[5b])

fraction ^{29}X of $^{14}N\ ^{15}N(g)$ is a measure of the desorption flux (isothermal). The relevant equation is

$$F = \left(\frac{2\mathcal{S}\mathcal{P}}{AkT_g}\right)[(^{29}X - {}^{29}X_0)/(1 - 2\ ^{29}X_0)], \qquad (8.4.4)$$

where F is the desorption flux and $^{29}X_0$ is the concentration of $^{14}N\ ^{15}N(g)$ at steady state in the apparatus when only the mass spectrometer is operating. Figure 17 shows that the calculated value of F is well behaved for different rates of the pumping speed $\mathcal{S}$ at different steady-state pressures. These isotopic desorption flux measurements are compared with conventional temperature-programmed desorption results in Fig. 18 and it may be seen that the agreement is excellent. The use of isotopic mixing methods for determining desorption kinetic parameters is a relatively unexplored area of research at present.

8.5. Theories of Thermal Desorption

8.5.1. The Mobile Precursor Model in Adsorption and Desorption

A number of chemisorption studies on surfaces have shown that the sticking coefficient remains nearly constant (and often near unity) over a wide coverage range. This observation has led to the postulate that a mobile precursor state to chemisorption is present, and that this state may sample both filled and empty adsorption sites as it migrates, until finally becoming chemisorbed. This behavior on single crystals has been reviewed by Schmidt[59] and is demonstrated by typical sticking coefficient data shown in Fig. 19. King[60] first proposed that one should also consider passage through a mobile precursor state in thermal desorption, and he computed the sizable influence that this process might have on the shape of thermal desorption spectra. Gorte and Schmidt[61] have formulated a desorption kinetic model involving passage through a precursor state which is illustrated for first-order desorption kinetics below, and the general idea presented here is applicable to more complex situations. A one-dimensional potential energy curve is shown in Fig. 20 for the hypothetical situation in which a mobile precursor state exists at the top of the barrier at energy E_d.

For an adsorbed molecule A(s), desorption proceeds through the reaction steps

$$A(s) \underset{k_a}{\overset{k_d}{\rightleftharpoons}} A^* + \text{site} \overset{k^*}{\longrightarrow} A(g), \qquad (8.5.1)$$

---

[59] L. D. Schmidt, *Top. Appl. Phys.* **4**, 64 (1975); see also L. D. Schmidt, *Catal. Rev.—Sci. Eng.* **9**, 115 (1974).

[60] D. A. King, *Surf. Sci.* **64**, 43 (1977).

[61] R. Gorte and L. D. Schmidt, *Surf. Sci.* **76**, 559 (1978).

8.5. THEORIES OF THERMAL DESORPTION

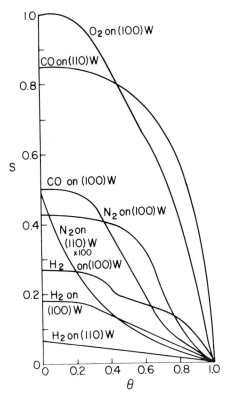

FIG. 19. Plot of sticking coefficient versus fractional coverage on several single crystal planes at 300 K. The shape of the curves (plateau region) in many cases is more significant than the absolute value of S.

where the k's represent first-order rate constants for the elementary steps shown.

If the precursor A* is in equilibrium with A(s) (at coverage θ_s) and is at a low steady-state concentration during desorption, one may write

$$d\theta_{A*}/dt = k_d\theta_s - k_a\theta_{A*}(1-\theta_s) - k^*\theta_{A*} = 0, \quad (8.5.2)$$

and hence

$$d\theta_s/dt = -k^*\theta_{A*} = -k^*k_d\theta_s/[k^* + k_a(1-\theta_s)]. \quad (8.5.3)$$

For $k^* \gg k_a$ we obtain normal first-order desorption kinetics, viz.,

$$d\theta_s/dt = -k_d\theta_s.$$

For the case in which $k^* \ll k_a$, then

$$d\theta_s/dt = -k^*k_d\theta_s/[k_a(1-\theta_s)] \quad (8.5.4)$$

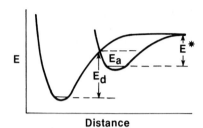

FIG. 20. One dimensional potential energy curves for chemisorbed species with precursor state at energy E_d.

and we introduce a factor $k^*/k_a(1 - \theta_s)$ into the rate expression. For this case, Eq. (8.5.4) becomes:

$$d\theta_s/dt = -[v^*v_d/v_a][\theta_s/(1 - \theta_s)]e^{-(E^*+E_d-E_a)/RT}. \quad (8.5.5)$$

The activation energy term in the exponential factor is just the total barrier height for the process going from A(s) to A(g), i.e., the heat of adsorption.

A comparison of theoretically calculated first-order temperature-programmed desorption behavior for a mobile precursor model compared to desorption via a strictly immobile activated complex is shown in Fig. 21. It can be seen that the $\theta/(1 - \theta)$ factor in Eq. (8.5.5) causes a significant broadening of the desorption peak as well as a shift to lower temperature. Thus, if a mobile precursor is involved in desorption, estimates of pre-exponential

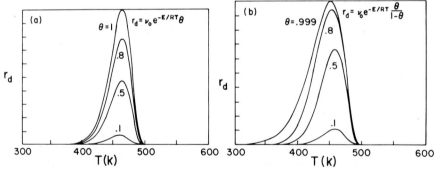

FIG. 21. Calculated thermal desorption spectra assuming $E_d = 26$ kcal mole^{-1}, $v_0^{(1)} = 10^{13}$ s^{-1} and $\beta = 100$ K s^{-1}. (a) Normal first order desorption. (b) Limiting case for single precursor [Eq. (8.5.5)]. (From Gorte and Schmidt.[61] Copyright North-Holland Physics Publishing, Amsterdam, 1978.)

8.5. THEORIES OF THERMAL DESORPTION

factors and desorption energies from peak shapes will be inexact[62] due to the dependence of these parameters on peak shapes, as was demonstrated by Yates et al.[62] in the case of H_2 desorption from Rh(111).

As shown by Gorte and Schmidt,[61] dissociative adsorption, followed by second-order desorption via a mobile precursor will introduce a $\theta^2/(1-\theta)^2$ term into Eq. (8.5.5). This will cause desorption peak broadening and shifting to lower temperatures also, and examples are given by Gorte and Schmidt.[61]

It is clear that mobility of the activated complex is an important factor to consider in any kinetic measurement of thermal desorption. As Gorte and Schmidt[61] point out, "the absence of a precursor in condensation at low temperature implies the absence of a precursor expression for desorption, and the presence of a precursor in condensation suggests that these expressions might be anticipated in desorption."

8.5.2. Statistical Thermodynamics of Adsorption and Desorption

The rates of adsorption and desorption processes on surfaces may be conceptualized within the framework of the Eyring theory of reaction rates.[63] This theory postulates that the reaction (adsorption or desorption) proceeds via an energetically activated state (or complex) that is intermediate in structure between the reactants and products in the process under consideration. It exists at the top of a potential energy barrier whose height is the activation energy for the reaction process. Passage over the energy barrier occurs by motion along a path called the reaction coordinate that describes the molecular configuration of the reactants and products. It is proposed that the activated complex at the top of the barrier exists in low concentration *in equilibrium* with the reactants, permitting one to apply statistical theory to the problem.

For adsorption on a uniform surface, let N^* be the equilibrium concentration of activated complexes (molecules cm^{-2}), N_s the number of adsorption sites per square centimeter, and N_g the number of gas-phase molecules per cubic centimeter. For an adsorption process in which a molecule is adsorbed without dissociation on a surface, one may write an equilibrium constant involving the reactants and activated complex as:

$$K^*(\text{eq}) = N^*/N_g N_s = f^*/f_g f_s, \qquad (8.5.6)$$

where the three f terms above are the *complete* partition functions for the species or the site, viz., $f = \sum_i g_i e^{-\epsilon_i/kT}$ with g_i being the degeneracy of the

---

[62] J. T. Yates, Jr., P. A. Thiel, and W. H. Weinberg, *Surf. Sci.* **84**, 427 (1979).

[63] S. Glasstone, K. J. Laidler, and H. Eyring, "The Theory of Rate Processes." McGraw-Hill, New York, 1941.

quantum state of energy ϵ_i. Thus,

$$N^* = f^{*\prime} N_g N_s / f_g f_s. \tag{8.5.7}$$

The rate of the reaction (adsorption in this case) is equal to the concentration of activated complexes N^* multiplied by the frequency of crossing of the barrier. Let us assume that the activated complex exists in a region of length δ along the reaction coordinate at the top of the barrier. For translation over the barrier the average velocity $\bar{v}$, determined from Maxwell–Boltzmann statistics for a one-dimensional problem, is

$$\bar{v} = (kT/2\pi m^*)^{1/2}, \tag{8.5.8}$$

where m^* is the effective mass of the complex. The average time τ of crossing the barrier is

$$\tau = \delta/\bar{v} = \delta(2\pi m^*/kT)^{1/2} \tag{8.5.9}$$

Therefore it follows that the rate of transmission over the barrier is

$$\mathcal{K}N^*/\tau, \tag{8.5.10}$$

where $\mathcal{K}$ is a transmission coefficient reflecting the probability that the activated complex will pass over the potential barrier to product. Thus,

$$-\frac{dN_g}{dt} = \mathcal{K}N^*/\tau = \mathcal{K}/\tau (f^{*\prime} N_g N_s / f_g f_s). \tag{8.5.11}$$

Now it is convenient to re-express $f^{*\prime}$, a complete partition function, as $f^{*\prime} = f^* f^*_{\text{trans},(1\text{-dim})}$ where f^* now has removed from it the one-dimensional translational partition function corresponding to motion along the reaction coordinate, over the barrier. From quantum mechanical considerations of a particle in a one-dimensional box of length δ,

$$f^*_{\text{trans},(1\text{-dim})} = (2\pi m^* kT)^{1/2} \delta/h. \tag{8.5.12}$$

Therefore, rewriting equation (21), we obtain

$$-dN_g/dt = \mathcal{K}kT/h (f^* N_g N_s)/f_g f_s). \tag{8.5.13}$$

It is convenient to extract from the partition function f^* the zero-point energy of the initial state of the system by making this energy the arbitrary zero reference energy, and redefining f^* on this basis. Thus,

$$-\frac{dN_g}{dt} = \mathcal{K} \frac{kT}{h} \frac{f^*}{f_g f_s} N_g N_s e^{-\epsilon_1/kT}, \tag{8.5.14}$$

where ϵ_1 is the difference in zero-point energy for the reactant and the activated complex, i.e., the activation energy for the process. A continuation

8.5. THEORIES OF THERMAL DESORPTION

of this treatment for the cases of immobile and mobile adsorption is given by Glasstone et al.[63]

By a similar argument, the rate of first-order desorption via an activated complex is given by

$$\frac{dN_g}{dt} = \mathcal{H} \frac{kT}{h} \frac{f^*}{f_a} N_a e^{-\epsilon_2/kT}, \tag{8.5.15}$$

where N_a and f_a refer to the adsorbed species, and ϵ_2 is the activation energy for desorption of a single molecule, referred to the zero-point energy of the adsorbed species. For simple first-order kinetics of desorption in which *both* the adsorbate species and the activated complex are considered to be immobile, the ratio $f^*/f_a \cong 1$, and we are left with a pre-exponential factor of $\mathcal{H}kT/h$ for the desorption process. Many experiments have shown that first-order desorption processes often exhibit pre-exponential factors of this order of magnitude.

We may now convert our formulation of the Eyring theory of activated desorption into thermodynamic terms by remembering that $K^*_{(eq)} \equiv e^{-\Delta G_0^*/RT}$ where $\Delta G_0^* \equiv \Delta H_0^* - T\Delta S_0^*$, where ΔH_0^* and ΔS_0^* are respectively the activation enthalpy and entropy for formation of the activated complex from the reactants. Therefore, for first-order desorption, with

$$K^*_{(eq)} = \frac{N^*}{N_a} = \frac{f^*}{f_a} e^{-\epsilon_2/kT}, \tag{8.5.16}$$

we rewrite Eq. (8.5.14) as

$$\frac{dN_g}{dt} = \mathcal{H} \frac{kT}{h} e^{\Delta S_0^*/R} e^{-\Delta H_0^*/RT} N_a. \tag{8.5.17}$$

For higher-order desorption, of order n, Eq. (8.5.16) may be rewritten as a dimensionless equilibrium constant,

$$K^*_{(eq)} = N^* N_s^{(n-1)}/N_a^n, \tag{8.5.18}$$

where N_s is the number of sites per square centimeter. This produces the general equation for the nth order desorption rate within the Eyring framework:

$$\frac{dN_g}{dT} = \frac{\mathcal{H}}{N_s^{(n-1)}} \frac{kT}{h} e^{\Delta S_0^*/R} e^{-\Delta H_0^*/RT} N_a^n. \tag{8.5.19}$$

Since $N_s \approx 10^{15}$ cm^{-2} for most adsorbents, we may calculate approximate values of the pre-exponential factor for desorption. For little or no activation entropy, $\Delta S_0^* \approx 0$, and for $\mathcal{H} \approx 1$, the pre-exponential factor $\nu_0^{(n)}$ is

$$\nu_0^{(n)} = kT/hN_s^{(n-1)}. \tag{8.5.20}$$

In the temperature range 50–2000 K, for first-order kinetics,

$$10^{12} \lesssim v_0^{(1)}(=kT/h) \lesssim 4 \times 10^{13} \text{ s}^{-1},$$

whereas for second-order kinetics,

$$10^{-3} \lesssim v_0^{(2)}\left(=\frac{kT}{hN_s}\right) \lesssim 4 \times 10^{-2} \text{ cm}^2 \text{ s}^{-1}.$$

We may ask what will happen to the pre-exponential factor for nth-order desorption, $v_0^{(n)}$, when the activated complex is given two degrees of *translational freedom*. The two-dimensional translational partition function for a single particle is

$$f^*_{(2D,\text{tr})} = \frac{2\pi mkT}{h^2} \mathcal{A} \tag{8.5.21}$$

where $\mathcal{A}$ is the total surface area available to the translating complex. The canonical ensemble partition function for N indistinguishable and independent particles is

$$\mathcal{F}^*_{(2D,\text{tr})} = \frac{1}{N!}(f^*_{(2D,\text{tr})})^N. \tag{8.5.22}$$

By definition, the Helmholtz free energy, $A^* \equiv E^* - TS^*$, is given by

$$A^* \equiv -kT \ln \mathcal{F}^* = -NkT \ln\left(\frac{2\pi mkT}{h^2} \frac{A e}{N}\right).$$

For a two-dimensional gas, $E^* = NkT$, and therefore

$$S^*_{(2D,\text{tr})} = \frac{E^* - A^*}{T} = 2Nk + Nk \ln\left(\frac{2\pi mkT}{h^2} \frac{A}{N}\right) \tag{8.5.23}$$

For a molecule of mass = 30 amu, and assuming

$$A/N \approx 10^{-15} \text{ cm}^2, \quad T = 300 \text{ K}, \quad \text{and}$$
$$Nk = R = 1.986 \text{ cal K}^{-1} \text{ mole}^{-1}, \tag{8.5.24}$$

the term $e^{\Delta S_0^*/R}$ in Eq. (8.5.19) will therefore cause a factor of $10^{2.9}$ increase in the pre-exponential factor due to complete two-dimensional translational freedom for the activated complex in desorption, if we assume an immobile adsorbate.

Additional enhancement of the pre-exponential factor in desorption can occur if other degrees of freedom are permitted for the activated complex in desorption. For example, for a hypothetical diatomic molecule that freely rotates in the activated complex but is restrained from rotation in the chemisorbed state,

$$f^*_{rot} = 8\pi^2 IkT/\sigma h^2, \qquad (8.5.25)$$

where I is the moment of inertia and σ the symmetry number, $=1$ for homonuclear diatomics and $=2$ for heteronuclear diatomics. Following the procedure outlined in Eqs. (8.5.21) through (8.5.23), and defining the canonical ensemble rotational partition function as

$$\mathscr{F}^*_{rot} = (f^*_{rot})^N \qquad (8.5.26)$$

we find that

$$S^*_{rot} = \Delta S^*_{0(rot)} = Nk \ln\left(\frac{8\pi^2 IkTe}{\sigma h^2}\right). \qquad (8.5.27)$$

For a homonuclear molecule of mass 30 and internuclear separation of 10^{-8} cm,

$$\Delta S^*_{0(rot)} = 9.60 \quad \text{cal K}^{-1} \text{ mole}^{-1}.$$

This will result in a further contribution of a factor of $10^{2.1}$ to the pre-exponential factor $e^{\Delta S^*_0/R}$ in Eq. (8.5.19) due to complete rotational freedom for the activated complex. Thus within the framework of the Eyring theory, in which an immobile adsorbate of mass $=30$ desorbs via a completely mobile and freely rotating activated complex,

$$10^{17} \lesssim \nu_0^{(1)} \lesssim 10^{18.6} \qquad (T = 50-2000 \text{ K})$$

and

$$10^2 \lesssim \nu_0^{(2)} \lesssim 10^{3.6} \qquad (T = 50-2000 \text{ K}).$$

Several examples of anomalously high pre-exponential factors in thermal desorption spectroscopy from single crystals have been observed. For example, physisorbed Xe on W(111) exhibits first-order desorption kinetics with a pre-exponential factor of 10^{15} s^{-1}, as reported in 1974 by Dresser et al.[64] The interpretation of these results involved the postulate of an activated complex with two degrees of translational freedom parallel to the surface, with an essentially immobile Xe adsorbate at 140 K. A comparison of the desorption data with theoretical desorption curves is shown in Fig. 13 for three Xe coverages.

Another example of an anaomalously high pre-exponential factor for first-order desorption comes from the work of Falconer and Madix[65] who

---

[64] M. J. Dresser, T. E. Madey, and J. T. Yates, Jr., *Surf. Sci.* **42**, 533 (1974).
[65] J. Falconer and R. Madix, *Surf. Sci.* **48**, 393 (1975).

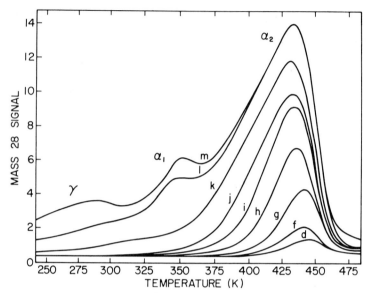

FIG. 22. Temperature programmed desorption spectra of CO from Ni(110). θ_{CO} ranges from 0.026 (curve d) to 1.00 (curve m). (From Falconer and Madix.[65] Copyright North-Holland Physics Publishing, Amsterdam, 1975.)

studied CO desorption from Ni(110). Although there was obvious overlap of desorption states (Fig. 22) a pre-exponential factor of 8.5×10^{15} s^{-1} was required to give the fit shown in the figure for first-order desorption of the high-temperature CO state. This high pre-exponential factor was confirmed by Helms and Madix[66] using molecular-beam methods.

Finally, Comrie and Weinberg[67] reported a disagreement between thermal desorption energetics measured by temperature-programmed desorption and by isosteric heat measurements for CO adsorbed on Ir(111). This difference was attributed to a high pre-exponential factor (2.4×10^{14} s^{-1}) for first-order desorption of the higher energy CO state.

In a very careful study, Pfnur et al.[55] have reported very high pre-exponential factors for the first-order desorption of a high-energy CO state from Ru(001). Values dependent on coverage range from 10^{16} to $10^{19.5}$ s^{-1} as shown in Fig. 23. It is not clear at present whether additional factors must be included to explain the very high $10^{19.5}$ s^{-1} value.

[66] C. Helms and R. Madix, *Surf. Sci.* **52**, 677 (1975).
[67] C. Comrie and W. H. Weinberg, *J. Chem. Phys.* **64**, 250 (1976).

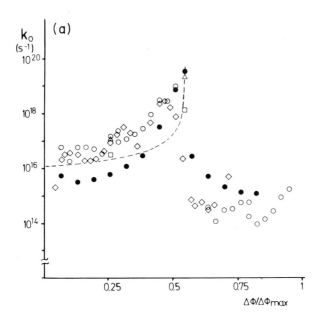
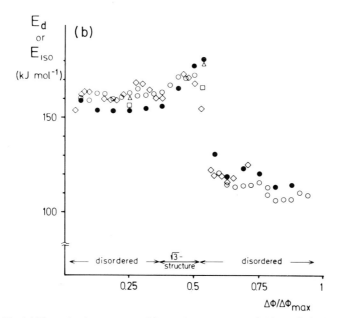

Fig. 23. (a) The activation energy and first order pre-exponential factor k_0 for CO desorption from Ru(001). (b) The range of existence of an ordered layer of CO under desorption conditions. The various types of data points represent measurements made by independent experimental methods. $\Delta\phi/\Delta\phi_{max}$ is a measure of the CO coverage. (From Pfnur et al.[55] Copyright North-Holland Physics Publishing, Amsterdam, 1978.)

8.6. Concluding Remarks

The use of thermal desorption methods for the study of species adsorbed on surfaces is widespread. The study of thermal desorption supplies useful information to basic science as well as information of importance to technology. As shown in Fig. 1 we have witnessed a doubling in rate of publication in this field approximately every 3 years during the past decade. A number of current reviews of the field are available for the reader who wishes to have additional information.[68-72] It is anticipated that, in the future, further refinement in experimental methods for studying desorption processes will be devised, so that thermal desorption may be understood on a more fundamental molecular basis. In addition, continued widespread application to technological problems is anticipated, particularly to high-area powdered substrates. These applications have not been treated in this article.

[68] V. N. Ageev and N. I. Ionov, *Prog. Surf. Sci.* **5,** Part 1 (1974).
[69] D. Menzel, *Top. Appl. Phys.* **4,** Part 4 (1975).
[70] L. D. Schmidt, *Catal. Rev.—Sci. Eng.* **9,** 115 (1974).
[71] D. A. King, *CRC Crit. Rev. Solid State Mater. Sci.* **7**(3) (1978).
[72] T. E. Madey and J. T. Yates, Jr., *Surf. Sci.* **63,** 203 (1977).

9. EXPERIMENTAL METHODS IN ELECTRON- AND PHOTON-STIMULATED DESORPTION

By Theodore E. Madey and Roger Stockbauer

Surface Science Division
National Bureau of Standards
Gaithersburg, Maryland

Since the early 1970s a number of review articles have surveyed both theoretical developments and experimental advances in electron-stimulated desorption (ESD) and photon-stimulated desorption (PSD).[1-7] In the present review effort, we shall not attempt to reproduce these considerable efforts. Rather, we shall concentrate on the basic experimental methodology of ESD and PSD: experimental techniques, applications and examples developed and employed during the past few years will be described. In addition, we shall discuss ESD/PSD as perturbing effects in surface analysis.

In Chapter 9.1 a survey of the present state of ESD and PSD theory will be presented. Since this is basically a "methods" chapter, this discussion will not be extensive, and the reader will be guided to the appropriate literature. In Chapter 9.2 experimental methods will be discussed in detail. The emphasis will be on ESD/PSD applications to adsorbed monolayers on surfaces but we shall close with a discussion of beam effects in surface analysis that involve "deeper" damage.

9.1. Theory and Mechanisms of Electron-Stimulated Desorption and the Relationship between ESD and PSD

9.1.1. Experimental Observations

When a surface containing an adsorbed layer is bombarded by electrons or photons at energies greater than some threshold value, usually $\gtrsim 10$ eV, electronic excitations in the adsorbed layer may result in the desorption of

---

[1] T. E. Madey and J. T. Yates, Jr., *J. Vac. Sci. Technol.* **8**, 525 (1971).
[2] D. Menzel, *Surf. Sci.* **47**, 370 (1975); *Top. Appl. Phys.* **4**, 101 (1975).
[3] D. Menzel, *J. Vac. Sci. Technol.* **20**, 538 (1982).
[4] V. N. Ageev and N. I. Ionov, *Prog. Surf. Sci.* **5**, Part 1, 1 (1974).

ions and neutral fragments (both ground state and metastable species). In addition, dissociation and/or polymerization processes may be initiated by electron or photon bombardment. For electrons or photons in the range 10–1000 eV incident upon metal surfaces containing monolayer quantities of adsorbed gases, the general experimental observations are as follows.[2,3,6]

(1) The cross sections for most ESD/PSD processes are smaller than comparable gas-phase processes involving electron- or photon-induced dissociation and ionization of molecules. For 100-eV electrons, typical gas-phase dissociative-ionization cross sections for small molecules are $\sim 10^{-16}$ cm^2. Cross sections for ESD of most adsorbates lie in the range $10^{-18}-10^{-23}$ cm^2, with both lower and higher values observed for certain systems. Also, the cross sections for neutral desorption are typically larger than the cross sections for desorption of ions. Although ions frequently comprise a small fraction of the desorbing species, the ease of experimental detection of ions means that their desorption characteristics are the most frequently studied ESD processes.

(2) Mostly positive atomic ions are seen, with H^+, O^+, F^+, and Cl^+ being the most abundant. Negative ions of each of these four atoms have also been detected,[8,9] but their yields are generally about $\frac{1}{100}$ of that for positive ions. The most common molecular ions reported from ESD of adsorbed monolayers are CO^+, OH^+, and OH^-.[1-3,10] Electron-stimulated desorption of multilayers can result in desorption of more complex molecular ions (e.g., $C_6H_{12}^+$ from a multilayer of C_6H_{12},[11] $H^+(H_2O)_n$ from thick films of ice[12])

(3) Cross sections for ESD and PSD are very sensitive to the mode of bonding. In general, cross sections for rupture of an internal bond in a weakly adsorbed molecule (the C–O bond in adsorbed CO, the H—O bond in adsorbed H_2O) are higher than the cross sections for rupture of a metal–atom bond (low coverages of hydrogen or oxygen on tungsten).

(4) The binding energies of most adsorbates are sufficiently large that direct momentum transfer between electron and adsorbate does not provide sufficient energy to cause desorption of neutral species. Power densities in most ESD studies are sufficiently low that thermal heating of the substrate

---

[5] M. J. Drinkwine, Y. Shapira, and D. Lichtman, in "Radiation Effects on Solid Surfaces" (M. Kaminsky, ed.), Advances in Chemistry Series, No. 158, p. 171. Am. Chem. Soc., Washington. D.C., 1976; see also M. J. Drinkwine and D. Lichtman, Prog. Surf. Sci. 8, 123 (1977).
[6] T. E. Madey and J. T. Yates, Jr., Surf. Sci. 63, 203 (1977).
[7] E. Bauer, J. Electron Spectrosc. Relat. Phenom. 15, 119 (1979).
[8] J. L. Hock, J. H. Craig, and D. Lichtman, Surf. Sci. 85, 101 (1979).
[9] M. L. Yu, Phys. Rev. B 19, 5995 (1979).
[10] M. L. Knotek and P. J. Feibelman, Phys. Rev. Lett. 40, 964 (1978).
[11] T. E. Madey and J. T. Yates, Jr., Surf. Sci. 76, 397 (1978).
[12] R. H. Prince and G. R. Floyd, Chem. Phys. Lett. 43, 326 (1976).

does not induce desorption. These observations, coupled with the fact that ions are generally desorbed having most probable kinetic energies in the range 0–10 eV, indicate that ESD and PSD proceed via an electronic excitation mechanism.

(5) Positive ions liberated from surfaces by ESD and PSD often desorb in sharp cones of emission in specific directions related to the symmetry of the substrate and the structure of adsorbed molecules and molecular complexes. The observation of ESD ion angular distributions (ESDIAD) has resulted in the development of a new method for surface structure characterization.[6,13–15]

9.1.2. Mechanisms of Ion Formation and Desorption

In the following paragraphs, we present a brief introduction to the theoretical basis of ESD and PSD. For further details, the interested reader is referred to a number of recent articles describing the current status of theory.[16–19]

We begin with a comparison of ESD and PSD, then discuss models for ion formation and desorption, and close with a discussion of the experimental basis for the ESDIAD method for determining surface molecular structure.

In most cases, the electronic excitations that cause desorption of surface species can be initiated by either electron or photon bombardment. The equivalence of the electron and photon excitation processes has been well documented by the agreement of desorption thresholds, ion energy and angular distributions, and the nature of the surface species from which desorption occurs.[20–21] Nevertheless, differences do appear in the spectral shape of the ion yield versus excitation energy under photon or electron

---

[13] T. E. Madey and J. T. Yates, Jr., *Chem. Phys. Lett.* **51**, 77 (1977); see also T. E. Madey and J. T. Yates, Jr., *Proc. 7th Int. Vac. Congr., Vienna*, p. 1183 (1977).

[14] J. C. Czyzewski, T. E. Madey, and J. T. Yates, Jr., *Phys. Rev. Lett.* **32**, 777 (1974).

[15] T. E. Madey, in Inelastic Ion–Surface Collisions" (W. Heiland and E. Taglauer, eds.), p. 80. Springer-Verlag, Berlin and New York, 1981; *Surf. Sci.* **79**, 575 (1979).

[16] N. H. Tolk, M. M. Traum, J. C. Tully, and T. E. Madey, eds., "Desorption Induced by Electronic Transitions," (Springer-Verlag, Heidelberg, 1983) and references therein.

[17] (a) D. R. Jennison, p. 26, (b) J. C. Tully, p. 31, (c) R. Gomer, p. 40, (d) P. J. Feibelman, p. 61, and (e) D. E. Ramaker, p. 70, in "Desorption Induced by Electronic Transitions" (N. H. Tolk, M. M. Traum, J. C. Tully, and T. E. Madey, eds.), (Springer-Verlag, Berlin and New York, 1983).

[18] D. E. Ramaker, *J. Vac. Sci. Technol.* **A1**, 1137 (1983).

[19] D. R. Jennison and D. Emin, *Phys. Rev. Lett.* **51**, 1390 (1983).

[20] (a) M. L. Knotek, in "Desorption Induced by Electronic Transitions" (N. H. Tolk, M. M. Traum, J. C. Tully, and T. E. Madey, eds.) p. 139. (Springer-Verlag, Berlin and New York, 1983). (b) M. L. Knotek, V. O. Jones, and V. Rehn, *Phys. Rev. Lett.* **43**, 300 (1979).

[21] T. E. Madey, R. Stockbauer, J. F. van der Veen, and D. E. Eastman, *Phys Rev. Lett.* **45**, 187 (1980).

excitation. These different spectral dependences reflect the differences in the physics of the initial excitation process itself.

In general, there is a much slower rise in the electron excitation cross section just above threshold relative to the photon cross section results, because the primary electron must have an empty final state to occupy in the solid. In addition, an electron beam acts essentially as a *white radiation* source for exciting both dipole allowed and dipole forbidden excitations over a wide spectral range. Thus PSD excited by monochromatic photons exhibits much sharper thresholds than ESD, and the spectral line shape may in some instances be expected to exhibit directly the final state density information. ESD has much weaker thresholds, with the spectrum spread out over a much larger energy and with the small dipole-forbidden contributions superimposed on the larger dipole-allowed "step-function" contributions. This makes it more difficult to interpret and extract quantitative information from ESD spectra than from PSD spectra.[20,22]

The stimulated desorption process can be described in a very general way as a sequence of three steps:[22,23]

(1) a fast initial electronic excitation ($\sim 10^{-16}$s) leading to a repulsive excited state,

(2) decay of the excited state by conversion of electronic energy to nuclear motion but in competition with other decay mechanisms that redistribute the electronic energy ($10^{-15}-10^{-14}$s), and

(3) a modification of the desorbing species (e.g., energy, charge state, trajectory) as it recedes from the surface ($10^{-14}-10^{-13}$s).

Our knowledge of desorption processes is based, in large part, on examination of the desorbed species:[22,23] their identities, angular and energy distributions, charge states, and electronic and vibrational energy distributions. All of these quantities can provide clues to the desorption mechanisms, but modifications of the desorbing species (step 3) can obscure the interpretation of the interesting dynamics which occurs in step 2.

Electron- and photon-stimulated desorption are usually discussed in terms of specific models, but they all are consistent with the above general picture. The first of these is known as the Menzel–Gomer–Redhead model,[24,25] illustrated in Fig. 1. In this model the primary process is a Franck–Condon excitation or ionization to a repulsive neutral or ionic state. Although the actual nature of the repulsive state is not usually speci-

---

[22] T. E. Madey, D. E. Ramaker, and R. Stockbauer, *Ann. Rev. Phys. Chem.* **35**, 215 (1984).

[23] J. C. Tully, *in* "Desorption Induced by Electronic Transitions" (N. H. Tolk, M. M. Traum, J. C. Tully, and T. E. Madey, eds.) p. 31. (Springer-Verlag, Berlin and New York, 1983.)

[24] D. Menzel and R. Gomer, *J. Chem. Phys.* **41**, 3311 (1964); **41**, 3329 (1964).

[25] P. A. Redhead, *Can. J. Phys.* **42**, 886 (1964).

9.1. THEORY AND MECHANISMS OF STIMULATED DESORPTION

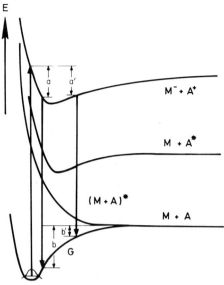

FIG. 1. Potential energy curves for adsorbate levels relevant for ESD. Abscissa: distance between adsorbate and substrate. G: adsorbate ground state; (M + A)*: antibonding adsorbate state; M + A*: excited adsorbate state; M⁻ + A⁺: ionized adsorbate state. Transitions are shown for ionic desorption only. Depending on the distance at which retunneling occurs, it leads to recapture in the ground state ($b > a$) or to desorption as a neutral particle ($b' < a'$). (From Menzel.[2] Copyright North-Holland Physics Publishing, Amsterdam, 1975.)

fied, the initial proposal was in the context of a valence excitation followed by reneutralization. As such, this view is supported by desorption thresholds corresponding to valence electron excitations and by the dominance of the neutral yield over the ion yield. Since the detailed nature of the electronic excitations is not specified, this model does not have predictive capabilities concerning the initial excited state or about the surface atomic arrangements required for desorption to occur.[22] A variation of the MGR desorption model has been proposed by Antoniewicz.[26] In this picture, the excited ion begins its trajectory by moving toward the surface where it is neutralized. After neutralization, the atom exists on a strongly repulsive potential curve leading to desorption most likely as a neutral. Evidence for this mechanism has been found in ESD of rare gases from metals.[27]

Another basic model for desorption has been formulated by Knotek and

---

[26] P. Antoneiwicz, *Phys. Rev. B* **21**, 3811 (1980).
[27] Q.-J. Zhang, R. Gomer, and D. R. Bowman, *Surf. Sci.* **129**, 535 (1983).

Feibelman (KF)[10,28,29] to explain ESD of positive ions from maximal valency transition metal oxides (i.e., oxides in which the metal cation is oxidized to the rare-gas configuration, e.g., Ti^{4+} in TiO_2). The model is based on ionization of a metal core level as the primary process (Fig. 2). The interatomic Auger decay of the core hole creates a positive anion at an initially negative ion site ($O^{--} \rightarrow O^+$ in Fig. 2). The expulsion of the positive ion results from the reversal of the Madelung potential. This mechanism explains the thresholds observed in many ESD/PSD experiments, as well as the large charge transfer occurring in these processes.

The Auger process has been demonstrated to be of importance in desorption of ions from both covalent systems and nonmaximal valency systems, but its role is somewhat different from the KF model. In covalent systems the Auger process must create two holes localized in a bonding orbital.[17-19] The expulsion of the positive ion still results from the unshielded nuclear–nuclear repulsion (also called hole–hole repulsion), but an additional condition must exist; the holes must remain localized for a sufficiently long time ($\sim 10^{-14}$ s) in the bonding orbital that initially held the atom to the surface so that electronic energy can be converted to nuclear motion. This mechanism is more appropriately called the Auger-stimulated desorption (ASD) model. It has been shown to describe desorption from nonmaximal valency surface

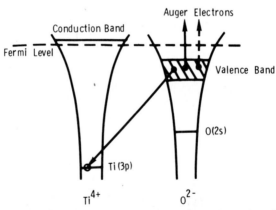

FIG. 2. Auger decay model for stimulated desorption of ions. Formation of a core hole in the 3p shell of maximal valency TiO_2 is followed by interatomic Auger decay leading to formation of an O^+ ion. (From Knotek and Feibelman.[28] Copyright North-Holland Physics Publishing, Amsterdam, 1979.)

[28] M. L. Knotek and P. J. Feibelman, *Surf. Sci.* **90**, 78 (1979).
[29] M. L. Knotek, *Phys. Scripta* **T6**, 94 (1983).

9.1. THEORY AND MECHANISMS OF STIMULATED DESORPTION

oxides[30] as well as molecular adsorbates (i.e., dissociation of OH on Ti, Cr, and Cu) as a result of metal core-hole excitation.[31] Multiple electron excitations resulting from Auger decay following deep core ionization of surface molecules (C 1s, O 1s, N 1s in adsorbed CO, N_2, NO) also result in ion desorption from adsorbed molecules.[32,33]

The measurement of ESD ion angular distributions (ESDIAD) have proven to be useful for determining the structures of surface molecules.[15,34] ESD and PSD ions desorb in discrete cones of emission, in directions determined by the orientation of the bonds that are ruptured by the excitation. For example, ESD of CO bound in a standing-up configuration on a metal surface will result in desorption of O^+ in the direction of the surface normal, whereas ESD of O^+ from "inclined" CO, or of H^+ from H_2O adsorbed via the O atom will result in desorption in off-normal directions (Fig. 3).

Based on a large body of experimental data,[34] it appears that for all molecular systems for which we have *a priori* knowledge of the surface structure, the ion-desorption angle observed in ESDIAD is related to the surface bond angle. In particular, it is found that the expected azimuthal angle of the surface bond is preserved in ESDIAD, but the polar angle is increased for the ion trajectory, due largely to image charge effects. To date, no exceptions have been found. Thus, measurements of the patterns of ion desorption

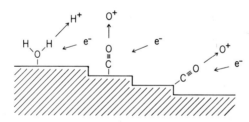

FIG. 3. Schematic bonding configuration showing relationship between surface bond angle and ion desorption angle in ESDIAD.

---

[30] R. Stockbauer, D. E. Ramaker, E. Bertel, R. L. Kurtz, and T. E. Madey, *J. Vac. Sci. Technol.* **A2**, 1053 (1984).

[31] E. Bertel, R. Stockbauer, R. L. Kurtz, D. E. Ramaker, and T. E. Madey, in preparation.

[32] R. Franchy and D. Menzel, *Phys. Rev. Lett.* **43**, 855 (1979); P. Feulner, R. Treichler, and D. Menzel, *Phys. Rev. B* **24**, 7427 (1981).

[33] J. E. Houston and T. E. Madey, *Phys. Rev. B* **26**, 554 (1982); T. A. Carlson and M. O. Krause, *J. Chem. Phys.* **56**, 3206 (1972).

[34] T. E. Madey, F. P. Netzer, J. E. Houston, D. M. Hanson, and R. Stockbauer, *in* "Desorption Induced by Electronic Transitions" (N. H. Tolk, M. M. Traum, J. C. Tully, and T. E. Madey, eds.), p. 120. (Springer-Verlag, Berlin and New York, 1983); T. E. Madey, D. L. Doering, E. Bertel, and R. Stockbauer, *Ultramicroscopy* **11**, 187 (1983); T. E. Madey and J. T. Yates, Jr., *Surf. Sci.* **63**, 203 (1977).

provide direct information about the geometrical structure of surface molecules in the adsorbed layer. ESDIAD is sensitive to the *local molecular structure,* so that long-range order is not necessary for determination of adsorbate geometry.

In support of the picture that ion-desorption angles are related to surface bond angles, Clinton[35] has shown that an ESD ion excited to a spherically symmetrical repulsive final state will experience an initial impulse in the direction of the original bond. Also, within the framework of the KF model and the multielectronic excitation model of ESD/PSD, the initial repulsive interaction is largely coulombic in origin, arising from hole–hole repulsion. Hence, the initial desorption direction is along the bond that is ruptured by the excitation. In addition, ion-desorption times ($\sim 10^{-14}-10^{-15}$ s) and electron tunneling times for reneutralization ($\sim 10^{-15}$ s) are so rapid with respect to molecular vibration times ($\sim 10^{-12}-10^{-13}$ s) that one would not expect significant molecular rearrangements to occur during desorption, i.e., the ion-desorption angle should reflect the surface bond angle.

There are, however, final state effects that can affect the ion-desorption trajectories and the resultant ESDIAD patterns. Some of these effects include anisotropy in the ion neutralization rate, "focusing" effects due to curvature in the final state potential, and deflection of the escaping ions by the electrostatic image potential.[34] In general, we have no detailed knowledge of the final state potentials, although model calculations have been performed for specific systems.[36,37] There are recent calculations[38-40] in which the influence of the image force and of ion neutralization processes on ion-desorption trajectories have been estimated. These calculations[38-40] have shown that the image potential invariably causes an increase in the polar desorption angle θ of an ion leaving a planar metal surface; the image potential does not influence the azimuthal angle ϕ. The amount of distortion of the ion trajectory by the image potential is directly related to the parameter $|V_I|/E_0$, which defines the "strength" of the image potential. (The quantity V_I is the image potential at the initial ion–surface separation, and E_0 is the initial kinetic energy.) Larger values of $|V_I|/E_0$ lead to large distortion of the trajectories. Furthermore, for monoenergetic, monodirectional ions there is a critical angle for desorption θ_c, which depends on $|V_I|/E_0$. For initial desorption polar angles $\theta_0 > \theta_c$, the image potential will bend the ions

---

[35] W. L. Clinton, *Phys. Rev. Lett.* **39**, 965 (1977).
[36] E. Preuss, *Surf. Sci.* **94**, 249 (1980).
[37] R. Janow and N. Tzoar, *Surf. Sci.* **69**, 253 (1977).
[38] W. L. Clinton, *Surf. Sci.* **112**, L791 (1981).
[39] Z. Miskovic, J. Vukanic, and T. E. Madey, *Surf. Sci.* **141**, 285 (1984).
[40] Z. Miskovic, J. Vukanic, and T. E. Madey, *Symp. Phys. Ionized Gases, Yugoslavia, 1984.*

back to the surface, and escape is impossible. For $\theta_0 = \theta_c$, the polar angle of the final trajectory is $\theta = 90°$, parallel to the surface.

Finally, it is hoped that a comparison of the angular distributions of neutral species and negative ions with those of positive ions will provide insight into the role of neutralization processes in determining the direction of desorption.

9.2. Experimental Methods in ESD and PSD

The experimental techniques useful for ESD studies can be categorized as (a) methods based on direct detection of the desorbing species, and (b) methods based on detecting changes in the surface being bombarded by electrons. Each will be considered in turn. We shall close with a discussion of electron-beam-induced damage in surface analysis.

9.2.1. Direct Detection of Desorbing Species

9.2.1.1. Detection of Positive Ions: Total Yields, Mass and Energy Analysis, Thresholds.
The simplest way to detect positive ions produced by ESD from an adsorbed layer on a surface is simply to surround the sample by a hemispherical collector and measure the ion current produced during electron bombardment. This is the method used by Redhead,[25] Yates et al.,[41] and Madey et al.[42] to examine ESD of ions from adsorbed layers on Mo and W. By appropriately biasing the sample, ion energy distributions were also measured. The use of grids between the sample and the collector electrode facilitated measurements of energy distributions and prevented electrons from the bombarding filament from reaching the collector. In addition, one could discriminate against ions formed in the gas phase by electron bombardment.

The next stage of sophistication in ESD ion measurement was the use of mass spectrometers to detect and mass analyze the desorbing ion signal. Moore,[43] Lichtman and McQuistan,[44] and Menzel[45] employed magnetic sector mass spectrometers for their early ESD measurements. The target to be studied was bombarded with electrons from a hot filament or an electron gun, and the ESD ions were accelerated into the mass spectrometer, analyzed, and detected. In the early 1970s, it became more common to use

---

[41] J. T. Yates, Jr., T. E. Madey, and J. K. Payn, *Nuovo Cimento, Suppl.* **5,** 558 (1967).
[42] T. E. Madey, J. T. Yates, Jr., D. A. King, and C. J. Uhlaner, *J. Chem. Phys.* **52,** 5215 (1970).
[43] G. E. Moore, *J. Appl. Phys.* **32,** 1241 (1961).
[44] D. Lichtman and R. B. McQuistan, *Prog. Nucl. Energy, Anal. Chem.* **4,** Part 2, 95 (1965).
[45] D. Menzel, *Ber. Bunsenges. Phys. Chem.* **72,** 591 (1968).

quadrupole mass analyzers[46,47] for ESD ion detection. Since quadrupole instruments are mass filters rather than momentum analyzers, they can be used to mass analyze ions over a wide energy range. Ion energy analysis is accomplished using retarding grids at the input of the quadrupole mass filter.[48]

All of the mass spectrometers used for ESD ion detection employ multiplier detectors and are highly sensitive, particularly in the pulse-counting mode. However, they invariably have a narrow solid angle of acceptance, the ion transmission properties and sensitivities of the analyzer–detector system are not usually known, and the angular distribution of ESD ions is not generally known, so that absolute ion desorption yields cannot be measured using these instruments.

Mass and energy analysis, as well as absolute ion yields, can be measured with an instrument that combines a mass spectrometer with a hemispherical analyzer having a large-solid-angle collector. Madey and Yates,[47] and Ashcroft et al.[49] built instruments in which the collector has a hole that allows a small fraction of the ion current to enter a quadrupole mass filter situated behind the collector. Figure 4 illustrates the analyzer of Ashcroft et al.[49] along with the configuration of bias potentials. Spherical geometry was maintained in the grids G_1, G_2, and G_3, the ion collector C, and the guard hemisphere H. The influence of field distortion due to nonspherical geometry at the target and the presence of the electron gun certainly limited the ultimate resolution; the outside of the gun drift tube could, in principle, have been covered with a resistive coating[50] and biased appropriately to minimize the field distortion between the grids.

Any electrostatic analyzer designed for electron analysis can be used for ion analysis by simply reversing the polarity of the voltages applied to the electrodes. Ions are analyzed as easily as electrons since all electrostatic analyzers measure particle energy-to-charge ratio that is independent of particle mass.

Cylindrical-mirror analyzers (CMAs) have been used to energy analyze desorbing ions by Niehus and Bauer.[51] The advantage of the CMA is its high resolution and high sensitivity. Bauer and co-workers used a quadrupole mass spectrometer for mass analysis in the same chamber in which the CMA

---

[46] D. R. Sandstrom, J. H. Leck, and E. E. Donaldson, *J. Chem. Phys.* **48**, 5683 (1968).

[47] T. E. Madey and J. T. Yates, Jr., *J. Vac. Sci. Technol.* **8**, 39 (1971).

[48] J. W. Coburn, *Surf. Sci.* **11**, 61 (1968).

[49] K. W. Ashcroft, J. H. Leck, D. R. Sandstrom, B. P. Stimpson, and E. M. Williams, *J. Phys. E* **5**, 1106 (1972).

[50] H. D. Hagstrum, personal communication (1979).

[51] H. Niehus and E. Bauer, *Rev. Sci. Instrum.* **46**, 1275 (1975); S. Prigge, H. Niehus, and E. Bauer, *Surf. Sci.* **75**, 635 (1978).

9.2. EXPERIMENTAL METHODS IN ESD AND PSD

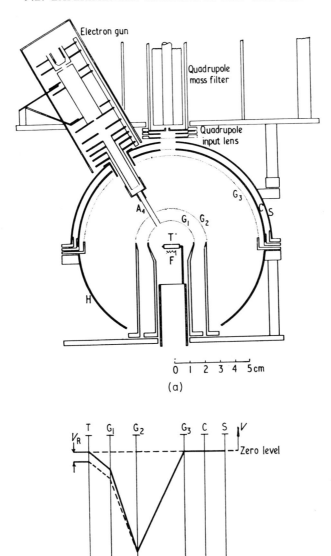

FIG. 4. (a) Experimental system for studies of electron-stimulated desorption. (b) Voltage configuration at electrodes. See text for discussion of apparatus. (From Ashcroft et al.[49] Copyright 1972 The Institute of Physics.)

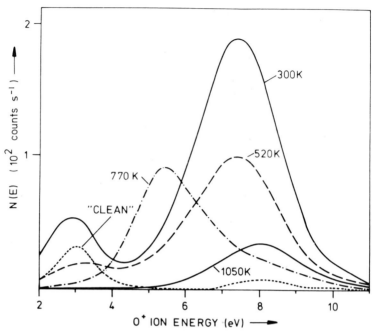

FIG. 5. Energy distributions of O^+ ions desorbed normal to a surface covered with 1 ML oxygen at 300 K with $E_p = 160$ eV and $i_p = 0.7$ μA for various annealing temperatures shown as curve parameters. (From Prigge et al.[51] Copyright North-Holland Physics Publishing, Amsterdam, 1978.)

was used for energy analysis; an example of the energy distributions measured using the CMA is given in Fig. 5.[51] Nishijima and Propst[52] employed a high-resolution cylindrical magnetic spectrometer for mass and energy analysis of ESD ions.

Ageev and Kutsenko[4,53] were the first to apply time-of-flight (TOF) techniques to the study of ESD processes. More recently, Traum and Woodruff[54] performed TOF measurements with a commercial double pass CMA to obtain simultaneous energy distributions and mass determinations for ESD and PSD ions. The dispersive character of the analyzer together with the use of pulse counting for ion detection permits ion energy spectra to be obtained with incident electron-beam currents from the concentric electron gun of

[52] M. Nishijima and F. M. Propst, *Phys. Rev. B* **2**, 2368 (1970).
[53] V. N. Ageev and E. N. Kutsenko, *Zh. Tekh. Fiz.* **39**, 1725 (1969).
[54] (a) M. M. Traum and D. P. Woodruff, *J. Vac. Sci. Technol.* **17**, 1202 (1980); (b) D. P. Woodruff, P. D. Johnson, M. M. Traum, H. H. Farrell, N. V. Smith, R. L. Benbow, and Z. Hurych, *Surf. Sci.* **104**, 282 (1981).

only a few nanoamperes. A rotable aperture in the CMA permits studies of the angular dependences of ion emission (ESDIAD) along with masses and kinetic energies.

Figure 6 shows a schematic of a similar CMA instrument used by Hanson et al.[55] for ESD ion-mass analysis. The electron beam is deflected onto the sample for a short period of time (~ 1 μs) and as the desorbed ions pass through the CMA, ions of different mass separate in time, the heavier ions taking longer to reach the detector. The time interval between the electron pulse on the sample and detection of an ion is measured by a time-to-amplitude converter (TAC) and a pulse-height analyzer (PHA). The instrument has extremely low noise since there is no direct line of sight from the sample

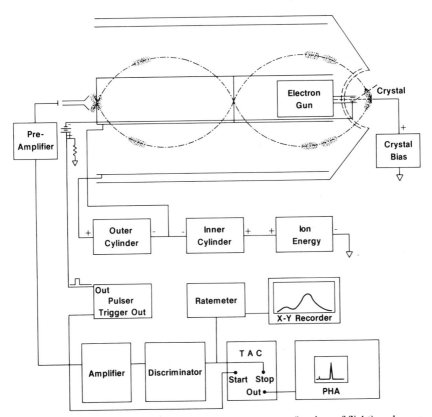

FIG. 6. Schematic of a CMA used for simultaneous mass (by time-of-flight) and energy analysis of ESD ions. (From Hanson et al.[55])

[55] D. M. Hanson, R. Stockbauer, and T. E. Madey, *Phys. Rev. B* **24**, 5513 (1981).

to the detector and has high sensitivity since the TAC-PHA system analyzes ions of all masses simultaneously.

Other circuits used to measure ion flight times are shown in Fig. 7. Part (a) illustrates an inexpensive circuit consisting of a delay and gate generator connected to the gate of a scaler. Ion pulses are counted by the scaler only while the gate is on. The TOF mass spectrum is obtained by recording the number of counts as a function of the delay setting. The disadvantage of this circuit is its inefficiency, since it is able to count ions only during the time when the gate is on.

In contrast, the TAC-PHA method illustrated in Fig. 7 is able to measure the flight time of each ion detected. A pulse starts the TAC when the electron beam is deflected on the sample. The detected ion signal stops the TAC which then outputs a pulse whose height is proportional to the time interval between the start and stop signals. This output pulse in turn is analyzed by

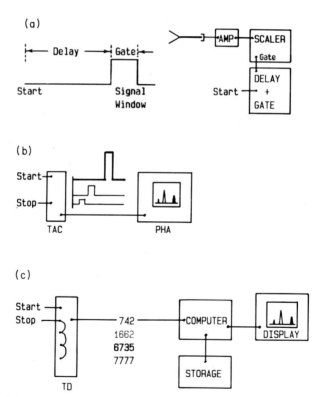

FIG. 7. Timing circuits for measuring ion flight times. (a) Delay and gate circuits, (b) TAC-PHA, and (c) time digitizer.

9.2. EXPERIMENTAL METHODS IN ESD AND PSD

the PHA, which displays the number of pulses versus pulse height which is the TOF mass spectrum.

The TAC–PHA combination can be replaced by a time digitizer as illustrated in Fig. 7c. This output from the time digitizer is a digital number that is proportional to the ion flight time and can be read directly by a computer.

One of the most important pieces of experimental data needed to understand ion-desorption mechanisms is the ion yield curve. This is the measurement of the ion emission as a function of electron or photon energy. The thresholds in the ion yield can then be correlated with either core or valence excitation cross sections. This correlation or lack thereof provides insight into the ion-desorption mechanism.

The ESD threshold measurements of Knotek and Feibelman[10] were made using a small electron gun (~ 2 cm long) mounted on the entrance housing of a quadrupole mass analyzer. A schematic of Knotek's apparatus is shown in Fig. 8. The electron energy is scanned at constant emission current and the

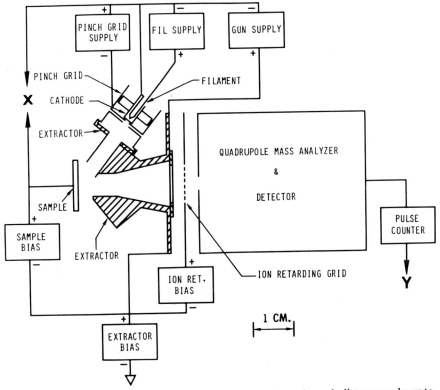

FIG. 8. ESD apparatus of Knotek. Electron gun, extractors, and sample distance are drawn to scale.

ion counting rate is monitored as a function of electron energy; the data are stored in a multichannel scaler. The thresholds so measured can be correlated with loss features observed in low-energy electron loss spectra, as shown in Fig. 9.

Knotek et al.[20b] and Jaeger et al.[56] used time-of-flight mass spectrometry to identify ions desorbed by photons. Figure 10 shows the drift tube analyzer used by Jaeger et al.[56] and the O^+ yield curve from O on Mo(100). They used synchrotron radiation at the Stanford Synchrotron Radiation Laboratory (SSRL), which comes in 0.3 ns pulses every 780 ns and provides a well-defined trigger pulse and a broad time window for mass determination. To increase the light output, the synchrotron can be operated in a multiple bunch mode in which more than one bunch of electrons are circulated simultaneously. This, however, decreases the time between light pulses and causes the TOF mass spectra to overlap. To overcome this difficulty, one bunch can be left out of the multiple bunch sequence so that the resulting light pulse train is nonperiodic. This nonperiodicity is reflected in each of the superimposed TOF mass spectra and can be used to disentangle them.

Finally, the reader is referred to two reviews, the first a comprehensive article on the transport, dispersion, and detection of electrons, ions, and

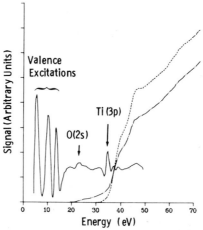

FIG. 9. Low-energy electron loss (second-derivative spectra) and H^+ and O^+ ESD from an annealed TiO_2 surface. Desorption thresholds for H^+ are the O 2s and Ti 3p core-level ionization potential and O^+ at the Ti 3p threshold. ---, O^+ ESD; - · -, H^+ ESD; ——— LEELS. (From Knotek.[55a] Copyright North-Holland Physics Publishing, Amsterdam, 1980.)

[55a] M. L. Knotek, Surf. Sci. **91**, L17 (1980).
[56] R. Jaeger, J. Stöhr, J. Feldhaus, S. Brennan, and D. Menzel, Phys. Rev. B **23**, 2102 (1981).

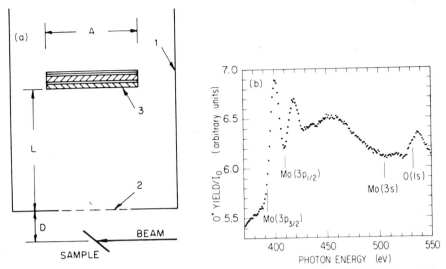

FIG. 10. (a) Drift tube time-of-flight arrangement used by Jaeger et al.[56] 1. Stainless steel cylindrical tube. 2. 90% transmission mesh. 3. Double microchannel-plate assembly. (b) O^+ yield from oxygen on Mo(100). Note the correlation of the ion thresholds with the indicated energy of the Mo and O core levels.

neutrals by Granneman and van der Wiel[57]; this paper covers charged particle optics, analyzer characteristics, and particle detectors. The second is on the instrumentation involved in the detection of ions from the photon-stimulated desorption by Stockbauer[57a] and covers time-of-flight and electrostatic analyzers.

9.2.1.2. Special Problems in Measurements of PSD Ion Yields; Monochromator Corrections. In PSD data analysis one encounters several unique problems. The most serious of these is the ion signal due to the second-order light from the photon monochromator. This is due to the fact that any grating monochromator, in addition to transmitting light of energy $h\nu$ to which it is tuned, also passes some amount of second-order light of energy $2h\nu$ and perhaps third-order light of energy $3h\nu$. Normally, this is not a problem in UPS experiments where the second-order light produces a photoelectron energy distribution that is displaced far enough from the primary distribution that the two do not overlap. This arises from electron kinetic energy being a linear function of photon energy. However, this is not

---

[57] E. H. A. Granneman and M. J. van der Wiel, in "Handbook on Synchrotron Radiation," Vol. 1 (E. E. Koch, ed.) p. 367. (North-Holland, Amsterdam, 1983).

[57a] R. Stockbauer, Nucl. Instrum. Methods 222, 284 (1984).

the case for photodesorbed ions whose kinetic energy is independent or at best a slowly varying function of photon energy. The contributions to the ion signal from the different orders of light overlap and must be separated to obtain an ion yield versus photon energy curve. This problem is particularly acute, as pointed out by Woodruff et al.,[54b] in the threshold region for which the ion yield approaches zero. If there is significant ion yield and light at the second-order energy, a large ion signal will be observed that can completely obscure features in the threshold of the yield curve. Hence, the second-order contributions must be properly removed from the data.

In addition the observed ion count rate must be normalized to the photon flux to obtain the ion yield curve. This correction requires knowledge of the monochromator transmission, i.e., the amount of light incident on the sample as a function of photon wavelength.

The monochromator first-, second-, and higher-order light output then must be measured. The methods used to make these measurements will be given below followed by the method used to correct the PSD ion signal. In the following discussion, it has been assumed that only second-order light makes any significant contribution, i.e., third and higher orders have been neglected. It is a simple matter to extend the derivation to include contributions from higher-order light.

The first method described below requires two devices to measure the monochromator transmission function. These are a calibrated photodiode that gives a known amount of current per incident light flux and an electron energy analyzer. There is a technique available using only a calibrated photodiode, which involves operating the synchronous light source under different conditions to obtain different but known distributions of radiation. This techique has been developed by Saloman.[58] However, it will not be described here, since most synchrotrons are not able to operate under the required conditions. The technique described below assumes a calibrated photodiode and electron energy analyzer are available.

We proceed by defining the following quantities:

m_λ = monochromator wavelength setting
$F(m_\lambda)$ = intensity of first-order light with wavelength λ when monochromator is set at m_λ
$S(m_\lambda)$ = intensity of second-order light with wavelength $\lambda/2$ when monochromator is set at m_λ
$I(m_\lambda)$ = measured current from photodiode
$D(\lambda)$ = response of photodiode to light of wavelength λ

---

[58] E. B. Saloman, *Appl. Opt.* **14**, 1391 (1975).

9.2. EXPERIMENTAL METHODS IN ESD AND PSD

The function $F(m_\lambda)$ is the monochromator transmission function required for the PSD data analysis.

The calibrated photodiode is placed at the exit slit of the monochromator and its photocurrent is measured as the monochromator is scanned through its spectral range. The measured photocurrent will consist of contributions from both the first- and second-order light:

$$I(m_\lambda) = D(\lambda)F(m_\lambda) + D(\lambda/2)S(m_\lambda). \quad (9.2.1)$$

To eliminate $S(m_\lambda)$ from this equation, the quantity

$$P(m_\lambda) = S(m_\lambda)/F(m_{\lambda/2}) \quad (9.2.2)$$

is defined as the ratio of the second-order light at a monochromator setting of m_λ to the first-order light of the same wavelength, i.e., measured at $m_{\lambda/2}$. The ratio $P(m_{\lambda/2})$ is measured by using the electron energy analyzer to obtain UPS spectra of *any* surface measured as a function of m_λ. The ratio of any peak height observed in the second-order spectrum at m_λ to its height in the first-order spectrum at $m_{\lambda/2}$ gives $P(m_\lambda)$. Since both spectra are obtained with the same wavelength of light, the ratio is independent of photoemission cross section, electron energy analyzer transmission, and pass energy. Substituting (9.2.2) into (9.2.1) and rearranging, we obtain

$$F(m_\lambda) = \frac{I(m_\lambda) - D(\lambda/2)P(m_\lambda)F(m_{\lambda/2})}{D(\lambda)}. \quad (9.2.3)$$

Since F appears on both sides of the equation, it must be calculated using a bootstrap technique. Use is made of the fact that toward the short-wavelength end of the monochromator range, the second-order light contribution is negligible, so that $P(m_\lambda)$ and, hence, the second term on the right in (9.2.3) is zero.

The bootstrap to calculate $F(m_\lambda)$ involves starting at the short-wavelength limit of the monochromator, where only $I(m_\lambda)$ and $D(\lambda)$ need to be known and working toward longer wavelengths. As the range in which $P(m_\lambda)$ becomes nonzero is approached, $F(m_{\lambda/2})$ will have already been calculated and can be used to obtain $F(m_\lambda)$. This continues until the long-wavelength limit of the monochromator is reached so that $F(m_\lambda)$ is determined throughout its range.

A similar bootstrap technique is used to extract the ion yield curve from the measured ion count rate as illustrated in the following derivation. The quantity $C(m_\lambda)$ is defined as the measured ion count rate at a monochromator setting of m_λ and will be the sum of contributions from both first- and second-order light, i.e.,

$$C(m_\lambda) = \sigma(\lambda)F(m_\lambda) + \sigma(\lambda/2)S(m_\lambda), \quad (9.2.4)$$

where $\sigma(\lambda)$ is the desired ion yield at wavelength λ. Using (9.2.2) and rearranging, we find

$$\sigma(\lambda) = [C(m_\lambda) - \sigma(\lambda/2)P(m_\lambda)F(m_{\lambda/2})]/F(m_\lambda) \qquad (9.2.5)$$

As with F in (9.2.3), σ appears on both sides of the expression and is calculated starting at the short-wavelength end for which the second-order contribution can be neglected and moving to longer wavelengths using $\sigma(\lambda/2)$ for which the second-order light contributes.

The advantage of this type of analysis is that $P(m_\lambda)$ is independent of the electron analyzer transmission function and the photoemission cross section of the surface used to determine it. The only serious drawback arises for those monochromators that have a second-order contribution at m_{λ_0} but no first-order light at $m_{\lambda_0}/2$. In this case $P(m_{\lambda_0})$ becomes infinite. There is little recourse in this event. Although it is possible to determine $F(m_{\lambda_0})$ and $S(m_{\lambda_0})$ without using $P(m_{\lambda_0})$, they are of little use since $\sigma(\lambda_0/2)$ in (9.2.5) cannot be measured when there is no light at $\lambda_0/2$. Nonetheless, if σ can be extrapolated into the unknown wavelength range or obtained from measurements on a different monochromator, the data can be analyzed using $F(m_\lambda)$ determined directly without using $P(m_\lambda)$.

The function $F(m_\lambda)$ can be determined directly if the relative photoemission cross section of a sample and analyzer transmission function are known. UPS spectra are obtained for the sample surface and the following ratio is measured as a function of wavelength setting:

$$R(m_\lambda) = P_S(m_\lambda)/P_F(m_\lambda), \qquad (9.2.6)$$

where $P_F(m_\lambda)$ and $P_S(m_\lambda)$ are the peak heights of a UPS feature due to the first- and second-order light at the same monochromator wavelength setting m_λ. Peak height $P_F(m_\lambda)$ is due to light of wavelength λ, whereas $P_S(m_\lambda)$ is due to light of wavelength $\lambda/2$. These peak heights are given by

$$P_F(m_\lambda) = S(m_\lambda)\gamma(\lambda/2)A(T_{\lambda/2})$$

and
$$\qquad (9.2.7)$$

$$P_S(m_\lambda) = S(m_\lambda)\gamma(\lambda/2)A(T_{\lambda/2})$$

where $\gamma(\lambda)$ is the relative photoemission cross section for the observed UPS feature and $A(T_\lambda)$ is the transmission of the analyzer for electrons of kinetic energy T_λ. Substituting (9.2.7) into (9.2.6) and solving for $S(m_\lambda)$ we find

$$S(m_\lambda) = R(m_\lambda)F(m_\lambda)\gamma(\lambda)A(T_\lambda)/\gamma(\lambda/2)A(T_{\lambda/2}). \qquad (9.2.8)$$

This can be substituted into (9.2.1) to solve that expression for $F(m_\lambda)$.

The photoemission cross sections γ and electron analyzer transmission A enter explicitly. To a first approximation A can be calculated from known characteristics of the analyzer. The cross section γ, however, is more difficult to obtain. It is a function not only of λ but also of the angle with which the electron leaves the surface as well as the angle of incidence and the polarization of the light. Hence, instruments of different geometry, i.e., different angular apertures and different angles of photon incidence and electron emission will give different peak intensities.

A third method makes use of a thin filter of known transmission and a calibrated photodiode. The photodiode is used to measure the output of the monochromator with and without the filters. The resulting photocurrents are given by

$$I(m_\lambda) = D(\lambda)F(m_\lambda) + D(\lambda/2)S(m_\lambda) \qquad (9.2.9)$$

$$I^f(m_\lambda) = D(\lambda)T(\lambda)F(m_\lambda) + D(\lambda/2)T(\lambda/2)S(m_\lambda), \qquad (9.2.10)$$

where $I^f(m_\lambda)$ is the measured photocurrent with the filter in place and $T(\lambda)$ is the transmission of the filter at wavelength λ. Note that equation (9.2.9) is identical to equation (9.2.1). Solving for $F(m_\lambda)$ and $S(m_\lambda)$ one obtains

$$F(m_\lambda) = \frac{I^f(m_\lambda) - I(\lambda)T(\lambda/2)}{D(\lambda)[T(\lambda) - T(\lambda/2)]} \qquad (9.2.11)$$

and

$$S(m_\lambda) = \frac{I(m_\lambda) - D(\lambda)F(m_\lambda)}{D(\lambda/2)} \qquad (9.2.12)$$

If an electron energy analyzer is available, the transmission of the filter can be obtained by measuring the intensity of a UPS feature with and without the filter assuming that the scattering of the light through the filter is minimal, the transmission is given by the ratio of the two intensities.

The above derivations are given in terms of the photon wavelength since most monochromators are calibrated in wavelength rather than energy. Since it is customary to plot PSD ion yield waves in terms of energy, it is necessary to convert from photon wavelength to photon energy at some state in the data analysis.

9.2.1.3. Measurements of Ion Angular Distributions: ESDIAD and Angle-Resolved PSD. Two types of detection schemes have been employed in ESDIAD studies of adsorbates on metals. Madey and Yates[6] have used a display-type aparatus in which the ion angular distribution patterns are visually displayed on a fluorescent screen. Niehus[59] and Jaeger and Menzel[60] employed narrow-aperture detectors that necessitated spatial scanning to measure the angular distribution patterns. The relative merits of each method will be discussed below.

[59] H. Niehus, *Surf. Sci.* **78**, 667 (1978); **80**, 245 (1979); **87**, 567 (1981).
[60] R. Jaeger and D. Menzel, *Surf. Sci.* **93**, 71 (1980).

A schematic of the relevant components of the stainless steel, ion-pumped ultrahigh-vacuum chamber built at NBS for ESDIAD studies[6] is shown in Fig. 11. The single crystal sample S is mounted on a manipulator that allows translations in the x, y, and z directions as well as rotary motion ($\pm 180°$) about the axis perpendicular to the plane of the drawing. The sample crystal is heated resistively or by electron bombardment and can be cooled to ~90 K using liquid nitrogen; temperatures are determined using a W–3%-Re versus W–25%-Re thermocouple.

A focused low-energy electron beam (30–1500 V) is generated using the electron gun, passes through the drift tube, and strikes the sample S. Typical electron currents are $\lesssim 2 \times 10^{-7}$ A at 200 eV with a beam diameter of $\lesssim 0.5$ mm. Deflection plates on the gun allow the beam to be easily scanned over the surface of the sample.

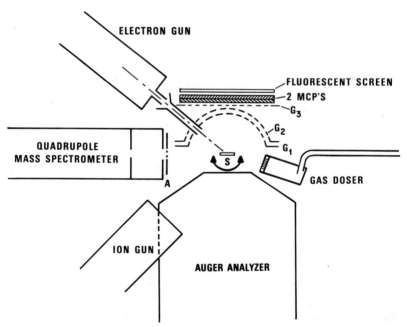

FIG. 11. Schematic of ultrahigh-vacuum NBS ESDIAD apparatus. The radius of curvature of G_1 is 2.0 cm and the active area of each MCP has a diameter of 4.0 cm. See text for details. For most ESDIAD measurements, typical potentials are: $G_1 = G_2 = 0$ V, $G_3 = -70$ V, MCP entrance $= -700$ V, MCP midpoint $= 0$ V, MCP exit $= +700$ V, fluorescent screen $= +3800$ V. Electron-gun filament potential $V_f = 70-1500$ V, crystal potential $V_B = 0$ to $+100$ V. Electron energy $E_e = e(|V_f| + |V_B|)$. (From Madey and Yates.[6])

9.2. EXPERIMENTAL METHODS IN ESD AND PSD

The grid-microchannel plate (MCP) optical system[61] is used to display visually the ESDIAD patterns. Electron stimulated desorption ions liberated from the crystal pass through G_1 and G_2 and are accelerated between G_3 and the entrance of the first MCP. Electrons liberated from the output of the second MCP are accelerated to the fluorescent screen, where the proximity-focused ESDIAD patterns are observed and photographed. The gain of the MCP array is usually in the range 10^6-10^7. By changing the bias of the MCP entrance so as to accelerate electrons and repel positive ions, the elastic low-energy electron diffraction (LEED) pattern from the surface can be observed. Switching from the LEED to the ESDIAD mode can be accomplished in a few seconds. The LEED and ESDIAD patterns are photographed at $f/2.8$ using Tri-X film (ASA 400) with 2–10 s exposures. Because of the high angles of emission of some of the ion cones causing the ESDIAD patterns, a bias potential V_B is frequently applied to the crystal to "compress" the patterns for all photography.[6] Figure 12 contains some ESDIAD patterns obtained using this apparatus.

The quadrupole mass spectrometer (QMS) is used for residual gas analysis and for detection and mass analysis of ESD ions. By rotating the sample S so as to face the QMS rather than the ion optics, the angular profile of ESDIAD beams can be determined. Electrode A contains a 3-mm-diameter hole covered with a high-transparency mesh (40 lines cm^{-1}) and is used for crude retarding potential energy analysis of the ion beams. The full cone angle of ESDIAD ion beams accepted by the QMS is 4°.

The gas handling system used for the gas doser is a more versatile version of the system described previously.[62] To insure a uniform dose of molecules onto the surface of S, the end of the doser contains a microcapillary array (length-to-radius for each channel = 40) through which the molecules effuse. The ratio of direct flux on the crystal to random background flux of the dosed gas is estimated to be > 40. The doser has been used for a variety of gases (O_2, CO, H_2O, NH_3, C_3H_6, C_6H_{12}, C_8H_{16}). The gas purity at the sample is high, with little difficulty from wall reactions or displacement processes in the ion pumps.

Surface cleanliness of the sample is verified using the Auger spectrometer. In most cases, samples are cleaned by repeated heating in oxygen followed by heating in vacuum, but often it is necessary to use the argon ion gun for sputter cleaning.

All of the ESDIAD measurements made to date in this apparatus have been dc measurements, i.e., continuous collection of all ions during electron bombardment of the sample. In principle, however, it is possible to pulse the

---

[61] D. J. Ruggieri, *IEEE Trans. Nucl. Sci.* **NS-19**, 74 (1972).
[62] T. E. Madey, *Surf. Sci.* **33**, 355 (1972).

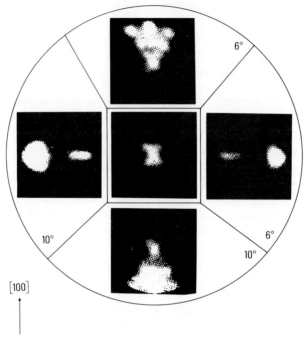

FIG. 12. ESDIAD patterns for desorption of O^+ ions from a fractional monolayer of oxygen (exposure 4 L) adsorbed on a multifaceted tungsten single crystal at 100 K. The central facet is oriented with its surface parallel to the (110) plane and the outer facets are inclined by 6° and 10° with respect to the (110) plane. The terrace widths on the 6° and 10° stepped surfaces are 13 Å and 22 Å, respectively. In all the patterns, the spot in the center of the picture corresponds to O^+ ions desorbing normal to the surface. The additional beams are due to off-normal desorption. (From Madey.[15])

electron beam and time-gate the microchannel-plate detector for time-of-flight analysis of the ions that give rise to the ESDIAD patterns; such measurements are currently underway in our laboratory. Panitz[63] has described in detail the pulse circuitry and procedures for time-gating microchannel plates.

A vidicon camera parallel-detection system that was designed and built originally for angle-resolved electron spectroscopy[63a] has been used by Rowe, Traum, and colleagues[63b] for ESDIAD studies. The display-type en-

[63] J. A. Panitz, *Prog. Surf. Sci.* **8**, 219 (1978).

[63a] S. P. Weeks, J. E. Rowe, S. B. Christman, and E. E. Chaban, *Rev. Sci. Instrum.* **50**, 1249 (1979).

[63b] J. E. Rowe and M. Traum, private communication (1981).

9.2. EXPERIMENTAL METHODS IN ESD AND PSD

ergy analyzer and detector is similar to that of Fig. 11, but an imaging and digitizing system permits rapid electronic acquisition and storage of data in two dimensions. The imaging-digitizing system consists of a commercially available optical multichannel analyzer and a vidicon camera coupled to a minicomputer. The ESDIAD patterns are displayed as digital pictures, plotted in the form of intensity versus ion-desorption angle for a fixed kinetic energy window.

The apparatus used by Niehus[59] for ESDIAD measurements is shown in Fig. 13. The electron gun produces either a continuous or a pulsed beam of electrons (100–500 eV, 1×10^{-8} to 5×10^{-6} A). The ion collector is a shielded channeltron multiplier (aperture angle of 3°) attached to a motor-driven goniometer that can be positioned within 0.1° over two thirds of the halfspace in front of the target. During ESDIAD experiments, the settings of the goniometer and data collection are both computer controlled. An ion angular distribution for O^+ from a W(111) crystal obtained using this apparatus is shown in Fig. 14.[59]

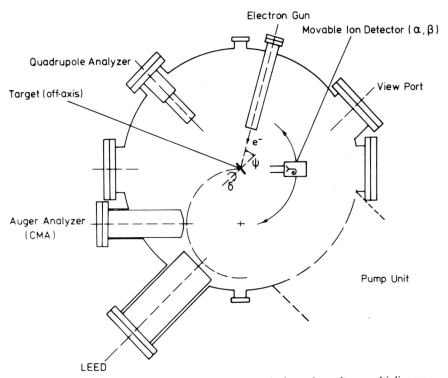

FIG. 13. Apparatus used by Niehus[59] for ESDIAD, employing a channeltron multiplier as a movable ion detector.

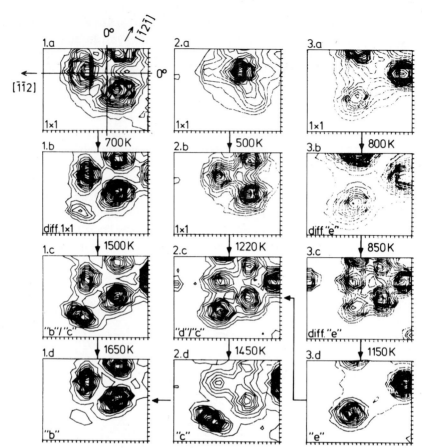

FIG. 14. ESDIAD patterns of oxygen adsorption layers. The intensity contour lines are plotted in a linear scale as a function of azimuth and polar angles of the desorbing O^+ ions; angular increment 4°; intensity increment 1/20 of the maximum value. All patterns are measured at room temperature and show about one quarter of the halfspace in front of the specimen. The surface normal (0°, 0°) as well as the azimuthal orientation of the crystal have not been changed during the measurements of all patterns and is shown in 1a. The indication of LEED pattern is given on the left side of each ESDIAD pattern; "diff." stands for diffuse. The first column 1a to 1d shows the development of ESDIAD patterns after an oxygen exposure of 2×10^{-6} Torr s as a function of subsequent annealing for 30 s at the indicated temperature. Exposure of 3.5×10^{-6} Torr s leads to the sequence shown in column 2. Annealing at 1650 K of a surface layer represented in pattern 2d results in a transition to 1d. Column 3 shows the data for an exposure of 100×10^{-6} Torr s. Patterns 3d changes to 2c upon annealing at 1220 K. (From Niehus.[59] Copyright North-Holland Physics Publishing, Amsterdam, 1978.)

9.2. EXPERIMENTAL METHODS IN ESD AND PSD

The kinetic energy distributions of desorbed ions are determined using a time-of-flight technique. The primary electron pulse is 0.5 μs, and the mean flight time of O^+ ions is about 15–20 μs. Signal averaging techniques are used to record the time of flight spectrum, which is then transformed into an energy spectrum.

Niehus and Krahl-Urban[63c] have shown that the background signal in the ESDIAD patterns obtained using the scan device (Fig. 13) can be significantly reduced by using a chopped electron beam in connection with a gated detector system. This procedure effectively eliminates the major source of spurious background signal, viz., photoelectron emission at the entrance of the electron multiplier detector due to photons emitted by electron impingement on the target, and/or by ESD ions generated at the shielding grids by electrons scattered from the target. The gating technique suppresses this "ion/photon" peak, which occurs with shorter flight times than the ESDIAD signal. The advantages of the improved procedure are increased sensitivity, low background, high angular resolution and ion energy determination at any point in the ESDIAD pattern. In principle, the signal to noise ratio of patterns obtained in the display apparatus (Fig. 11) can also be improved by chopping the electron beam and gating the high voltage applied to the channel plate.

The ability to measure relative ion beam intensities and the angular profile of individual ion beams in a field-free condition is a major advantage of the Niehus apparatus. However, because of the time necessary to record an ESDIAD pattern of the type shown in Fig. 14 (up to 1 hr depending on signal intensity), more "fragile" adsorbates can easily be destroyed by the electron beam before the pattern is obtained; e.g., the H_2O "halo" ESDIAD pattern[13] is destroyed in ~20 seconds at 1×10^{-7} A. In such cases, the microchannel-display ESDIAD system, in which the pattern is recorded photographically in a few seconds, offers an advantage.

Jaeger and Menzel[60] characterized the ESDIAD behavior of O^+ from CO on W(100) using a mass spectrometer detector. The electron gun used for ESD was mounted at a fixed position with respect to the sample normal, and moved with the sample as the sample was rotated in front of the entrance aperture to the mass spectrometer. Care was taken to operate at low current densities, and a scan of O^+ ion current from the W(100) crystal as function of polar angle (along a single azimuth) took 100 s.

Similar techniques, which use synchrotron radiation, have been employed to study angle-resolved PSD. One problem concerns the low counting rates expected due to the relatively low ion yields; another difficulty is the background signal in channeltrons and microchannel plates due to scattered

---

[63c] H. Niehus and B. Krahl-Urban, *Rev. Sci. Instrum.* **52**, 56 (1981).

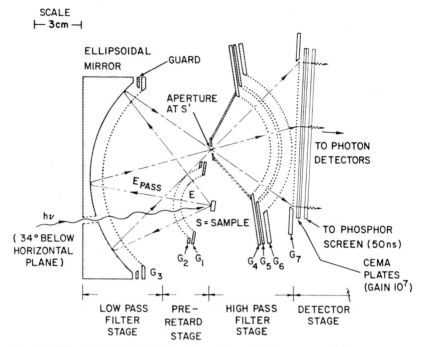

FIG. 15. Ellipsoidal mirror display analyzer designed by Eastman et al. for angle resolved PSD, ESD, and photoemission. See Table I for specifications. (From Eastman et al.[64])

high-energy photons. Eastman et al.[64] have built an ellipsoidal mirror analyzer with a microchannel array detector having large acceptance angle; it is being used for studies of angle-resolved photoelectron spectroscopy using synchrotron radiation. For UPS measurements, electrons emitted from the sample are reflected from an ellipsoidal mirror, refocused on an aperture, and displayed visually on the detector (Fig. 15). The aperture reduces the flux of scattered photons on the microchannel-plate detector. Van der Veen et al.[65] have reversed potentials on this analyzer, and detected PSD O^+ ion emission from single crystal V_2O_5 in the direction normal to the surface; large onsets in O^+ emission were observed at the V core level ionization energies, as expected on the basis of the Knotek–Feibelman[10] mechanism. Since the analyzer operates as a bandpass filter, ion energy distributions are

[64] D. E. Eastman, J. J. Donelon, N. C. Hien, and F. J. Himpsel, *J. Nucl. Instrum. Methods* **172**, 327 (1980).
[65] J. F. Van der Veen, F. J. Himpsel, D. E. Eastman, and P. J. Heimann, *Solid State Commun.* **36**, 99 (1980).

easily measured. Ion masses are determined by time-gating the grids, using a time-of-flight method. Madey et al.[21] have used this analyzer to demonstrate the equivalence between ESDIAD and angle-resolved PSD for a chemisorption system yielding multiple ion beams: oxygen on W(111). It is a well-designed analyzer with excellent operating characteristics for ESDIAD/PSD studies. An example of angle resolved PSD patterns of O^+ from W(111) is shown in Fig. 16.

Engelhardt, Menzel, and colleagues[65a] have described another energy-dispersive analyzer system suitable for measuring ion angular distributions in ESD or PSD. The analyzer consists of a toroidal prism and a truncated conical lens, and the system employs a position-sensitive detector to display simultaneously the ion (or electron) signal emitted from a surface over the range of polar angles $0° < \theta < 90°$ at a fixed azimuthal angle θ. Total $N(E,\theta,\phi)$ spectra are obtained by mechanical variation of the azimuth of the sample and by accumulating and processing spectra with a minicomputer.

9.2.1.4. Detection of Negative Ions. There have been several literature reports of negative ESD ions (H^-, O^-, F^-, Cl^-).[8,9,66] Lichtman and co-workers[8,66] have modified the potentials on their magnetic sector mass spectrometer so as to detect negative ions. The input to the channeltron detector was grounded, and the output potential was biased positively (+2900 V). The output was capacitively coupled to the pulse-counting circuitry. Yu[9] has used a quadrupole mass spectrometer with the multiplier detector biased to detect negative ions. He found that the ratio of counting rates O^-/O^+ for

TABLE I. Specifications for Display Type Analyzer

Grids	Radius	Voltage
G_1	2.54 cm	$V_{SAMPLE}(SCAN)$
G_2	3.18	GND
G_3	9.21	GND
G_4	4.93	GND
G_5	5.72	$-V_t = E_p - \Delta E$
G_6	6.76	GND
G_7	FLAT	GND
Mirror	—	$-V_r \approx E_p$
Guard	—	$0.5 V_r$

[65a] H. A. Engelhardt, W. Bäck, D. Menzel, and H. Liebl, *Rev. Sci. Instrum.* **52**, 835 (1981); H. A. Engelhardt, A. Zartner, and D. Menzel, *Rev. Sci. Instrum.* **52**, 1161 (1981).
[66] J. L. Hock and D. Lichtman, *Surf. Sci.* **77**, L184 (1978).

494   9. ELECTRON- AND PHOTON-STIMULATED DESORPTION

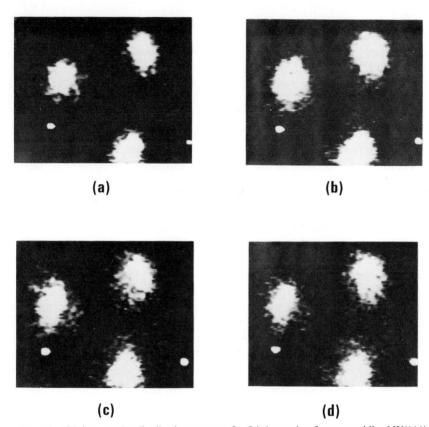

FIG. 16. PSD ion angular distribution patterns for O^+ desorption from an oxidized W(111) surface, measured using the apparatus of Fig. 15. $h\nu$ (eV) = (a) 55.1, (b) 44.3, (c) 35.4, (d) 31.0. (From Madey et al.[21])

oxygen on Mo is about 0.01. The O^-/O^+ ratio is a function of temperature, and increases with decreasing surface work function when Cs and oxygen are coadsorbed.

Traum and Woodruff[54a] have suggested that their CMA time-of-flight method should be useful for separating the ESD negative ion signal from the secondary-electron signal from the sample, but this has not yet been demonstrated.

It is anticipated that studies of ESD thresholds and ESD ion angular distributions for negative ions will shed new light on the mechanisms of ion formation and the nature of the electronically excited states involved in the ESD process.

9.2.1.5. Detection of Metastables.   Electron stimulated desorption of

CO adsorbed on transition metals is known to result in release of O^+, CO^+, and CO neutrals. In addition, Redhead[25] reported the possible release of an excited neutral species. Definitive evidence for the ESD of excited neutrals in ESD of CO on W(100) was reported by Newsham et al.[67,68] They used a time-of-flight technique to distinguish between ions and excited neutrals, as well as photons emitted by the sample. They suggested that the excited species corresponded to the $a^3\Pi$ state of CO.

The basis for detection[69] of metastable (excited neutral) molecules is via the secondary electrons ejected during their de-excitation at a collector surface. If the lifetime of the excited state is long enough for the molecule to reach the detector, and if the energy of the excited state of the molecule (or atom) is greater than the work function of the collector, then there is a high probability that an electron will be emitted during de-excitation. A few atoms and molecules that have relatively long-lived metastable states (> 1 ms) and excitation energies > 4 eV are $CO(a^3\Pi)$ and $N_2(A^3\Sigma^+)$, both ~ 6 eV, metastable states of all the rare gases, O(1S) at ~ 4 eV, NO ($^4\Pi$), and $H_2 c(2p)$ $^3\Pi_u$. Of these, singlet oxygen would appear to be a good candidate for a future ESD study.

The Auger effect detector used by Stern[70] in his molecular-beam studies of metastable CO consisted of a Ta ribbon detector which was operated while heated to a few hundred degrees Celsius. Metastables striking the detector released secondary electrons that were accelerated to the input of an electron multiplier. For their ESD studies Newsham et al.[67,68] simply allowed the metastables to strike the input end of a channeltron multiplier and detected the secondary-electron pulses.

Another method used for detecting excited species is via fluorescence decay of short-lived species. Prince et al.[71] have observed a fluorescence spectrum during bombardment of a thin film of ice at 77 K by a 200-eV electron beam. The data suggest that the excited OH radical is formed at the surface of the film during electron bombardment, and the radiation occurs during de-excitation of the excited OH in the gas phase. Electron stimulated desorption of excited OH from a TiO_2 surface has also been found by Bermudez and Hoffbauer.[71a] Radiative decay of electronically excited species at metal surfaces has been detected by Artamonov and colleagues, and re-

---

[67] I. G. Newsham, J. V. Hogue, and D. R. Sandstrom, *J. Vac. Sci. Technol.* **9**, 596 (1972).
[68] I. G. Newsham and D. R. Sandstrom, *J. Vac. Sci. Technol.* **10**, 39 (1973).
[69] T. C. English and J. C. Zorn, in "Molecular Physics," Part B (D. Williams, ed.), 2nd Ed., Methods of Experimental Physics, Vol. 3, p. 669. Academic Press, New York, 1974.
[70] R. Stern, Ph.D. Thesis, Harvard Univ., Cambridge, Massachusetts, 1968.
[71] R. H. Prince, G. N. Sears, and F. J. Morgan, *J. Chem. Phys.* **64**, 3978 (1976).
[71a] V. Bermudez and R. Hoffbauer, *Phys. Rev. B* (1984). In press.

viewed by Artamonov and Samarin.[72] Tolk and colleagues[73] have detected fluorescence radiation from electronically excited alkali neutrals released by ESD of an alkali halide crystal.

9.2.1.6. Detection of Neutral ESD Products. There have been few reports of the *direct* detection of ground state neutral species desorbed from a surface by ESD. Petermann[74] and Menzel[75] reported small increases in CO partial pressure upon electron bombardment of CO covered surfaces. Additional references to measurements of neutral desorption via partial pressure changes are given by Drinkwine and Lichtman.[5] Most other evidence for neutral desorption has been via measurements of a decrease in ion current during ESD, or by monitoring *in situ* changes in surface coverage during electron bombardment using Auger spectroscopy or other methods (the detection of ESD damage in adsorbed layers will be discussed in Section 9.2.3).

Feulner et al.[75a] and Zhang and Gomer[75b] have made substantial advances in detection of ESD neutrals. The system that was used for detection of desorbing neutrals by Feulner et al.[75a] had a glass enclosure around the mass spectrometer ion source,[75c] which served as a stagnation volume. An assembly consisting of a filament between two grids was placed in front of the entrance aperture of the glass enclosure; the single crystal sample was positioned about 1 mm in front of this device. Desorbing ions were repelled from the entrance to the mass spectrometer by suitable potentials. The neutral desorption signal was separated from the large background by lock-in detection at 20 Hz. Electron current densities were kept below 10^{-5} A cm^{-2} so that beam heating was negligible. Continuous thermal desorption due to radiation only added to the noise level. By averaging over multiple scans, between which the sample was cleaned and recovered to avoid beam-induced changes, a desorption signal as low as 1/50 of the background (corresponding to $\sim 10^{-11}$ Pa) in the enclosure could be measured. Even so, considerable noise was encountered, which made the direct determination of thresholds difficult. To overcome this difficulty, the integral of the difference of signals with and without the adsorbate layer was taken; this procedure effectively removed the noise.

---

[72] O. M. Artamonov and S. N. Samarin, *Radiat. Eff.* **40**, 201 (1979).

[73] N. H. Tolk, L. C. Feldman, J. S. Kraus, R. J. Morris, M. M. Traum and J. C. Tully, *Phys. Rev. Lett.* **46**, 134 (1981); N. H. Tolk, M. M. Traum, J. S. Kraus, T. R. Pian, W. E. Collins, N. G. Stoffel and G. Margaritondo, *Phys. Rev. Lett.* **49**, 812 (1982).

[74] L. A. Petermann, *Nuovo Cimento, Suppl.* **1**, 601 (1963).

[75] D. Menzel, P. Kronauer, and W. Jelend, *Ber. Bunsenges. Phys. Chem.* **75**, 1074 (1971).

[75a] P. Feulner, R. Treichler, and D. Menzel, *Phys. Rev. B* **24**, 7427 (1981).

[75b] Z.-J. Zhang and R. Gomer, *Surf. Sci.* **109**, 567 (1981).

[75c] P. Feulner and D. Menzel, *J. Vac. Sci. Technol.* **17**, 662 (1980).

9.2. EXPERIMENTAL METHODS IN ESD AND PSD

Using a system similar to this, Feulner et al.[75d] measured the angular distributions of neutral CO desorbed by electron impact from chemisorbed CO on Ru(001), and of neutral N_2O and N_2 from chemisorbed and physisorbed N_2O on Ru(001).

Molecular-beam-scattering studies of surfaces have detector requirements similar to those in neutral ESD/PSD; detectors that have been used in molecular-beam work include electron bombardment detectors (usually coupled with a mass spectrometer), surface ionization detectors, and calorimetric methods. These and other techniques are discussed by English and Zorn[69] and Valyi.[76] Electron bombardment detectors offer the advantage of being "universal" detectors, but at the expense of low detection efficiency and sensitivity to background gas molecules. Surface ionization detectors, in which atoms or molecules are thermally ionized on collision with a heated metal or oxide surface, are highly sensitive but limited to a relatively small class of atoms and molecules. Zandberg and Ionov[77] have provided a comprehensive discussion of the theoretical and practical aspects of surface ionization and have listed the elements that have been detected using the surface ionization method (Table II).

A sensitive superconducting thin film bolometer detector has been tested

TABLE II. List of Atomic Species That Have Been Detected Using a Surface Ionization Method[a]

Element	Emitter	Element	Emitter
Li	W	Al	W
Na	$W + O_2$	Ga	$W + O_2$
K	W	In	$W + O_2$
Rb	W	Tl	$W + O_2$
Cs	W	F	$W + Th$
Mg	$W + O_2$	Cl	W
Ca	$W + O_2$	Cl	$W + Th$
Sr	$W + O_2$	Br	$W + Th$
Ba	W	I	$W + Th$
Ba	$W + O_2$	Pr	Mo

[a] From Zandberg and Ionov.[77]

[75d] P. Feulner, W. Riedl, and D. Menzel, *Phys. Rev. Lett.* **50**, 986 (1983).
[76] L. Valyi, "Atom and Ion Sources." Wiley, New York, 1977.
[77] E. Y. Zandberg and N. I. Ionov, "Surface Ionization" (transl. from Russ. Isr. Program Sci. Transl., Jerusalem, 1971; also available from U.S. Dep. Commer., NTIS, Springfield, Virginia.

by Cardillo et al.[78] in molecular-beam scattering from surfaces. The thermal energy released on condensation of the beam on a liquid-He-cooled detector is measured; sensitivities to thermal beams having fluxes of 10^{12} particles sec^{-1} are measured with a signal/noise ratio greater than 100.

Other potentially useful techniques for detecting certain molecules and molecular fragments (CH_2, NO, CN, OH, . . .) that may possibly be desorbed by ESD under appropriate conditions are laser-excited fluorescence[79] and multiphoton ionization.[80] These highly sensitive laser techniques are sensitive to single atoms or molecular fragments.

We expect studies of threshold and yields for neutral ESD to be of great importance in clarifying the mechanisms of the electronic excitations.

9.2.2. Evaluation of ESD and PSD Data: Cross Sections and Electron-Induced Surface Damage

9.2.2.1. Calculation of Cross Sections for Electron-Induced Surface Processes.

Moore[43] demonstrated a number of years ago for CO on Mo that the ESD ion current is a linear function of electron bombardment current over 5 orders of magnitude of current measurement; the proportionality was found to be true for several values of electron energy. Furthermore, it is well known that the ESD ion yield varies with electron energy at constant electron current.

Such experimental observations substantiate the hypothesis that ESD results can be analyzed in terms of isolated desorption processes; that is, the ESD ion yield from a surface monolayer at a fixed electron energy can be written as a first-order rate equation[1-7]

$$i^+ = I_e Q^+ N, \qquad (9.2.13)$$

where i^+ is the ion current in amperes, I_e the electron bombardment current in amperes, Q^+ the cross section for desorption of ions in square centimeters, and N the surface coverage (atoms, or molecules per square centimeter) in the binding state affected by ESD. The quantities i^+ and I_e are easily measured in an apparatus of the type described in sub-section 9.2.1.1. (Fig. 4). If N is known from other measurements, then Q^+ can be computed. For some

---

[78] M. J. Cardillo, G. E. Becker, G. D. Kubiak, V. Narayanamurti, and M. A. Chin, *J. Vac. Sci. Technol.* **17**, 130 (1980).

[79] R. N. Zare and P. J. Dagdigian, *Science (Washington, D.C.)* **185**, 739 (1974); G. S. Hurst, M. G. Payne, S. D. Kramer, and J. P. Young, *Rev. Mod. Phys.* **51**, 767 (1979).

[80] N. Winograd, J. P. Baxter, and F. M. Kimock, *Chem. Phys. Lett.* **88**, 581 (1982); F. M. Kimock, J. P. Baxter, P. H. Kobrin, D. L. Pappas, and N. Winograd, *Proc. Soc. Photo-Opt. Instrum. Eng.* **426**, 24 (1983).

9.2. EXPERIMENTAL METHODS IN ESD AND PSD

adsorption systems,[15,81] i^+ varies linearly with the coverage N at the initial stages of adsorption, as given in Eq. (9.2.13). In other cases,[59,82] N is the concentration of species in a particular binding state that appears at high coverages and may be a small fraction of the total quantity of adsorbate; this makes unequivocal determination of N (and Q^+) difficult.

The time rate of change of coverage N under electron bombardment can be written as

$$-dN(t)/dt = n_e QN, \qquad (9.2.14)$$

where Q is the *total* cross section and can include one or more processes that result in decrease in population of those species affected by ESD. They may include desorption of ions, ground state neutrals, metastables, and conversion from one binding mode to another.[1-7] The electron bombardment flux n_e is given by $n = I_e/A\epsilon$, where A is the sample area in square centimeters bombarded by the electron beam of current I_e, and ϵ is the charge on the electron in coulombs.

Integration of Eq. (9.2.14) yields

$$N(t)/N_0 = \exp - (I_e Qt/A\epsilon) \qquad (9.2.15)$$

(The discussion in Section 9.2.3 includes a variety of ways to measure $N(t)$, such as surface work function and Auger electron spectroscopy.)

For a simple ESD desorption process, a plot of $\ln[N(t)/N_0]$ versus t will yield a straight line whose slope varies directly with I_e. If competing processes occur (e.g., desorption and conversion from one binding mode to another), such plots may exhibit curvature.

In certain cases (particularly for weakly bound physisorbed layers or insulating substrates), there is a possibility that electron bombardment heating can induce thermal desorption. Equation (9.2.15) provides a test for this. If the slope of the $\ln[N(t)/N_0]$ versus t plot varies directly with I_e, the coverage change is due to ESD. If the slope increases faster than a linear dependence on I_e, there are likely to be thermal contributions to the desorption process caused by beam heating (i.e., Eq. (9.2.14) may have a thermal desorption term added to the ESD term).

If we differentiate Eq. (9.2.13) and substitute Eq. (9.2.14), we obtain

$$di^+/dt = -nQi^+ = I_e Q i^+/A\epsilon. \qquad (9.2.16)$$

We then integrate (9.2.16) to obtain

$$i^+(t)/i_0 = \exp - (I_e Qt/A\epsilon) \qquad (9.2.17)$$

---

[81] T. E. Madey, *Surf. Sci.* **36**, 281 (1973).
[82] S. W. Bellard and E. M. Williams, *Surf. Sci.* **80**, 450 (1979).

Equation (9.2.17) has the same functional form as Eq. (9.2.15), so that Q can be determined from the dependence of $i^+(t)$ during electron bombardment. Equation (9.2.17) must be used with care, however. For example, the O^+ ion yield from oxygen on W(100) is dominated by the β_1 state, which comprises, at most, a few percent of the total oxygen coverage.[59,81,82] The value of Q for β_1 oxygen only relates to this minority state; the total cross section for the majority of the adsorbed species has been determined from Eq. (9.2.11) using Auger spectroscopy.[82]

Table III lists some representative values of ESD cross sections for various adsorbates and substrates; more complete tables are given by Madey and Yates,[1] Menzel,[2] and Shek et al.,[83] and Redhead et al.[84]

In principle, the above relations can be used to determine PSD cross sections for specific values of photon energy; simply substitute photon flux n_{ph} for electron bombardment flux $n_e = I_e/A\epsilon$.

9.2.2.2 Assessment of Damage by Electron Bombardment of a Surface Monolayer.

We now address the question: how long can an adsorbed monolayer be bombarded by electrons before detectable damage will occur?

From Eqs. (9.2.15) and (9.2.17), we can define a lifetime, τ (s), of a surface layer equal to the time in which $N(t)$ or $i^+(t)$ drops to $1/e$ of its initial value as

$$\tau = A\epsilon/I_e Q. \qquad (9.2.18)$$

(Note that for a 5-μA beam 1 mm in diameter — typical LEED conditions — and for "worst case" values of $Q \sim 10^{-16}$ cm^2, τ is about 3 s). The "critical electron exposure" $D_{1/e}$ necessary to reduce the concentration to the $1/e$ value is the product of current density times τ,

$$D_{1/e} = (I_e/A)\tau \quad \text{or} \quad D_{1/e} = \epsilon/Q. \qquad (9.2.19)$$

The units of $D_{1/e}$ are C cm^{-2}. For $Q \sim 10^{-16}$ cm^2, $D_{1/e} \simeq 1.6 \times 10^{-3}$ C cm^{-2}. As will be seen below, numbers of this magnitude appear frequently in discussions of surface analysis of fragile or unknown materials.

For most surface science applications, a more reasonable criterion would be the dose D_d necessary to cause *detectable* damage to the monolayer, i.e., $\Delta N/N = 0.1$ (i.e., $N(t)/N_0 = 0.9$). Thus, from Eq. (9.2.15), the electron dose required to cause detectable damage D_d is

$$D_d = 0.11\epsilon/Q.$$

We should emphasize that beam damage may also include chemical or physical changes in the surface layer (e.g., dissociation without desorption) that do not change N.

[83] M.-L. Shek, S. P. Withrow, and W. H. Weinberg, *Surf. Sci.* **72**, 678 (1978).
[84] P. A. Redhead, J. P. Hobson, and E. V. Kornelson, "The Physical Basis of Ultrahigh Vacuum." Chapman & Hall, London, 1968.

TABLE III. Some Representative Cross Sections for ESD of Adsorbed Species

Adsorbate	Substrate	Binding State	Ion	Ionic Cross Section Q^+ (cm²)	Electron Energy (eV)	Total Cross Section Q (cm²)	Dissociation–Desorption Ratio	Threshold (eV)	IED Peak (eV)	References
CO	Ir(111)	$\theta = 1/3$	—	—	86	0.8×10^{-17}–1.7×10^{-17}	≤0.01–0.02	—	—	83
CO	Ir(111)	$\theta = 1/3$	—	—	2500	1×10^{-7}	—	—	—	83
CO	Pt(111)		—	—	1500	5.4×10^{-18}	0.04	—	—	85
CO	W	Virgin	—	—	80	3×10^{-19}	—	—	—	24
CO	W	α	—	—	80	3×10^{-18}	—	—	—	24
CO	W	β	—	—	80	3×10^{-21}	—	—	—	24
CO	W	1	CO⁺	1–2×10^{-20}	100	3×10^{-19}	—	15.1	1	86c
CO	W	2	O⁺	1–2×10^{-20}	100	3×10^{-18}	—	18.7	7	86c
O_2	Te film	10 Å TeO_2	—	—	2000	3×10^{-18}	—	—	—	86a
O_2	Ti	—	—	—	300	10^{-18}–10^{-19}	—	—	2	86b
H_2	W(100)	$\theta \leq 0.2$	H⁺	2×10^{-23}	100	—	—	—	—	86e
H_2	W	$\theta < 1$	H⁺	3×10^{-23}	100	—	—	—	—	86f
C_6H_{12}	Ru(001)	$\theta < 1$	H⁺	—	150	8×10^{-17}	—	—	—	119
NH_3	Ni(111)	$\theta < 1$	—	—	100	1.6×10^{-16}	—	—	—	86d
Xe	Ni, Pt	$\theta < 1$	—	—	1000–3000	1×10^{-17}	—	—	—	87

[85] R. M. Lambert and C. M. Comrie, *Surf. Sci.* **59**, 33 (1976).
[86] (a) R. G. Musket, *Surf. Sci.* **74**, 423 (1978); (b) P. H. Dawson, *Surf. Sci.* **65**, 41 (1977); (c) P. A. Redhead, *Nuovo Cimento, Suppl.* **5**, 586 (1967); (d) C. Seabury, T. N. Rhodin, R. J. Purtell, and R. P. Merrill, *Surf. Sci.* **93**, 117 (1980); (e) T. Madey, *Surf. Sci.* **36**, 281 (1973); (f) M. Nishijima and F. M. Propst, *J. Vac. Sci. Technol.* **7**, 420 (1970).
[87] B. G. Baker and B. A. Sexton, *Surf. Sci.* **52**, 353 (1975).

In a following subsection 9.2.4, we shall discuss the critical electron exposures necessary to observe beam damage in thin films and solids.

9.2.3. Methods Based on Changes in Surface Properties: Detection of ESD and PSD Damage in Adsorbed Monolayers

A number of different experiments have been reported in which changes in some surface property were used to deduce that electron stimulated desorption, cracking, or conversion from one binding state to another has occurred. Some of these methods have been discussed previously[1-4] and include field emission and retarding-potential work function measurements, thermal desorption measurements, changes in low-energy electron diffraction (LEED) patterns, changes in electron reflection coefficient, and decreases in Auger peak intensities. In each case, the cross sections Q for the ESD processes are determined by using Eq. (9.2.15) described above,

$$N(t)/N_0 = \exp - (I_e Qt/A\epsilon). \tag{9.2.15}$$

The ratio $N(t)/N_0$ is determined by assuming that the magnitude of the physical or chemical change being measured is proportional to surface coverage.

Some interesting applications of other surface characterization methods to studies of ESD processes will be discussed in more detail. Although most of the examples will involve ESD of adsorbed CO, this is *not* the only molecule subject to electron damage! Hooker and Grant[88] applied Auger electron spectroscopy to a study of CO on Ni(110). Carbon monoxide is known to adsorb molecularly on Ni, but under electron bombardment by the Auger beam, electron-stimulated desorption and dissociation processes occur. Figure 17 illustrates the changes in both Auger line shape and peak intensity during the measurement. The Auger line shape is known to be an indicator of the local chemical state of an atom,[89] and the carbon is clearly transformed into a surface carbidelike species after prolonged electron bombardment. A fraction of the carbon is desorbed, probably as molecular CO.

Martinez and Hudson[90] have also applied Auger spectroscopy to an examination of the adsorption and decomposition of CO on Pt(111). The Pt surface was saturated with a monolayer of molecular CO, the electron beam was turned on, and the oxygen and carbon Auger peaks were monitored as a function of time. The data are plotted in Fig. 18, and are explained by using a model in which the CO is excited by the electron beam and subsequently

---

[88] M. P. Hooker and J. T. Grant, *Surf. Sci.* **55**, 741 (1976).
[89] R. R. Rye, T. E. Madey, J. E. Houston, and P. H. Holloway, *J. Chem. Phys.* **69**, 1504 (1978).
[90] J. M. Martinez and J. B. Hudson, *J. Vac. Sci. Technol.* **10**, 35 (1973).

9.2. EXPERIMENTAL METHODS IN ESD AND PSD

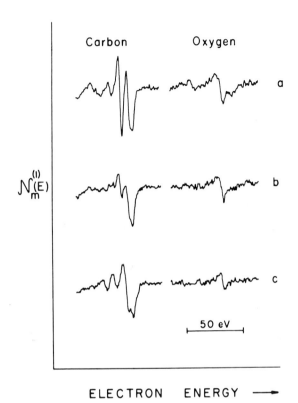

FIG. 17. AES line shape changes during ESD of CO/Ni (110) with a 1.5-keV, 1.5-μA electron beam for (a) 0 min, (b) 10 min, and (c) 40 min. (From Hooker and Grant.[88] Copyright North-Holland Physics Publishing, Amsterdam, 1976.)

decays by either desorption of molecular CO or by desorption of an O atom, leaving an adsorbed C atom. The cross section for this process is 2×10^{-19} cm^2.

Fuggle et al.[91] have used thermal desorption methods, ultraviolet photoemission spectroscopy (UPS), x-ray photoelectron spectroscopy (XPS), and x-ray excited Auger spectroscopy (XAES) to study the electron-induced decomposition of CO on Ru(001). The chemical shift in the XPS O 1s peak following electron bombardment clearly indicates that dissociation of molecular CO into adsorbed atomic C and O has occurred. A similar effect was also seen in a LEED study of CO on Ru(001).[91a]

Yates and King[92] used thermal desorption methods to measure the de-

---

[91] J. C. Fuggle, E. Umbach, P. Feulner, and D. Menzel, *Surf. Sci.* **64**, 69 (1977).
[91a] T. E. Madey and D. Menzel, *Proc. 2nd Int. Conf. Solid Surf., Kyoto*, p. 229 (1974).
[92] J. T. Yates, Jr. and D. A. King, *Surf. Sci.* **38**, 114 (1973).

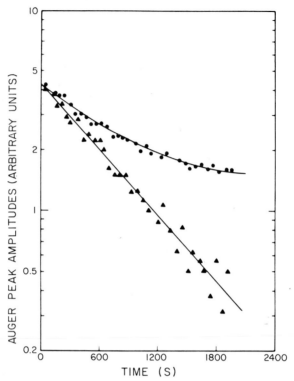

FIG. 18. Electron impact desorption and decomposition of CO on Pt(111). $E_p = 1570$ eV, $I_e = 11$ μA, $A = 1$ mm². (●) Carbon, (▲) oxygen. (From Martinez and Hudson.[90])

crease in coverage of the molecular α-CO state on W(100) as a function of electron bombardment time. The cross section they measured for ESD of CO (4×10^{-18} cm²) was in good agreement with that deduced from a measurement of the decrease in CO⁺ yield versus time (cf. Eq. (9.2.17)).

An excellent account of the use of LEED and Auger to characterize ESD processes in the case of CO on Ir(111) is given by Shek et al.[83] They discuss both experimental procedures and the methodology for extracting cross sections.

Electron stimulated desorption processes have plagued LEED investigators for years. Low-energy electron diffraction structure analyses of "fragile" adlayers require measurements of intensity versus electron energy of many individual LEED beams over a wide energy range. This is usually a time-consuming process, during which ESD can be a major problem. In an attempt to minimize the effect of beam damage on adsorbed layers during

LEED intensity measurements, Müller and co-workers[93] devised a technique for recording LEED intensities very rapidly using a photographic method.

Unusual electron-beam-induced *disordering* effects have been reported in a LEED study by Naumovets and Fedorus.[94] They observed that ordered submonolayer Li films adsorbed on a W(110) surface are disordered upon bombardment of the surface at 5 K by low-energy electrons. The phenomenon is nonthermal in nature, and characterized by a threshold energy of 54 eV — the Li K shell ionization energy. They suggest that the first stage in the disordering process is excitation of the Li adatom by creation of a vacancy in its K shell. This observation and interpretation predated the core-hole ionization model of Knotek and Feibelman.[10] A similar electron-beam induced disordering process[94] has been observed for ordered hydrogen layers on W(110) at 5 K, even for electrons having near zero kinetic energy!

It is safe to conclude that any method sensitive to surface coverage, chemical state, geometrical ordering, or electronic properties can be used to assess the extent of beam damage.

9.2.4. Measurements of Electron-Beam-Induced Damage in Thin Films and Bulk Samples

To this point, we have concentrated on those excitations that lead primarily to desorption during electron bombardment of a solid surface. However, in many cases the damage wrought by an electron beam can be quite deep. The electron penetration depth[95] in Al, for example, is ~ 5000 Å for a 5-keV beam. This is much greater than the mean free path for inelastic electron scattering,[96] typically 5 – 30 Å for electron energies < 5 keV. Deep in the bulk of solid samples, as well as in the surface layer, electron bombardment can result in excitation of core or valence electrons. Depending on the electronic structure of the solid, bonds can be broken, radicals and other products formed, diffusion initiated, defects produced, and even collapse of the crystal lattice brought about. Such processes can be particularly troublesome in surface analysis techniques such as, for example, Auger spectroscopy, which samples the subsurface region of a solid. For many nonmetallic samples, there can be more potentially damaging events (due to low-energy electron

---

[93] K. Heinz, E. Lang, and K. Müller, *Surf. Sci.* **87**, 595 (1979).
[94] A. G. Naumovets and A. G. Fedorus, *Zh. Eksp. Teor. Fiz.* **68**, 1183 (1975) [*Sov. Phys.—JETP (Engl. Transl.)* **41**, 587 (1976)]; V. V. Gonchar, O. V. Kanash, A. G. Naumovets, and A. G. Fedorus, *Pis'ma Zh. Eksp. Teor. Fiz.* **28**, 358 (1978) [*Sov. Phys.—JETP Lett. (Engl. Transl.)* **28**, 330 (1978)].
[95] J. I. Goldstein, *in* "Practical Scanning Electron Microscopy, Electron and Ion Microprobe Analysis" (J. I. Goldstein and H. Yakowitz, eds.), p. 49. Plenum, New York, 1976.
[96] M. P. Seah and W. A. Dench, *Surf. Interface Anal.* **1**, 2 (1979).

losses) than potentially information-producing events (brought about by core-hole ionization). For many biological molecules and organic layers, an average of 40 eV is transferred per inelastic event, enough energy to break most chemical bonds and produce fragments with several electron volts kinetic energy.[97] Corbett[98] has reviewed the status of radiation damage and defects in bulk metallic, covalent, and ionic systems; he has considered damage production processes, defect configurations, and defect migration processes in the vicinity of a surface.

Knotek and Feibelman[28] have provided criteria for the stability of ionically bonded surfaces in ionizing environments. They base their arguments on the core-hole Auger mechanism described in Section 9.1.3. The materials most likely to be decomposed under electron irradiation are certain maximal valency compounds. Such materials include TiO_2, V_2O_5, MoO_3, SiO_2, and Al_2O_3. The effect of covalency on decomposition is less clear, although creation of *deep* core holes may lead to an Auger cascade and coulomb explosion in some instances.[32,33]

In the following sections, a few examples of beam damage in surface analysis as well as "critical doses" for minimally perturbed analysis will be given, and a brief introduction to the literature of electron-beam lithography will be provided. The reader is referred to Pantano and Madey[99] for a more detailed discussion of electron-beam damage in Auger electron spectroscopy.

9.2.4.1. Electron- and Photon-Induced Damage in Surface Analysis. Van Oostrom[100] has published a review of Auger microanalysis, and has discussed such problems as charging, heating by the electron beam, and beam-induced damage. He considered the case of electron impact dissociation and desorption of Al_2O_3, and his data are shown in Fig. 19. When an electron beam strikes an oxidized Al target, decomposition occurs and results in two peaks in the Auger spectrum at 54 and 68 eV. These peaks correspond to Al in the oxidized and metallic state, respectively. Figure 19 is a plot of the Auger peak-to-peak heights for the low-energy Al and the oxygen peaks. It is seen that an electron dose of $< 10 \, C \, cm^{-1}$ is needed for a "correct" analysis of this surface. At higher doses, ESD processes affect the analysis. Pantano and Madey[99] and Fontaine *et al.*[101] have discussed other examples of beam damage in Auger microscopy.

---

[97] J. P. Langmore, *in* "Principles and Techniques of Electron Microscopy; Biological Applications" (M. A. Hayat, ed.) Vol. 9, p. 1. Van Nostrand/Reinhold, New York, 1978.
[98] J. W. Corbett, *Surf. Sci.* **90**, 205 (1979).
[99] C. G. Pantano and T. E. Madey, *Appl. Surf. Sci.* **7**, 115 (1981).
[100] A. van Oostrom, *Surf. Sci.* **89**, 615 (1979).
[101] J. M. Fontaine, J. P. Durand, and C. LeGressus, *Surf. Interface Anal.* **1**, 196 (1979).

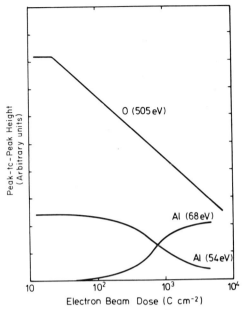

FIG. 19. Effect of electron bombardment dose (C cm^{-2}) on AES peak-to-peak height during electron bombardment of Al_2O_3. (From van Oostrom.[100] Copyright North-Holland Physics Publishing Amsterdam, 1979.)

Sasaki and co-workers[102] have used x-ray photoelectron spectroscopy (XPS) to examine electron irradiation damage in two "maximal valency" oxides, Li_2CrO_4 and Li_2WO_4. Single crystals were irradiated using 0.3–1.6 keV electrons *in situ* in an XPS spectrometer. Chemical shifts of core-level peaks in XPS spectra indicated formation of Cr III on the irradiated Li_2CrO_4 surface and of W IV and W V on the irradiated Li_2WO_4 surface. Their data indicated that substantial damage was observed on the Li_2CrO_4 for doses as low as 5×10^{-2} C cm^{-2}. The differential energy loss (the so-called "G value"—a figure of merit for measure of radiation damage, defined as the number of molecules produced or decomposed per 100 eV of radiation adsorbed) for the Cr compound is about 10^{-1}. Comparison of this G value with previous data suggests that solid surfaces of this compound are several orders of magnitude *more sensitive* to ionizing radiation than bulk crystals.

In a round-robin characterization of catalyst samples (SiO_2, Al_2O_3 and a Na-activated zeolite) using Auger spectroscopy,[103] large variations in peak

[102] T. Sasaki, R. S. Williams, J. W. Wong, and D. A. Shirley, *J. Chem. Phys.* **69**, 4374 (1978).
[103] T. E. Madey, C. D. Wagner, and A. Joshi, *J. Electron Spectrosc. Relat. Phenom.* **10**, 359 (1977).

height ratios were observed. The most probable cause of these variations was believed to be electron-beam-induced damage. Both desorption and electron-beam-induced diffusion[104] processes were suspected.

Auger line shape analysis, photoelectron spectroscopy, and thermal desorption spectroscopy have been used to study the effects of electron bombardment on condensed multilayers of $(CH_3)_2O$, CH_3OH, and H_2O.[105] The results show that electron doses as low as 5×10^{-4} C cm^{-2} (a 0.5-mm diameter, 1-μA beam for 1 s) can cause detectable damage. New chemical species are created in the condensed layer by the electron-beam interaction, with water and hydrocarbons being most abundant.

As a final example of electron-beam damage in surface analysis, we consider the work of Coad et al.,[106] in which XPS was used to examine beam damage caused in typical AES studies of "as received" oxidized metals. They point out that XPS and AES analyses of virgin surfaces are frequently in disagreement; for example, the oxygen and carbon levels appear much larger using XPS. By rastering an electron beam over the area analyzed by XPS, and thereby reducing the electron-beam exposure, they found that the XPS analysis was more comparable to that by AES. They conclude that AES can provide an unperturbed analysis of a virgin surface only at electron-beam exposures of about 2×10^{-3} C cm^{-2} or less! This exposure is less than that required for most commercial Auger systems by several orders of magnitude.

The x-ray photons used in typical XPS experiments can also introduce radiation damage by production of secondary electrons, as well as by the primary excitation process (production of core and valence holes). Evidence for such processes has been reported in XPS studies of inorganic compounds,[107] and has been reviewed by Copperthwaite.[108] In general, however, photon beams produce less damage per inner shell excitation than electron beams, provided that the photon energy is not too much greater than the excitation energy of the core-hole edge under investigation.[109,110]

9.2.4.2. Critical Doses for Beam Damage in Films and Solids. In Table IV, we tabulate a representative list of solids and films for which electron-beam-induced damage has been observed.[99] We loosely distinguish between

---

[104] C. G. Pantano, Jr., D. B. Dove, and G. Y. Onada, Jr., *J. Vac. Sci. Technol.* **13**, 414 (1976).
[105] P. H. Holloway, T. E. Madey, C. T. Campbell, R. R. Rye, and J. E. Houston, *Surf. Sci.* **88**, 121 (1979).
[106] J. P. Coad, M. Gettings, and J. C. Riviere, *Faraday Discuss. Chem. Soc.* No. 60, p. 269 (1975).
[107] R. G. Copperthwaite and J. Lloyd, *J. Electron Spectrosc. Relat. Phenom.* **14**, 159 (1978).
[108] R. G. Copperthwaite, *Surf. Interface Anal.* **2**, 17 (1980).
[109] A. Barrie, *J. Electron Spectrosc. Relat. Phenom.* **7**, 75 (1975).
[110] M. Isaacson and M. Utlaut, *Optik (Stuttgart)* **50**, 213 (1978).

TABLE IV. Critical Electron Exposures for Electron-Bombardment Damage of Molecular Solids and Films[a]

Sample	Method element(s) monitored	Electron energy (eV)	$D_{1/e}$ (C cm^{-2})	D_d (C cm^{-2})	Ref.
Oxidized Al	AES; O, Al	5000	—	10	100
Cu-, Fe, and metal-free pthalocyanines	{LEED {AES	15–200 1000	— —	≥1 >1	111
Cu pthalocyanine	ELS	8×10^4	0.8	—	112
SiO$_x$ (1.2 < x < 2)	AES; O, Si	2000	—	0.6	113
Si$_3$N$_4$	AES; O, N	2000	—	stable	113
Li$_2$WO$_4$	XPS; W	300–1600	—	0.5	102
NaF, LiF, NaCl	LEED	100	—	0.06	114
KCl	AES; Cl	1500	0.25	0.03	114
Li$_2$SO$_4$	XPS; O, S	1000	~0.5	0.05	115
Li$_2$CrO$_4$	XPS; Cr	300–1600	—	<0.05	102
Napthalene	LEED	>25	—	0.03	116
Oxidized Te	AES	2000	—	0.02	117
H$_2$O (ice film)	AES; O	1500	—	10^{-2}	105
H$_2$O (ice film)	LEED	80	≥10^{-2}		13
LiNO$_3$	XPS; N, O	1000	0.05		115
Nitrocellulose	ELS; N	7×10^4	2.3×10^{-3}		112
	ELS; O	7×10^4	5.7×10^{-3}		
Air-exposed metals	XPS; C, O	—	—	2×10^{-3}	106
Formvar film	ELS; O	7.5×10^4	—	2×10^{-3}	112
(CH$_3$)$_2$O film	AES; C, O	1500	—	1×10^{-3}	105
Na$_3$AlF$_6$	AES; F	3000	—	10^{-4}–10^{-3}	118
C$_6$H$_{12}$ film	LEED	100–200	3×10^{-4}	—	119
CH$_3$OH film	AES; C, O	1500	—	2.5×10^{-4}	105

[a] $D_{1/e}$ is the electron exposure necessary to reduce the measured concentration to about $1/e$ of its initial value, D_d is the dose at which damage is detected.

$D_{1/e}$ (electron dose necessary to reduce the concentration of one chemical state to $1/e$ of its initial value) and D_d (the electron dose at which damage is clearly detectable). In many of the references the dose estimates are not precise, so that these data have mainly order-of-magnitude significance. Nevertheless, it is of interest to note that many of the critical doses are in the

[111] J. C. Buchholz and G. A. Somorjai, *J. Chem. Phys.* **66**, 573 (1977).
[112] R. F. Egerton, *Proc. Ann. Meet., Electron Microsc. Soc. Am.* **37**, 514 (1979).
[113] Y. E. Strausser and J. S. Johannessen, *ARPA/NBS Workshop IV, Surf. Anal. Silicon Devices, Gaithersburg, Md., 1975,* published in *NBS Spec. Publ. (U.S.)* No. 400-23, p. 125 (1976).
[114] H. Tokutaka, M. Prutton, I. G. Higginbotham, and T. E. Gallon, *Surf. Sci.* **21**, 233 (1970).
[115] T. Sasaki, R. S. Williams, J. S. Wong, and D. A. Shirley, *J. Chem. Phys.* **68**, 2718 (1978).
[116] L. E. Firment and G. A. Somorjai, *Surf. Sci.* **55**, 413 (1976).
[117] R. G. Musket, *Surf. Sci.* **74**, 423 (1978).
[118] A. G. Knapp and J. R. Hughes, *Proc. 7th Int. Vac. Congr., Vienna,* p. 2161 (1977).
[119] T. E. Madey and J. T. Yates, Jr., *Surf. Sci.* **76**, 397 (1978).

range $10^{-2}-10^{-3}$ C cm^{-2}. That this range of values is consistent with electron–molecule processes will be shown below.

As discussed in Sections 9.2.2 and 9.2.3, procedures for determining cross sections for ESD processes in adsorbed monolayers are well known. However, the sensitivity of the surface analysis methods used to obtain the data in Table IV (AES, XPS, etc.) to several molecular layers of solid, as well as the complex chemistry in the bombarded region, makes this a less straightforward process for multilayers.[99]

We assume by analogy with ESD in monolayers that the concentration of undamaged species within the detection region at any time $X(t)$, obeys a first-order relation [cf. Eq. (9.2.15)]

$$X(t)/X_0 = \exp - (I_e Q't/A\epsilon). \qquad (9.2.20)$$

The quantities I_e, A, and ϵ have been defined previously. The quantity X_0 is the initial undamaged concentration (molecules cm^{-3}) within the analysis region; Q' (in square centimeters) is an *effective* cross section (due to the fact that the analysis method is sensitive to more than one monolayer) for the electron-stimulated decrease in the ratio $X(t)/X_0$. For AES or XPS, the analysis region is taken to be approximately three times the inelastic mean free path λ_e for electrons at the energy being analyzed. Within the analysis depth $d = 3\lambda_e$, the number of molecular "layers" is $n = 3\lambda_e/a$, where a is the appropriate molecular dimension. The cross section *per molecule* for electron damage is approximated by $Q = nQ'$, where n is typically 10–20 layers. Hence, following Eq. (9.2.20), the critical electron exposure for damage in a molecular solid is

$$D_{1/e} = \epsilon/Q' = n\epsilon/Q. \qquad (9.2.21)$$

If we assume Q to be of the order 10^{-16} cm^2, a typical value for dissociative ionization of small gas-phase molecules, then $D_{1/e} \approx 1.6 \times 10^{-3}$ C cm^{-2}. The fact that so many organic and inorganic solids exhibit critical doses in this range indicates that the primary ionization process dominates the beam damage mechanism, and that de-excitation and neutralization effects are not as important. Values of $D_{1/e}$ much greater than 10^{-3} C cm^{-2} are largely a result of reneutralization or "bond healing" processes, whereas much smaller values can be due to the multiple effects of secondary electrons and high damage cross sections. We emphasize that the above formalism and the numbers derived from it are to be viewed as useful approximations, not as a rigorous analysis. The cross sections Q depend on bonding and can be sensitive functions of the excitation energy (ignored for simplicity). We have also not explicitly considered the fact that for electron beams having energies of a few kilo-electron-volts, the probability of damage by low-energy secondaries can be comparable to or greater than the primary beam damage.

Generally speaking, inorganic compounds are more stable than organics, with saturated hydrocarbons being less stable than molecules with unsaturated carbon bonds.

9.2.4.3. Electron Lithography. In surface analysis, the objective is to make rapid analyses with minimal ESD effects. Electron lithography,[120-122] on the other hand, requires both negative and positive resists that have been designed to be highly sensitive to electron irradiation. It is beyond the scope of this article to discuss the methods, physics, and chemistry of electron lithography, other than to point out that the sensitivities (critical doses) typically found for 10-keV electrons interacting with resist materials are in the range $10^{-5}-10^{-7}$ C cm^{-2}. These doses are orders of magnitude smaller than the threshold doses for ESD damage, which are in the range $10^{-3}-10$ C cm^{-2} for "fragile" materials (cf. Table IV). The resist materials are very high-molecular-weight polymers, and bombardment by either primary or secondary electrons causes such radiation damage processes as crosslinking and/or chain scission to occur with a high probability. One damage event in a large molecule can have a profound effect on its chemistry; hence, the critical doses required are smaller than those observed in the AES studies of smaller molecules.

9.2.4.4. Specimen Damage in Electron Microscopy. Electron microscopists have long been concerned with electron-induced specimen damage, particularly for high primary beam energies (~ 80 keV), and some of the data in Table IV is taken from that literature. A more extensive discussion of experimental methods in electron microscopy and the electron microprobe is beyond the scope of this review, but the interested reader is referred to the literature.[97,99,110,112,123-126]

9.2.5. Electron-Stimulated Adsorption

There have been a number of reports using AES of electron-beam-induced adsorption processes in which enhanced adsorption on surfaces is observed to occur with simultaneous electron irradiation and gas exposure. For these cases, gas exposure in the *absence* of irradiation results in adsorption processes that are either slow or nonexistent. It has been suggested that the gas is first adsorbed into a weakly bound state, is then dissociated by the electron beam, and finally is either chemisorbed on the surface or diffuses into the

---

[120] M. J. Bowden and L. F. Thompson, *Solid State Technol.* **22**, 72 (1979).
[121] L. F. Thompson, L. E. Stillwagon, and E. M. Doerries, *J. Vac. Sci. Technol.* **15**, 938 (1976).
[122] L. F. Thompson and R. E. Kerwin, *Ann. Rev. Mater. Sci.* **6**, 267 (1976).
[123] S. J. B. Reed, "Electron Microprobe Analysis." Cambridge Univ. Press, London and New York, 1975.
[124] R. F. Egerton, *Phys. Status Solidi A* **37**, 663 (1976).
[125] R. M. Glaeser, in "Introduction to Analytical Electron Microscopy" (J. J. Hren, J. Goldstein, and D. C. Joy, eds.), p. 423. Plenum, New York, 1979.
[126] D. G. Howitt and G. Thomas, *Radiat. Eff.* **34**, 209 (1977).

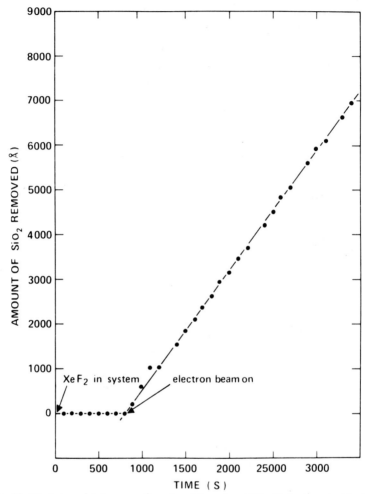

FIG. 20. Electron-assisted gas–surface chemistry using 1500-eV electrons incident on an SiO_2 surface exposed to gaseous XeF_2. Beam current, 45 μA; beam current density, 50 mA cm^{-2}. $P_{total} = 6 \times 10^{-4}$ Torr with most of the ambient gas being xenon. Neither exposure to XeF_2 nor an electron beam produces etching by itself. Simultaneous exposure produces an etch rate of ~ 200 Å min.$^{-1}$ (From Coburn and Winters[129]).

bulk. In the absence of the beam, the weakly bound molecules simply desorb from the surface. This interpretation is consistent with the electron-stimulated adsorption of N (from NH_3) on Ru(001),[127] and the beam-induced oxidation of polycrystalline Ni.[128]

[127] L. R. Danielson, M. J. Dresser, E. E. Donaldson, and J. T. Dickinson, *Surf. Sci.* **71**, 615 (1978).
[128] J. Verhoeven and J. Los, *Surf. Sci.* **58**, 566 (1976).
[129] J. N. Coburn and H. F. Winters, *J. Appl. Phys.* **50**, 3189 (1979).

A dramatic example of electron-assisted surface chemistry,[129] which appears to be initiated by electron-stimulated adsorption, is shown in Fig. 20. Exposure of an SiO_2 film to XeF_2 gas results in little or no adsorption or chemical attack on the film. When the electron beam is turned on, the SiO_2 can be etched away at a rate of several hundred angstroms per minute, as determined using a quartz microbalance. The electron beam dissociates weakly bound XeF_2, and the F atoms react with surface Si to form volatile SiF_4; the initial reduction of the SiO_2 to Si may also be a consequence of the bombardment. The chemistry involved may be more complex than this simple reduction and etching of the elemental Si, but this example provides a plausible mechanism.

An alternative mechanism for ESA was suggested by Neave and Joyce.[130] They observed that at the gas pressures typically used for adsorption ($\sim 10^{-6}$ Torr), an ion beam of significant intensity will be created by electron impact ionization, and the ions impinging on the substrate will have quite different adsorption characteristics from the parent neutral species. They provided evidence that the enhanced adsorption of oxygen on GaAs can be explained by such a process. They suggested that the enhanced adsorption could involve multiple effects: the oxygen ions can have a greater sticking probability than neutrals, and ion bombardment damage of the substrate might create active sites for neutral adsorption.

The clear way to minimize ESA effects is to avoid electron bombardment of surfaces in the presence of reactive gases at pressures greater than 10^{-10} Torr.

9.3. Conclusions

Experimental methods based on electron or photon irradiation of surfaces are myriad. In all cases, however, the experimentalist must be aware that the beams are capable of severely perturbing the surface under study. The accompanying ESD/PSD effects can be either a blessing of a curse, but they cannot be ignored. The intent of this chapter has been to provide a guide to the physics and chemistry of ESD/PSD processes, as well as to their measurement and control.

Acknowledgments

The authors are grateful to M. Knotek for providing Fig. 8. This work was supported in part by the Office of Naval Research.

[130] J. H. Neave and B. A. Joyce, *J. Phys. D* **10**, 243 (1977).

AUTHOR INDEX

Numbers in parentheses are reference numbers and indicate that an author's work is referred to although the name is not cited in the text.

A

Abbott, R. C., 370, 416(99, 100), 417
Adler, I., 230
Adler, J. G., 26, 101(49)
Adnot, A., 26, 27(47), 32, 47(67), 50(67), 122(47, 67), 123
Agamy, S. A., 313, 316(43)
Ageev, V. N., 432, 444, 464, 465, 476, 498(4), 499(4), 502(4)
Agrawal, D. C., 418
Ahern, A. J., 354
Aiferieff, M. E., 382
Aitken, K. L., 403
Allison, A. H., 82
Allred, A. L., 107
Allred, D. D., 347
Almén, O., 331
Anbar, M., 400
Andersen, H. H., 331
Anderson, C. R., 226
Anderson, J., 51, 138, 159, 161, 238
Anderson, P. A., 14
Anderson, R. J., 432
Anderson, S., 26, 32(31), 33, 34(31), 35(31), 62(31), 122, 344
Andersson, S., 26, 33, 34, 35(39, 81), 37, 54, 56, 57, 122(39, 81, 117, 254, 283), 123(117), 210
Anger, H. O., 287
Antoneiwicz, P., 469
Aono, M., 348
Appelbaum, J. A., 214
Appleton, B. R., 347
Arai, H., 107
Arakawa, E. T., 197
Aria, H., 105
Aris, R., 449
Artamonov, O. M., 495, 496
Ashcroft, K. W., 474, 475

Ashworth, F., 375
Aspnes, D. E., 81, 82
Ayrault, G., 393, 395(167)

B

Bach, P. W., 410(285)
Bäck, W., 493
Backx, C., 26, 32(45, 46), 33, 50(44, 45), 122(43, 44, 45, 46, 282)
Badgley, R. E., 361
Baer, A. D., 172, 173(46), 177(46), 179(46), 180(46)
Baer, Y., 225, 231
Bagus, P. S., 124
Bahadur, K., 380, 381
Baker, B. G., 501
Ball, D. J., 313, 316(44)
Barbour, J. P., 367
Barker, R. A., 254, 434
Barnes, G. J., 350
Barnes, M. R., 32, 47(68), 50(68), 122(68)
Barnes, R. F., 274, 275(64)
Barnett, E. F., 216
Baró, A. M., 59, 60, 62, 122(132)
Barofsky, D. F., 399
Baroody, E. M., 211
Barrie, A., 224, 508
Barthes, M. G., 447
Basset, J. M., 110
Bassett, D. W., 392, 393, 395
Batra, I. P., 124
Bauer, E., 313, 317(46), 465(5), 466, 474, 476(51), 498(7), 499(7)
Bauer, K., 22
Baun, W. L., 307
Baxter, J. P., 498
Bay, H. L., 331
Bearden, J. A., 189

AUTHOR INDEX

Beaven, P. A., 410, 415
Becker, G. E., 498
Becker, J. A., 350, 354, 372, 374, 428
Beckey, H. D., 400, 401
Bee, J. V., 410(286)
Beeby, J. L., 277
Begeman, S. H. A., 307
Behm, R. J., 62, 84, 122(182)
Behrisch, R., 330, 348
Belin, M., 95, 101, 102(213), 103(200)
Beni, J., 233
Belkin, N. V., 369
Bell, A. E., 361, 362, 367(57)
Bellard, S. W., 499
Benbow, R. L., 476, 482(54b)
Bennett, P. A., 284, 288, 297(86)
Bennette, C. J., 367
Benninghoven, A., 328, 339(76), 343, 346
Berger, A. S., 410
Bermudez, V., 495
Bernasek, S. L., 285
Bertel, E., 471
Bertolini, J. C., 26, 33(41, 42), 34, 122(42, 258, 262, 274, 275, 277, 278)
Bertoni, C. M., 155
Besocke, K., 443
Bettler, P. C., 367
Betty, K. R., 78
Beyer, J., 410(285)
Bhaduri, S., 74, 75
Bhattacharya, R. S., 332
Bienfait, M., 437
Biersack, J., 311
Bingham, J. K., 410(286)
Bird, P. H., 110
Birkner, G. K., 101(217), 102
Blakely, J. M., 238, 254(21), 255(21), 293(21)
Blauth, E., 221
Blauth, E. W., 317, 320
Bloch, J. H., 400, 401, 406, 410(257, 258)
Blott, B. H., 361
Blyholder, G., 65, 105
Boers, A. L., 306, 307, 311, 328(40)
Bohdansky, J., 321, 330(61), 333(61)
Bohn, G. K., 136, 317
Boling, J. L., 367
Bond, W. L., 289
Bonzel, H. P., 25, 70, 122(268)
Bowden, M. J., 511
Bowker, M., 444
Bowman, D. R., 469
Bowsen, W. M., 95, 98, 99, 101, 106(202), 108(202), 109(202), 110(210)
Boyes, E. D., 412
Bozio, R., 102
Bradshaw, A. M., 26, 27, 47(64), 59, 75(53, 54), 79, 81, 84, 85, 86, 87(184), 88, 89, 90, 91, 92, 93, 94, 117(54), 124, 440
Brako, R., 47, 48, 49(107), 50(107)
Brandes, R. G., 372
Brandon, D. G., 397, 398
Braundmeier, A. J., 197
Breig, E. L., 51
Brennan, S., 235, 480
Brenner, S. S., 414
Brent, D. A., 400
Briant, C. L., 418
Brice, D. K., 345, 347
Bril, A., 284
Brodie, I., 207, 359, 401
Bronckers, R. P. N., 310, 328(36)
Brongersma, H. H., 324, 325(64), 328(69)
Bruce, G., 331
Bruche, E., 354
Bruchmann, H. D., 59, 60(132), 62(132), 122(132)
Brundle, C. R., 25, 344
Buchholz, J. C., 509
Buck, T. M., 308, 313, 316, 319, 320(59), 324, 325(64)
Burland, G. N., 124
Burns, R. P., 382
Burr, A. F., 189
Burstein, E., 96, 97(207)
Burton, J. J., 418
Butz, R., 22, 432

C

Cade, P. E., 56
Cailler, M., 197
Caldwell, C. W., 238
Camp, W. J., 224
Campbell, C. T., 70, 508
Campuzano, J. C., 59, 65, 66, 124(127)
Candy, J. P., 61
Cardillo, M. J., 498
Cardona, M., 145, 166, 170(43)
Carette, J. D., 26, 27(47), 32, 34, 35(77), 37(76), 47(67), 50(67), 122(47, 67)
Carlson, T. A., 471
Caroli, C., 101(216), 102
Carter, G., 348

AUTHOR INDEX

Cartwright, P., 390
Castaing, R., 404
Catterick, T., 83, 124
Chaban, E. E., 140, 141, 184(19), 488
Chadwick, P. A., 285
Chambers, R. S., 414
Chan, C. M., 449
Chandrasekhar, B. S., 26, 101(49)
Chang, C. C., 25, 231
Chang, L. L., 277, 288(72)
Charbonnier, F. M., 367
Chen, M. H., 230
Chen, Y. S., 316
Chen, Y.-S., 319, 320(59)
Chesters, M. A., 26, 81(23)
Chiang, T.-C., 166, 170(44)
Chick, F. H. C., 416
Chin, M. A., 498
Chinn, M. D., 238, 254, 285, 286(8)
Christman, S. B., 140, 141, 150, 151(24, 25), 153, 154, 155(29), 157, 158(31), 184(19), 488
Christmann, K., 62, 84, 88
Chung, N. S., 197
Church, E. L., 189
Churchill, M. R., 110
Chutjian, A., 38, 40, 42(88, 89)
Chutjian, H., 53
Cini, M., 230
Ćirić, D., 299, 313(7, 8)
Citrin, P. H., 225, 231, 234
Clabes, J., 293
Clampitt, R., 403
Clark, G. J., 347
Clearfield, H. M., 248, 292, 293, 294
Clinton, W. L., 472
Coad, J. P., 508
Coburn, J. W., 329, 337(83, 84), 474, 512, 513(129)
Cocke, D. L., 401
Cohen, M. L., 151, 166, 169(41, 42), 182
Cohen, P. I., 213, 214, 264, 276, 284(57), 288(57), 295
Cohen, S. A., 345, 347
Colby, B. W., 403
Coleman, R. V., 26
Colligon, J. S., 348
Collins, W. E., 496
Collman, J. P., 68
Colmenares, C. A., 290
Combescot, R., 101(216), 102
Comrie, C., 462, 501

Conrad, H., 69, 89, 94(192), 440, 441
Cooper, C. B., 316
Cooper, E. C., 367
Copperthwaite, R. G., 508
Corbett, J. W., 506
Cornish, J. C. L., 124
Cotton, F. A., 64
Cowan, P., 395
Cowley, J. M., 277
Craig, J. H., 466, 493(8)
Cranstoun, G. K. L., 411(278)
Crasemann, B., 230
Crewie, A. V., 368
Cromwell, J. E., 239
Crossley, A., 65, 88(141), 124(141)
Crouser, L. C., 26
Crowell, A. D., 443
Culbertson, R. J., 403
Culver, R. V., 1, 3(6)
Cunningham, S. L., 101(219), 102
Czanderna, A. W., 400
Czyzewski, J. C., 467

D

Dagdigian, P. J., 498
Dahl, A. I., 434
Dahlstrom, R. K., 361
Daley, H. L., 403
Dalmai-Imelik, G., 26, 33(41, 42), 34, 122(42, 262, 274, 275, 278)
Danev, V., 368
Danielson, L. R., 512
Datz, S., 307
Davenport, J. W., 48, 52, 53, 54, 55, 56, 57, 122(117), 123(117)
Davies, J. A., 443
Davis, L. C., 101(215), 102
Davis, N., 110, 112(245)
Davis, W. D., 444
Davisson, C. J., 237, 238(1)
Dawson, P. H., 501
Dayton, B. B., 441
Debe, M. K., 434
de Bersuder, L., 284
de Cheveigne, S., 95, 103(200)
Defourneau, D., 95, 101, 102(213), 103(200)
de Khuizenaar, E. E., 329
Delage, A., 34, 37(76)
Delanaye, F., 47, 48, 59
Delargy, K. M., 410(282), 415
Delchar, T., 19

Delchar, T. A., 431
Deline, V. R., 337
Demuth, J. E., 122(276, 279, 280), 123
denBoer, M. L., 195, 197, 213, 214, 236
Dench, W. A., 198, 505
Dennis, R. L., 275
Derrick, P. J., 401
Desplat, J.-L., 18
de Vries, A. E., 332
de Wit, A. G. J., 310, 328(36)
DiChio, D., 40
Dickinson, J. T., 512
Dietz, E., 166, 170(45)
DiFoggio, R., 366
Dignam, M. J., 79, 80(169)
Dillon, J. A., 443
Dobretsov, L., 1, 2(4), 5(3), 11(3), 19(4)
Dobrzynski, L., 31, 47(66)
Doering, D. L., 471
Doerries, E. M., 511
Dolan, W. W., 350, 353, 367, 401(21)
Dolcetti, G., 68
Domke, M., 359
Donaldson, E. E., 474, 512
Donelon, J. J., 140, 492
Doniach, S., 149
Dorn, R., 26, 32, 122
Dove, D. B., 277, 288(72), 508
Down, D. S., 78, 124(166)
Doyen, G., 88
Drachsel, W., 406, 410(257, 258)
Dräger, W. M., 320
Drechsler, M., 350, 359
Dresser, M. J., 448, 461, 512
Drinkwine, M. J., 465(5), 466, 496, 498(5), 499(5)
Driscoll, R. K., 124
Dubois, L. H., 239, 443
Duke, C. B., 25, 96, 102, 105(206), 238, 271(14), 382
Duncan, M. T., 105
Dupp, G., 331
Durand, J. P., 506
Durig, J. R., 68
Dyke, W. P., 350, 353, 367, 368, 401(21)

E

Eastman, D. E., 140, 166, 170(44, 45), 225, 467, 492, 493(21), 494(21)
Eberhagen, A., 19

Eckstein, W., 306, 307, 313
Edelman, F., 221
Edelman, H. S., 155, 175
Efrima, S., 58, 65
Egerton, R. F., 509, 511
Ehrenberg, W., 238
Ehrhardt, H., 53
Ehrlich, G., 365, 392, 393, 395, 414, 428, 431, 441(2)
Einstein, A., 128, 394
Einstein, T. L., 236, 297
Eischens, R. P., 59, 61, 65(122), 105
Eisenberg, R., 68
Eisenberger, P., 234
Elam, W. T., 189, 210, 236
Ellis, W. P., 238, 313
Emin, D., 467, 470(19)
Engelhardt, H. A., 447, 451(55), 462(55), 463(55), 493
Englert, W., 299, 310(9), 313(9), 324, 325(63)
English, T. C., 495, 497
Erickson, N. E., 189, 225
Erickson, R. L., 305(18)
Erley, W., 122(257, 259, 281), 443
Ermrich, W., 367
Ernst, N., 399
Ertl, G., 25, 33, 62, 69, 84, 88, 89, 94(192), 122(182), 239, 440, 441(29)
Estrup, P. J., 51, 238, 239, 254, 434
Evans, C. A., Jr., 337, 403
Evans, E., 47, 48, 50(101), 51(101)
Evans, H. E., 95, 101, 106, 107(232), 108, 109, 110(203, 204, 241), 111, 112(203, 204, 205, 241)
Everhart, T. E., 197
Ewald, H., 317
Eyring, C. F., 349, 378
Eyring, H., 457, 459(63)

F

Fain, S. C., Jr., 238, 254, 285, 286(8)
Falconer, J., 462
Farnsworth, H. E., 238, 273(6), 443
Farrell, H. H., 166, 169(41), 239, 476, 482(54b)
Fedorus, A. G., 505
Feibelman, P. J., 214, 231, 466, 467, 470, 479, 492, 505, 506
Feldhaus, J., 480
Feldman, L. C., 496

AUTHOR INDEX

Fellner-Feldagg, H., 216
Felter, T. E., 434
Feltham, R. D., 68
Feuchtwang, T. E., 101(218), 102
Feuerbacher, B., 26, 32(45), 33(43, 44, 45, 46), 50(44), 122(43, 44, 45, 46)
Feulner, P., 68, 445, 447, 451(55), 462(55), 463(55), 471, 496, 497, 503
Firment, L. E., 509
Firsov, O. B., 311
Fischer, T. E., 70
Fisher, G. B., 122(264)
Fitton, B., 26, 32(45), 33(43, 44, 45, 46), 34(45), 50(44), 122(43, 44, 45, 46)
Flickner, M., 421, 423(325)
Floyd, G. R., 466
Foesch, J. A., 391
Foley, E. B., 329, 336(79), 337(79)
Fontaine, J. M., 506
Foster, C. A., 289, 432
Fouilloux, P., 61
Fowler, R. H., 7, 350
Franchy, R., 471, 506(32)
Franciosi, A., 175
Francis, S. A., 59, 65(122), 82
Franck, J., 198
Frank, O., 400
Freilich, A., 418
Froitzheim, H., 26, 27(32, 35), 28, 32, 33(32, 36), 34, 46(36, 78), 50(32, 33), 72(36), 122(32, 33, 34, 35, 36), 139
Frost, D. C., 285
Fuggle, J. C., 229, 503
Fukuda, Y., 189, 195, 210

G

Gadzuk, J. W., 62, 206, 230, 361, 362, 363(46)
Gallon, T. E., 509
Garbout, A., 84, 85(183), 91(183)
Garland, C. W., 105
Geiger, A. L., 26, 101
Gelius, U., 223
George, T. H., 412
Gerhard, W., 329, 337(85)
Gerlach, R. L., 194
Germer, L. H., 237, 238(1)
Gesley, M., 20
Gettings, M., 508
Ghiglia, D. C., 421, 423(324), 421(325)
Giaever, I., 374, 375, 416, 417, 421

Gibson, J. W., 21
Gierlich, H. H., 400
Giessmann, V., 400
Girlando, A., 102
Gland, J. L., 122
Glaeser, R. M., 511
Glasstone, S., 457, 459
Gobby, P. L., 172, 173(46), 177(46), 179(46), 180(46)
Godfrey, D. J., 304
Godfrey, T. J., 415
Golay, M. J. E., 78
Goldberg, H. A., 17(22)
Golden, D. E., 53
Golden, W. G., 78, 124(166)
Goldstein, J. I., 505
Gomer, R., 10, 21, 351, 359(18), 361, 362(50), 364, 366, 367, 372, 379, 381, 384, 397, 400, 402, 403, 435, 436, 437, 442, 467, 468, 469, 470(17), 496
Gonchar, V. V., 505
Goncil, B., 299, 313(7)
Good, R. H., Jr., 8, 352, 361, 362(45)
Goodman, D. W., 36, 67(87)
Gorte, R., 454, 456, 457
Gossling, B. S., 350
Goutte, R., 329
Graham, W. R., 392, 393
Grandke, T., 166, 170(43)
Granneman, E. H. A., 481
Grant, J. T., 502, 503
Green, T. S., 138, 221
Greene, J. E., 329
Greenler, R. G., 26, 27(25), 59, 65, 66, 124(127), 81, 82, 83, 107
Greenwald, A. C., 368
Grigson, C. W. B., 288
Grønlund, F., 331
Gronwald, K. D., 238, 287, 294
Gruen, D. M., 340
Grundner, M., 314, 316, 319, 336, 340(100), 342, 343(100)
Grunze, M., 124
Grupta, R. K., 124
Gudat, W., 174
Guilland, C., 329
Guinet, C., 95, 103(200)
Gunnarson, O., 229
Gurney, R. W., 378
Gurney, T., Jr., 418
Guseva, M. I., 331

H

Haas, G. A., 1, 19(1), 20, 21
Haas, H., 64
Haefer, R., 372
Haggmark, L., 311
Hagstrum, H. D., 305, 306(17), 307(17), 310(17), 474
Hahn, P. O., 293
Hair, M. L., 26
Hall, T. M., 410
Halpern, B., 364, 402
Hamann, D. R., 214
Hammaker, R. M., 59, 65(122)
Hammarqvist, H., 210
Hammond, D., 216
Hanke, G., 238
Hansen, G. R., 401
Hanson, D. M., 471, 472(34), 477
Hansma, P. K., 26, 29, 59, 95, 96(58), 101, 102, 103, 104, 105
Hardy, J. R., 101(219), 102
Haring, A., 332
Harrick, N. J., 26
Harrington, W. L., 304, 334, 339(98), 340(98), 341(98), 343(98)
Harris, L. A., 227
Harrison, D. E., 331
Harrison, W. W., 313, 315
Harrower, G. A., 220
Haydock, R., 399
Hayward, D. O., 365
Heiland, W., 299, 301, 302, 303, 304, 305, 306(17), 307, 308, 309, 310(9, 17), 313(9), 314, 316(12, 51), 319(51), 322, 324, 325, 326, 333, 336, 340(100), 342(100), 343(100), 348
Heilman, P., 238
Heimann, P. J., 492
Hein, N. C., 140
Heindric, A., 400
Heinen, H. J., 400
Heinz, K., 238, 505
Heinze, W., 21
Helms, C., 462
Henderson, J. E., 361
Hendra, P. J., 26
Henkel, E., 350
Henzler, M., 238, 239, 250(25), 287, 293, 294
Hermanson, J., 160, 172, 173(46), 177(46), 179(46), 180(46)
Heroz, R., 319
Herring, C., 1, 5(4)
Hertz, G., 198
Herzberg, G., 63
Herzog, R. F. K., 328
Hewitt, R. C., 234
Hickmott, T. W., 365
Hickson, D. A., 101, 102(212)
Hien, N. C., 492
Higginbotham, I. G., 509
Hill, M. P., 360
Hill, R. H., 189
Himpsel, F. J., 140, 225, 492
Hintenberger, H., 299
Hirose, H., 338
Ho, K. M., 166, 169(41, 42)
Ho, W., 32, 34, 38(82), 47(69, 70, 71), 48, 50, 51, 52(103), 53(103), 54(103), 55(103), 68(71), 122(69, 70, 71, 82), 123(82)
Hobbs, G. M., 349
Hobson, J. P., 444, 500
Hock, J. L., 466, 493
Hocker, L. O., 368
Hofer, W. O., 334, 334, 338
Hoffbauer, R., 495
Hoffmann, F. M., 27, 31, 47(63, 64), 75(53, 54), 76(55), 77, 78, 79, 81(63), 84, 85, 86, 87(184), 88, 89, 90, 91, 92, 93, 94, 117(54, 63), 118(63), 119(63), 120(63), 121(63), 122(266), 123, 124, 440
Hogg, A. M., 400
Hogue, J. V., 495
Hollins, P., 88, 124(185)
Holloway, P. H., 502, 508
Holloway, S., 277
Holscher, A. A., 15, 360
Honig, R. E., 304, 334, 339(98), 340(98), 341(98), 343(98)
Hooker, M. P., 502, 503
Hopster, H., 26, 32(36, 40), 33(36), 34, 35(40), 36(79), 46(36), 47(79), 72(36, 40), 122(36, 40, 79, 268), 344
Hörl, E., 372, 374
Horlich, G., 78
Horn, K., 26, 32, 76, 80, 81(160), 122, 124
Horr, K., 84, 85(183), 91(183)
Hosoi, K., 400
Houston, J. E., 194, 199, 203, 210, 224, 230, 238, 239(27), 255(27), 260, 263(56), 350, 471, 472(34), 502, 506(33), 508
Howitt, D. G., 511

Hren, J. J., 389
Huchital, D. A., 220
Hudda, F. G., 365, 392
Hudson, J. B., 502, 504
Hughes, J. R., 509
Hulm, J. K., 364
Hulpke, E., 299, 313(6)
Humphrey, L., 401
Hupkens, T. M., 310, 328(36)
Hulse, J. E., 59, 64(126), 65(126)
Hunt, R. H., 56
Hurst, G. S., 498
Hurych, Z., 476, 482(54b)
Hussain, M., 124
Hutchinson, R., 418, 419

I

Ibach, H., 25, 26, 27(32, 35), 28, 31, 32, 33, 34, 35(40), 36, 46, 47(37, 65, 79), 50(32, 33), 59, 60(132), 62(132), 72(36, 40), 84(37), 122, 139, 259, 267, 268, 269, 273, 276, 279, 280, 281
Ibbotson, D. E., 434
Iberl, F., 326
Imelik, B., 122(258, 278)
Inghram, M. G., 379, 381, 382, 384(126), 400, 402
Ionov, N. I., 432, 444, 464, 465, 476(4), 497, 498(4), 499(4), 502(4)
Isaacson, M., 368, 508, 511(110)
Ishibashi, M., 400
Itskovich, F. I., 1, 5(5)
Iwanaga, M., 400

J

Jach, T., 226
Jackson, D. P., 299, 310, 311(35), 313(9), 328(35)
Jaeger, R., 235, 480, 485, 491
Jaklevic, R. C., 26, 97
James, B. D., 110, 112
Janow, R., 472
Jason, A. J., 382, 402
Javan, A., 368
Javelas, R., 329
Jefferies, D. K., 403
Jeland, W., 496
Jennison, D. R., 467, 470(17, 19)
Jepsen, D. W., 238, 239, 271(15)

Jeziorowski, H., 326, 327(72)
Jiru, P., 105
Johannessen, J. S., 509
Johannson, L., 235
Johansson, J., 225
Johnig, G., 359
Johnson, B. F. G., 74(158), 75
Johnson, D., 368
Johnson, P. D., 476
Johnson, R. P., 353, 354(22), 482(54b)
Jona, F., 238, 239, 271(15)
Jones, E. R., Jr., 242, 243(47), 273, 274(63)
Jones, V. O., 199, 467, 480(20b)
Jonker, B. T., 214
Joshi, A., 507
Joyce, B. A., 513

K

Kaminska, T. J., 285
Kaminsky, M., 348
Kanash, O. V., 505
Karpuzor, D. S., 307
Kaska, W. C., 95, 103(198), 104(198), 105(201)
Kasper, H., 150, 151(24), 180, 181(49)
Katnani, A. D., 153, 154, 155
Kawashima, H., 112
Kay, E., 329, 337(83, 84)
Kellogg, G. L., 393, 395, 399, 406, 410(256, 259), 415
Kelly, R., 332
Kelsey, T., 369
Kem, R., 437, 438
Kemeny, P. C., 165
Kennelly, W. J., 112
Kerwin, R. E., 511
Kesmodel, L. L., 285
Keywell, F., 331
Khoshnevisan, M., 390
Kimock, F. M., 498
King, D. A., 26, 27(25), 59, 65, 88, 124, 395, 431, 433, 434, 446, 454, 464, 473, 503
Kingsman, D. R., 399
Kirk, C. F., 392
Kirkpatrick, A. R., 368
Kirschner, J. 199, 213, 313, 315, 320
Kirtley, J., 105
Kirtley, J. R., 59, 101(129), 102
Kivillis, V. M., 307, 308(26)
Klein, J., 95, 101, 102(213) 103

Klein, R., 365
Klier, K., 443
Knapp, A. G., 509
Knapp, J. A., 138, 172, 173(46), 177(46), 179(46), 180(46)
Knotek, M. L., 199, 466, 467, 469, 470, 479, 480, 492, 505, 506
Knözinger, H., 326, 327
Kobayashi, H., 107
Kobrin, P. H., 498
Koch, J., 88, 89, 94(192)
Kock, F. B., 347
Koestner, R. J., 239
Kohrt, C., 366, 435, 436, 437
Kolb, J. R., 112
Kölbel, H., 105
Koma, A., 32
Kondo, T., 338
Kornelson, E. V., 444, 500
Kortan, A. R., 297
Koshihawa, T., 197
Kowalczyk, S. P., 130, 189
Krahl-Urban, B., 432, 491
Kraihanzel, C. S., 64
Kramer, S. D., 498
Kraus, J. S., 305(19), 496
Krause, M. O., 191, 192, 471
Krauss, A. R., 340
Krebs, H. J., 79, 124
Krefft, H., 199
Krishnaswamy, S. V., 382, 405, 410, 412, 413
Krocker, R. M., 95, 105
Kröger, F. A., 284
Krohn, V. E., 340, 402, 403(234)
Kronauer, P., 496
Ku, R., 25
Kubiak, G. D., 498
Kulkarmi, S., 68
Kummler, D., 400
Kunz, C., 174
Küppers, J., 25, 69, 88, 239
Kuroda, T., 410
Kurtz, R. L., 471
Kutsenko, E. N., 476
Kuyatt, C. E., 17, 35, 38(85, 86), 40, 41, 45(86), 139, 362

L

Lagally, M. G., 238, 239, 247, 248, 253(26), 255(27), 256, 257, 260, 265(55), 266(55), 269, 270(59), 271, 273(9), 274, 275(64, 65), 276, 277(41), 280, 281(45), 286(9), 291(9), 292, 293, 294(92), 295(26, 53), 296(74)
Lagreid, N., 331
Laider, K. J., 457, 459(63)
Laine, R. M., 95, 103(198, 199), 104(198)
Lam, K., 403
LaMarche, P. H., 403
Lambe, J., 26, 97
Lambert, R. M., 501
Lampton, M., 287, 390
Lanczos, C. Z., 378
Lander, J. J., 227, 230, 238, 239, 242, 243(47)
Lang, E., 238, 505
Langan, J. D., 102, 103
Langhans, L., 53
Langmore, J. P., 506, 511(97)
Langmuir, I., 427
Lapeyre, G. J., 138, 159, 161, 172, 173(46), 177(46), 179, 180
Laramore, G. E., 224, 235, 236
Larsen, P. K., 138, 139(14), 166, 167(40), 168(40), 169
Lasser, R., 229
Latta, E. E., 69, 89, 94(192), 440, 441(29)
Lauritsen, C. C., 350
Lea, C., 361, 362(50)
Leck, J. H., 474, 475(49)
Lederer, D., 101(216), 102
Lee, M. J. G., 17(22)
Lee, P. A., 233
Lee, R. N., 210, 226
Léger, A., 95, 101, 102(213), 103(200)
LeGressus, C., 506
Lehmann, W. D., 400
Lehwald, S., 26, 27(32, 35), 32, 33(32, 36), 46(36), 50(32, 33), 72(36), 122(32, 33, 34, 35, 36, 256, 259, 267, 269, 273, 276, 281)
Leidheiser, H., Jr., 443
Leisch, M., 410(283)
LeLay, G., 437, 438
Lenac, Z., 47, 48, 49(107), 50(107)
Lent, C. S., 264, 276, 277(68), 281(68), 284(57), 288(57, 68), 295
Levi-Setti, R., 403
Lewis, B. F., 26
Lewis, R. J., 75, 410
Ley, L., 165, 166, 170(43), 189
Liau, Z. L., 410
Lichtman, D., 74(158), 465(5), 466, 473, 498(5), 499(5)

Liebl, H., 313, 315, 317, 320, 328, 331, 334, 338, 339, 346, 347(104), 493, 496
Lin, C. C., 51
Lin, L., 230
Lindau, I., 235
Linder, F., 53
Lipari, N. O., 102
Little, L. H., 26
Little, R. G., 368
Livingston, W., Jr., 370, 416(99)
Lloyd, J., 508
Loebach, E., 328, 339(76)
Los, J., 512
Lu, T.-M., 238, 239, 248, 253(26), 255(27), 260, 265(55), 266(55), 269, 270(59), 277(41), 280, 293, 295(26), 296(74)
Lucas, A., 47, 48
Lucas, A. A., 59, 382
Ludeke, R., 32, 277, 288(72)
Lundqvist, S., 96, 97(207)
Lundy, R., 364
Lüth, H., 26, 32, 79, 122, 124
Lytle, F. W., 232

M

McAllister, W. A., 68
McCarroll, B., 444
McCrocken, G. M., 348
MacDonald, R. J., 348
McFeely, F. R., 189
McGuire, E. J., 192, 231
Machlin, E. S., 418
McIntyre, J. D. E., 81, 82
McKay, K. G., 194
McKee, C. S., 238, 395
Mackenzie, K. V., 361
Mackeown, S., 349, 378(7)
Mackie, W., 15(18)
McKinney, J. T., 242, 243(47), 273, 274(63), 414
McKinstry, D., 397
McLane, S. B., 383, 410
McLean, W., 290
McNair, D., 313, 316(44)
McNeely, D. R., 20
McQuistan, R. B., 473
McRae, E. G., 238, 239, 271(13)
Madden, H. H., 84
Maderlechner, G., 330
Madey, T. E., 36 67(87), 92, 94, 115, 116(253), 189, 225, 428, 431, 432, 433, 434, 435, 440, 441, 442, 444, 448, 452, 453, 461(64), 464, 465, 466, 467, 468, 469(22), 471, 472, 473, 474, 477, 485, 486, 487, 488, 493(21), 494, 498(1, 6), 499, 501, 502, 503, 506, 507, 508, 509, 510(99), 511(99)
Madix, R. J., 107, 444, 462
Maekawa, T., 112
Maekawa, Y., 400
Magee, C. W., 334, 339(98), 340(98), 341, 343(98), 347
Mahl, H., 354
Mahan, G. D., 47, 48, 59, 65(124), 163
Mahoney, J. F., 403
Makowiecki, D. M., 290
Manley, B. W., 390
Manneville, M., 437, 438
Marcus, P. M., 238, 271(15)
Marcus, S. M., 101(214), 102
Margaritondo, G., 128, 138, 139(12), 142(2), 150, 151, 155, 157, 158(31), 159, 166, 167(40), 168(40), 175, 180, 181, 496
Marks, T. J., 112
Martin, E. E., 367
Martin, J. A., 238, 239, 274(46)
Martin, R. L., 189
Martinez, J. M., 502, 504
Martinka, M., 412, 413
Mashkova, E. S., 307
Massardier, J., 122(275)
Matsuda, T., 112
Matthews, J. A. D., 229
Matthews, L. D., 443
May, J. W., 239
Mayer, J. W., 410
Mayer, L., 22
Melchior, W., 303, 304(12), 316(12)
Melmed, A. J., 367, 371, 372, 373, 374, 410(275), 417
Menadue, J. F., 277
Menzel, D., 31, 47(63, 64), 68, 81(63), 115, 116(253), 117(63), 118(63), 119(63), 120(63), 121(63), 124(63, 64), 326, 367, 445, 447, 451(55), 462(55), 463(55), 464, 465, 466(2, 3), 468, 469, 471, 473, 480, 485, 491, 493, 496, 497, 498(2, 3), 499(2, 3), 500, 502(2, 3), 503, 506(32)
Mercer, E. E., 68
Meriaux, J.-P., 329
Merrill, R. P., 501
Messier, R., 410(280, 284)

Metiu, H., 58, 65
Meyer, C. D., 68
Mies, F. H., 56
Mihelich, J. W., 189
Miller, G. L., 319, 320(59)
Miller, M. K., 410, 415
Millikan, R. A., 349, 350, 378(7)
Mills, D. L., 47, 48, 50(101), 51(101)
Mills, P. F., 410(286), 415
Minnucci, J. A., 368
Miscovic, Z., 472
Mitchell, K. A. R., 285
Miyazaki, H., 400
Molchanov, V. A., 306, 307(23, 29)
Molière, G., 311
Montague-Pollack, H., 370, 416(98)
Moore, G. E., 473
Moore, R. D., 403
Moore, W. J., 331
Morar, J. F., 210, 214, 226, 236
Morgan, F. J., 495
Morris, R. J., 496
Morrison, J., 238, 239, 242, 243(47)
Morrow, B. A., 107
Moseley, H. G. J., 187
Mosesman, M., 26
Moskovits, M., 59, 64(126), 65(126)
Mul, P., 324, 328(69)
Müller, E. W., 8, 15, 349, 352, 354, 356, 359, 361, 365, 370, 372, 374, 376, 378, 379, 380, 381, 382, 383, 384, 385, 387(129), 388(126), 392, 397, 399, 404, 405, 406, 410(248, 251, 252, 253, 254, 287, 288, 299), 412, 413, 417
Müller, J., 68
Müller, K., 231, 238, 505
Müller, N., 320
Murato, Y., 196
Murphy, E. L., 361, 362(45)
Musket, R. G., 290, 501, 509

N

Nakamura, S., 393, 410
Narayanamurti, V., 498
Natali, S., 40
Naumovets, A. G., 505
Neave, J. H., 513
Neff, S. N., 305(19)
Neff, L. D., 105

Nelson, G. C., 324
Netzer, F. P., 471, 472(34)
Neumann, M., 33, 88
Newman, R. W., 389
Newns, D. M., 47, 48(99), 50(99)
Newsham, I. G., 495
Ng, Y. S., 410
Ngoc, T. C., 271
Nibbering, N. M. M., 400
Nichols, M. H., 1, 5(4)
Niehus, H., 313, 317(46), 474, 476(51), 485, 489, 490, 491, 499(59), 500(59)
Nieuwenhuys, B. E., 359
Nishigaki, S., 393, 406, 410(257, 258)
Nishijima, M., 476, 501
Nordheim, L. W., 350
Norman, C. E., 331
Norton, J. R., 68
Norton, P. R., 443
Nottingham, W. B., 208
Nozieres, P., 101(216), 102
Nyberg, C., 210

O

Oechsner, H., 329, 337(85)
Ohno, Y., 410
Ohtani, S., 196
Okawa, L., 62, 63, 64(137)
Oliver, J. H., 192
Olszewski, G. B., 285
Onada, G. Y., Jr., 508
Ono, M., 36, 67(87)
Onsgaard, J., 322, 324(62)
Oppenheimer, J. R., 378
Orloff, J. H., 402
Ortega, A., 31, 47(63, 64), 78, 81(63), 84, 85(183), 86, 87, 91(163, 183), 117(63), 118(63), 119(63), 120(63), 121(63), 124(63, 64, 163, 184)
Ottenberger, W., 321, 330(61), 333(61)
Overeijnder, H., 332
Ozin, G. A., 112

P

Padva, A., 102
Palmberg, P. W., 84, 134, 136, 317, 439
Palmer, R. L., 248
Pandey, K. C., 231

AUTHOR INDEX

Panin, B. V., 299
Panitz, J. A., 357, 369, 380, 383, 386, 387, 388, 389, 390, 391, 404, 405, 408, 409, 410(47, 247, 248, 250, 261, 290), 414, 415, 416, 417, 418, 421, 422, 423
Pankey, T., 21
Pantano, C. G., 506, 508, 510(99), 511(99)
Pappas, D. L., 488, 498
Paresce, F., 287
Parilis, E. S., 307
Park, R. L., 25, 84, 189, 194, 195, 199, 203, 210, 213, 214, 224, 226, 236, 238, 260, 263(56), 273(16, 18), 297
Parrott, S. L., 107
Parsely, M. J., 392
Paul, W., 340
Payn, J. K., 473
Payne, M. G., 498
Payzant, J. D., 400
Pearce, H. A., 81
Pecile, C., 102
Pendry, J. B., 25, 238, 239(16), 271(16), 291(16)
Penn, D., 11
Perel, J., 403
Peria, W. T., 227
Perović, B., 299, 313(8)
Perry, D. L., 238
Persson, B. N. J., 47, 50(102)
Petermann, L. A., 496
Peterson, D. T., 152
Pethica, B. A., 360
Petry, R. L., 199
Petzow, G., 289
Pfnür, H., 31, 47(63, 64), 81(63), 117(63), 118, 119, 120, 121, 124, 447, 451, 462, 463
Pian, T. R., 496
Pianetta, P., 235
Pillon, J., 197
Piper, T. C., 26, 122, 123
Pirug, G., 122(268)
Plagge, A., 88
Plative, P. G., 392
Pliskin, W. A., 61, 105
Plummer, E. W., 11, 32, 34(70, 71), 35, 38(86), 41, 45(86), 47(69, 70, 71), 48(69, 70, 71), 50(69), 51(70, 71), 122(69, 70, 71), 361, 362, 363(46), 392
Plyler, E. K., 56
Poelsema, B., 306, 311, 328(40)

Powell, C. J., 189, 190, 225, 226, 230
Preuss, E., 432, 472
Prigge, S., 474
Primet, M., 61
Prince, R. H., 466, 495
Pritchard, J., 26, 76, 80, 81, 82, 83, 88, 125
Proca, G. A., 138, 221
Propst, F. M., 26, 30(38), 34(38), 122, 123, 476, 501
Prutton, M., 509
Pukite, P. R., 264, 267, 277(68), 281(68), 284, 288(57, 68), 295
Purcell, E. M., 223
Purtell, R. J., 501
Pyke, D. R., 410(278)

Q

Quicksall, C. O., 114
Quinby, M. S., 68
Quinn, C. M., 360
Quinn, J. J., 195, 196

R

Rálek, M., 105
Ralph, B., 410
Ramaker, D. E., 467, 468, 469(22), 470(17, 18), 471
Rao, B., 79, 80(169)
Ratajczykowa, I., 26, 27(25)
Rawlinson, W. F., 214, 215
Redhead, P. A., 428, 429, 430, 431, 441(4), 444, 447, 448, 468, 473, 495, 500, 501
Reed, D. A., 393, 418, 419
Reed, S. J. B., 511
Rehn, V., 199, 467, 480(20b)
Reidl, W., 497
Reifenberger, R., 17
Reimann, A. L., 1, 6(7)
Rendulic, K. D., 418
Reynolds, W. D., 401
Rhead, G. E., 447
Rhodin, T. N., 10, 392, 501
Rice, J. K., 40
Richardson, N. V., 59
Richardson, O. W., 205, 350
Rigden, J. D., 220
Ringers, D. A., 230
Ringo, G. R., 402, 403(234)
Ritchie, R. H., 197

Rivière, J. C., 1, 19(2), 508
Roberts, M. W., 238, 360, 395
Robertson, G. H., 403
Robertson, J. A. B., 401
Robinson, C. A. H., 319, 320(59)
Robinson, H. H., 193, 214, 215
Robinson, J. E., 313, 316(43)
Robinson, M. T., 330, 331
Robinson, W. L., 68
Rochow, E. G., 349
Rochow, T. G., 349
Roelofs, L. D., 236, 297
Rogers, J. W., Jr., 107
Rollgen, F. W., 400
Roos, G., 338
Roptin, D., 197
Rose, D. J., 359, 370(31)
Rosenberg, D., 331
Rosebury, F., 289
Rosengren, A., 225
Roth, J., 321, 330(61), 333(61), 334
Rothschild, W. G., 105
Roundy, V., 48
Rousseau, J., 26, 33(41, 42), 34, 122(42), 262, 274, 277, 278)
Rowe, J. E., 128, 140, 141, 142(2), 150, 151(24, 25), 153, 154, 155(29, 184), 157, 158(31), 159, 166, 167(40), 168(40), 180, 181(49), 488
Roy, D., 34, 35(76, 77)
Rüdenauer, F. G., 338
Rudge, M. R. H., 197
Ruggieri, D. J., 487
Rye, R. R., 502, 508

S

Sachtler, W. M. H., 359
Saint-James, D., 101(216), 102
St. John, G. A., 400
Saito, Y., 107
Sakata, T., 393
Sakurai, T., 403, 406, 410(254)
Saloman, E. B., 482
Saloner, D. A., 256, 295(53)
Samarin, S. N., 496
Sandstrom, D. R., 434, 474, 475(49), 495
Sangster, M. J. L., 101, 102(213)
Sar-El, H. Z., 221

Sasaki, T., 507, 509
Saski, N., 197
Savage, D. E., 264
Sayers, D. E., 232
Scalapino, D. J., 59, 101(129, 214), 102
Schäffler, H. G., 309
Scharmann, A., 331
Schattke, W., 101(217), 102
Scheffler, M., 59, 65, 82
Scherzer, B. M. U., 330
Schlüter, M., 151, 166, 167(40), 168(40), 169(41), 182
Schmeitz, M., 59
Schmidt, H., 53
Schmidt, L. D., 432, 433, 434, 446, 454, 456, 457, 464
Schmidt, W. A., 400
Schmitt, S., 68
Schober, O., 88
Schönhammer, K., 229
Schottky, W., 350
Schreiner, D. G., 260, 263(56)
Schrieffer, J. R., 48, 52(103), 53(103), 54(103), 55(103)
Schrott, A. G., 285
Schulten, H. R., 400, 401
Schulz, G. J., 48, 53(105, 106)
Schuster, F., 303, 304(12), 316(12)
Schwartz, S. B., 197
Schwarz, J. A., 101, 102(212)
Schwinger, J., 141, 142(20), 143(20)
Schwoebel, R. L., 238
Seabury, C., 501
Seah, M. P., 198, 505
Sears, G. N., 495
Semancik, S., 254
Seidman, D. N., 410
Seliger, R. D., 403
Sexton, B. A., 34, 35(80), 36(79), 47(79), 122(79, 80, 260, 263, 265), 123, 501
Shapira, Y., 465(5), 466, 498(5), 499(5)
Sharma, S. P., 308
Shaw, C. G., 285
Shek, M.-L., 500, 504
Shelef, M., 326
Sheline, R. K., 64
Shelton, H., 11
Sheppard, N., 81
Shepherd, F. R., 285
Shibata, S., 400

AUTHOR INDEX

Shigeishi, R. A., 59, 65(123), 88(123), 124
Shih, A., 21
Shimizu, R., 197
Shimp, L. A., 112
Shirai, E., 400
Shirley, D. A., 189, 229, 507, 509
Shockley, W., 353, 354(22)
Shulga, V. I., 307
Shurwell, H. F., 112
Sichtermann, W. K., 343
Siegbahn, K., 189, 216, 219
Siegbahn, M., 189
Siegel, B. M., 401
Siekhaus, W. J., 290
Sigmund, P., 329, 332(87), 333(87), 348
Simms, D. L., 329, 336(79), 337(79)
Simonson, M. G., 26
Simpson, J. A., 35, 38(84), 139
Sims, M. L., 26, 82(22)
Smith, A. K., 110
Smith, B. E., 110, 112
Smith, D. P., 299, 302, 305(18), 316, 317, 324
Smith, G. D. W., 410, 415
Smith, N. V., 135, 138, 139, 166, 167(40), 168(40), 169(41), 476, 482(54b)
Smith, R. J., 161
Smith, S. T., 368
Snoek, C., 307
Šokčević, D., 47, 49(107), 50
Sokoloff, D. R., 368
Soma, M., 62, 63(137), 64(137)
Somorjai, G. A., 122(284), 123, 239, 285, 443, 509
Southern, A. L., 331
Sovitsky, A., 78, 380, 390, 410(293), 412, 415
Spangenberg, K. R., 41, 259
Sparnaay, M. J., 324, 325(64)
Speer, D. A., 372
Spicer, W. E., 129, 130(4), 149
Spindt, C. A., 207, 401
Spiro, T. G., 114
Sproull, W. T., 238
Srivastana, S. K., 53
Staib, P., 199, 213
Stair, P. C., 238, 285
Staudenmaier, G., 331, 338
Stayer, R. W., 15, 367
Steiger, W., 338
Steinwedel, H., 340
Steit, K. M., 33

Stenzel, W., 84, 85(183), 91(183)
Stephan, C. H., 390
Stern, E. A., 197, 232
Stern, R. C., 444, 495
Stern, T. E., 350
Stillwagon, L. E., 511
Stimpson, B. P., 474, 475(49)
Stobie, R. W., 79, 80(169)
Stockbauer, R., 467, 468, 469(22), 471, 472(34), 477, 481, 493(21), 494(21)
Stoffel, N. G., 138, 139(12), 153, 154(28), 155, 159, 175, 496
Stöhr, J., 235, 480
Stolt, K., 393
Stozier, J. A., Jr., 239
Strangler, F., 372, 374
Strausser, Y. E., 509
Strehlow, W. H., 324
Streit, K. M., 88
Strobel, H. A., 79
Sugata, E., 367
Sunjïc, M., 47, 382
Suzuki, M., 400
Swanson, L. W., 15(18), 20, 26, 358, 361, 362, 367, 397, 402
Szoke, A., 368
Szymoński, M., 332

T

Taglauer, E., 207, 299, 301, 302, 303, 304, 305, 306(17), 307(17), 308, 309, 310, 311(35), 313(9), 314, 316(12, 51), 319(51), 320, 322, 324(62), 325(63), 326, 328(35), 329, 333, 336, 337(84), 340(100), 342(100), 343(100), 348
Takeda, K., 367
Takezawa, N., 107
Tamm, P. W., 432, 433, 434, 446
Tamura, H., 338
Taylor, G. I., 403
Taylor, J. B., 427
Taylor, J. L., 44, 434, 449, 450, 451
Taylor, T. N., 313
Teller, W., 394
Terada, K., 196
Terada, M., 112
Terzić, I., 299, 313(7, 8)

Thiel, P. A., 34, 36(83), 46, 67, 70, 71, 72(146), 73(146), 74, 84, 117(146), 122(145, 146, 182, 266), 123, 443, 457
Thomas, E. W., 336, 337(101)
Thomas, G. E., 28, 31, 34(57), 39, 43, 44, 47(61, 62), 48, 52, 62(62), 63, 64(62), 67, 68, 114, 115, 116, 122, 329, 511
Thomas, R. E., 1, 19(1), 20, 21
Thompkins, F. C., 1, 3(6)
Thompson, L. F., 511
Thompson, L. W., 107
Thum, F., 331
Thurstans, R. E., 395
Tice, D. R., 392
Todd, C. J., 10
Tokutaka, H., 509
Tolk, N. H., 305, 306(17), 307(17), 310(17), 329, 336(79), 337, 467, 496
Tominaga, H., 105
Tommet, T. N., 285
Tompkins, F. C., 19
Tompkins, H. G., 26
Tompsett, M. F., 288
Tong, S. Y., 25, 238, 239(17), 254(17), 271(17), 291(17)
Tongson, L. L., 316
Topping, J., 4
Torrens, I. M., 311
Toth, R. A., 56
Toyuki, H., 68
Tracy, J. C., 84, 136, 238, 254(21), 255(21), 293(21), 317, 443
Trajmar, S., 53
Traum, M. M., 135, 138, 139(14), 166, 467, 476, 488, 496, 482(54b), 494
Treichler, R., 471, 496
Trolin, J. K., 350, 367
Tsang, T., 230
Tsaur, B. Y., 410
Tsong, T. T., 376, 379, 381, 383, 384(126), 385(126), 388(126), 392, 393, 395, 397, 406, 407, 410, 412(126)
Tuck, R. A., 21
Tucker, C. W., Jr., 238, 271(14)
Tully, J. C., 305, 306(17), 307(17), 310(17), 314(21), 467, 468, 470(17), 496
Turaev, N. Y., 307, 308(26)

Turner, P. J., 390
Tzoar, N., 472

U

Uchida, K., 62, 63(137), 64(137)
Uhlaner, C. J., 473
Ulehla, M. V. K., 122(261)
Ulmer, K., 221
Umbach, E., 68
Unterwald, F., 238
Unwin, R., 84, 85(183), 91(183)
Ustinov, Y. K., 444
Utlaut, M., 508, 511(110)
Uva, E., 40

V

Valyi, L., 497
Vanderslice, T. A., 444
van der Veen, J. F., 225, 467, 492, 493(21), 494(21)
van der Weg, W. F., 316, 336, 337 (101)
van der Wiel, M. J., 481
Van Hove, J. M., 264, 271, 276, 277(68), 284(57), 288(57, 68), 295
Van Hove, M. A., 62, 84, 122(182), 238, 239, 254(17), 271(17), 291 (17)
Van Oostrom, A., 390, 410(289), 506, 507
Van Oostrom, A. G. J., 352, 360, 361(20), 369(20), 390, 410(289)
Vastel, J., 338
Vaughan, R. W., 105
Venables, J. A., 437
Verbeek, H., 306, 307(23), 313
Verbraak, C. A., 410(285)
Verhey, L. K., 306, 311, 328(40)
Verhoeven, J., 512
Vernickel, H., 367
Viehböck, F. P., 328
Vierle, O., 26
von Ardenne, M., 313
Vorburger, T. V., 11
Voss, D. E., 345, 347
Vukanić, J., 299, 313(7), 472

AUTHOR INDEX

W

Wachs, I. E., 107
Wagener, S., 21
Wagner, A., 410
Wagner, C. D., 507
Wagner, H., 22, 122(257, 259, 281), 443
Walko, R. J., 392, 393, 412
Wallbridge, M. G. H., 110, 112(245)
Walls, J. M., 368, 395
Walther, H., 77
Walther, V., 299
Wang, C., 366, 442
Wang, G.-C., 238, 239, 255(27), 269, 270(59), 280, 293, 296
Watanabe, T., 400
Watson, D. J., 74(158), 75, 416
Watson, P. R., 285
Watts, A. J., 410
Waugh, A. R., 403, 410(293), 412, 415
Weaver, J. H., 152, 159
Webb, M. B., 239, 242, 243(34, 47), 271, 273, 274, 275, 276, 284, 290
Weber, R. E., 227
Weeks, S. P., 140, 488
Weheeler, M. A. Z., 326
Wehner, G. K., 329, 331
Weinberg, W. H., 26, 28, 31, 34(57), 39, 43, 44, 47(59, 61, 62), 48, 52, 62, 63, 64(62), 65, 67, 68, 70, 71, 72(146), 73, 84, 95, 96, 98, 99, 101, 102, 105, 106, 107(232), 108(202), 109(202), 110(203, 205, 210, 241), 111,112(203, 204, 205, 241), 114, 115, 116, 117(146), 122, 123(266), 434, 449, 450(58), 451(58), 457, 462, 500, 504(83)
Weissler, G. L., 7
Weissmann, R., 231, 330
Welkie, D. G., 238, 239, 247, 248, 257, 273(9), 277(41), 286(9), 291(9), 292
Welter, L. M., 368
Wendelken, J. F., 26, 30(38), 34(38), 122(38, 261), 289, 432
Wertheim, G. K., 225, 231
Westerberg, E. R., 401
Wheatley, G. H., 313, 316, 319, 320(59)
Whelan, J. M., 329
Whitcutt, R. D. B., 361
White, C. W., 305, 306(17), 307(17), 310(17), 336, 337, 347
White, J. M., 70, 107
Whitton, J. L., 334, 335
Williams, D. W., 368
Williams, E. D., 65, 105, 115(143), 116(143), 443
Williams, E. M., 474, 475(49), 499
Williams, M. W., 197
Williams, P., 337
Williams, P. R., 410(282), 415
Williams, R. H., 165
Williams, R. S., 507, 509
Williams, W. T., 368
Willis, R. F., 26, 27(46), 32, 33(43, 44, 45, 46), 34(70, 71), 47(68, 69, 70, 71), 48(69, 70, 71), 50(44, 68, 69), 51(70, 71), 122(43, 44, 45, 46, 68, 69, 70, 71, 282)
Willis, W. R., 331
Wilson, K., 311
Winograd, N., 498, 500, 504(83)
Winters, H. F., 512, 513(129)
Withrow, S. P., 289, 432, 434
Witmore, R. W., 107
Wittmaack, K., 328, 329, 332(86), 337(86), 340(77)
Wolf, P., 372
Wolfram, T., 29, 47(60), 95(60), 96(60)
Wong, J. W., 507, 509
Wood, J. V., 410(286)
Woodruff, D. P., 138, 304, 476, 482, 494
Worley, S. D., 105
Wortman, R., 364

Y

Yahiku, A. T., 403
Yamato, Y., 400
Yang, A. C., 105
Yao, H. C., 105, 326
Yates, J. T., Jr., 25, 26, 27(25), 36, 67, 70(145, 146), 71(145, 146), 72(146), 73(145, 146), 92, 94(195, 196, 197), 105, 117(146), 122(145, 146), 428, 431, 432, 433, 440, 443, 444, 446, 448, 452, 453, 457, 461(64), 464, 465, 466, 467, 471, 473, 474, 485, 486,

487(6), 498(1, 6), 499(1, 6), 500,
502(1), 503, 509
Yermakov, Y. I., 110
Yonas, G., 368
Yoneda, Y., 107
Yonts, O. C., 331
Yoshida, R., 400
Yoshida, T., 400
Yoshida, Y., 400
Young, J. P., 498
Young, R. D., 361
Yu, M. L., 466, 493
Yurasova, V. E., 307

Z

Zandberg, E. Y., 497
Zakharov, V. A., 110
Zare, R. N., 498
Zartner, A., 300, 301, 305(20), 307(10), 314(20), 493
Zettlemoyer, A. C., 443
Zhang, Q.-J., 469, 496
Zhao, L.-H., 238, 239(27), 255(27)
Ziskoven, Z., 400
Zorn, J. C., 495, 497
Zuccaro, C., 74(158), 75

SUBJECT INDEX

A

Absorption
  coefficient, 176, 179
  edge, 175
Absorption infrared spectroscopy, 26
Abundance, 408
Adsite symmetry, 57
Adsorbate diffusion, 363
Adsorption, 3, 356, 359
  covalent, 4
  dipole layer, 3
  heat of, 366, 456
  ionic, 4
  isosteric heat of, 438, 440
  rate, 457
  on reduced metal clusters, 103
  states, 152, 154
  statistical thermodynamics of, 457
  van der Waals, 4
AES, see Auger electron spectroscopy
Aiming error, 414
Alkali metals, 403
Aluminum, 155
Angle-resolved energy distribution curve, 171
Angle-resolved photoemission, 133, 135, 159, 160, 161
Angle-resolved photoemission spectra, 170
Angle-resolved photon-stimulated desorption, 485, 491, 493
Angle-resolved spectrometers, 140
Angular distribution, 333, 473
  measurements, 272, 279
  of neutrals, 497
  of photoelectrons, 164
Angular-profile measurements, 286
  LEED, 291
  RHEED, 275, 295
Antiphase domains, 254
  rotational, 254
  translational, 254
Antiphase islands, 254
  rotational, 254
  translational, 254

Appearance-potential spectroscopy, 198
  resolution, 205
Application of various techniques in work function measurements, 21
Arrhenius equation, 430, 451
  plot, 395
ASD, see Auger-stimulated desorption
Atomic mixing, 333
Atomic potential, 311
Atom-probe, 404, 407
  10-cm, 414
  imaging, 415
  pulsed laser, 406
  voltage-pulsed, 404
Atom-probe mass spectroscopy, 404
Atom-probe spectroscopy, 404
Attenuation, 244
Attenuation coefficient, 242
Au, see Gold
Auger analysis, 231
Auger decay, 148
Auger effect detector, 495
Auger electron appearance-potential spectroscopy, 202, 210
Auger electron spectroscopy, 33, 45, 154, 226, 324, 496, 500, 504, 507, 508
  detector, 495
Auger emission, 150, 227
Auger line shape, 230, 508
Auger microscopy, 506
Auger neutralization, 304
Auger peak, 179, 502
Auger process, 148, 174, 470
Auger spectrometer, 487
Auger-stimulated desorption, 470
Auger transition, 227
Auger yield, 190

B

Band structure mapping, 165
  technique, 166
  three-dimensional, 169
  two-dimensional, 165, 168, 181

SUBJECT INDEX

Band structure of adsorbed overlayer, 168
Barium, 374, 376
Bayard–Alpert gauge, 444
Beam-current problems, 320
Beam lines, 144, 150, 179
Beam rastering, 313
Benzene, 421
Best image
  field, 383
  voltage, 383, 387
Binary scattering, 300, 302
Binding energy, 189, 466
Binding states, 428, 429
Biological molecule, 416
  deposited from aqueous solution, 416
BIV, see Best image, voltage
Bolometer detector, 497
Bragg-intensity-versus-temperature measurement, 279
Breakdown, 368, 369
  voltage, 368
Bremsstrahlung, 199, 202, 207
Broadening, 247
Brown–Lien grazing-incidence monochromator, 144

C

C, see Carbon
Calibrated photodiode, 485
Calibration, 226
Capacitance methods, 17, 19
  contact potential, 17, 18, 19
  Fermi-level, 17
  Kelvin–Zisman (vibrating capacitor), 17
  reference surface, 19
Capacitor analyzer, 220
Carbon, 154, 176, 184
Carbon monoxide, 154
  on Pd, 95
  on Pd(100), 84
  on Pd(111), 88
  on Pd(210), 91
  on Ru(001), 66, 113, 117
  on W(100), 324
Catalyst surface, 326
CFS, see Constant final state
Channel electron multiplier, 286

Channel-plate electron multiplier, 140, 184
  Chevron, 287
Channeltron multiplier, 320
Characteristic loss, 196, 229
Characteristic time, 430
Chemical potential, 3
Chemical shift, 223, 224
Chemisorbed layer, 364
Chemisorbed species, 366
Chemisorption, 131, 154, 155, 157, 169, 181, 182, 365, 454
Chlorine, 154, 155, 157, 158, 169, 181, 182
CIS, see Constant initial state
CISS, see Constant-initial-state spectroscopy
Cl, see Chlorine
Clausius–Clapeyron equation, 439
CMA, see Cylindrical-mirror analyzer
Collimation, 143
Collision cascades, 329, 330, 333
Commensurate overlayer, 278
Composition analysis, 338
Condensed multilayer, 508
Conditioning, 369
Constant final state, 172, 173, 174, 175, 176, 177, 179, 183, 185
Constant-final-state energy, 173
Constant-final-state-energy spectroscopy, 173
Constant initial state, 172, 177, 179, 180, 181, 182, 183, 185
Constant-initial-state mode, 177
Constant-initial-state spectroscopy, 177, 179
Contact potential, 12, 17, 18
Contour slice images, 421
Conventional photon sources, 141, 147
Cooperative diffusion, 392
Copper, 170
Copper-phthalocyanine, 370
Core electron binding energy, 215
Core hole
  excitation, 199
  lifetime, 191, 200
Coster–Kronig transition, 192
Coulomb potentials, 311
Coverage measurement
  absolute, 441, 442, 443
  during programmed desorption, 451
  relative, 441
Crater effects, 338
Critical electron exposure, 509

SUBJECT INDEX

Critical wavelength, 143
Cross-section, 244, 303, 304, 311, 330
  desorption, 322
  elastic scattering, 244
  electron-induced surface processes, 498
  electron-stimulated desorption, 466, 500, 510
  electron-stimulated desorption, adsorbed species, 501
  inelastic scattering, 244
  ion scattering, 303, 304, 330
  low-energy electron scattering, 244
  photoemission, 484
  photon-stimulated desorption, 466
Cylindrical-deflector analyzer, 127°, 34, 35, 139
Cylindrical-mirror analyzer, 136, 159, 169, 221, 474, 476, 477, 494

D

Daly-type converters, 320
Damage
  desorption, 321
  electron-beam-induced, 505
  electron bombardment, 500
  ion beam, 321
  radiation, 173, 333
  sputtering, 321
de Broglie wavelength, 358, 379
Debye–Waller effect, 213, 235
Debye–Waller factor, 274, 275, 296, 297
Defects, 247
Density
  of final states, 175, 177, 179
  of initial states, 144
  of occupied states, 132, 149, 150, 175, 183
  of states, 145, 147, 150, 151, 154, 171, 173, 200, 202
  of unoccupied states, 149, 181, 183
Depth
  analysis, 327, 347
  profiling, 313
  scale, 410
Desorption, 321, 322, 356, 376, 385, 408
  activation energy, 438, 456
  Auger-stimulated, 470
  characteristic time, 430
  cross section, 322
  experimental data treatment, 447

  isosteric heat of, 439
  mechanisms, 467
  model, 469
  from polycrystalline substrates, 427
  rate, 430, 457
  states, 434
  statistical thermodynamics of, 457
Desorption field, 408
Desorption images, 412
Desorption kinetics
  chemical methods, 453
  using gas evolution measurement, 447, 449
  kinetics parameter, 448
Detector, 283, 286
  bolometer, 497
  Faraday cup, 283, 287
  fluorescent screen, 283, 284, 287
  multiplier, 474
  photoelectric, 208
  position-sensitive pulse, 287
  solid-state, 210
  surface barrier, 210
  surface ionization, 497
  synchronous, 285
Differential pumping, 321
Differentiation of spectrum, 203
Diffraction, 212, 372
  from surfaces, 242
  pattern, 267
  theory, 240
Diffractometer
  components, 280
  sources of instrument broadening, 260
Diffusion, 364, 378, 384, 389
  activation energy, 365
  of chemisorbed layer, 364
  cooperative, 392
  surface diffusion activation energy, 395
Diffusion coefficient, 364, 394
Diffusion constant, 364
Diffusivity, 395
Dipolar (multipolar) scattering, 46, 56, 60
Dipolar theory, 56
Dipole attraction, 383
Dipole–dipole interaction, 62, 64
Dipole energy, field-induced, 380, 384
Dipole moment, 360
Dipole-normal selection rule, 47
Dipole selection rule, 230
Disappearance-potential spectroscopy, 213

Disappearing-phase transformations, 279
Dispersion-compensated electrostatic analyzer, 217
Dispersive analyzer, 218
Display analyzers, 140
DNA, 423
Double-focusing mass spectrometers, 328, 340
Drift region, 405
Duoplasmatron ion source, 347
Dynamic range, 284
Dynamic screening, 231

E

EAPFS, *see* Extended appearance-potential fine structure
EDC, *see* Energy distribution curve
EDJDOS, *see* Energy distribution of the joint density of states
EELS, *see* Electron energy loss spectroscopy
Elastic scattering, 244
Elastic yield, 213
Electrochemical potential, 2
Electron-beam-induced damage, 505
Electron beam method, for work function measurement, 14
   adsorption, 15
   application, 14
   contact potential, 15
Electron emission methods, for work function measurement, 5
   field-electron emission, 8
   photoelectric emission, 7
   thermionic method, 5
Electron-energy analyzers, 134, 182
Electron-energy distribution, 11
Electron-energy loss spectrometer, 33
Electron-energy loss spectroscopy, 26, 28, 66, 75, 344
   resolution, 38
   sensitivity, 30
   tabulation of results, 121
Electron gun, 206, 281, 366, 476, 486
   field emission source, 282
   low-energy, 282
   magnetically focused, 282
   for RHEED, 283

Electronic excitation, 331
Electron-impact ion source, 313
Electron-induced desorption, 209
Electron microscopy, 511
Electron penetration depth, 505
Electron spectroscopy, 218
Electron-stimulated adsorption, 511, 513
Electron-stimulated desorption, 367, 446, 467, 471, 477, 487, 493
Electron-stimulated-desorption ion angular distribution, 467, 471, 477, 485, 493
Electron tunneling, 350, 378
Electrostatic analyzer, 219
Electrostatic potential, 155
Electrostatic stress, 417
Ellipsoidal mirror analyzer, 492
Emission methods, for work function measurement, 4
   electronic band structure effect, 5
   patch field, 5
   temperature dependence, 5
Emission spectrum, 197
Emission techniques, for work function measurement, 20
   band structure, 20
   field emission, 20
   free-electron limit, 20
   photoemission, 20
   thermionic emission, 20
Emitted current density, 6
Emitted electrons, 15
   energy spread, 15
   Fermi-energy, 15
End form, 385
Energy analysis, 314, 316, 340, 474
Energy analyzer, 316
Energy distribution, 137, 332, 473, 476
   of electrons, 129
   of field-emitted electrons, 361
   of field-ionized gas atoms, 380
   of ions, 492
   of photoelectrons, 130, 164
   of secondary particles, 332
Energy distribution curve, 130, 137, 140, 144, 147, 154, 159, 165, 167, 170, 173, 175, 179, 181, 183
   interpretation of, 147
Energy distribution of collected electrons, 16
Energy distribution of the joint density of

SUBJECT INDEX

states, 145, 146, 148, 172, 173, 177, 178, 180
Energy resolution, 303, 314, 317, 318
Energy spread, 405
Energy-window analyzers, 136
Equilibrium position determinations, 291
ESA, see Electron-stimulated adsorption
Escape depth, 127, 130, 131, 152, 179
ESD, see Electron-stimulated desorption
ESDIAD, see Electron-stimulated-desorption ion angular distribution
Ewald construction, 241
   grazing-angle x-ray diffraction, 241
   LEED, 241
   penetrating radiation, 241
   RHEED, 241
EXAFS, see Extended-x-ray-absorption fine structure
Extended appearance-potential fine structure, 232, 235, 236
Extended-range "grasshopper" monochromator, 144
Extended-x-ray-absorption fine structure, 297
Extinction distance, 245
Extra-atomic relaxation, 224
Extrinsic Auger satellites, 229
Eyring theory of reaction rates, 457, 459, 461

F

Faraday cup detector, 283, 287
Fast-Fourier-transform infrared spectroscopy, 78
FEED, see Field-emission energy distribution
FEEM, see Field-electron emission microscope
FEM, see Field-emission microscope
Fermi level, 19
Fermi sea, 8
   current density, 8
FERP, see Field emission retarding-potential method
Ferritin, 417, 421
FeTi, see Iron-titanium
FFTS, see Fast-Fourier-transform infrared spectroscopy
Fiber-optic faceplate, 389

Field desorption, 385, 398
   mass spectrometry, 400
   microscopy, 412
Field-electron-emission microscope, 349, 353, 354, 359
   pressure in, 354
Field-electron-emission microscopy, 349, 363
   application, 363
Field-electron energy distribution, 362
Field emission, 8, 20, 351, 356, 368
   electron source, 206
   energy distribution, 26
   microscope, 21
Field-emission-referenced electron spectroscopy, 226
Field emission retarding-potential method, 15
   field emission cathode, 15
   work function, 15
Field emission source, 282
Field-emitted electrons, 353
   energy distribution, 361
Field evaporation, 385, 387, 397, 399, 404, 410
   initiation, 404
Field-induced dipole energy, 380, 384
Field-ion current, 402
Field-ion images, 379
Field ionization, 378, 401
Field-ionization ion source, 347
Field-ionization mass spectroscopy, 400
Field-ion microscope, 376, 384, 385, 387, 392, 412
Field-ion source, 339, 401
   liquid metal, 403
Field-ion tomography, 421
Field strength, 410
Filter, 485
   monochromators, 485
FIM, see Field-ion microscope
Final density of states, 172
Final-state correlation effects, 231
Final-state effects, 150, 169, 179
Final-state formation, 337
Finite-size effects, 246
First-order desorption kinetics, 425, 448, 454, 459, 460, 462
Flash desorption, 428

Flavanthrene, 372
Flicker noise, 202, 259, 402
Fluorescence
  decay, 495
  laser-excited, 498
Fluorescent-screen detector, 283, 287
Forbidden zone, 383
Fountain analyzer, 221
Fourier transform, 165
Fowler–Nordheim equation, 350, 352, 353, 369
Fowler–Nordheim formula, 8
Fowler–Nordheim slope, 359
Fowler–Nordheim theory, 351, 352
Franck–Condon excitation, 468
Franck–Condon ionization, 468
Frequency shift, 62
  dipole–dipole interaction, 62, 64
  local bonding variation, 62, 65
  vibrational coupling, 62
Free-electron lasers, 184
Free-electron wave, 164
FRESCA, see Field-emission-referenced electron spectroscopy

G

GaAs, see Gallium arsenide
Gallium, 403
  source, 403
Gallium arsenide, 170, 171
  GaAs(001), 295
  GaAs(110), 291
  sputter etched, 292
Gallium selenide, 150, 165, 167, 180
Gas doser, 487
GaSe, see Gallium selenide
Gas evolution measurement, 447, 449
Gas jet, 339
GDMS, see Glow-discharge mass spectrometry
Germanium, 155
  Ge(111), 154, 155
Globar source, 77
Glow-discharge mass spectrometry, 329
Gold, 176, 403
Goniometer, 288
Grid microchannel plate, 487, see also Microchannel plate detector, Channel-plate electron multiplier
G value, 507

H

Heat of adsorption, 366, 456
Hemispherical deflector, 180°, 34, 35
  lens, 35, 40, 41
  resolution, 40, 41
Hermanson Rule, 159, 161
Heterojunction discontinuities, 155
High-brightness source, nearly monochromatic electrons, 368
High-voltage pulse, 404
Hopping gas model, 383
Hydrogen, 154
  physisorbed, 402

I

IETS, see Inelastic electron tunneling spectroscopy
Image charge, 351
Image compression factor, 357, 358
Image-force theory, 397, 398
Image intensification, 378
Image potential, 351, 379, 398
Imaging atom-probe, 415
Imaging
  of single atom, 374
  of single molecule, 370
Incommensurate overlayer, 256
Indium, 154, 155
Indium selenide, 167
Inelastic electron scattering, 200
Inelastic electron tunneling spectroscopy, 26, 28, 59, 95, 97
  experimental, 96
  resolution, 30, 81
  result tabulation, 121
  sensitivity, 30
  spectra, 97
  theoretical, 96
  tunnel junction fabrication, 98
  tunneling current, 96
Inelastic scattering, electrons, 193, 244
  mean free path, 198, 244
Infrared catastrophe, 207
Infrared laser, 77
Infrared reflection–absorption spectroscopy, 27, 30, 31, 75
  detector, 79
  experimental, 76

SUBJECT INDEX

modulation systems, 79
resolution, 30, 78, 81
result tabulation, 117, 121
sensitivity, 30
theoretical considerations, 81
Infrared reflection
　from clean metal surface, 82
　from metal surface with adsorbate, 82
Infrared source, 76
　diffraction grating, 77, 78
　Globar source, 77
　infrared laser, 77
　intensity, 77
　Nernst glower, 77
Initial density of states, 171
Inner potential, 233
Inorganic cluster compound, 108
InSe, *see* Indium selenide
Instrument broadening in diffractometers, 260
Instrument response, 263
　function, 260
Integrated-intensity-versus-energy measurement, 268, 284, 291
Intensity-temperature measurement, 296
Intensity-versus-energy measurement, 269, 286
　in LEED, 269
　in RHEED, 270
Interatomic spacing, 232, 234
Interference function, 241, 250
Internal electron distribution curve
　$L(T)$, 130
　$N(T)$, 132, 134, 135, 137, 144, 145
Internal energy distribution, 144
Intra-atomic relaxation, 224
Intrinsic plasmon, 229
Ion
　angular distribution, 485
　current, 401
　energy distribution, 492
Ion beam damage, 321
Ion desorption trajectory, 472
Ion formation mechanism, 467
Ionization
　efficiency, 410
　probability, 379
　of residual gas, 369
Ion microprobe, 338, 347
Ion-neutralization process, 472

Ion sources, 313, 337, 403
　duoplasmatron, 347
　electron-impact, 313
　field-ion, 339
　field-ionization, 347
　liquid-gallium, 403
　liquid-metal, 402, 403
　liquid-metal field desorption, 403
　plasma, 314
　plasma-discharge, 337
Ion yield, 303, 334, 474, 482
　curve, 483
　negative-ion, 337
　positive-ion, 337
IRAS, *see* Infrared reflection–absorption spectroscopy
Iron-titanium, 150, 151, 152
Isochronal focusing, 405
Isosteric heat
　of adsorption, 438, 440
　of desorption, 439
Isothermal desorption, 436

J

Jason effect, 382
Jason peak, 382
JDOS, *see* Joint density of states
Johnson noise, 202
Joint density of states, 145, 146, 174
　energy distribution, 145, 146

K

KCl, *see* Potassium chloride
Kelvin probe, 22
KF model, *see* Knotek and Feibelman model
Kinetic energy, 467, 482
　distribution, 491
　window, 175
Kinetic parameters, 448
Knotek and Feibelman mechanism, 492
Knotek and Feibelman model, 469, 472, 505
Kuyatt–Simpson electron spectrometer, 35, 40

L

Laplace's equation, 350
Laser, 406

SUBJECT INDEX

Laser-excited fluorescence, 498
Lateral resolution, 339
Laue atlas, 432
Laue back reflection technique, 432
Laue conditions, 241
Layer compound, 167, 168, 169, 180
LEED, see Low-energy electron diffraction
Lifetime broadening, 190, 192, 216
Lifetime of the excited state, 171
Liquid-gallium source, 403
Liquid metal, 402, 403
Liquid-metal field desorption source, 403
Lithography, 511
Local density of states, 154
Locally enhanced magnification, 370
Loss spectrum, 195
Low-energy electron diffraction, 227, 241, 324, 326, 344, 487, 500, 502, 504, 505
   detector, 283
   theories, 271
Low-energy electron gun, 282
Low-energy electron loss spectra, 480
Low-temperature field-ion microscope, 384, 389
Luminosity, 220

M

Madelung potential, 223
Magnetically focused gun, 282
Magnetic sector mass spectrometer, 444, 473, 493
Magnetic spectrometers, 219
Magnification, 354
   of cylindrical microscope, 354
   of field-electron emission microscope, 357
   of field-ion image, 388
   of field-ion microscope, 385, 389
   of point projection microscope, 357
Mahan cones, 163, 164
   primary, 163, 164
   secondary, 163, 164
Mapping of band structure, see Band structure mapping
Mass analysis, 473
Mass resolution, 303, 344, 405
Mass separation, 314, 338
Mass spectrometer, 404, 444
   double-focusing, 328, 340
   magnetic-sector, 444, 473, 493

   Omegatron, 44
   quadrupole, 444, 487, 493
   time-of-flight, 404, 444
Mass spectrometry, 340, 398
Mass spectrum, 408
Mass-to-charge ratio, 405
Material balance equation, 430
Matrix effect, 339, 346
Matrix element effect, 147, 149, 169, 179
Matrix elements, 147, 150, 151, 156, 157, 169, 180, 230
MCP, see Microchannel-plate detector
Mean free path, 176, 245, 505
   high-energy electron, 245
   inelastic-electron, 198, 244
   inelastic scattering, 192
   x-ray, 245
Measurement of diffracted beam intensity
   channel electron multiplier, 286
   Faraday cup detector, 283, 287
   fluorescent screen detector, 283, 284, 287
   parallel detection, 285
   photography, 285
   synchronous detection, 285
   vidicon camera, 285, 488
Measurement of phase transitions, 279
Menzel–Gomer–Redhead model, 468
Mercury, 403
Metal–semiconductor interfaces, 155
Metastable species, 494, 495
Microchannel-plate detector, 390, 391, 487, 492
Microscope, 21, 353
   cathode-emission, 21
   electron emission, 21
   field-electron-emission, 349, 353, 359
   field-emission, 21
   field-ion, 376, 384, 387, 392, 412
   low-temperature field-ion, 384, 389
   point projection, 357
   reflected electron, 22
Miller indices, 355
Minimum angle of resolution 263, 264
Mirror symmetry, 168
M/n, see Mass-to-charge ratio
Mobile-precursor model, 454
Molecular-beam scattering, 497
Molecular imaging, 370, 416
Molière approximation, 311
Monochromatic electrons, 368

SUBJECT INDEX

Monochromator, 32, 33, 76, 78, 144, 150, 176, 178, 179, 184, 481
   Brown–Lien grazing-incidence "grasshopper," 144
   extended-range "grasshopper," 144
   toroidal-grating, 144
Monochromator transmission function, 483
Monoenergetic ions, 401
Mosaic structure, 248
Multilayer images, 412
Multiphoton ionization, 498
Multiple ionization, 216
Multiple scattering
   electron, 234
   ion, 307, 308, 310
Multiplier detector, 474

N

Negative ion resonance scattering, 52, 53, 54
Negative ions, 466, 493
Nernst glower, 77
Neutralization, 303, 304, 307, 310, 326, 473
   Auger neutralization, 304
   quasiresonance neutralization, 304–305
   resonance neutralization, 304
Neutral-particle detection, 320
Neutral species, 496
Nitrogen, 184
Nitrogen adsorption, 360
Nitrous oxide, 66, 75
   on Ru(001), 66, 75
Noise
   flicker, 202, 259, 402
   Johnson, 202
   shot, 259
N-th order desorption kinetics, 460
Nuclear reactions, 347
Nucleation, 366

O

O, *see* Oxygen
Omegatron spectrometer, 444
Opening angle, 143
Optical absorption, 174
Optical absorption coefficient, 145, 179
Optical excitation, 148, 163
Order–disorder transitions, 279
Oscillator strength, 233

Overlayer phase diagram, 297
Oxidation state, 223
Oxygen, 154, 155, 184
   activation energy for diffusion, 365
   on Ag(100), 326
   on Cu(110), 328
   on W(110), 365

P

Palladium, 378
Palladium thimble, 389
Parallel detection, 285
Parallel-plate analyzer, 220
Parity, 160, 183
Partial yield, 179
Parital-yield spectroscopy, 173, 179, 183
Particle detection, 320
Path of electrons, 244
Penetrating radiation, 241
PHA, *see* Pulse-height analyzer
Phase diagram, 278
Phase shift, 233
Photocurrent, 483
Photodiode, 482, 485
Photoelectric detectors, 208
Photoelectric emission, 7
   Fowler analysis, 7
   patchy surface, 7
   photoelectric yield, 7
   threshold frequency, 7
Photoelectric method, 8
Photoelectron parameters, 133
Photoelectron spectroscopy, 133, 508
   angle-resolved, 133, 135
Photoemission, 20, 21
   angle-resolved, 160, 161
   cross section, 484
   spectra, 170
Photon monochromator, 481, *see also* Monochromator
Photon parameters, 133
Photon polarization, 158
   in angle-resolved photoemission, 159
   selection rules, 156
Photon source, 133
   conventional, 141, 147
Photon-stimulated desorption
   angle-resolved, 185, 491, 493
   cross section, 466

Physisorbed layer, 364
Physisorption process, 131
Plane-mirror analyzer, 138
Plasma discharge ion source, 337
Plasmon, 195, 196, 229
    decay, 197
    gain satellites, 229
    losses, 150
Point-projection microscope, 357
Point source, 368
Polarization, 133, 143, 156, 159, 160, 171, 179, 180, 183, 360, 485
Polarization modulation, 80
Polarized-photon photoemission, 156, 158
Position-sensitive pulse detector, 287
Post ionization, 337, 399
Potassium chloride, 179, 180
Potential modulation, 203, 220
Prebreakdown current, 369
Preferential sputtering, 324, 333
Primary electrons, 132, 148, 174
Primary Mahan cones, 163, 164, 166
Primary photoelectrons, 177
Probe hole, 359, 366, 414
Protein multilayers, 417
PSD, see Photon-stimulated desorption
Pulse-counting mode, 474
Pulsed-laser atom-probe, 406
Pulse duration, 405
Pulse-height analyzer, 477, 478
PYS, see Partial-yield spectroscopy

Q

QMS, see Quadrupole mass spectrometer
Quadrupole mass analyzer, 474
Quadrupole mass filter, 328, 340
Quadrupole mass spectrometer, 444, 487, 493
Quantitative Auger analysis, 231
Quantum efficiency, 209
Quasi-atomic behavior, 231
Quasi-resonance neutralization, 304
Quenched surface states, 154

R

Radiation damage, 177, 333
Radiative decay, 495

Radius of curvature, field emitter tip, 358
Random strain, 247
Reciprocal lattice, 241, 243, 244, 245, 247
    antiphase island, 254, 256
    of defects, 247
    $p(1 \times 1)$ structure, 257
    superlattice, 253, 254
    terrace, 252
Reciprocal-lattice rods, 244, 247, 250, 254
    broadening, 247
Reference energy, 149
Reference surface, 18
Reflected-electron microscope, 22
    contact potential, 22
    patches, 22
    scanning, 22
Reflection high-energy electron diffraction, 241, 264, 283
    detector, 283, 285, 288
    measurement of angular profile, 295
Reflection infrared spectroscopy, 28, 29, 75
Relaxation energy, 224, 229
Reliability factor, 271
Reneutralization, 510
Resistive-anode encoder, 287
Resolution
    of appearance-potential spectroscopy, 205
    of cylindrical microscope, 354
    of field-ion microscope, 384, 385
    mass, 303, 344, 405
    of point-projection microscope, 358, 359
Resolving power, 258, 263, 286, 295
    in LEED, 264
    in RHEED, 264
Resonance, 382
Resonance neutralization, 304
Retarding-potential analyzer, 227, 362
Retarding-potential method, 11, 20, 21
    band structure, 20
    emission current density, 11
    energy distribution, 20
    experimental arrangement, 20
    field-emission retarding potential technique, 20
    nonuniform surface, 22
    reflection mechanism, 20
    scanning arrangement, 21
    scanning low-energy probe technique, 20, 21

SUBJECT INDEX

Retarding-potential method (*continued*)
  Shelton triode, 11
  standard surface, 22
RHEED, *see* Reflection high-energy electron diffraction
Richardson equation, 5, 11
  temperature-dependent, 5

S

Sample preparation, 289
Scanning ion micrographs, 403
Scanning low-energy electron probe, 20, 21
Scattering cross-section, 303, 304
Scattering probability, 49
Schottky-barrier height, 155
Schottky effect, 6
Schottky hump, 397, 398
Screening, 229, 231
Secondary, *see* Secondary electron
Secondary electron, 131, 148, 150, 153, 174, 176, 178, 179
Secondary-electron emission, 193, 202, 331
Secondary-electron multiplier, 320
Secondary-electron yield, 210
Secondary-emission spectrum, 227
Secondary-ion mass spectroscopy, 339, 403
  static, 339
Secondary Mahan cone, 163, 164
Secondary neutral mass spectrometry, 329
Second-order desorption kinetics, 425, 457, 460
Second-order light, 481
Selective ionization, 410
Sensitivity, 258, 286
SEXAFS, *see* Surface-extended x-ray-absorption fine structure
Seya–Namioka monochromator, 144
Shadow cone, 310
Shadowing coefficient, 304
Shadow technique, 418
Shelton method, 13, 14
  application, 14
  retarding-potential, 13
  thermionic emission, 13
Shot noise, 259
Si, *see* Silicon
Signal gating, 313
Signal-to-noise ratio, 285, 292
Signal-to-shot-noise ratio, 259

Silicon, 154, 157, 175, 181, 182
Si(111), 154, 157, 169, 182
Si(111)–Al, 155
Si(111)–Cl, 157, 158, 169, 181
Si(111) (2 × 1) surface, 153, 175
Si(111) (7 × 7) surface, 153, 155
SIMS, *see* Secondary-ion mass spectroscopy
Single molecule, 370
SLEEP, *see* Scanning low-energy electron probe
SNMS, *see* Secondary neutral mass spectrometry
S/N ratio, *see* Signal-to-noise ratio
Soft x-ray appearance-potential spectroscopy, 202, 207
Solid-state detector, 210
Space charge buildup, 6
Spatial ionization, 383
Spatial resolution, 287
Spectrometers
  angle-resolved, 140
  Kuyatt–Simpson electron, 35, 40
  magnetic, 219
Spherical-capacitor analyzer, 223
Spherical electrostatic analyzer, 317
Spherical-grid retarding potential analyzer, 219, 223
Spherical-sector analyzer, 139
Spin–orbit splitting, 216
Sputter cleaning, 487
Sputter etching, 290
Sputtering, 321, 328, 366
  rate, 367
  yield, 321, 330, 334, 367
Static capacitor, 19
Statistical thermodynamics of desorption, 457
Stefan–Boltzmann law, 76
Step height, 251
Sticking-coefficient determination, 442
Sticking probability, 324
Storage ring, 141, 142
Streak length in RHEED, 276
String model, 309, 310
Subgrain structure, 248
Superlattice, 253, 254
Surface adsorbates, 356
Surface analysis, 313
Surface barrier, 162, 171
Surface-barrier detector, 210

SUBJECT INDEX

Surface binding states, 428
Surface chemical shift, 224
Surface cleaning, 324, 354
Surface composition, 231, 333
Surface coverage, 363
Surface crystallography measurements, 267
Surface defects, 245
Surface diffusion, 363, 392, 393, 395
  activation energy, 395
Surface-extended x-ray-absorption fine structure, 232, 233, 234, 235, 236
Surface ionization detectors, 497
Surface plasmons, 196, 382
Surface potential, 2
  correlation energy, 2
  dipole layer, 2
  ion-core potential, 2
Surface roughness, 367
Surface segregation, 324
Surface self-diffusion, 367
Surface sensitivity, 131, 152
Surface-state quenching, 152
Surface state, 152, 153
  studies in electron distribution curve mode, 152
Surface steps, 248
Surface structure, 328, 333
Synchronous detector, 285
Synchrotron radiation, 127, 128, 141, 144, 147, 171, 183, 184, 217, 225, 480, 491
  source, 156, 171

T

TAC, *see* Time-to-amplitude converter
Taylor cone, 403
Temperature-programmed desorption, 428, 431, 446
  from single crystal, 432
Terrace
  distribution of sizes, 252
  reciprocal-lattice rod, 252
  structure factor, 250
Thermal accommodation, 383
Thermal activation, 354, 406
Thermal desorption, 354, 430, 464, 499, 503
  apparatus, 436
  data treatment, 447
  detector, 444

surface binding state, 428
theory, 454
Thermal desorption from polycrystalline substrates, 427
surface binding states, 428, 429
Thermal desorption from single crystal, 431
  preparation and mounting of single crystal, 432
Thermal-desorption spectroscopy, 508
Thermal-diffuse scattering, 274
Thermal spikes, 330
Thermal vibrations, 247
Thermionic emission, 5, 20, 21, 205
  current density of electrons, 5
Thermionic method, 6
Thomas–Fermi potential, 311
Three-dimensional band structure mapping, 169
Three-step model, 127, 129, 130
Threshold energies for sputtering, 367
Threshold measurement, 479
Time digitizer, 479
Time-of-flight, 328, 404
  method, 494
  system, 319
  technique, 476, 491, 495
Time-of-flight atom-probe, 404
Time-of-flight mass spectrometer, 404, 444
Time-of-flight mass spectrometry, 480
Time-of-flight spectrum, 478
Time structure, 143
Time-to-amplitude converter, 477, 478
Tip apex, 355
Tip profile, 356
TOF, *see* Time-of-flight
Toroidal-grating monochomator, 144
Toroidal prism, 493
Total yield spectrum, 211
Transition density, 201
Transition probabilities, 146
True secondary electrons, 194
Tungsten
  W(001), 161
  W(100), 291
  W(110), 296
Tunneling, 382, 384
Two-dimensional angular profile, 292
Two-dimensional band structure mapping, 165, 166, 168, 181

SUBJECT INDEX

Two-dimensional crystals, 165
Two-dimensional layered compound, 169
Two-dimensional ordered systems, 166
Two-dimensional systems, 168
Twofold symmetric molecule, 372

U

Ultraviolet photoelectron spectroscopy, 33, 484, 485, 503
Undulators, 184
UPS, *see* Ultraviolet photoelectron spectroscopy

V

Vacuum level, 128
Vibrational coupling, 62
  intermolecular, 64
  intramolecular, 62
Vidicon camera, 285, 488

W

Width of energy distribution for field-ionized hydrogen, 381
Wigglers, 184
WKB approximation, 352, 379
Wood's metal, 403
Work function, 2, 8, 14, 19, 21, 128, 137, 141, 145, 148, 155, 167, 206, 350, 356, 359, 375, 397
  adsorbate, 9
  adsorption, 3, 5, 9, 21
  band structure effects, 11
  crystallographic direction, 9
  different planes, 21
  dipole layer, 3
  dipole moment discreteness, 10
  geometric factor, 9
  individual crystal planes, 9
  ordering, 21
  patch effects, 5
  single crystal, 21
  versus surface coverage, 439
  temperature dependence, 5
  theory, 2
  thermal desorption, 5
  true, 10
  variation, 9
  variation with temperature, 14
Work function measurement, 15
  capacitance method, 17
  electron beam method, 14
  emission method, 4
  field-emission retarding-potential method, 15
  retarding-potential method, 11
  technique comparison, 19
  thermionic method, 6

X

XAES, *see* X-ray excited Auger spectroscopy
XPS, *see* X-ray photoelectron spectroscopy
X-ray absorption, 189
X-ray emission, 189
X-ray excited Auger spectroscopy, 503
X-ray photoelectron spectroscopy, 189, 214, 503, 507
X-ray regime, 150

Z

Zero-order desorption kinetics, 425, 437, 438
Zinc-Phthalocyanine, 372